R

DESK REFERENCE FOR ORGANIC CHEMISTS

Michael B. East, Ph.D.
FAR Research
A Division of Societe Protex
Palm Bay, Florida

and

David J. Ager, Ph.D.
NSC Technologies
A Division of the NutraSweet Company
Mt. Prospect, Illinois

KRIEGER PUBLISHING COMPANY
MALABAR, FLORIDA
1995

Original Edition 1995

Printed and Published by
KRIEGER PUBLISHING COMPANY
KRIEGER DRIVE
MALABAR, FLORIDA 32950

Library of Congress Cataloging-In-Publication Data

East, Michael B.
Desk reference for organic chemists / Michael B. East and David J. Ager.
p. cm.
ISBN 0–89464–818–7 (acid-free)
1. Chemistry, Organic—Handbooks, manuals, etc. I. Ager, David J. II. Title.
QD257.7.E28 1995
547—dc20 93–33239
CIP

10 9 8 7 6 5 4 3 2

CONTENTS

Chapter 1

Chemical Acronyms

This chapter contains an alphabetical list of chemical acronyms. A few words are needed to explain the conventions used in the list. Named reagents have not been included, and should be accessed by the list in Chapter 5.

Any numbers, parentheses, Greek letters, hyphens, or stereochemical descriptors have not been considered for the purposes of alphabetizing. Stereochemical descriptors have been considered independent of the main acronym if a hyphen is used as a separator. Thus, 2,4-DDT, *o,p*-DDT, and (DDT) would all be found as if the entry were DDT. Alphabetization is based on the acronym entry, not the definition. When alternative nomenclature is available, as in the case of 2,4-DDT and *o,p*-DDT, the numerical system has been used to name the compound rather than *ortho, para*, etc. The exceptions are where the stereochemical descriptor is an integral part of the acronym. Amino acids naturally occur as the L-isomer, and this is inferred if no stereochemical descriptor is given. The D-isomer must, therefore, be differentiated by inclusion of the stereochemical descriptor. For explanations of the stereochemical descriptors see the glossary (Chapter 3). Otherwise, if alternative nomenclature is available for a compound, the different forms have been noted, separated by a semi-colon.

Capitalized entries will always precede lower case entries. When there is more than one way of representing the same compound, and the acronyms only vary by letter case, alternatives are given in parentheses as in, for example, DOPA (Dopa, dopa). It is common practice to use lower case acronyms and abbreviations when the compound is a ligand on a metal. In some cases, the reader is left to piece together the structure of the complete molecule when the acronym only refers to a specific ligand or fragment of the whole molecule.

Note, that in many cases, one acronym can refer to a number of compounds. It is for the reader to decide which one is correct in that specific context. In some cases, an acronym may have been used incorrectly in the literature. The interpretation of these acronyms are noted, often without comment, so care should be taken to ensure that the original author's meaning for this acronym is interpreted correctly from the context. Such an example is HMPA, which has often been confused with HMPT.

1122	1,1-dichloro-2,2-difluoroethene
A	actin
A	adenine
A	adenosine
A	alanine
A-5′-P	adenosine 5′-monophosphate
AA	acetylacetone; 2,4-pentanedione
AA	amino acids
AA	anisylacetone
AAA	acetoacetanilide
AAAF	2-(*N*-acetoxyacetylamino)fluorene
AABPH	acetonylacetone-bisphenylhydrazone
Aad	L-2-aminoadipic acid
7-AAD	7-amino-actinomycin D
AADEA	aminoacetaldehyde diethyl acetal
AADH	α-amino acid dehydrogenase
AADMA	aminoacetaldehyde dimethyl acetal
AAF	2-acetylaminofluorene
AAMX	acetoacet-*m*-xylidide; *m*-acetoacetoxylidide
AAO	acetaldehyde oxime; acetaldoxime
AAOA	acetoacet-*o*-anisidide; *o*-acetoacetanisidide
AAOC	acetoacet-*o*-chloroanilide; *o*-acetoacetochloranilide
AAOT	acetoacet-*o*-toluidide; *o*-acetoacetotoluidide
9-AAP1	1-([(2-[9-acridinyl]amino)-1-methylethyl]amino)-3-(1-naphthaleneoxy)-2-propanol
AAT	alanine-aspartate transferase
ABA	abscisic acid
ABEI	*N*-(4-aminobutyl)-*N*-ethylisoluminol
ABH	azidobenzoyl hydrazide
ABL	2-acetylbutyrolactone; α-acetyl-γ-butyrolactone
ABS	acrylonitrile-butadiene-styrene copolymer
ABS	alkyl benzene sulfonates
ABTS	2,2′-azinobis(3-ethylbenzothiazoline-6-sulfonic acid)
AC	acrylic
Ac	acetate
Ac (ac)	acetyl
A_2C	2-(2-methoxyethoxy)ethyl 8-(*cis*-2-*n*-octylcyclopropyl)-octanoate
Ac_2O	acetic anhydride
ACAA	adenocarcinoma associated antigen
ACAC	2,4-pentanedione; acetylacetone
acac	acetylacetonate; acetylacetonato
acaen	*N*,*N*′-bis(1-methyl-3-oxobutylidene)ethylenediamine
ACC	1-aminocyclopropane-1-carboxylic acid; amino cyclopropylcarboxylic acid

AcCh	acetylcholine
AcChE	acetylcholinesterase
ACDP	acetone cyclic diperoxide
ACDPI	5-amino-methyl N^3-carbamoyl-1,2-dihydro-3*H*-pyrrolo[3,2-e]-indole-7-carboxylate
ACE	adrenal cortical extract
ACE	angiotensin converting enzyme
ACE-Cl	1-chloroethyl chloroformate
ACES	*N*-(2-acetamido)-2-aminoethanesulfonic acid; *N*-(carbamoylmethyl)taurine
acet	acetone
ACh	acetylcholine
AChE	acetylcholinesterase
ACHPA	(3*S*,4*S*)-4-amino-5-cyclohexyl-3-hydroxypentanoic acid
Acm	acetamidomethyl
ACMP	*O*-anisylcyclohexyl methylphosphine
ACN	acetonitrile
ACN	acrylonitrile
ACND	acenaphthylenedione
AcOEt	ethyl acetate
AcOH	acetic acid
AcOOH	peracetic acid
ACP	acyl carrier protein
AcP	acetyl phosphate
t-ACPD	*trans-dl*-1-amino-cyclopentanedicarboxylic acid
ACR	9*H*,10*H*-acridine; acridine
ACS	antireticular cutotoxic serum
Act-Al	sodium polyhydroxyaluminum monocarbonate hexitol complex
ACTE	agkistrodon contortrix thrombin-like enzyme
2,3,5-ACTF	2-amino-3-chloro-5-(trifluoromethyl)pyridine
2,6,4-ACTF	2-amino-6-chloro-4-(trifluoromethyl)pyridine
ACTH	adrencorticotrophin; adrenocorticotropin; adrenocorticotropic hormone
ACV	acyclovir
acyl CoA	acyl derivative of coenzyme A
Acylase I	aminoacylase
Ad	1-adamantyl; adamantyl
AD/here	methyl 2-cyanoacrylate
ADA	adenosine deaminase
ADA	*N*-(2-acetamido)iminodiacetic acid; *N*-(carbamoylmethyl)iminodiacetic acid
ADAM	9-anthryldiazomethane
7-ADCA	7-aminodesacetoxycephalosporanic acid
ADD	1,1′-(azodicarbonyl)dipiperidine

Adda	(2*S*,3*S*,8*S*,9*S*)-3-amino-9-methoxy-2,6,8-trimethyl-10-phenyldeca-4,6-dienoic acid
ADDC	ammonium diethyldithiocarbamate
Ade	adenine
ADH	alcohol dehydrogenase
ADH	antidiuretic hormone
ADMA	alkyldimethylamine
ADMCS	allyldimethylchlorosilane
Ado	adenosine
Ado-5′-PPP[S]	adenosine-5′-[γ-thio]triphosphate tetralithium salt
Ado-5′-PP[S]	adenosine-5′-[β-thio]diphosphate trilithium salt
Ado-5′-P[S]	adenosine-5′-[α-thio]monophosphate dilithium salt
Adoc	1-adamantyloxycarbonyl
AdoCbl	5′-deoxyadenosylcobalamin
AdoHcy (adoHcy)	5-(5′-adenosyl)-L-homocysteine; *S*-adenosylhomocysteine
AdoMet (adoMet)	adenosylmethionine; *S*-adenosylmethionine
ADP	adenosine 5′-diphosphate
5′-ADP	adenosine-5′-diphosphoric acid
ADP[α,β-CH_2]	adenosine 5′-[α,β-methylene]diphosphoric acid
ADPGPP	adenosine diphosphate glucose pyrophosphorylase
Adpoc	1-(1-adamantyl)-1-methylethoxycarbonyl
ADPRP	2′-monophosphoadenosine 5′-diphosphoribose
ADP-β-S	adenosine-5′-[β-thio]diphosphate trilithium salt
AEBSF	4-(2-aminoethyl)-benzenesulfonyl fluoride
AEC	1-amino-2-ethylaminocyclopropane-1-carboxylic acid
AEC	3-amino-9-ethylcarbazole
AEDANS	(2-aminoethylamino)-1-naphthalenesulfonic acid
AEP	1-(2-aminoethyl)piperazine; aminoethylpiperazine
AEP	(2-aminoethyl)phosphonic acid
$AESe^{ox}$	selenocystamine hydrochloride
AET	2-aminoethylisothiouronium bromide hydrobromide; S-(2-aminoethyl)isothiouronium bromide hydrobromide
AFP	α-fetoprotein
AGP	acid glycoprotein from human plasma
AGPA	α-amino-β-guanidino propionic acid
AGU	anhydroglucose units
AHC	alkali halide clusters
AHF	aminohexahydrofluorine
AHs	aromatic hydrocarbons
Aib	α-aminoisobutyric acid; $C_{\alpha,\alpha}$-dimethylglycine
AIBN	2,2′-azobisisobutyronitrile; 2,2′-azobis(2-methylpropionitrile); azobis(isobutronitrile)
AICA	4-amino-5-imidazolecarboxamide; 5-aminoimidazole-4-carboxamide

AIP	aluminum isopropoxide
AIR	5-aminoimidazole ribonucleotide
AITC	2,3,4-tri-*O*-acetyl-α-D-arabinopyranosyl isothiocyanate
ALA	aminolevulinic acid; δ-aminolevulinic acid
ALA-O	aminolevulinic acid, dehydrates
Ala	alanine
Ala (ala)	alanyl
β-Ala	β-alanine
Ala(P)	1-aminoethylphosphonic acid
ALAD	5-aminolevulinic acid dehydratase
AlaPHOS	1-diphenylphosphino-2-dimethylaminopropane
alc	alcohol
Alk	alkali
All	allyl
All	β-D-allose
Alloc	allyloxycarbonyl
Aloc	allyloxycarbonyl
Alt	D-altrose
AM	actomyosin
Am	amyl; pentyl
AMA	octyl-dodecyl ammonium methanearsonate
p-AMB	*p*-aminobenzophenone
AMBA	3-amino-4-methoxybenzanilide
AMC	7-amido-4-methylcoumarin
AMCA-HPDP	*N*-[6-(7-amino-4-methylcoumarin-3-acetamido)hexyl]-3′-(2′-pyridyldithio)propionamide
AMCA-NHS	succinimidyl-7-amino-4-methylcoumarin-3-acetate
AMEO	3-aminopropyltriethoxysilane
AM-ex-OL®	4-chloro-2-phenylquinazoline
Amino G Acid	7-amino-1,3-naphthalenedisulfonic acid
Amino H Acid	8-amino-1,3,6-naphthalenetrisulfonic acid
Amino J Acid	6-amino-1,3-naphthalenedisulfonic acid
Amino R Acid	3-amino-2,7-naphthalenedisulfonic acid
Ammediol	2-amino-2-methyl-1,3-propanediol
AMMO	2-aminopropyltrimethoxysilane
AMNT	aminopropanedinitrile 4-methylbenzenesulfonate; aminomalononitrile *p*-toluenesulfonate; ammoniopropanedinitrile *p*-toluenesulfonate
AMO	2-isopropyl-4-dimethylamino-5-methyl-phenyl-1-piperidinecarboxylate methyl chloride
AMO	actinomycin D
$(AMOC)_2O$	di-*tert*-amyl-dicarbonate
2-AMP	2-(aminoethyl)pyridine
3′(+2′)-AMP	adenosine-3′(+2′)-monophosphoric acid monohydrate

3′-AMP	adenosine 3′-monophosphate
5′-AMP	adenosine-5′-monophosphoric acid
AMP	2-amino-2-methyl-1-propanol hydrochloride
AMP	adenosine 5′-monophosphate; adenosine 5′-phosphate
AMP-CP	adenosine 5′-[α,β-methylene]diphosphoric acid
AMP-CPP	adenosine 5′-[α,β-methylene]triphosphate tetralithium salt
AMPD	2-amino-2-methyl-1,3-propanediol
AMP-NH_2	adenosine-5′-monophosphoramidate sodium salt
AMP-PCP	adenosine 5′-[β,γ-methylene]triphosphate tetralithium salt
AMP-PNP	adenosine 5′-[β,γ-imido]triphosphate tetralithium salt dihydrate
AMP-S	adenosine 5′-[α-thio]monophosphate dilithium salt
AMPS	2-acrylamido-2-methylpropanesulfonic acid
AMPSO	3-[(1,1-dimethyl-2-hydroxyethyl)amino]-2-hydroxypropanesulfonic acid
AMPT	2-amino-2-methyl-1-propanethiol
AMPT	α-methyl-D,L-tyrosine methyl ester hydrochloride
AMS	ammonium sulfamate
AMSO	*N*-acetyl (*S*)-methionine (*R,S*)-sulfoxide
Amt	*tert*-amyl; 1,1-dimethylpropyl
AMTCS	amyltrichlorosilane
AmtOH	*t*-amyl alcohol; 2-methyl-2-butanol
AN	acetonitrile
2-AN	2-acetonaphthone
An	anisyl
3-ANA	3-aminonocardicinic acid
ANB-NOS	azido-2-nitrobenzoyloxysuccinimide
14-Ane-S4	1,4,8,11-tetrathiacyclotetradecane
ANI	arylnitrenium ions
ANL	*Aspergillus niger* lipase
ANM	*N*-(4-anilino-1-naphthyl)maleimide
ANP	atrial natriuretic peptides
ANPO	2-(1-naphthyl)-5-phenyloxazole
ANPP	4-azido-2-nitrophenyl phosphate
ANS (1,8-ANS)	anilino-8-naphthalenesulfonic acid
Ans	ansyl
ANSA	8-anilino-1-naphthalenesulfonic acid
ansyl	8-anilino-1-naphthalenesulfonate
ANTI	3-acetoxy-*N*-trimethylanilinium iodide
ANTU	1-(1-naphthyl)-2-thiourea
AOAA	aminooxyacetic acid
AOC	allyloxycarbonyl
p-AOM	*p*-anisyloxymethyl; 4-(methoxyphenoxy)methyl
AOT	bis(2-ethylhexyl)sulfosuccinate sodium salt

4-AP	4-aminopyridine
6-APA	6-aminopenicillanic acid
APAAP	alkaline phosphatase-anti-alkaline phosphatase
APAD	3-acetylpyridine adenine dinucleotide
APANS	thorin
APAP	4-acetamidophenol; *N*-acetyl-*p*-aminophenol
AP-Cl	2-acetoxypropionyl chloride
APDC	1-pyrrolidinecarbodithioic acid, ammonium salt; ammonium 1-pyrrolidinecarbodithioate
APDC	pyrrolidinedithiocarbamate
APDP	4-(*p*-azidosalicylamido) butyl-3′-(2′-pyridyldithio)propionamide
APDTC	1-pyrrolidinecarbodithioic acid, ammonium salt; ammonium pyrrolidinedithiocarbamate
APE	alkylphenol ethoxylates
APF	animal protein factor
APG	alkyl polyglycosides
APG	*p*-azidophenylglyoxal hydrate
APH	acetone phenylhydrazone
AP-M	leucine aminopeptidase
APM	aminoglycoside 3′-phosphotransferase
APME	*N*-acetyl phenylalanine methyl ester
APMSF (p-APMSF)	4-amidinophenylmethanesulfonyl fluoride
Apo-A-1	Apolipoprotein A-1
APP	cocarboxylase tetrahydrate
APPEI	2-{[*N*-(1-phenylethyl)imino]ethyl}pyridine
APPM	*N*-acetyl-4-diphenylphosphino-2-diphenylphosphinomethyl pyrrolidine
APPS	adenosine 3′-phosphate-5′-phosphosulfate
APRTase	adenine phosphoribosyltransferase
APS	adenosine 5′-phosphosulfate
APTP	*N*-(4-azidophenylthio)phthalimide
APV	2-amino-5-phosphonovaleric acid
Ar	aryl
Ara	arabinose
Ara-5-P	arabinose 5′-phosphate
Ara-A	adenine 9-β-D-arabinofuranoside
Ara-C (ara-C)	cytosine β-D-arabinofuranoside; cytosine arabinoside
araCTP	cytosine arabinoside 5′-triphosphate
Arg	arginine
Arg	arginyl
ArgP	arginine phosphate
5-ASA	5-aminosalicylic acid
Asa	β-carboxyaspartic acid

ASB-n	ammonium sulfobetaine-n, where n = 1 to 4
ASBA	4-(*p*-azidosalicylamido)butylamine
ASC	*p*-acetylaminobenzenesulfonyl chloride; *N*-acetylsulfanilyl chloride
ASIB	1-(*p*-azidosalicylamido)-4-(iodoacetamido)butane
ASLA-Cl	acetyl (*S*)-lactyl chloride
Asn	asparagine
Asn	asparaginyl
ASO	allele-specific oligonucleoside
Asp	aspartate
Asp	aspartic acid
Asp	aspartyl
Asx	aspargine or aspartic acid
ATA	aminotriazole
ATA	anthranilamide
ATA	aurintricarboxylic acid
ATBH	acetone *tert*-butylhydrazone
ATCase	aspartate transcarbamoylase
ATCh-I	acetylthiocholine iodide
ATEE	*N*-acetyl-L-tyrosine ethyl ester
ATENOLOL	4-[2′-hydroxy-3′-(isopropylamino)propoxy]-phenylacetamide
2,4-ATF	2-amino-4-(trifluoromethyl)pyridine
ATP	adenosine 5′-triphosphate; adenosine triphosphate
ATPase	adenosine triphosphatase
ATP[α,β-CH_2]	adenosine-5′-[α,β-methylene]triphosphate tetralithium salt
ATP[β,γ-CH_2]	adenosine-5′-[β,γ-methylene]triphosphate tetralithium salt
ATP[β,γ-NH]	adenosine-5′-[β,γ-imido]triphosphate tetralithium salt
ATPP	adenosine tetraphosphate
ATP-γ-S	adenosine-5′-[γ-thio]triphosphate tetralithium salt
2-ATT	2-acetylthiazole
AVE	allyl vinyl ether
AVG-HCl	L-α-(2-aminoethoxyvinyl)glycine hydrochloride
AVP	arginine vasopressin; vasopressin
1-aza-12-crown-4	1,4,7-trioxa-10-azacyclodecane
1-aza-15-crown-5	1,4,7,10-tetraoxa-13-azacyclopentadecane
1-aza-18-crown-6	1,4,7,10,13-pentaoxa-16-azacyclooctadecane
AZIDO-FDA (Azido-FDA)	azidofluorescein diacetate; 4(5)-azidofluorescein diacetate
Azol	*p*-aminophenol
AZT	3′-azido-3′-deoxythymidine
AZTTP	3′-azido-2′-deoxythymidine
AzUR	6-azauridine

B	nucleoside base (adenine, cytosine, guanine, thymine, or uracil)
B-β-pinanyl-9-BBN	*B-beta*-pinanyl-9-borabicyclo[3.3.1]nonane
B. cereus	*Bacillus cereus*
B. subtilis	*Bacillus subtilis*
BA	6-benzlaminopurine; benzylaminopurine
BA	benzyladenine
BAA	*N*α-benzoyl-L-arginineamide hydrochloride
t-BAA	*tert*-butyl acetoacetate
BAB	branched alkylbenzene
BAC	*N*,*N*'-bis(acryloyl)cystamine
1,3-BAC	1,3-bis(aminomethyl)cyclohexane; 1,3-cyclohexanebis-(methylamine)
BACH	*N*-biotinyl-6-aminocaproic hydazide
BACO	1,4-diazabicyclo[2.2.2]octane
BADEA	bromoacetaldehyde diethyl acetal
BADMA	bromoacetaldehyde dimethyl acetal
BAEE	*N*α-benzoyl-L-arginine ethyl ester
BAK	benzalkonium chloride
BAL	2,3-dimercapto-1-propanol; British anti-Lewisite
BAME	*N*α-benzoyl-L-arginine methyl ester
BAMPITC	4-(benzyloxycarbonyl-aminomethyl)phenyl isothiocyanate
BANA	*N*α-benzoyl-DL-arginine-2-naphthylamide
BANI	*N*α-benzoyl-DL-arginine.4-nitroanilide
BAO	2,5-bis(4-aminophenyl)-1,3,4-oxadiazole
BAP	6-benzylaminopurine
BAP	1,4-bis(acryloyl)piperazine
BAP	bacterial alkaline phosphatase
BAP (BaP)	benzo[a]pyrene
BAPA	*N*α-benzoyl-DL-arginine-4-nitroanilide hydrochloride
BAPABA	*N*α-benzoyl-L-arginyl-4-aminobenzoic acid hydrochloride
BAPNA	*N*α-benzoyl-DL-arginine-4-nitroanilide hydrochloride
BAPTA	1,2-bis(2-aminophenoxy)ethane-*N*,*N*,*N*'*N*'-tetraacetic acid; bis(*O*-aminophenoxy)ethane-*N*,*N*,*N*',*N*'-tetraacetic acid
BAPTA/AM	bis(*O*-aminophenoxy)ethane-*N*,*N*,*N*',*N*'-tetraacetic acid, tetra(acetoxymethyl)-ester
Base L	lycofawcine
BASED	*bis*-[β-(4-azidosalicylamido)ethyl] disulfide
BAT	tropine hydrochloride
BB	bibenzyl
p-BB	*p*-benzylbiphenyl

BB powder	potassium sorbate
BBB	filixic acids
BBC	α-bromobenzyl cyanide
BBD	4-benzylamino-7-nitrobenzofurazan; 7-benzylamino-4-nitrobenz-2-oxa-1,3-diazole
BBD	2,5-bis-(4-biphenylyl)-1,3,4-oxadiazole
BBEDA	*N,N'*-bis(benzylidene)ethylenediamine
9-BBN	9-borabicyclo[3.3.1]nonane
9-BBN	9-borabicyclo[3.3.1]nonyl
9-BBN-H	9-borabicyclo[3.3.1]nonane
BBO	2-(4-biphenylyl)-5-phenyloxazole
BBO	2,5-bis(4-biphenylyl)oxazole
BBOD	2,5-bis(4-biphenylyl)-1,3,4-oxadiazole
BBOT	2,5-bis(5-*tert*-butyl-2-benzoxazolyl)thiophene; 2,5-bis-(5-*t*-butylbenzoxazolyl-[2'])-thiophene
BBP	benzyl butyl phthalate
BBQ	7*H*-benzimidazo[2,1-*a*]benz[*de*]isoquinoline-7-one
BCA	2,2'-biquinoline-4,4'-dicarboxylic acid disodium salt dihydrate
BCA	bicinchoninic acid
BCA	*N*-benzylcyclopropylamine
BCB	bromocresol blue
BCC	benzylchlorocarbene
BCCN	*trans,trans*-4'-butyl[1,1'-bicyclohexyl]-4-carbonitrile
BCDC	*N*-benzylcinchonidinium chloride
BCECF	2',7'-bis(2-carboxyethyl)-4(5)-carboxyfluorescein
BCECF	2',7'-bis(2-carboxyethyl)-5(6)-carboxyfluorescein
BCECF/AM	2',7'-bis(carboxyethyl)-5(6')-carboxyfluorescein, pentaacetoxymethyl ester
BCG	bromocresol green
BCH	2-amino-2-norbornanecarboxylic acid
BCIP	5-bromo-4-chloro-3-indolyl phosphate
BCM	mannomustine
BCNC	*N*-benzylcinchonidinium chloride
BCNU	1,3-bis(2-chloroethyl)-1-nitrosourea
BCP	bromocresol purple
BCP	butyl carbitol piperonylate
BCPB	bromochlorophenol blue
BCPC	*sec*-butyl *N*-(3-chlorophenyl)carbamate
BCS	β-chlorostyrene
Bct	*N*ε-biotinyl-L-lysine
Bd-*t*-BOC	1,1-dimethyl-2,2-dibromoethoxycarbonyl
bda	benzylideneacetone
1,3-BDB	1,3-bis(α-diazobenyl)benzene

BDC-OH	4,4′-bis(dimethylamino)diphenyl carbinol
BDCP	tris(2,4,6-tribromophenoxy)dichlorophosphorane
T-BDCP	*trans*-1,2-bis[diphenylphosphino)methyl]cyclopropane
BDCPH	*trans*-1,2-bis(diphenylphosphinoxy)cyclohexane
BDCS	*tert*-butyldimethylchlorosilane
t-BDEA	*tert*-butyldiethanolamine
BDMA	benzyldimethylamine; *N*,*N*-dimethylbenzylamine
BDMAP	1,6-bis(dimethylamino)pyrene
BDMS	*tert*-butyldimethylsilyl
BDMSi	*tert*-butyldimethylsilyl
BDOG	β-dodecyl glucopyranoside
BDPA	α,γ-bisdiphenylene-β-phenylallyl, free radical
BDPM	*p*-biphenylyldiphenylmethane
BDPOP	2,4-bis[(diphenylphosphino)oxy]pentane
BDPP	bis(diphenylphosphino)pentane
BDT	1,3-benzodithiol-2-yl; 1,3-benzodithiolan-2-yl
BDTF	1,3-benzodithiolylium tetrafluoroborate
2,4-BDTP	2,4-bis(5,6-diphenyl-1,2,4-triazin-3-yl)pyridine
2,4-BDTPS	2,4-bis(5,6-diphenyl-1,2,4-triazin-3-yl)-pyridine tetrasulfonic acid tetrasodium salt
BEDT-TTF	bis(ethylenedithio)tetrathiafulvalene
BEHP	bis(2-ethylhexyl) phthalate
BEMP	2-(*tert*-butylamino)-2-(diethylamino)imino-1,3-dimethylperhydro-1,3,2-diazaphosporine
BES	*N*,*N*-bis(2-hydroxyethyl)-2-aminoethanesulfonic acid
BEST	2-(trimethylsilyl)ethanethiol
BETE	tropine hydrochloride
BF	benzoylformate
1,2-BF	11*H*-benzo[*a*]fluorene
2,3-BF	11*H*-benzo[*b*]fluorene
3,4-BF	7*H*-benzo[*c*]fluorene
BFA	brefeldin A
BFE	bromotrifluoroethylene
BFO	benzofurazan-1-oxide
BFT	2,2-bis(trifluoromethyl)ethylene-1,1-dicarbonitrile
BGE	butyl glycidyl ether
BGIG	5-bromo-4-chloro-3-indolyl-β-D-glucuronic acid
BHA	3-*tert*-butyl-4-hydroxyanisole; butylated hydroxyanisole
BHC	lindane; 1,2,3,4,5,6-hexachlorocyclohexane (incorrectly named as benzene hexachloride)
γ-BHC	lindane pesticide
BHMF	2,5-bis(hydroxymethyl)furan; 2,5-furandimethanol
BHMT	bis(hexamethylene)triamine
BHPA	1-(β-hydroxyethyl)piperazine

BHT	2,6-di-*t*-butyl-4-methylphenol; butylated hydroxytoluene; 2,6-di-*tert*-butyl-*p*-cresol
BHYMA	bis(hydroxymethyl)acetaldehyde
BiBLEs	bibracchial lariat ethers
Bic	5-benzisoxazolylmethoxycarbonyl
BICINE (Bicine) (bicine)	*N*,*N*-bis(2-hydroxyethyl)glycine
4-BIG	bis[4(5)imidazoyl] glycolic acid
BIGCAP	*N*,*N*′-[(octanoylimino)bis(trimethyllene)]-bis-(D-gluconamide)
BIGCHAP (Big CHAP) (BigCHAP)	*N*,*N*-bis[3-(D-gluconamido)propyl]cholamide
BINAL-H	lithium (1,1′-binaphthalene-2,2′-diolato)-(ethanolato)hydridoaluminate
BINAP	2,2′-bis(diphenylphosphino)-1,1′-binaphthalene
BINAP	2,2′-bis(diphenylphosphino)-1,1′-binaphthyl
BINAPO	2,2′-bis(diphenylphosphono)-1,1′-binaphthyl
BINOL	1,1-bi-2-naphthol
BIO-11-UTP	5-(*N*-[*N*-biotinyl-ε-aminocaproyl]-3-aminoallyl)uridine 5′-triphosphate
BIO-14-ATP	8-(*N*-[*N*-biotinyl-ε-aminocaproyl]-8-aminohexylamino)adenosine 5′-triphosphate
BIO-17-ATP	N^6-(*N*-[*N*-biotinyl-ε-aminocaproyl]-6-aminohexylcarbamoylmethyl) adenosine 5′-triphosphate
BIO-4-dUTP	5-([*N*-biotinyl]-3-aminoallyl)-2′-deoxyuridine 5′-triphosphate
BIO-4-UTP	5-([*N*-biotinyl]-3-aminoallyl)uridine 5′-triphosphate
Biotin-BMCC	1-biotinamido-4-[4′-(maleimidoethyl)cyclohexane-carboxamido]butane
Biotin-HPDP	*N*-[6-(biotinamido)hexyl]-3′-(2″-pyridylidithio) propionamide
Biotin-LC-DPPE	*N*-(6-((biotinoyl)amino)hexanoyl)dipalmitoyl-L-α-phosphatidylethanolamine, triethylammonium salt
BIPM	*N*-[4-(2-benzimidazoyl)phenyl]maleimide
BIPY	2,3-bipyridine
bipy	2,2′-bipyridine
bipy	2,2′-bipyridyl; bipyridyl
BIS	bis-acrylamide, methylene
bis-AMP	*N*-bis(hydroxyethyl)-2-amino-2-methyl-1-propanol
bis-ANS	bis(5,5′)-8-anilino-1-naphthalenesulfonic acid
BISBI	2,2′-bis[(diphenylphosphino)methyl]-1,1′-biphenyl
Bis-DHA	dehydro-L(+)-ascorbic acid, dimer

bis(HOMOTRIS) [Bis-homotris] [bis(homotris)]	tris(3-hydroxypropyl)aminomethane; 4-amino-4-(3-hydroxypropyl)-1,7-heptanediol
BIS-MSB (Bis-MSB) (bis-MSB)	1,4-bis(2-methylstyryl)benzene
bis-NAD	N^2,2-*N*-adipodihydrazido-bis-(N^6-carbonylmethyl)-nicotinamide adenine dinucleotide
bispicen	*N,N'*-bis(2-pyridylmethyl)-1,2-ethanediamine
bispichxn	*N,N*-bis(2-pyridylmethyl)-1,2-cyclohexanediamine
bispictn	*N,N*-bis(2-pyridylmethyl)-1,3-propanediamine
bispin	*N,N*-bis(2-pyridylmethyl)-2,3-butanediamine
BIS-TRIS (bis-Tris)	2-[bis(2-hydroxyethyl)amino]-2-(hydroxymethyl)-1,3-propanediol; [bis(2-hydroxyethyl)amino]tris-(hydroxymethyl)methane; 2,2-bis(hydroxymethyl)-2,2′,2″-nitrilotriethanol
BIS-TRIS Propane	1,3-bis(tris(hydroxymethyl)methylamino)propane
BLM	bleomycin
BLO	γ-butyrolactone
BLUO-GAL	5-bromoindolyl-β-D-galactopyranoside
BMC	4-bromomethyl-7-methoxycoumarin
BMC	butadiene-maleic acid copolymer
BMC (BMC-P) (BMC-S)	sulfuryl chloride, disulfur dichloride, and aluminum trichloride mixtures
BMDMCS	(bromomethyl)dimethylchlorosilane
BMDT-TTF	bis(methylenedithio)tetrathiafulvalene
BMH	bismaleimidohexane
BMME	bis-maleimidomethyl ether
Bmpc	2,4-dimethylthiophenoxycarbonyl
Bmpm	1,1-bis(4-methoxyphenyl)-1′-pyrenylmethyl
BMPP	benzylmethylphenylphosphine
BMS	borane dimethylsulfide complex; borane-methyl sulfide complex
BN	benzonitrile
Bn	benzyl
BNAH	1-benzyl-1,4-dihydronicotinamide; *N*-benzyl-1,4-dihydronicotinamide
BNB	2,4,6-tri-*tert*-butylnitrosobenzene
BNEN	bicyclo[4.3.0]non-1(6)-en-2-one
BNHS	(+)-biotin *N*-hydroxysuccinimide ester
BNOA	2-naphthoxyacetic acid
BNP	(+)-biotin 4-nitrophenyl ester
BNPP	bis-(*p*-nitrophenyl)phosphate

BNPS-SKATOLE (BNPS-Skatole) (BNPS-skatol)	2-(2-nitrophenylsulfenyl)-3-methyl-3′-bromoindolenine; 3-bromo-3-methyl-2-(2-nitrophenylmercapto)3*H*-indole;
BnZ (Bnz)	benzyl
Bo-XAN	xanthophyll
BOA	benzoxazolinone
BOA	*o*-butoxyacetanilide
BOC (Boc) (t-Boc)	*tert*-butoxycarbonyl; carbo-*tert*-butoxy; *t*-butoxycarbonyl
BOC-BMI	1-(*tert*-butoxycarbonyl)-2-*tert*-butyl-3-methyl-4-imidazolidinone
$(BOC)_2NH$	di-*tert*-butyl-iminodicarboxylate
$(BOC)_2O$	di-*tert*-butyl-dicarbonate
BOC-ON	2-(*tert*-butoxycarbonyloxyimino)-2-phenylacetonitrile
BOC-O-PCP	*tert*-butyl pentachlorophenyl carbonate
BOC-OSU	*N*-*tert*-butoxycarbonyloxysuccinimide
BOC-OTCE	1,2,2,2-tetrachloroethyl-*tert*-butylcarbonate
BOC-OTCP	*tert*-butyl 2,4,5-trichlorophenyl carbonate
BOC-OX	*tert*-butyl 2,3-dihydro-2-oxo-3-oxazolecarboxylate
BOC-S	*S*-*tert*-butoxycarbonyl-2-mercapto-4,6-dimethylpyrimidine
BOE	ethyl biscoumacetate
BOM	(benzyloxy)methyl
BOM	*tert*-butoxymethyl
BON	3-hydroxy-2-naphthoic acid; β-oxynaphthoic acid
BOP	(benzotriazolyloxy)tris(dimethylamino) phosphonium hexafluorophosphate; benzotriazol-1-yloxytris-(dimethylamino)phosphonium hexafluorophosphate; benzotriazolyl-*N*-oxy-tris(dimethylamino)phosphonium hexafluorophosphate
BOP-Cl	bis(2-oxo-3-oxazolidinyl)phosphinic chloride; *N*,*N*-bis(2-oxo-3-oxazolidinyl)phosphinic chloride
BORONAL	1-hydroxy-4-(*p*-toluidino)anthraquinone
BOT	borane-1,4-thioxane
BP	benzophenone
B-PABA	*N*-(+)-biotinyl-4-aminobenzoic acid
BPAS	4-benzamidosalicylic acid, sodium salts
BPB	bromophenol blue
BPBG	butyl phthalyl butyl glycolate
BpBr	4-bromobiphenyl
BPC	*n*-butylpyridinium chloride
BPCC	2,2′-bipyridinium chlorochromate
BPDODP	[(diphenylphosphino)oxy]-1,3-diphenylpropane
BPDT	2-*p*-biphenylyl-1,3-dithiane
BPE	bis(phosphino)ethanes
BPEA	9,10-bis(phenylethynyl)anthracene

bpea	*N,N*-bis(2-pyridylmethyl)ethylamine
BPEN	5,10-bis(phenylethynyl)naphthacene
BPEN	5,12-bis(phenylethynyl)naphthacene
9-BpFl	9-biphenylylfluorene
BPG	2,3-bisphosphoglycerate
1,3-BPG	1,3-bisphosphoglycerate
BPM	bipiperidyl mustard
BPMC	2-*sec*-butylphenyl-*N*-methylcarbamate
BPME	*N*-benzyl phenylalanine methyl ester
BPO	2-(4-biphenylyl)-5-phenyloxazole
BPOC reagent	2-(4-biphenyl)-prop-2-yl 4′-methoxycarbonylphenyl carbonate
Bpoc	1-methyl-1-(4-biphenyl)ethoxycarbonyl; biphenylisopropyloxycarbonyl
BPPFA	*N,N*-dimethyl-1,1-(1,2-bis(diphenylphosphino)-ferrocenyl)ethylamine; *N,N*-dimethyl{1-[2-(diphenylphosphino)ferrocenyl]ethyl}amine
BPPFA-IP	*N,N*-dimethyl-1-[1′,2-bis(diphenylphosphino)ferrocenyl]-2-methylpropylamine
BPPFA-Ph	*N,N*-dimethyl-1-[1′,2-bis(diphenylphosphino)ferrocenyl] benzylamine
BPPFOH	1-[1′,2-bis(diphenylphosphino)ferrocenyl]ethanol
BPPM	*N-tert*-butoxycarbonyl-4-diphenylphosphino-2-diphenylphosphinomethylpyrrolidine; *N*-(butoxy-carbonyl)-4-[(diphenylphosphino)methyl]pyrrolidine
BPPOH	1-[1′,2-bis(diphenylphosphino)ferrocenyl]ethanol
BPR	bromophenol red
BPST	2-(2′-benzothiazolyl)-5-styryl-3-(4′-phthalhydrazidyl)-tetrazolium
BPTI	bovine pancreatic trypsin inhibitor
bpy	2,2′-bipyridine
bpy	2,2′-bipyridyl
BQ	3-benzoyl-quinoline-2-carboxaldehyde
BQ	1,4-benzoquinone; *p*-benzoquinone
BQC reagent	2,6-dibromoquinone-4-chlorimide
BR	polybutadiene rubber
bR	bacteriorhodopsin
BRA	β-resorylic acid
BrAntr	bromoanthracene
Br-DMEQ	3-bromomethyl-6,7-dimethoxy-1-methyl-2(1*H*)-chinoxalinone
BrDNFB	5-bromo-2,4-dinitrofluorobenzene
5-BrdU	5-bromo-2′-deoxyuridine
Br-MB	3-bromomethyl-7-methoxy-1,4-benzoxazin-2-one

Br-Mmc	4-bromomethyl-7-methoxycoumarin; 7-methoxy-4-bromomethylcoumarin
Bromo-PADAP	2-(5′-bromo-2′-pyridylazo)-5-(diethylamino)phenol
BRP	"bis-roof" porphyrin
5-Br-PSAA	2-(5-bromo-2-pyridylazo)-5-(*N*-propyl-3-sulfo-propylamino)aniline sodium salt
Bs	brosylate; *p*-bromobenzenesulfonate
BS^3	bis(sulfosuccinimidyl) suberate
BSA	benzenesulfonic acid
BSA	bovine serum albumin
BSA	*N*,*N*-bis(trimethylsiyl)acetamide
BSA	*N*,*O*-bis(trimethylsilyl)acetamide
BSC	*N*,*O*-bis(trimethylsilyl) carbamate
BSF	*N*,*N*-bis(trimethylsilyl)formamide
BSH	benzenesulfonyl hydrazide
BSLDH	*Bacillus stearothermophilius* lactate dehydrogenase
BSO	buthionine-sulfoximine
BSOCOES	bis[2(succinimidooxycarbonyloxy)ethyl] sulfone
BSPT	2-(2′-benzothiazolyl)-5-styryl-3-(4′-phthalhydrazidyl)tetrazolium
BSS	bis(trimethylsilyl)sulfate
BST	bovine somatotropin
BST	2-(2′-benzothiazolyl)-5-styryl-3-(4′-phthalhydrazidyl)tetrazolium
BSTFA	bis(trimethylsilyl)trifluoroacetamide; *N*,*O*-bis(trimethylsilyl)trifluoroacetamide
BSU	*N*,*N*′-bis(trimethylsilyl)urea
BT	blue tetrazolium
Bt	benzotriazol-1-yl; benzotriazolyl
BTA	benzoyltrifluoroacetone
BTAC	benzyltrimethylammonium chloride
BTAF	benzyltrimethylammonium fluoride
BTAM	benzyltrimethylammonium hydroxide
BTB	bromothymol blue
BTBA-BDTD Zn Complex	bis(tetra-*n*-butylammonium)bis(1,3-dithiole-2-thione-4,5-dithiolato) zinc complex
BTC	benzalkonium chloride
BTC	*N*,*N*′-bis(trimethylsilyl)carbodiimide
BTC	tetrazolium blue chloride
BTCEAD	bis(2,2,2-trichloroethyl)azodicarboxylate
BTDA	3,3′,4,4′-benzophenonetetracarboxylic dianhydride
BTE	tropine benzylate
Btea	benzyltriethylammonium (cation)

BTEAC	benzyltriethylammonium chloride
BTEE	*N*-benzoyl-L-tyrosine ethyl ester
BTF	2,2-bis(trifluoromethyl)ethylene-1,1-dicarbonitrile
BTF	benzotrifluoride
BTFA	bis(trifluoroacetamide)
BTFP	2-bromotrifluoropropene
BTI	[bis(trifluoroacetoxy)iodo]benzene
BTMA	benzyltrimethylammonium tetrachloroiodate
BTMA (Btma)	benzyltrimethylammonium (cation)
BTMS	bromotrimethylsilane
BTMSA	bis(trimethylsilyl)acetylene
BTMSA	*N*,*O*-bis(trimethylsilyl)acetamide
BTMSBD	1,4-bis(trimethylsilyl)-1,3-butadiyne
BTMSPO	bis(trimethylsilyl)peroxide
BTS	bithionol (sulfoxide); 2,2′-bis(4,6-dichlorophenol) sulfoxide
Bu	butyl
Bum	*t*-butoxymethyl
t-Bumeoc	1-(3,5-di-*t*-butylphenyl)-1-methylethoxycarbonyl
BuOH	butanol
Butyl-PBD	2-(4-biphenylyl)-5-(4-*tert*-butylphenyl)-1,3,4-oxadiazole; 2-(4-*tert*-butylphenyl)-5-(4-biphenylyl)-1,3,4-oxadiazole
BVU	bromisovalum
BY	baker's yeast
Bz	benzyl
Bz (bz)	benzoyl
Bzac	benzoylacetone
BZL (Bzl) (bzl)	benzyl
BzMS	benzyl methyl sulfide
BzN	benzonitrile
BzOH	benzoic acid
BzOOH	perbenzoic acid
BZPPM	*N*-benzoyl-4-diphenylphosphino-2-diphenylphosphinomethyl pyrrolidine
BZQ	benquinamide
C	cysteine
C	cytidine
C	cytosine
C Acid	3-amino-1,5-naphthalenedisulfonic acid
c	cyclo
CA	α-bromobenzyl cyanide
CA	carbonic anhydrase
Ca-ATPase	calcium pump
CAB	cellulose-acetate-butyrate (polymer)

CAB	chiral (acyloxy)borane
CAB	cholesteryl 4-(2-anthryloxy)butyrate
CADEA	chloroacetaldehyde diethyl acetal
CADMA	chloroacetaldehyde dimethyl acetal
caged GTP	guanosine-5′-triphosphate P^3-[1-(2-nitrophenyl)ethyl ester] disodium salt
CAI	carbonic anhydrase inhibitor
CaM	calmodulin
Cam	camphanoyl
Cam	carboxamidomethyl
CAMP	cyclohexylanisylmethylphosphine
Camp	camphor
cAMP (c-AMP)	adenosine cyclic 3′,5′-phosphate; adenosine 3′,5′-cyclic phosphate; adenosine 3′,5′-phosphate; cyclic AMP
CAMPOS	1,3-bis[(diphenylphosphino)methyl]-1,2,2-trimethylcyclopentane
CaMU	cauliflower mosaic virus
CAN	ceric ammonium nitrate; ammonium cerium(IV) nitrate
CAP	calf intestine alkaline phosphatase
CAP	catabolite activator protein
CAP	cellulose acetate phthalate
CAP	cellulose-acetate-propionate (polymer)
CAP-Li_2	carbamoyl phosphate, dilithium salt
CAPP	carbomyl-4-diphenylphosphino-2-diphenylphosphinopyrrolidines
CAPPM	*N*-formamido-4-diphenylphosphino-2-diphenylphosphinomethyl pyrrolidine
CAPS	3-cyclohexylamino-1-propanesulfonic acid
CAPSO	3-(cyclohexylamino)-2-hydroxy-1-propanesulfonic acid
CAS	ceric ammonium sulfate
CAT	2-chloro-4,6-bis(ethylamino)-*s*-triazine
CAT	carnitine acetyltransferase
CAT	chloramphenicol acetyltransferase
cat	catalytic
CatBH	catecholborane
Cathyl	ethoxycarbonyl; carbethoxy
CB	4-carboxybenzophenone
CB	catecholborane
Cb	benzyloxycarbonyl; carbobenzoxy
CBA (P-CBA) (p-CBA)	4-carboxybenzaldehyde
CBA	chlorobenzoic acid
CBC	carbomethoxybenzenesulfonyl chloride; methyl-2-(chlorosulfonyl)benzoate

CBI	1,2,9,9a-tetrahydrocyclopropa[1,2-c]benz[1,2-e]indol-4-one
Cbl	cobalamin
CBMIT	1,1′-carbonylbis(3-methylimidalolium) triflate
CBn	benzyloxycarbonyl; carbobenzoxy
CBQ	3-(4-carboxybenzoyl)-quinoline-2-carboxaldehyde
CBZ (Cbz)	benzyloxycarbonyl; carbobenzoxy; carbobenzyloxy
CBz-HONB	*N*-benzyloxycarbonyloxy-5-norbornene-2,3-dicarboximide
CCC	2-dichloroethyltrimethylammonium chloride
CCC	calcium cyanamide citrated
CCCP	carbonylcyanide-3-chlorophenylhydrazone
β-CCE	ethyl 9*H*-pyrido[3,4-*b*]indole-3-carboxylate
CCH	cyclohexylidenecyclohexane
CCK	cholecystokinin
CCL	*Candida cylindracea* lipase
CCNU	1-(2-chloroethyl)-3-cyclohexyl-1-nitrosourea
CcO	cytochrome *c* oxidase
CCP	cytochrome *c*-peroxidase
CD	cyclodextrin
α-CD	α-cyclodextrin
β-CD	β-cyclodextrin
γ-CD	γ-cyclodextrin
CDA	cytidine deaminase
CDAA	2-chloro-*N*,*N*-di-2-propenylacetamide; chlorodiallylacetamide; *N*,*N*-diallyl-2-chloroacetamide
CDC	cycloheptaarylose-dansyl chloride complex
CDD	chlorinated dibenzo-*p*-dioxin
CDEC	2-chloroallyl *N*,*N*-diethyldithiocarbamate; diethyldithiocarbamic acid, 2-chloroallyl ester
CDF	chlorinated dibenzofurane
CDFBA	2-chloro-4,5-difluorobenzoic acid
CDH	ceramide dihexosides
CDHPDAB	cetyl(2,3-dihydroxyethyl)dimethylammonium bromide
CDI	1,1′-carbonyldiimidazole; carbonyldiimidazole
CDNA (cDNA)	complementary DNA
CDNT	4-chloro-3,5-dinitrobenzotrifluoride
CDP	cytidine 5′-diphosphate
CDPI	methyl N^3-carbamoyl-1,2-dihydro-3*H*-pyrrolo[3,2-e]-indole-7-carboxylate
CDPP	cyclohexyldiphenylphosphine
CDT	1,1′-carbonyldi(1,2,4-triazole)
CDTA	*trans*-1,2-diaminocyclohexane-*N*,*N*,*N*′,*N*′-tetraacetic acid
CE	cholesterol esterase

CE	cyanoethyl
CEA	2-cyanoethyl *N,N*-diisopropylphosphoramidite
CEA	carcinoembryonic antigen
Cee	1-(2-chloroethoxy)ethyl
CEEA	*N*-(2-cyanoethyl)-*N*-ethylamine
CEEMT	*N*-(2-cyanoethyl)-*N*-ethyl-*m*-toluidine
CEMA	*N*-(2-cyanoethyl)-*N*-methylaniline
CEPEA	*N*-(2-hydroxyethyl)-*N*-(2-cyanoethyl)aniline
CETAB	hexadecyltrimethylammonium bromide
CF	5(6)-carboxyfluorescein
5-CF	5-carboxyfluorescein
6-CF	6-carboxyfluorescein
CFC-11	trichlororofluoromethane
CFC-111	1,1,1,2,2-pentachloro-2-fluoroethane
CFC-112	1,1,1,2,-tetrachloro-2,2-fluoroethane
CFC-113	trichlorotrifluoroethane
CFC-114	dichlorotetrafluoroethane
CFC-115	chloropentafluoroethane
CFC-12	dichlorodifluoromethane
CFC-13	chlorotrifluoromethane
CFC-211	heptachlorofluoropropane
CFC-212	hexachlorodifluoropropane
CFC-213	pentachlorotrifluoropropane
CFC-214	tetrachlorotetrafluoropropane
CFC-215	trichloropentafluoropropane
CFC-216	dichlorohexafluoropropane
CFC-217	chloroheptafluoropropane
CFCs	chlorofluorocarbons
5-CFDA	5-carboxyfluorescein diacetate
6-CFDA	6-carboxyfluorescein diacetate
CFX™	poly(carbon monofluoride)
CGA	cytosylglucuronic acid
CGRP	calcitonin gene-related peptide
CH	cyclohexenone
CH	cycloheximide
CHA	cyclohexylamine
CHAPS	3-[(3-cholamidopropyl)dimethylammonio]-1-propanesulfonate
CHAPSO	3-[(3-cholamidopropyl)dimethylammonio]-2-hydroxy-1-propanesulfonate
1,4-CHD	1,4-cyclohexadiene
CHDA	cyclohexanedicarboxylic acid
CHDP	cyclohexanone cyclic diperoxide
ChE	cholinesterase

CHECK tungsten 7077	cetylpyridinium tetrakis(diperoxotungsten)phosphate; tetrahexylammonium tetrakis(diperoxotungsten)phosphate
CHEDAB	cetyl(2-hydroxyethyl)dimethylammonium bromide
CHES	2-(cyclohexylamino)ethanesulfonic acid
CHIRAPHOS (chiraphos)	bis(diphenylphosphino)butane; 2,3-bis(diphenylphosphino)butane
Chl	chlorophyll
chl	chloroform
Chloro-BPEA	chloro-9,10-bis(phenylethynyl)anthracene
CHLORO-IPC (Chloro IPC)	isopropyl-*N*-(3-chlorophenyl)carbamate; *m*-chlorocarbanilic acid, isopropyl ester
CHMI	5-(carboxymethyl)-2-hydroxymuconate isomerase
CHP	1-cyclohexyl-2-pyrrolidinone; *N*-cyclohexylpyrrolidinone
CHPDAB	cetyl(2-hydroxypropyl)dimethylammonium bromide
CHS	chalcone synthase
CHT	cycloheptatriene
Chx (cHx)	cyclohexyl
Chx_2BH	dicyclohexylborane
CIAP	calf intestinal alkaline phosphatase
CIPC	isopropyl *m*-chlorocarbanilate; *m*-chlorocarbanilic acid, isopropyl ester
Cipca	ascorbic acid
5-ClA	5-chloroisatoic anhydride
CLAYCOP	clay-supported cupric nitrate
CLB	*p*-chlorobenzoate
2-ClBP	2-chlorobiphenyl
3-ClBP	3-chlorobiphenyl
4-ClBP	4-chlorobiphenyl
m-Cl-CCP	carbonylcyanide-3-chlorophenylhydrazone
CLE	*Candida lipolytica* esterase
CLECs	cross-linked enzyme crystals
7-Cl-KYNA	7-chloro-4-hydroxyquinoline-2-carboxylic acid
CLIP	corticotropin-like intermediate lobe peptide
CIPO	bis(2-carboisopentyloxy-3,5,6-trichlorophenyl)oxalate
ClQ	2-chloroquinoline
CM	calmodulin
CM	carboxymethyl (as in CM-cellulose)
CM-D	carboxymethyl dextran
CMA	calcium magnesium acetate
CMA	calcium methanearsonate
CMA	carbomethoxymaleic anhydride
CMC	1-cyclohexyl-3-(2-morpholinoethyl)carbodiimide
CMC	carboxymethylcellulose

CMC	chloromethylcarbene
CMC	sodium carboxymethyl cellulose
CMDMCS	(chloromethyl)dimethylchlorosilane; chloro(chloromethyl)dimethylsilane
CMDO	chloromethyl-1,3-dioxolane
CMNT	4-chloro-3-nitrobenzotrifluoride
CMOS	(*S*)-[(carboxymethyl)oxy]succinic acid
CMP	cytidine 5′-monophosphate; cytidine 5′-phosphate
3′(+2′)-CMP	cytidylic acid
CMP-NeuAc	cytidine 5′-monophospho-*N*-acetylneuraminic acid
CMPI	2-chloro-1-methylpyridium chloride
CMPO	octyl(phenyl)-*N*,*N*-diisobutylcarbamoylmethylphosphine oxide
CMPP	2-(2-methyl-4-chlorophenoxy)propionic acid; 2-(4-chloro-*o*-tolyloxy)propionic acid
CMU	3-(*p*-chlorophenyl)-1,1-dimethylurea
CN	1-cyanonaphthalene
CN-Cbi	vitamin B_{12}
CNAH	4-chloro-2-nitroaniline
p-CNB	*p*-cyanobenzophenone
CNN	ceramide monohexosides
CNT	cyanotoluene
CNT	*m*-tolunitrile
CNT	*o*-tolunitrile
CNT	*p*-tolunitrile
CO	cholesterol oxidase
Co I	coenzyme I
Co II	coenzyme II
CoA (CoASH)	coenzyme A
Coc	cinnamyloxycarbonyl
COD (cod) (1,5-COD)	1,5-cyclooctadiene; *cis*,*cis*-1,5-cyclooctadiene; cyclooctadiene
coe	cyclooctene
COHED	5-(carboxymethyl)-2-oxo-3-hexene-1,6-dioate decarboxylase
COMT	catechol-*O*-methyl transferase
CoQ	coenzyme Q; ubiquinone
COT	carnitine octanoyltransferase
COT	cyclooctatetraene
CP	cyclopropene
CP	cyclopentene
3-CP	2-(*m*-chlorophenoxy)propionic acid
6-CP	6-chloropurine
Cp	cyclopentadiene

Cp (cp)	cyclopentadienyl; η^5-cyclopentadienyl
Cp* (cp*)	η^5-pentamethylcyclopentadienyl
CPA	carboxypeptidase A
CPA	cyclopropylamine
4CPA (4-CPA)	4-chlorophenoxyacetic acid
CPC	cetylpyridinium chloride
CPCBS	*p*-chlorophenyl-*p*-chlorobenzene sulfonate
CPD	cyclopropeniumyldiazonium dication
CPI	7-methyl-1,2,8,8a-tetrahydrocyclopropa[1,2-*c*]pyrrolo[3,2-e]indol-4(5*H*)-one
CPI	cyclopropanimines
CPK	creatine phosphokinase
CPMC	*O*-chlorophenyl-methyl-carbamate
CPO	chloroperoxidase
CPO	cyclopropanone
CPP	carboxypeptidase P
CPP	copalyl pyrophosphate
CPP	cyclopental[*cd*]pyrene
CPPM	*N*-cholesteryloxycarbonyl-4-diphenylphosphino-2-diphenylphosphinomethyl pyrrolidine
CPPO	bis(2-carbopentyloxy-3,5,6-trichlorophenyl) oxalate
CPR	chlorophenol red
CPRG	chlorophenol red-β-D-galactopyranoside
CPS-I	carbamyl phosphate synthetase-I
CPS-II	carbamyl phosphate synthetase-II
CPT	carnitine palmityltransferase; carnitine acyl-transferases
CPTn	carnitine palmityltransferase, where n = 1 or 2
CPTA	2-(4-chlorophenylthio)trimethylamine chloride
CPTA	*N*,*N*-diethylaminoethyl 4-chlorothiophenyl ether
CPTEO	3-chloropropyltriethoxysilane
CPTMO	3-(chloropropyl)trimethoxysilane
CPTPase	carboxypeptidase transpeptidase
CPTr	4,4′,4″-tris(4,5-dichlorophthalimido)triphenylmethyl
CPVC	chlorinated polyvinyl chloride
CPX	1,3-dipropyl-8-cyclopentylxanthine
Cr	creatine phosphate
CRA	complex reducing agent
CRF	corticotropin releasing factor
CRH	corticotrophic releasing hormone; corticotropin releasing factor
CRL	*Candida rugosa* lipase
cRNA	complementary RNA
12-Crown-4 (12-crown-4)	1,4,7,10-tetraoxacyclododecane

15-Crown-5 (15-crown-5)	1,4,7,10,13-pentaoxacyclopentadecane
18-Crown-6 (18-crown-6)	1,4,7,10,13,16-hexaoxacyclooctadecane
CRP	C-reactive protein
CRP	camp receptor protein
CS	citrate synthase
CS	cumenesulfonate
CSA	camphorsulfonic acid
CSA	chondroitin 4-sulfate; chondroitin sulfate A
CSF	colony-stimulating factor
CSH	cysteamine hydrochloride
CSI	chlorosulfonyl isocyanate
CSP	calf spleen phosphodiesterase
CSSC	cysteamine hydrochloride
C-Stuff	hydrazine hydrate and methanol
α-CT	α-chymotrpsin
CTA	citraconic anhydrlde
CTAB	cetyltrimethylammonium bromide; cetrimonium bromide; hexadecyltrimethylammonium bromide
CTABr	cetyltrimethylammonium bromide
CTACl	cetyltrimethylammonium chloride
CTACN	cetyltrimethylammonium cyanide
CTAOH	cetyltrimethylammonium hydroxide
CTc	chlorotetracycline
CTEA	*N*-cyano-*N*,*N*,*N*-triethylammonium tetrafluoroborate
2,3-CTF	2-chloro-3-(trifluoromethyl)pyridine
2,4-CTF	2-chloro-4-(trifluoromethyl)pyridine
2,5-CTF	2-chloro-5-(trifluoromethyl)pyridine
2,6-CTF	2-chloro-6-(trifluoromethyl)pyridine
3,5-CTF	3-chloro-5-(trifluoromethyl)pyridine
CTFE	chlorotrifluoroethylene
CTFE	chlorotrifluoroethylene polymer
CTH	ceramide trihexosides
CTMP	1-[(2-chloro-4-methyl)phenyl]-4-methoxypiperidin-4-yl
CTMS	chlorotrimethylsilane
CTP (5-CTP)	cytidine 5′-triphosphate; cytidine triphosphate
CTPS	*sym* collidinium toluene-4-sulfonate
CVD	β-carboxyvinyldiazonium ions
CVL	*Chromobacterium viscosum* lipase
Cy	cyclohexyl
Cy^{3+}	*N*,*N*′-bis[3-(trimethylammonio)propyl]thiadicarbocyanine tribromide
CYAP	*O*,*O*-dimethyl *O*-(*p*-cyanophenyl) phosphorothioate

cyclam	1,4,8,11-tetraazacyclotetradecane
Cyclic AMP (cyclic AMP)	adenosine 3′:5′-cyclic phosphate; adenosine 3′,5′-cyclic monophosphate
cyclic GMP	guanosine 3′,5′-cyclic monophosphate
Cyd	cytidine
CYP	*p*-cyanophenyl ethyl phenylphosphonothioate
Cys	cysteine
Cys	cysteinyl
CySH	cysteine
CySSCy	cystine
Cyt	cytosine
cyt-c	cytochrome c
cytRNA	cytoplasmic RNA
D	2,2′-dithiodibenzoic acid
D	aspartic acid
2,4-D	(2,4-dichlorophenoxy)acetic acid; 2,4-dichlorophenoxyacetic acid
d	2′-deoxyribo
D600	methoxyverapamil
DA	dehydroxylated alumina
DA	dopamine
dA	2′-deoxyadenosine
DAA	diacetone acrylamide
DAA	diacetone alcohol
DAB	3,3′-diaminobenzidine tetrahydrochloride
DAB	diaminobenzidine (usually 3,3′)
DAB	*p*-dimethylaminoazobenzene
10-DAB III	10-deacetylbaccatin III
DABAN	4,4-diaminobenzanilide
DABCO (Dabco)	1,4-diazabicyclo[2.2.2]octane; triethylenediamine
DABIA	*N*-(4-dimethylaminoazobenzene-4′)iodoacetamide
DABITC	4-(*N*,*N*′-dimethylamino)azobenzene-4′-isothiocyanate; dimethylaminoazobenzene isothiocyanate
DABMI	4-dimethylaminophenylazophenyl-4′-maleimide
DABN	2,2′-diamino-1,1′-binaphthalene
DABS-Cl (DABSYL Chloride) (Dabsyl Chloride)	4-(*N*,*N*-dimethylamino)azobenzene-4′-sulfonyl chloride
DAC	α,α-dicyanoethyl acetate
3,5-DACB	3,5-diaminochlorobenzene
DACH	1,2-diaminocyclohexane
DACM	*N*-(7-dimethylamino-4-methyl-3-coumarinyl)maleimide
DACM-3	*N*-(7-dimethylamino-4-methyl-3-coumarinyl)maleinimide

DACPD	2,3-diamino derivative
DAD	diethyl azodicarboxylate
DADA	diisopropylamine dichloroacetate
DADF	diacetyldihydrofluorescein
dADP	deoxyadenose 5′-diphosphate
DAG	diacylglyceride
DAG	diacylglycerol
DAHP	3-deoxy-D-*arabino*-heptulosonic acid 7-phosphate
DAM	di(4-methoxyphenyl)methyl
DAM	diarylmethanol
DAMA-10	didecylmethylamine
DAMN	diaminomaleonitrile
DAMO	*N*-aminoethylaminopropyltrimethoxysilane; diaminotrimethoxysilane; *N*-[3-(trimethoxysilyl)propyl]ethylenediamine
DAMP	diethyl(diazomethyl)phosphonate
dAMP	2′-deoxyadenosine 5′-phosphate; deoxyadenose 5′-monophosphate
DAMSM	4-(dimethylamino) *N*-methyl-4-stilbazolium methylsulfate
DAN	2,3-diaminonaphthalene
DANITO	4-dimethylamino-1-naphthyl isothiocyanate
DANSYL	5-dimethylaminonaphthalene-1-sulfonyl
dansyl	8-(dimethylamino)-1-naphthalenesulfonate
Danthron	1,8-dihydroxyanthraquinone
DAP	ammonium phosphate, dibasic
DAP	diallyl phthalate
DAP	diallyl phthalate (polymer)
DAP	diaminopimelic acid
DAP	diammonium phosphate
β-DAP	β-dimethylaminopropiophenone hydrochloride
DAPI	4′,6-diamidino-2-phenylindole
DAPN	1-([2-(5-dimethylamino)naphthalene-1-sulfonylamino-ethyl]-amino)-3-(1-naphthaleneoxy)-2-propanol
DAPSONE	diaminodiphenyl sulfone
DAS	4,4′-diaminostilbene-2,2′-disulfonic acid
DAS	acetyl sulfide
DAS-Na	9,10-dimethoxyanthracene-2-sulfonic acid sodium salt
DAST	(diethylamino)sulfur trifluoride
DATC	*S*-(2,3-dichloroallyl)diisopropylthiocarbamate
DATD	*N*,*N*′-diallyltartardiamide
2,6,4-DATF	2,6-diamino-4-(trifluoromethyl)pyridine
DATMP	diethylaluminum-2,2,6,6-tetramethylpiperidine
dATP	deoxyadenosine 5′-triphosphate
DAVA	4,5-diaminovaleric acid

DAVLBH	desacetylvinblastine hydrazide
DB	dibromobimane
2,4-DB (4[2,4-DB])	4-(2,4-dichlorophenoxy)butyric acid; 2,4-dichlorophenoxybutyric acid
DBA	9,10-dibromoanthracene
DBA	dibenz[a,*h*]anthracene
DBA (dba)	dibenzylideneacetone; 1,5-diphenyl-1,4-pentadien-3-one
DBB	4,4′-di-*tert*-butylbiphenyl; di-*tert*-butylbiphenylide
DBB	*p*-diazobenzoyl biocytin
3,6-DBCat	3,6-di-*tert*-butylcatecholato
DBC·Br2	dibenzo-1,4,7,10,13,16-hexaoxacyclooctadecane, bromine complex
DBCP	1,2-dibromo-3-chloropropane
DBD	1,2,3-tris(4-carbomethoxyphenyl)propane
DBD-F	4-(*N*,*N*-dimethylaminosulfonyl)-7-fluoro-2,1,3-benzoxadiazole
DBD-Tmoc	2,7-di-*t*-butyl[9-(10,10-dioxo-10,10,10,10-tetrahydrothioxanthyl)]methylcarbonyl
DBDPO	decabromodiphenyl oxide; pentabromophenyl ether
DBDS	dibutyl disulfide
DBDU	5-(dihydroxyboryl)uridine
DBED	*N*,*N*′-dibenzylethylenediamine
DBF	dibenzofuran
DBH	1,3-dibromo-5,5-dimethylhydantoin
DBHA	dibutylated hydroxyanisole
DBIC	dibutylindolocarbazole
DBK	dibenzyl ketone
DBM	dibenzoylmethanato
DBM	*p*,*p*′-dinitrobenzhydryl
DBMC	4,6-di-*tert*-butyl-*m*-cresol
DBMIB	dibromomethylisopropylbenzoquinone
DBN	1,5-diazabicyclo[4.3.0]non-5-ene
2,6-DBN	2,6-dichlorobenzonitrile
DBNE	*N*,*N*-di-*n*-butylnorephedrine
DBP	(*E*)-2,3-dibromo-1-(phenylsulfonyl)-1-propene
DBP	di-*n*-butyl phthalate; dibutyl phthalate
DBPC	2,6-di-*tert*-butyl-4-methylphenol; 2,6-di-*tert*-butyl-*p*-cresol
DBPCl	dibenzylchlorophosphonate
DBPO	di-*tert*-butyl peroxy oxalate
DBPP	tris(2,4-di-*t*-butylphenyl)phosphite
DBS	dibenzosuberyl
DBS	dibutyl sebacate
3,6-DBSQ	3,6-di-*tert*-butylsemiquinone
DBT	dibenzyltriazone

DBTO	di(1-benzotriazolyl)oxalate
DBU	1,8-diazabicyclo[5.4.0]undec-7-ene
DCA	9,10-dicyanoanthracene
DCA	β,β-dicyanoacrylate
DCA	deoxycorticosterone acetate
DCA-O-PCP	pentachlorophenyl dichloroacetate
DCAc	dichloroacetyl
2,4-DCAD	2,4-dichlorobenzaldehyde
DCAF	2′,4′-bis[di(carboxymethyl)aminomethyl]fluorescein; 2,4-bis-[*N*,*N*′-di(carboxymethyl)-aminomethyl]fluorescein
DCB	1,2-dichlorobenzene
DCB	1,3-dicyanobenzene; dicyanobenzene
DCB	1,4-dichlorobutane
2,4-DCBA	2,4-dichlorobenzoic acid
2,4-DCBC	2,4-dichlorobenzyl chloride
2,4′-DCBP	2,4′-dichlorobenzophenone
DCBS	dichloroborane-methyl sulfide
2,4-DCBTF	2,4-dichlorobenzotrifluoride
3,4-DCBTF	3,4-dichlorobenzotrifluoride
DCC	dicyclohexylcarbodiimide; *N*,*N*′-dicyclohexylcarbodiimide
DCCI; (DCCi)	dicyclohexylcarbodiimide; *N*,*N*′-dicyclohexylcarbodiimide
DCDC	2,4-dichlorodichlorotoluene
DCDMS	dichlorodimethylsilane
dCDP	2′-deoxycytidine 5′-diphosphoric acid
DCE	dichloroethane
DCEE	dichloroethyl ether
D-CFZ	(3-chloro-2-hydroxypropyl)trimethylammonium chloride
DCH	2,3-dicyanohydroquinone
DCHA	dicyclohexylamine
DCHBH	dicyclohexylborane
DCHGF	1,2:5,6-di-*O*-cyclohexylidene-D-glucofuranose
DCI-HCl	1-(3′,4′-dichlorophenyl)-2-isopropylaminoethanol hydrochloride
DCK	dichloroketene
DCM	dichloromethane
dcm	*d*,*d*-dicampholylmethanate; *d*,*d*-dicampholylmethanato
DCMA®	*N*-[3,4-dichlorophenyl]methylacrylamide
DCME	α,α-dichloromethyl methyl ether
DCMME	dichloromethyl methyl ether
DCMO	5,6-dihydro-2-methyl-1,4-oxathiin-3-carboxanilide
dCMP	2′-deoxycytidine 5′-monophosphoric acid
dCMP	deoxycytidine 5′-monophosphate
DCMU®	3-[3,4-dichlorophenyl]1,1-dimethylurea
DCN	1,4-dicyanonaphthalene

DCNA®	2,6-dichloro-4-nitroaniline
DCNQ	*N,N'*-dicyano-1,4-naphthoquinone diimine
DCNQI	*N,N'*-dicyano-*p*-quinodiimine
DCNTT	2,5-bis(cyanoimino)-2,5-dihydrothieno[3,2-*b*]thiophene
DCOC	2,4-dichlorobenzoyl chloride
DCP	1,1-dimethylpiperidinium chloride
DCP	1,3-dichloropropane
DCP	2,3-dichlorophenol
DCP	3,9-dicyanophenanthrene
DCP	dodecamethoxy[1_4]orthocyclophane
DCPA	dimethyl tetrachloroterephthalate; dimethyl 2,3,5,6-tetrachloroterephthalate
DCPC	4,4'-dichloro-α-methylbenzhydrol
DCPD	dicyclopentadiene
DCPH	dicyclohexylphosphine
DCPP	dicyclohexyphenylphosphine
DCPTA	2-(3,4-dichlorophenoxy)triethylamine; 2-diethylaminoethyl-3,4-dichlorophenylether; *N,N*-diethylaminoethyl 3,4-dichlorophenyl ether
2,4-DCT	2,4-dichlorotoluene
3,4-DCT	3,4-dichlorotoluene
2,4-DCTC	2,4-dichlorobenzotrichloride
3,4-DCTC	3,4-dichlorobenzotrichloride
2,3,5-DCTF	2,3-dichloro-5-(trifluoromethyl)pyridine
2,5,3-DCTF	2,5-dichloro-3-(trifluoromethyl)pyridine
2,6,3-DCTF	2,6-dichloro-3-(trifluoromethyl)pyridine
2,6,4-DCTF	2,6-dichloro-4-(trifluoromethyl)pyridine
dCTP	2'-deoxycytidine 5'-triphosphoric acid
dCTP	deoxycytidine 5'-triphosphate
DCU	*N,N*-dichlorourethane
DCX	α,α'-dicyanoxylene
DD	dibenzo-*p*-dioxin
DD mixture	1,3-dichloropropene
DDA	4,4'-dichlorodiphenylacetic acid
DDA (ddA)	2',3'-dideoxyadenosine
DDAA	didehydroamino acids
DDAB	didecyldimethylammonium bromide
DDAO	*N,N*-dimethyldodecylamine-*N*-oxide
DDATHF	5,10-dideaza-5,6,7,8-tetrahydrofolic acid
ddATP	2',3'-dideoxyadenosine-5'-triphosphate
DDB	2,3-dimethoxy-1,4-bis(dimethylamino)butane
DDB	5-sulfo-4'-diethylamino-2,2'-dihydroxyazobenzene
DDC (ddC)	2',3'-dideoxycytidine
ddCTP	2',3'-dideoxycytidine 5'-triphosphate

DDD	2,2′-dihydroxy-6,6′-dinaphthyl disulfide; 6,6′-dithiodi-2-naphthol
DDD	1,1-dichloro-2,2-bis(*p*-chlorophenyl)ethane; 2,2-bis(*p*-chlorophenyl)-1,1-dichloroethane
m,p′-DDD	1,1-dichloro-2-(*m*-chlorophenyl)-2-(*p*-chlorophenyl)ethane
o,p′-DDD	1-(2-chlorophenyl)-1-(4-chlorophenyl)-2,2-dichloroethane; 1,1-dichloro-2-[*o*-chlorophenyl]-2-[*p*-chlorophenyl]ethane
p,p′-DDD	1,1-dichloro-2,2-bis(*p*-chlorophenyl)ethane; 2,2-bis(*p*-chlorophenyl)-1,1-dichloroethane
7,9-DDDA	7,9-dodecadien-yl acetate
8,10-DDDA	8,10-dodecadien-yl acetate
8,10-DDDOL	8,10-dodecadien-1-ol
o,p′-DDE	1,1-dichloro-2-[*o*-chlorophenyl]-2-[*p*-chlorophenyl]ethylene; l-(*o*-chlorophenyl)-1-(*p*-chlorophenyl)-2,2-dichloroethylene
p,p′-DDE	2,2-bis(4-chlorophenyl)-1,1-dichloroethylene; 1,1-dichloro-2,2-bis(*p*-chlorophenyl)ethylene
D_2dea	diethanolamine
DDED	dimethyl 2,2-dicyanoethylene-1,1-dicarboxylate
ddGTP	2′,3′-dideoxyguanosine 5′-triphosphate
DDH	1,3-dibromo-5,5-dimethylhydantoln
DDI (ddI)	2′,3′-dideoxyinosine
ddino	2′,3′-dideoxyinosine
ddITP	2′,3′-dideoxyinosine 5′-triphosphate
DDM	4,4′-dichlorodiphenylmethane
DDM	diphenyldiazomethane
DDMU	4,4′-dichlorodiphenyl-2-chloroethylene
ddNTP	dideoxynucleotide
DDOH	4,4′-dichlorodiphenylethanol
7-DDOL	7-dodecen-1-ol
DDP	dichlorodiammineplatinum; dichlorodiammineplatinum(II)
DDP	didehydropeptides
DDPC	1,2-dimyristoylamino-1,2-dideoxyphosphatidylcholine
DDPH	1,1-diphenyl-2-picrylhydrazyl (free radical)
DDQ	2,3-dichloro-5,6-dicyano-1,4-benzoquinone
DDS	4,4′-diaminodiphenyl sulfone
DDS	dihydroxydiphenyl sulfone
DDSA	dodecenylsuccinic anhydride
DDT	1,1-bis(*p*-chlorophenyl)-2,2,2-trichloroethane; 1,1,1-trichloro-2,2-bis(*p*-chlorophenyl)ethane
o,p′-DDT	1,1,1-trichloro-2-[*o*-chlorophenyl]-2-[*p*-chlorophenyl]ethane; 1-(*o*-chloropheny1)-1-(*p*-chlorophenyl)-2,2,2-trichloroethane

p,p′-DDT	1,1,1-trichloro-2,2-bis(*p*-chlorophenyl)ethane; 1,1-bis(*p*-chlorophenyl)-2,2,2-trichloroethane
DDTTA	4a,8a-diaza-2,6-dioxa-3,4,7,8-tetrahydro-4,4,8,8-tetramethylanthracene-1,5-dione
ddTTP	3′-deoxythymidine 5′-triphosphate
DDU	2′,3′-dideoxyuridine
DDVP	dimethyl 2,2-dichlorovinyl phosphate; phosphoric acid, 2,2-dichlorovinyl dimethyl ester
DDZ	α,α-dimethyl-3,5-dimethoxybenzyloxycarbonyl
DEA	*N*,*N*-diethylaniline
DEAA	*N*,*N*-diethylacetoacetamide
DEAA	*N*,*N*-diethylallylamine
DEAC	diethylaluminum chloride
DEAD	diethyl acetylenedicarboxylate
DEAD	diethyl azodicarboxylate
DEAD-TPP	diethyl azodicarboxylate-triphenylphosphine
DEADC	diethyl azodicarboxylate
DEAE	(diethylamino)ethyl (as in DEAE-cellulose)
DEAEA	2-(diethylamino)ethylamine
DEAH	diethylaluminum hydride
DEAI	diethylaluminum iodide
Deamino DPN	nicotinamide hypoxanthine dinucleotide
Deamino DPNH	nicotinamide hypoxanthine dinucleotide, reduced form
Deamino NAD	nicotinamide hypoxanthine dinucleotide; nicotinic acid adenine dinucleotide
Deamino NADH	nicotinamide hypoxanthine dinucleotide, reduced form
Deamino NADP	nicotinamide hypoxanthine dinucleotide phosphate
Deamino NADPH	nicotinamide hypoxanthine dinucleotide phosphate, reduced form
Deamino TPN	nicotinamide hypoxanthine dinucleotide phosphate
Deanol	*N*,*N*-dimethylethanolamine
DEAP	2,2-diethoxyacetophenone
DEASA	*N*,*N*-diethylaniline-3-sulfonic acid
DEBM	diethyl bromomalonate
DEC	2-diethylaminoethyl chloride hydrochloride
Dec	decalin
DEDA	7,7-dimethyleicosadienoic acid
DEDM	diethyl diazomalonate
DEDMPO	3,3-diethyl-5,5-dimethyl-1-pyrroline 1-oxide
DEDTC	diethyldithiocarbamic acid
Dedyl	diisopropylamine dichloroacetate
DEET (Deet)	*N*,*N*-diethyl-*m*-toluamide
DEF	diethyl fumarate
DEG	diethylene glycol

DEHP	bis(2-ethylhexyl) phthalate
DEHS-BSTFA	*N*,*O*-bis(diethylhydrogensilyl)trifluoroacetamide
DEII	diethylindoloindole
DEIPS	diethylisopropylsilyl
DEM	demeclocycline
DEM	diethyl maleate
DENA	*N*,*N*-diethylnicotinamide
DEOA	*N*,*N*-diethylethanolamine
Deoxy-BIGCHAP (deoxy-BigCHAP)	*N*,*N*′-bis(3-D-gluconamidopropyl)-deoxycholamide
DEP	diethyl phthalate
DEP	diethyl pyrocarbonate
DEPC	diethyl cyanophosphonate
DEPC	diethyl pyrocarbonate
DEPC	diethylphosphoryl cyanide; *O*,*O*′-diethyl phosphorocyanidate
DEPHA	di-(2-ethylhexyl)phosphoric acid
DERA	2-deoxyribose-5-phosphate aldolase; deoxyribose aldolase
DES	diethylstilbestrol
DESS	diethyl succinylsuccinate
DET	diethyl tartrate
DET	diethyltoluamide
DETAPAC	diethylenetriaminepentaacetic acid
DF	dibenzofuran
DFDD	1,1-dichloro-2,2-bis(*p*-fluorophenyl)ethane
DFDNB	1,5-difluoro-2,4-dinitrobenzene
DFDNB	difluoro-2,4-dinitrobenzene
DFDT	1,1,1-trichloro-2,2-bis(*p*-fluorophenyl)ethane
DFP	diisopropyl fluorophosphate
DG	diacylglycerol
dG	2′-deoxyguanosine
DGDG	digalactosyl diglyceride
dGDP	deoxyguanosine 5′-diphosphate
dGMP	deoxyguanosine 5′-monophosphate
dGTP	deoxyguanosine 5′-triphosphate
DHA	dehydro-L(+)-ascorbic acid, dimer
DHA	dehydroacetic acid
DHA	dihydroasparagusic acid
DHA	dihydroxyacetone
DHA (9,10-DHA)	9,10-dihydroanthracene
Dha	dehydroalanine
DHAB	2,2′-dihydroxyazobenzene

DHAP	dihydroxyacetone phosphate
DHBA	3,4-dihydroxybenzylamine hydrobromide
DHBP	3,4-dihydroxybutyl-1-phosphate
DHBP	3,4-dihydroxybutyl-1-phosphonic acid
DHBP	dihydroxybenzophenone (usually 4,4′)
DHEA	dehydroisoandrosterone
DHEBA	1,2-dihydroxyethylene-bis-acrylamide
DHET	dihydroergotoxine
DHF	dihydrofolate
$DHFH_2$	folate
DHFR	dihydrofolate reductase; dihydrofolic reductase
DHN	5,12-dihydronaphthacene
DHP	dehydropeptidase
DHP	diheptyl phthalate
DHP	dihydropyran
DHPG	9-(dihydroxypropoxymethyl)guanine
3,6-DHPN	3,6-dihydroxyphthalonitrile
DHPPEI	(1-phenylethyl)-2-pyridylmethylamine
DHQ	dehydroquinate
DHQ	hydroquinine
DHQD	hydroquinidine
$(DHQD)_2$-PHAL	1,4-bis-(9-*O*-dihydroquinidine)phthalazine
DHT	dihydrotachysterol
5,6-DHT	5,6-dihydroxytryptamine
5,7-DHT	5,7-dihydroxytryptamine creatinine sulfate
DHU	dihydrouridine
DIAD	diisopropyl diazodicarboxylate; diisopropyl azodicarboxylate
DIAMIDE (Diamide)	azodicarboxylic acid bis[dimethylamide]; diazine-dicarboxylic acid bis[*N*,*N*-dimethylamide]; 1,1′-azobis(*N*,*N*-dimethylformamide)
DIB	1,3-diphenylisobenzofuran
m-DIB	*m*-diisopropenylbenzene
p-DIB	*p*-diisopropenylbenzene
dib	1,4-diisocyanobenzene
DIBAC	diisobutylaluminum chloride
DIBAH	diisobutylaluminum hydride
DIBAL (Dibal)	diisobutylaluminum hydride
DIBAL-H	diisobutylaluminum hydride
DIBOA	2,4-dihydroxy-1,4-benzoxazin-3-one
DIBOC	di-*tert*-butyl dicarbonate
5,5′-dibromo-BAPTA	1,2-bis(2-amino-5-bromophenoxy)ethane-*N*,*N*,*N*′,*N*′-tetraacetic acid

1,8-dichloro-BPEA	1,8-dichloro-9,10-bis(phenylethynyl)anthracene
DIC	(dimethylamino)isopropyl chloride hydrochloride; 2-dimethylaminoisopropyl chloride hydrochloride
DIC	diisopropylcarbodiimide
DIDA	diisodecyl adipate
DIDP	diisodecyl phthalate
DIDS	4,4′-diisothiocyanatostilbene-2,2′-disulfonic acid
DIEA	diisopropylethylamine; *N*,*N*-diisopropylethylamine
dien	diethylenetriamine
DI-ET	*N*,*N*-diethyl-*p*-phenylenediamine monohydrochloride
DIF	diiodofluorone
DIF	diphenyliodonium fluoride
DIFP	diisopropyl fluorophosphate; diisopropylphosphofluoridate
DIHETE	dihydroxyeicosatetraenoic acid
5,12-DIHETE	5,12-dihydroxy-6,8,10,14-eicosatetraenoic acid
DiHPhe	2,5-dihydroxyphenylalanine
Dim	[2-(1,3-dithianyl)methyl]
DIMAD	dimethyl acetylenedicarboxylate
DIMBOA	2,4-dihydroxy-7-methoxy-1,4-benzoxazin-3-one
DIMEB	dimethyl-β-cyclodextrin
Dimethyl POPOP	1,4-bis-2(4-methyl-5-phenyl-2-oxazolyl)benzene
DIMF	2,4-diiodo-6-methoxy-3-fluorone
dimsyl	methylsulfinylmethylide (anion)
DINOL	2,2-dimethyl-α,α,α′,α′-tetrakis (β-naphthyl)-1,3-dioxolan-4,5-dimethanol
DINOL	2,3-*O*-isopropylidene-1,1,4,4-tetra(2-naphthyl)threitol
DIOP (diop)	1,4-bis(diphenylphosphino)-1,4-dideoxy-2,3-*O*-isopropylidenethreitol; 1,4-bis(diphenylphosphino)-2,3-*O*-isopropylidene-2,3-butanediol; (2,3-*O*-isopropylidene-2,3-dihydroxy-1,4-butylene)bis (diphenylphosphine); 4,5-bis(diphenylphosphino)methyl-2,2-dimethyldioxolan
Diox	dioxane
DIP	(*E*)-2,3-diiodo-1-(phenylsulfonyl)-1-propene
DIP	diisopropyl ether
DIP Chloride™	*B*-chlorodiisopinocamphenylborane
DIPA	diisopropylamine
DIPA	diisopropylamine dichloroacetate
DIPAMP	1,2-bis[(2-methoxyphenyl)phenylphosphino]ethane
DIPAMP	1,2-di(phenylarylphosphino)ethane
DIPB	di-isopinocampheylborane
DIPC	2-dimethylaminoisopropyl chloride hydrochloride
DIPCB	di-isopinocampheylborane
DIPCDI	*N*,*N*′-diisopropylcarbodiimide

DIPE	diisopropyl ether; isopropyl ether
DIPEA	diisopropylethylamine
DIPGF	1,2:5,6-di-*O*-isopropylidene-α-D-glucofuranose
DIPHIN	tetrahydrofuran-3,4-diyl-bis(diphenylphosphinite)
DIPHOS (diphos)	ethylenebis(diphenylphosphine); 1,2-bis(diphenylphosphino)ethane
diphos-4	1,4-bis(diphenylphosphino)butane
DIPMA	diisopropylmethylamine
DIPSO	3-[*N*,*N*-bis(2-hydroxyethyl)amino]-2-hydroxy-propanesulfonic acid; *N*,*N*-bis(2-hydroxyethyl)-3-amino-2-hydroxypropanesulfonic acid
DIPT (DiPT)	diisopropyl tartrate
Di-SNADNS	2,7-bis(4-sulfo-1-naphthylazo)-1,8-dihydroxynaphthalene-3,6-disulfonic acid
DIT	diiodotyranose
DITC	1,4-phenylene diisocyanate
dITP	2′-deoxyinosine 5′-triphosphate
DIURON	3-(3,4-dichlorophenyl)-1,1-dimethylurea
DK	β-diketone
DKP	potassium phosphate, dibasic
DLC	terephthalic acid bis[4-(methoxycarbonyl)phenyl] ester
DLP	polypropylene (isolactic form)
DLPC	1,2-dilauroyl-*sn*-glycero-3-phosphocholine
DLPE	1,2-dilauroyl-*sn*-glycero-3-phosphoethanolamine
DLPS	1,2-dilauroyl-*sn*-glycero-3-phospho-L-serine
DMA	dimethyl adipimidate
DMA	dimethylacetamide; *N*,*N*-dimethylacetamide
DMA	dimethylmorpholinium bromide
DMA	disodium methyl arsenate
DMA	dimethylamine
DMA	*N*,*N*-dimethylaniline
11DMA	1,1-dimethylallene
13DMA	1,3-dimethylallene
2,6-DMA	2,6-dimethylanisole
DMA-DEA	*N*,*N*-dimethylacetamide diethyl acetal
DMAA	*N*,*N*-dimethylacetoacetamide
DMAA	*N*,*N*-dimethylallylamine; *N*-allyldimethylamine
DMAB	borane-dimethylamine complex; dimethylamine borane
DMAC (DMAc)	*N*,*N*-dimethylacetamide; dimethylacetamide
DMAD	dimethyl acetylenedicarboxylate
DMAE	2-dimethylaminoethanol; dimethylaminoethanol; *N*,*N*-dimethylaminoethanol
DMAEMA	2-dimethylaminoethyl methacrylate
DMAP	4-dimethylaminopyridine; 4-*N*,*N*-dimethylaminopyridine

DMAP	dimethylaminopropylamine
p-DMAPhFl	9-(*p*-(dimethylamino)phenyl)fluorene
DMAPMA	dimethylaminopropyl methacrylamide
DMAPP	dimethylallyl diphosphate; dimethylallyl pyrophosphate
DMAT	dimethylallyltryptophan
DMB	4,4′-dichloro-α-methylbenzhydrol; 4,4′-dichloro-α-methylbenzylhydrol
DMB	2,3-dimethylbuta-1,3-diene
DMB	2,3-dimethylbutane
DMB	2,6-dimethoxybenzyl
DMB	4,4-dimethylmorpholinium bromide
DMB	hydroquinone dimethyl ether
DMBA	7,12-dimethylbenz[*a*]anthracene
DMBA	9,10-dimethylbenz[*a*]anthracene
DMBA	*N*-benzyldimethylamine
DMBAS	2,5-dimethoxy-4′-aminostilbene
DMBB	5-(1,3-dimethyl)-5-ethylbarbituric acid
dmbbmp	6-methyl-2-(1-(3,5-dimethoxybenzyl)benzimidazol-2-yl)pyridine
DMBI	1,3-dimethyl-2-phenylbenzimidizoline
DMBZ	5,6-dimethylbenzimidazole
DMC	2-(dimethylamino)ethyl chloride
DMC	4,4′-dichloro-α-methylbenzhydrol
DMC	penicillamine
DMC	dimethylcarbene
DMCH	4,4-dimethyl-2-cyclohexenone
DMCS	chlorodimethylsilane; dimethylchlorosilane
DMCTMS	trimethylsilyl *N,N*-dimethylcarbamate
DMD	3,3-dimethyldiazirine
DMD	dimethyldioxirane
DMDAAC	dimethyldiallylammonium chloride
DMDCS	dimethyldichlorosilane
DMDF	2,5-dimethoxy-2,5-dihydrofuran
DMDS	dimethyl disulfide
DMDT	2,2-dichloro-1,1-difluoroethyl methyl ether
DMDT®	1,1,1-trichloro-2,2-bis[*p*-methoxyphenyl]ethane
DME	dimethoxy
DME	dimethoxyethane; glyme
DME	dimethyl ether
DMECS	dimethylethylchlorosilane
dmepa	[(6-methyl-2-pyridyl)methyl][2-(2-pyridylethyl)](2-pyridylmethyl)amine
DMEU	1,3-dimethyl-2-imidazolidinone; *N,N*′-dimethylethyleneurea

DMF	dimethyl fumarate
DMF (dmf)	*N*,*N*-dimethylformamide; dimethylformamide
DMFA	*N*,*N*-dimethylformamide; dimethylformamide
DMF-DMA	dimethylformamide dimethyl acetal; *N*,*N*-dimethylformamide dimethyl acetal
DMFM	dimethyl fumarate
dmg	dimethylglyoximato(2^-)
DMG	N^2,N^2-dimethylguanosine
DMGG	1,1-dimethylbiguanide
DMH	*N*,*N*′-dimethyl-*N*,*N*′-bis(mercaptoacetyl)hydrazine
DMH	*N*,*N*-dimethylhydrazone
DMH^{ox}	*trans*-*N*,*N*′-dimethyl-*N*,*N*′-bis(mercaptoacetyl)-1,2-dithiane
DMI	1,3-dimethyl-2-imidazolidinone; *N*,*N*′-dimethylimidazalone
DMIPS	dimethylisopropylsilyl
DMIPSCl	isopropyldimethylchlorosilane
DMLDH	spiny dogfish muscle lactate dehydrogenase
DMMP	dimethyl methylphosphonate
DMNFHSCl	dimethyl(3,3,4,4,5,5,6,6,6-nonafluorohexyl)chlorosilane
DMOA	*N*,*N*-dimethyloctylamine
Dmoc	dithianylmethoxycarbonyl
DMP	2,2-dimethoxypropane
DMP	2,6-dimethylphenol
DMP	dimethyl phthalate
DMP	dimethyl pimelimidate
DMP	dimethyl pyrocarbonate
DMP	*N*,*N*-dimethylpyrrolidinium
2,2-DMP	2,2-dimethoxypropane
2,6-DMP	2,6-dimethylphenol
Dmp	dimethylphosphinyl
DMP-30	2,4,6-tris(dimethylaminomethyl)phenol
DMPA	1,2-dimyristoyl-*sn*-glycero-3-phosphatidic acid
DMPA	2,2-dimethoxy-2-phenylacetophenone
DMPA	*O*-(2,4-dichlorophenyl) *O*-methyl isopropylphosphoramidothioate
DMPA®	2,2-bis(hydroxymethyl)propionic acid
DMPADC	*N*,*N*-dimethylphosphoramic dichloride
DMPC	3-dimethylaminopropyl chloride hydrochloride
DMPC	1,2-dimyristoyl-*rac*-glycero-3-phosphocholine; 1,2-dimyristoyl-*sn*-glycero-3-phosphocholine
DMPD	4-amino-*N*,*N*-dimethylaniline
DMPD	1,4-dimethyl-piperazine-2,3-dione
DMPD	4,5-dimethyl-1,2-phenylenediamine

DMPE (dmpe)	1,2-bis(dimethylphosphino)ethane
DMPE	1,2-dimyristoyl-*sn*-glycero-3-phosphoethanolamine
DMPG	1,2-dimyristoyl-*sn*-glycero-3-phospho-*rac*-glycerol
DMPG	1,2-dimyristoyl-*sn*-glycero-3-phospho-*sn*-1-glycerol
$DMPH_4$	6,7-dimethyl-5,6,7,8-tetrahydropterine
DMPM	3,4-dimethoxybenzyl
DMPO	1,3-dimethyl-5-pyrazolone
DMPO	2,2-dimethyl-3,4-pyrroline *N*-oxide
DMPO	5,5-dimethyl-1-pyrroline-1-oxide; 5,5-dimethyl-1-pyrroline *N*-oxide
DMPO	5,5-dimethyl-1-pyrroline-*N*-oxyl
DMPO-OH	5,5-dimethyl-2-hydroxypyrrolidone-1-oxyl radical
DMPOX	5,5-dimethyl-2-pyrrolidone-1-oxyl radical
DMPP	1,1-dimethyl-4-phenylpiperazinium iodide
DMPP	dimenthylphenylphosphine
DMPPDA	4-amino-*N*,*N*-dimethylaniline
DMPS	1,2-dimyristoyl-*sn*-glycero-3-phospho-L-serine
DMPS	2,3-dimercapto-1-propanesulfonic acid (sodium salt)
DMPS	dimethylphenylsilyl
DMPSCl	phenyldimethylchlorosilane
DMPU	1,3-dimethyl-3,4,5,6-tetrahydro-2(1*H*)-pyrimidinone; *N*,*N*′-dimethyl-*N*,*N*′-propyleneurea
DMQ (1,4-DMQ)	1,4-dimethyl-2-quinolone
2,6-DMQ	2,6-dimethyl-2-quinolone
2,7-DMQ	2,7-dimethyl-2-quinolone
DMS	4,6-dimethoxybenzene-1,3-disulfonyl chloride
DMS	dimethyl suberimidate
DMS	dimethyl sulfide
DMSA	*meso*-2,3-dimercaptosuccinic acid
DMSB	dimethyl sulfide borane
DMSO (dmso)	dimethyl sulfoxide
$DMSO_2$	dimethyl sulfone
DMSS	dimethyl succinylsuccinate
DMT	4,4′-dimethoxytriphenylmethyl chloride; 4,4′-dimethoxytrityl chloride
DMT	dimethoxytriphenylmethyl
DMT	dimethyl terephthalate
DMT	*N*,*N*-dimethyltrytamine
dmt	4,4′-dimethoxytrityl
DMTD	2,5-dimercapto-1,3,4-thiadiazole; dimercaptothiadiazole
DMTr	di(*p*-methoxyphenyl)phenylmethyl
DMTSF	dimethyl(methylthio)sulfonium fluoroborate; dimethyl(methylthio)sulfonium tetrafluoroborate
DMTT®	tetrahydro-3,5-dimethyl-2*H*-1,3,5-thiadiazine-2-thione

DMU	dimethylurea
DMU®	3-(3,4-dichlorophenyl)1,1-dimethylurea
DmXM	bis(2,4-dimethylphenyl)methane
DN®	4,6-dinitro-*o*-cresol
DNA	deoxyribonucleic acid
DNAF	10-(2′,4′-dinitrophenylazo)-9-phenanthrol
DNAP	4-(2′,4′-dinitrophenylazo)-9-phenanthrol
DNAP	4-(2′,4′-dinitrophenylazo)phenol
DNase	deoxyribonuclease
DNB	dinitrobenzene
DNB	dinitrobenzoyl
DNB	*p,p*′-dinitrobenzhydryl
m-DNB	*m*-dinitrobenzene
DNBC	3,5-dinitrobenzoyl chloride
DNBF	4,6-dinitrobenzofuroxan
DNBPG	(*S*)-*N*-3,5-dinitrobenzoyl glycine
DNBP®	2,4-dinitro-6-*sec*-butylphenol
DNBS	2,4-dinitrobenzenesulfonic acid
DNBSC	2,4-dinitrobenzenesulfenyl chloride
DNC	4,4′-dinitrocarbanilide
DNC®	4,6-dinitro-*O*-cresol
2,4-DNCB	2,4-dinitrochlorobenzene
DNF	2,4-dinitrofluorobenzene
DNFA	2,4-dinitro-5-fluoroaniline
DNFB	1-fluoro-2,4-dinitrobenzene; 2,4-dinitrofluorobenzene
DNFB	2,4,6-trinitrobenzene-1-sulfonic acid
DNMBS	4-(4′,8′-dimethoxynaphthylmethyl)benzenesulfonyl
dNMP	deoxynucleotide 5′-monophosphate
DNOC	4,6-dinitrocarbanilide
DNOCHP	2-cyclohexyl-4,6-dinitrophenol
DNP (Dnp)	2,4-dinitrophenyl
DNP	2,4-dinitrophenylhydrazine
DNP	2,4-dinitrophenylhydrazone
DNP	deoxynucleoprotein
DNP	dinonyl phthalate
DNP®	2,6-diiodo-4-nitrophenol
2,4-DNP	2,4-dinitrophenol
2,6-DNPC	2,6-dinitro-4-cresol
Dnp-F	2,4-dinitrofluorobenzene
DNPA	2,4-dinitrophenyl acetate
DNPBA	3,5-dinitroperoxybenzoic acid
DNPDTC	2,4-dinitrophenyl *O*-ethyl dithiocarbonate
DNPF	2,4-dinitrofluorobenzene
DNPH	*O*-(2,4-dinitrophenyl)hydroxylamine

DNPMC	2,4-dinitrophenyl methyl carbonate
DNPO	bis(2,4-dinitrophenyl)oxalate
DNPTA	2,4-dinitrophenyl thiolacetate
DNPTC	*O*-ethyl *S*-(2,4-dinitrophenyl) thiocarbamate; *S*-(2,4-dinitrophenyl) *O*-ethyl thiocarbonate
DNQX	6,7-dinitroquinoxaline-2,3-dione
DNS	4,4′-dinitrostilbene-2,2′-disulfonic acid, disodium salt
DNS	5-dimethylamino-1-naphthalenesulfonic acid
DNS	dansyl chloride
DNS (Dns)	dansyl
2,4-DNS	2,4-dinitrobenzenesulfenyl chloride
DNS-BBA	*N*-dansyl-3-aminobenzeneboronic acid
DNS-F	5-dimethylaminonaphthalene-1-sulfonyl fluoride
DNSA	5-dimethylaminonaphthalene-1-sulfonamide
DNSAPITC	4-(5-dimethylaminonaphthalene-1-sulfonylamino)phenyl isothiocyanate
DNSCl	5-dimethylaminonaphthalene-1-sulfonyl chloride
DNTC	4-dimethylamino-1-naphthyl isothiocyanate
DNT-Cl	4-chloro-3,5-dinitrobenzotrifluoride
DNTP	*O*,*O*-diethyl-*O*-*p*-nitrophenyl phosphorothionate
DOA	dioctyl adipate
Dobz	*p*-(dihydroxyboryl)benzyloxycarbonyl
DOC	deoxycorticosterone
DOC	3,3′-diethyloxacarbocyanine
DOCA (Doca)	deoxycorticosterone acetate
DODA	dioctadecylamine
DODAC	dioctadecyldimethylammonium chloride
DODC	3,3′-diethyloxadicarbocyanine iodide
DOG	*sn*-1,2-dioctanoylglycerol
Dol	dolichol
DON	6-diaza-5-oxo-L-norleucine
DOP	dioctyl phthalate
DOPA	1,2-dioleoyl-*sn*-glycero-3-phosphoric acid
DOPA	L-α-dioleoylphosphatidic acid
DOPA (Dopa) (dopa)	3-(3,4-dihydroxyphenyl)alanine; 3,4-dihydroxyphenylalanine
D-DOPA	3-(3,4-dihydroxyphenyl)-D-alanine
DOPAC	3,4-dihydroxyphenylacetic acid
DOPAL	3,4-dihydroxyphenylacetaldehyde
DOPAMINE	3,4-dihydroxyphenylethylamine hydrochloride
DOPC	1,2-dioleoyl-*sn*-glycero-3-phosphocholine
DOPE	1,2-dioleoyl-*sn*-glycero-3-phosphoethanolamine
DOPET	3,4-dihydroxyphenethyl alcohol
DOPG	1,2-dioleoyl-*sn*-glycero-3-phospho-*rac*-glycerol

DOPG	1,2-dipalmitoyl-*sn*-glycero-3-phospho-*rac*-glycerol
DOPP	dioctyl phenylphosphonate
DOPS	1,2-dioleoyl-*sn*-glycero-3-phospho-L-serine
DOPS	3,4-dihydroxyphenylserine
DOPS	DL-*threo*-3,4-dihydroxyphenylserine
DOS	2,2-dibutyl-1,3,2-dioxastannolane
DOT	di-*o*-thymotide
DOTC	3,3′-diethyloxatricarboxyanine iodide
DoTM	di-*o*-tolylmethane
DOX	doxytetracycline
DOXYL	4,4-dimethyl-3-oxazolidinyloxy
2,4-DP	4-(2,4-dichlorophenoxy)propionic acid; 2,4-dichlorophenoxypropionic acid
DP3	1,3-diphenylpropene
DP5	1,5-diphenyl-1,3-pentadiene
DPA	9,10-diphenylanthracene
DPA	diphenylacetylene
DPA	diphenylamine
DPB	1,4-diphenyl-1,3-butadiene
DPB	1,4-diphenylbutane
DPBA	1,1-diphenylboronic acid
DPBF	diphenylisobenzofuran
DPC	dimethylpiperidium chloride
DPD	*N*,*N*-diethyl-*p*-phenylenediamine
DPDM	diphenyl diazomalonate
DPDPB	1,4-di-[3′-(2′-pyridyldithio)propionamido]butane
DPE	1,1-diphenylethene; 1,1-diphenylethylene
DPEP	deoxophylloerythroetioporphyrin
DPG (2,3-DPG)	2,3-diphosphoglycerate
DPH	1,6-diphenyl-1,3,5-hexatriene; diphenylhexatriene
DPH	*O*-(diphenylphosphinyl)hydroxylamine
DPIBF	1,3-diphenylisobenzofuran
DPIP	dichlorophenol-indophenol
DPM	2,2,6,6-tetramethylheptandione
DPM (dpm)	bis(diphenylphosphino)methane
DPM	di-π-methane
DPM	diphenylmethane
DPM	diphenylmethyl
DPM	dipivaloylmethane
dpm	2,2,6,6-tetramethyl-3,5-heptanedionate
DPMA-H	(diphenylmethylene)amine
DPMCl	diphenylchloromethane
DPMDS	diphenylmethane-4,4′-disulfonylchloride
DPMH	diphenylmethane

DPMS	diphenylmethylsilyl
DPMSCl	diphenylmethylchlorosilane
DPN	diphosphopyridine nucleotide; nicotinamide adenine dinucleotide
DPNH	α-nicotinamide adenine dinucleotide reduced form
DPNH	β-nicotinamide adenine dinucleotide reduced form
DPO	2,5-diphenyloxazole
DPO	diphenyl oxide
DPP	dipeptidyl peptidase
Dpp	diphenylphosphinyl
2,3-dpp	2,3-bis(2-pyridyl)pyrazine
DPP-Cl	diphenylphosphinyl chloride
DPPA	1,2-dipalmitoyl-*sn*-glycero-3-phosphatidic acid
DPPA	diphenyl phosphoroazidate; diphenylphosphoryl azide
DPPB (dppb)	1,4-bis(diphenylphosphine)butane; 1,4-diphenylphosphinobutane
DPPC	1,2-dipalmitoyl-*sn*-glycero-3-phosphocholine; α-dipalmitoyl phosphatidylcholine; dipalmitoylphosphatidylcholine
DPPD®	*N,N*-diphenyl-*p*-phenylenediamine
DPPE (Dppe) (dppe)	1,2-bis(diphenylphosphino)ethane; ethylenebis(diphenylphosphine)
DPPE	1,2-dipalmitoyl-glycero-3-phosphoethanolamine
DPPF	1,1′-bis(diphenylphosphino)ferrocene
dppf	1,2′-bis(diphenylphosphino)ferrocene
DPPG	1,2-dipalmitoyl-*sn*-glycero-3-phospho-*sn*-1-glycerol
DPPH	2,2-di(4-*tert*-octylphenyl)-1-picrylhydrazyl
DPPH	2,2-diphenyl-1-picrylhydrazyl; diphenylpicrylhydrazyl
Dppm	(diphenyl-4-pyridyl)methyl
DPPP (dppp)	1,3-bis(diphenylphosphino)propane
DPPS	1,2-dipalmitoyl-*sn*-glycero-3-phospho-L-serine
DPPZ	dipyrido[3,2-a:2′,3′c]phenazine
DPS	dimethylphenylsiloxy
DPS	*trans-p,p′*-diphenylstilbene
DPSO	3-[*N*-bis(hydroxyethyl)amino]-2-hydroxypropanesulfonic acid
DPT	cocarboxylase
DPTA	di-*O,O′*-pivaloyltartaric acid
DPTMDS	1,3-diphenyl-1,1,3,3-tetramethyldisilazane
DRB	5,6-dichloro-1-β-D-ribofuranosyl-benzimidazole; 5,6-dichlorobenzimidazole-1-β-D-ribofuranoside
DR-Et	Davy-reagent ethyl
DR-Me	Davy-reagent methyl
DR-T	Davy-reagent *p*-tolyl

DS Acid	2,3-dihydroxynaphthalene-6-sulfonic acid
DSAH	disuccinimidyl (*N*,*N*′-diacetylhomocysteine)
DSC	*N*,*N*′-disuccinimidyl carbonate
DSE	disodium ethylene bis[dithiocarbamate]
DSG	disuccinimidyl glutarate
DSMA	disodium methyl arsenate; disodium methanearsonate
DSO	di-(*N*-succinimidyl)oxalate
DSP	dithiobis(succinimidyl propionate)
DSP	sodium phosphate, dibasic
DSP-4	*N*-(2-chloroethyl)-*N*-ethyl-2-bromobenzylamine
DSPA	1,2-distearoyl-*sn*-glycero-3-phosphatidic acid
DSPC	1,2-distearoyl-*sn*-glycero-3-phosphocholine
DSPE	1,2-distearoyl-*sn*-glycero-3-phosphoethanolamine
DSPG	1,2-distearoyl-*sn*-glycero-3-phospho-*rac*-glycerol; 1,2-distearoyl-*sn*-glycero-3-phospho-*sn*-1-glycerol
DSS	2,2-dimethyl-2-sila-5-pentanesulfonate; 2,2-dimethyl-2-silapentane-5-sulfonate
DSS	dioctyl sodium sulfosuccinate
DSS	disuccinimidyl suberate
DST	disuccinimidyl tartrate
DT	1,3-dithiane
D4T	1-(2,3-dideoxy-β-D-glycero-pent-2-enofuranosyl)thymine; 2′,3′-didehydro-2′,3′-dideoxythymidine
dT	thymidine
DTA	2,5-dimercapto-*N*,*N*,*N*′,*N*′-tetramethyladipamide
DTA	di-2-thenoylacetylene
DTAF	5-([4,6-dichlorotriazin-2-yl]amino)fluorescein
DTAN	2,2′-dithio-di(1-naphthylamine); 2,2′-dithiobis-(1-aminonaphthalene)
DTBMS	di-*t*-butylmethylsilyl
DTBP	di-*tert*-butyl peroxide
DTBP	dimethyl 3,3′-dithiopropionimidate dihydrochloride
2,6-DTBP	2,6-di-*tert*-butylpyridine
DTBPA	4,4′-dithio-bis-(phenylazide)
DTBQ	3,5-di-*tert*-butyl-1,2-benzoquinone
DTBS	di-*t*-butylsilylene
$DTBSCl_2$	di-*tert*-butyldichlorosilane
DTBT	di-*tert*-butyl tartrate
DTC	dibutyltin chloride
DTCl	dodecyltrimethylammonium chloride
DTDC	3,3′-diethylthiadicarbocyanine iodide
dTDP	thymidine 5′-diphosphate
DTE	1,4-dithioerythritol; dithioerythritol
DTF	9-dicyanomethylene-2,4,7-trinitrofluorene

DTGS	deuteriated triglycerine sulfate
DTMC	4,4′-dichloro-α-(trichloromethyl)benzhydrol
dTMP	thymidine 5′-monophosphate
DTNB	5,5′-dithiobis(2-nitrobenzoic acid)
DTNP	2,2′-dithiobis(5-nitropyridine)
DTPA	diethylenetriaminepentaacetic acid
DTPA	diethylenetriaminepentaacetic anhydride
DTPA	{[(carboxymethyl)imino]bis(ethylenenitrilo)}tetraacetic acid
DTPP	diethoxytriphenylphosphorane
Dts	dithiasuccinimidyl
DTSSP	3,3′-dithiobis(sulfosuccinimidylpropionate)
DTT	1,4-dithiothreitol; dithiothreitol
DTTC	3,3′-diethylthiatricarbocyanine iodide
DTT^{ox}	*trans*-4,5-dihydroxy-1,2-dithiane
dTTP	thymidine 5′-triphosphate
DU	5-diazouracil
dUDP	deoxyuridine 5′-diphosphate
dUMP	deoxyuridine 5′-monophosphate
DuPhos	1,2-bis(2,5-diphenyl-1-phosphocyclopentyl)benzene
DUPONOL-C®	sodium lauryl sulfate
DUR	durene; 1,2,4,5-tetramethylbenzene
dUTP	deoxyuridine 5′-triphosphate
DVB	divinylbenzene
m-DVB	*m*-divinylbenzene
p-DVB	*p*-divinylbenzene
DVTMDS	1,3-divinyl-1,1,3,3-tetramethyldisilazane
DXE	dixylylethane
E	ether
E	glutamic acid
E^+	electrophile
E. coli	*Escherichia coli*
EA	ethanolamine
EA	ethyl acetate
EAA	ethyl acetoacetate
EAA	*N*-ethylanthranilic acid
EACA	ε-aminocaproic acid
EADC	ethylaluminum dichloride
EAK	ethyl amyl ketone; 3-octanone
Eapine	*B*-iso-2-ethylapopinocampheyl-9-borabicyclo[3.3.1]nonyl
EASC	ethylaluminum sesquichloride
EB	enterotoxin B
EBA	*N*-ethyl-*N*-benzylaniline
EBAC	ethyl-4-((4-ethoxybenzylidene)amino)cinnamate
EBASA	*N*-ethyl-*N*-benzylanillne-4-sulfonlc acid

EBBA	4′-ethoxybenzylidene-4-*N*-butylaniline; *N*-(4-ethoxybenzylidene)-4-butylaniline
EBDC	ethylene bis-dithiocarbamate
EBI	ethylene-1,2-bis(η^5-1-indenyl)
EBS	ethylbenzenesulfonate
EBSA	*p*-ethylbenzenesulfonlc acid
EBTHI	ethylene-1,2-bis(η^5-4,5,6,7-tetrahydro-1-indenyl)
ECEA	*N*-ethyl-*N*-chloroethylaniline
ECGS	endothelial cell growth supplement
EcoR1	EcoR1 restriction endonuclease
ED	ethynodiol
EDA	ethylenediamine
EDAC	1-(3-dimethylaminopropyl)-3-ethylcarbodiimide; 1-ethyl-3-(3-dimethylaminopropyl)carbodiimide
EDANS	2-aminoethylamino-1-naphthalenesulfonlc acid (1,5 or 1,8)
EDB	1,2-dibromoethane; ethylene dibromide
EDC	1-(3-dimethylaminopropyl)-3-ethylcarbodiimide; 1-ethyl-3-(3-dimethylaminopropyl)carbodiimide (hydrochloride); *N*-(3-dimethylaminopropyl)-*N*′-ethylcarbodiimide (hydrochloride)
EDC	1,2-dichloroethane; ethylene dichloride
EDCI	1-(3-dimethylaminopropyl)-3-ethylcarbodiimide; 1-ethyl-3-[3-(dimethylamino)propyl]carbodiimide (hydrochloride)
edda	ethylenediamine diacetate
EDDP	*O*-ethyl *S*,*S*-diphenyl dithiophosphate
EDDS	(*S*,*S*)-ethylenediamine-*N*,*N*′-disuccinic acid
EDPM	ethylene-propylene-diene rubber
EDT	ethanedithiol
EDTA	ethylenediamine-*N*,*N*,*N*′,*N*′-tetraacetic acid; ethylenediaminetetraacetic acid
EDTA (edta)	ethylenediaminetetraacetato
EDTN	1-ethoxy-4-(dichloro-*s*-triazinyl)naphthalene
EDTP	ethylenediamine tetrapropanol; *N*,*N*,*N*′,*N*′-tetrakis(2-hydroxypropyl)ethylenediamine
EE	1-ethoxyethyl; ethoxyethyl
EE	diethyl ether
EEDQ	1-ethoxycarbonyl-2-ethoxy-1,2-dihydroquinoline; 2-ethoxy-1-ethoxycarbonyl,1,2-dihydroquinoline; ethyl 1,2-dihydro-2-ethoxy-1-quinoline-carboxylate
EEE	2-(2-ethoxyethoxy)ethanol
EF(x)	elongation factor (x)
EFE	ethylene forming enzyme
EG	ethylene glycol

EGDMA	ethylene glycol dimethylacrylate
EGF	epidermal growth factor
EGS	ethylene glycol bis(succinimidyl succinate)
EGT	ethylene glycol bis(trichloroacetate)
EGTA	1,2-di(2-aminoethoxy)ethane-*N*,*N*,*N'*,*N'*-tetraacetic acid; ethylene glycol-bis(β-aminoethyl ether)-*N*,*N*,*N'*,*N'*-tetraacetic acid; ethylenebis(oxyethylenenitrilo)-tetraacetic acid
EGTA/AM	ethylene glycol-bis(β-aminoethyl ether)-*N*,*N*,*N'*,*N'*-tetraacetoxymethyl ester
EHPG	ethylenediamine-di(*O*-hydroxyphenylacetic acid
eIF(n)	eucaryotic initiation factor, where n = 2,3,4, or 6
ELH	egg laying hormone
EMC	ethyl mercuric chloride
EMCS	*N*-succinimidyl 6-maleimidocaproate
EME	ethyl methyl ether
EMP	ethyl 3-methylpropynoate
EMU	erythromycin
en	ethylenediamine
ENK	enkephalin
ENU	ethylnitrosourea
EO	ethylene oxide
EOC	ethoxycarbonyl
E4P	erythrose 4-phosphate
EPA	*cis*-5,8,11,14,17-eicosapentaenoic acid
EPA	diethyl ether, isopentane, and ethanol (5:5:2)
EPDM	ethylene-poly(propylene)-diene monomer
EPN	*O*-ethyl *O*-(*p*-nitrophenyl)thiobenzenephosphate; *O*-ethyl-*O*-*p*-nitrophenyl phenylphosphonothioate
EPO	erythropoietin
EPP	ethyl 3-phenylpropynoate
EPPS	4-(2-hydroxyethyl)-1-piperazinepropanesulfonic acid; 4-(2-hydroxyethyl)piperazine-1-propanesulfonic acid; 1-(2-hydroxyethyl)piperazine-4-(3-propanesulfonic acid)
EPR	ethylene-propylene rubber
EPSP	5-enolpyruvoylshikimate 3-phosphate
EPSPS	5-enolpyruvoylshikimate 3-phosphate synthase
EPT	ethylene-poly(propylene)ter polymer
EPTC	*S*-ethyl dipropylthiocarbamate
ET	bis(ethylenedithiolo)tetrathiafulvalene
ET	ethylene
Et	ethyl
ETA	ethylenediaminetetraacetic acid

EtAc	ethyl acetate (should be EtOAc)
ETAC-1	[2-(4-nitrophenyl)allyl]trimethylammonium iodide
ETBE	ethyl *tert*-butyl ether
$EtDBTh^+$	*S*-ethyldibenzothiophenium ion
ETFE	ethylene-tetrafluoroethylene
eth	diethyl ether
ETI	5,8,11-eicosatriynoic acid
ETIP	diethyl dithiolisophthalate
EtOAc	ethyl acetate
EtOH	ethanol
ETP	epidithiadioxopiperazine
ETPB	4-ethyl-2,6,7-trioxa-1-phosphabicyclo[2.2.2]octane
ETSA	ethyl trimethylsilylacetate
ETT	ethyl-1,3-thiazolidine-2-thione
ETTA	2,2′,2″,2‴-[1,2-ethanediylidenetetrakis(thio)tetrakisacetic acid
ETYA	5,8,11,14-eicosatetraynoic acid
EVA	ethylene/vinyl acetate copolymer
EVK	ethyl vinyl ketone
F	phenylalanine
F	formyl
F Acid	2-naphthol-7-sulfonic acid
FA	fatty acid
FA	furfuryl alcohol
FA	(2-furyl)acryloyl
facam	3-(trifluoromethylhydroxymethylene)-*d*-camphorato
FAD	flavin adenine dinucleotide; flavin-adenosine dinucleotide
$FADH_2$	flavin adenine dinucleotide (reduced form)
FAM	*N*-(3-fluoranthyl)maleimide
FAME	fatty acid methyl esters
FAMSO	methyl methylsulfinylmethyl sulfide
F-B	free base
FBP (F1,6-BP)	fructose 1,6-bisphosphate
F2,6-BP	fructose 2,6-bisphosphate
FBPase	fructose 1,6-bisphosphatase
FBPase-1	fructose-1,6-bisphosphatase
FBPase-2	fructose-2,6-bisphosphatase
Fc	ferrocene
FCAH	ferrocene carboxylic acid
FCCP	carbonyl cyanide 4-trifluoromethoxyphenylhydrazone
FCLA	3,7-dihydro-6-[4-[2-(*N*′(5-fluoresceinyl)thioureido)-ethoxy]phenyl]-2-methylimidazo[1,2-*a*]pyradin-3-one
$Fcm-NR_2$	*N*-ferrocenylmethyl

Fcm-SR	*S*-ferrocenylthiomethyl
FCPN	*N*,*N*-dimethylamino [2-bisphenylphosphinyl)ferrocenylmethane
FCs	fluorocarbons
Fd	ferredoxin
FDAA	1-fluoro-2,4-dinitrophenyl-5-L-alaninamide; N_α-(2,4-dinitro-5-fluorophenyl)-L-alaninamide
FdC	5-fluoro-2′-deoxycytidine
FDH	formate dehydrogenase
FDMA	perfluoro-*N*,*N*-dimethylcyclohexylmethylamine
FDNB	1-fluoro-2,4-dinitrobenzene; 2,4-dinitro-1-fluorobenzene
FDNDEA	5-fluoro-2,4-dinitro-*N*,*N*-diethylaniline
FDP	fructose 1,6-diphosphate
FDPase	fructose 1,6-diphosphatase
5-FdU	5-fluoro-2′-deoxyuridine
FEP	Fe(III) perchlorate
FEP	fluorinated ethylene-poly(propylene) resin; fluorinated ethylene-propylene (resin)
FETFE	fluoroelastomer with special tetrafluoroethylene additives
FFA	free fatty acids
FFDNB	1,5-difluoro-2,4-dinitrobenzene
FFN	fibroblast fibronectin
FGF	fibroblast growth factor
FH	fluorene
FH_4	tetrahydrofolic acid
FHD	decafluoroheptanedione
FHD-3	3-bromo-1,1,2,2-tetrafluoropropane
FHZ	ferritin hydrazide
FIGLU	formiminoglutamate
3-F-Ins	3-deoxy-3-fluoro-D-*myo*-inositol
Fl	flavin
Fl	fluorene
FLEC®	1-(9-fluorenyl)ethyl chloroformate
FLT	3′-α-fluoro-2′,3′-dideoxythymidine
FITC	fluorescein isothiocyanate
5-FITC	5-fluorescein isothiocyanate
FLUO	(1-[2-amino-5-(2,7-dichloro-6-hydroxy-3-oxy-9-xanthenyl)-phenoxyl]-2-[2-amino-5-methylphenoxy] ethane-*N*,*N*,*N*′,*N*′-tetraacetic acid
Fm	9-fluorenylmethyl
FMA	fluoroscein mercuric acetate
fMET	formylmethionine
fMet	*N*-formylmethionine
FMN	flavin mononucleotide; riboflavin 5′-monophosphate

$FMNH_2$	flavin mononucleotide (reduced form)
FMOC	9-fluorenylmethyl chloroformate
Fmoc	9-fluorenylmethoxycarbonyl; 9-fluorenylmethyloxycarbonyl
FMOC-Cl	9-fluorenylmethyl chloroformate
FNAP	4-fluoro-3-nitrophenyl azide
FNPS	4-fluoro-3-nitrophenyl sulfone; bis(4-fluoro-3-nitrophenyl) sulfone
FOD (fod)	1,1,1,2,2,3,3-heptafluoro-7,7-dimethyl-4,6-octanedionato; 6,6,7,7,8,8,8-heptafluoro-2,2-dimethyl-3,5-octanedionato
FP (Fp)	flavoprotein
Fp	dicarbonyl(η^5-cyciopentadienyl)iron(I)
F6P (F-6-P)	fructose 6-phosphate
FPA	1-formylpiperazine
FPD	1-phenyl-1,2-propanedione
FPME	*N*-formyl phenylalanine methyl ester
FPP	farnesyl diphosphate; farnesyl pyrophosphate
FPPM	*N*-formyl-4-diphenylphosphino-2-diphenylphosphinomethyl pyrrolidine
FQ	3-(2-furoyl)-quinoline-2-carboxaldehyde
FQUMP	fluorodeoxyuridylate
FREON	1,1,2-trichloro-1,2,2-trifluoroethane
Fru	fructose
FS	dipotassium nitrosodisulfonate; Fremy's salt
FSBA	5′-(4-fluorosulfonylbenzoyl)adenosine hydrochloride
FSF	fibrin
FSH	follicle stimulating hormone
FTCNQ	tetrafluorotetracyanoquinodimethane
FTN	perfluoro-1,3,7-trimethylbicyclo[3.3.1]nonane
F20TPP	*meso*-tetrakis(pentafluorophenyl)-porphinato iron(III) chloride
5-FU	5-fluorouracil
5-FU	5-fluorouridine
Fua	*N*-[3-(2-furfuryl)acryloyl]
FUDR	5-fluorodeoxyuridine; 2′-deoxy-5-fluorouridine
FUra	5-fluorouracil
Fura 2	1-(2-[5-carboxyoxazol-2-yl]-6-aminobenzofuran-5-oxy)-2-(2′-amino-5′-methylphenoxy)ethane-*N*,*N*,*N*′,*N*′-tetra-acetic acid
Fura 2-AM	1-(2-[5-carboxyoxazol-2-yl]-6-aminobenzofuran-5-oxy)-2-(2′-amino-5′-methylphenoxy)ethane-*N*,*N*,*N*′,*N*′-tetra-acetoxymethyl ester acid
Furacin	5-nitro-2-furfural semicarbazone

FUrd	5-fluorouridine
FUTP	5-fluorouridine triphosphate
G	glycine
G	guanine
G Acid	2-naphthol-6,8-disulfonic acid
G SALT	2-naphthol-6,8-disulfonic acid dipotassium salt
GA (GA_3)	gibberellic acid
GA	glutaraldehyde
GA	gramicidin A
GAA	glacial acetic acid
GABA (Gaba)	4-aminobutyric acid; γ-aminobutyric acid
GABOB	4-amino-3-hydroxybutyric acid
GAC	granular activated carbon
GAG	glyoxal bis(guanylhydrazone)
Gal	galactose
GalN	D-galactosamine
GalNAc	*N*-acetyl-D-galactosamine; *N*-acetylgalactosamine
GalT	galactosyltransferase
Gamma Acid	2-amino-8-naphthol-6-sulfonic acid
GAP	glyceraldehyde-3-phosphate
GAPDH	glyceraldehyde-3-phosphate dehydrogenase
GARFT	glycinamide ribonucleotide formyltransferase
GAS	galium arsonide
GC	glassy carbon
GDH	glutamate dehydrogenase
GDH	glycerol-3-phosphate dehydrogenase
GDP	guanosine 5′-diphosphate; guanosine diphosphate
GDP-Fuc	guanosine 5′-diphosphofucose
GDP-Man	guanosine 5′-diphosphomannosamine
GDPαS	guanosine 5′-*O*-(1-thiodiphosphate)
GDPβS	guanosine 5′-*O*-(2-thiodiphosphate)
GDPαSβS	guanosine 5′-*O*-(1,2-dithiodiphosphate)
GDPH	glycerol-3-phosphate dehydrogenase
GEDTA	glycoletherdiamine-tetraacetic acid
GEMSA	(2-guanidinoethylmercapto)succinic acid
GGPP	geranylgeranyl pyrophosphate
GH	growth hormone
GIP	6-*O*-(2-amino-2-deoxy-α-D-glucopyranosyl)inositol 1-phosphate
GIP	gastric inhibitory polypeptide
GK	glycerol kinase
Glc	glucose
GlcA	D-gluconic acid
GlcN	D-glucosamine

GlcNAc	*N*-acetyl-D-glucosamine
GlcUA	D-glucuronic acid
GLDH	glutamate dehydrogenase
GleNAc-P	*N*-acetylglucosamine phosphate
Gln	glutaminyl
Gln (gln)	glutamine
Glu	glutamate
Glu	glutamic acid
Glu	glutamyl
GLUPA	γ-L-glutamyl-4-nitroanilide
Glx	glutamine or glutamic acid
Gly	glycyl
Gly (gly)	glycine
glyc	glycerol
Glycin	*N*-(4-hydroxyphenyl)glycine
GlyDH	α-glycerophosphate dehydrogenase
GLYMO	3-glycidyloxypropyltrimethoxysilane
GMBS	*N*-(γ-maleimidobutyryloxy)succinimide; *N*-succinimidyl 4-maleimidobutyrate
GMBS	*N*-γ-maleimidobutyryrloxysuccinimide ester
GMDP	*N*-acetyl-D-glucosaminyl-β(1-4)-*N*-acetylmuramyl-L-alanyl-D-isoglutamine
GMP	guanosine 5′-monophosphate; guanosine monophosphate
5′-GMP	guanosine 5′-monophosphoric acid
G-3-5-MP	guanosine 3′,5′-cyclic monophosphate
GMP-PCP	guanosine 5′-[β,γ-methylene]triphosphate
GMP-PNP	guanosine-5′-[β,γ-imido]triphosphate
GnRH	gonadotropic releasing hormone
GOD (GOd)	glucose oxidase
GOT	glutamate-oxaloacetate transaminase
GP	glycerophosphate
G-1-P	α-D-glucose 1-phosphate; glucose 1-phosphate
G3P (G-3-P)	glyceraldehyde 3-phosphate
G-5′-P	guanosine 5′-monophosphate
G6P (G-6-P)	glucose 6-phosphate
α-GP	α-glycerophosphate
G3PDH	glyceraldehyde 3-phosphate dehydrogenase
G-6-PDH (G-6-P-dh)	glucose-6-phosphate dehydrogenase
α-GPDH	α-glycerophosphate dehydrogenase
GPI	glycosyl phosphatoyl inosityl; glycosylphosphoinositides; 6-*O*-(2-amino-2-deoxy-α-D-glucopyranosyl)-inositol 1-phosphate
GPO	glycerol 3-phosphate oxidase

GPP	geranyl pyrophosphate
GPSG	glycidoxypropyl silica gel
GPT	glutamate-pyruvate transaminase
GR	glutathione reductase
GR-I	government rubber I
GR-M	neoprene
GRN	government rubber nitrile
GRS	government rubber styrene
GSH	glutathione, reduced
GSH-Px	glutathione peroxidase
GSL	glycosphingolipid
GSSG	glutathione disulfide; glutathione, oxidized
GTC	guanidine isothiocyanate
GTP (5′-GTP)	guanosine 5′-triphosphate; guanosine triphosphate
GTPαS	guanosine 5′-*O*-(1-thiotriphosphate)
GTPβS	guanosine 5′-*O*-(2-thiotriphosphate)
GTP-γ-S-Li_4	guanosine 5′-[γ-thio]triphosphate tetralithium salt
GTPαSβS	guanosine 5′-*O*-(1,2-dithiotriphosphate)
GTPαSγS	guanosine 5′-*O*-(1,2-dithiotriphosphate)
GTX	gonyautoxin
Gua	guanine
GUM	guaiacolmethyl
Guo	guanosine
H	histidine
H Acid	1-amino-8-naphthol-3,6-disulfonic acid
H-7	1-(5-isoquinolinylsulfonyl)-2-methylpiperazine
H-8	*N*-[2-(methylamino)ethyl]isoquinoline-5-sulfonamide dihydrochloride
H-89	*N*-[2-((3-(4-bromophenyl)-2-propenyl)-amino)ethyl]-5-isoquinolinesulfonamide
H-9	*N*-(2-aminoethyl)-5-isoquinolinesulfonamide
HA	hemagglutinin
HA	hyaluronic acid
HA 100	1-(5-isoquinolinylsulfonyl)piperazine
HA 1004	*N*-(2-guanidinoethyl)-5-isoquinolinylsulfonamide
HA 1077	1-(5-isoquinolinylsulfonyl)homopiperazine
HAB	hexakis(alkanoyloxy)benzene
HABA	2-(4′-hydroxyazobenzene)benzoic acid; 2′-(4-hydroxyphenylazo)benzoic acid; 4-hydroxyazobenzene-2′-carboxylic acid
HABBA	2-(4′-hydroxyazobenzene)benzoic acid
HAC	(2*S*,6*R*)-6-(carboxymethyl)-2-hydroxy-2,4,4-trimethylmorpholinium chloride
HAC	hemiacetylcarnitinium

Hacac	acetylacetone
hANF	human atrial natriuretic peptide
HAV	hepatitis A virus
Hb	hemoglobin
β-HBA	D-β-hydroxybutyrate
HbA	adult hemoglobin
Hba	benzoylacetone
HBABA	2-(4′-hydroxyazobenzene)benzoic acid
HBC	(2*S*,6*R*)-6-(carboxymethyl)-2-hydroxy-4,4-dimethyl-2-phenylmorpholinium chloride
HBC	hemibenzoylcarnitinium
HbCO	carbon monoxide hemoglobin
HBD	bis(tri-*n*-butyltin)oxide; hexabutyldistannoxane
3-HBDH	3-hydroxybutyrate dehydrogenase
HbF	fetal hemoglobin
Hbg	biguanide
HBI	iminostilbene
HbO_2	oxyhemoglobin
HBPS	1′,3′,3′-trimethyl-6-hydroxyspiro[2*H*-1-benzopyran-2,2′-indoline]
HBT	hydrazinobenzothiazole
HBT	hydroxybenzotriazole
HBTU	2-(1*H*-benzotriazol-1-yl)-1,1,3,3-tetramethyluronium hexafluorophosphate; *O*-benzotriazol-1-yl-*N*,*N*,*N*′,*N*′-tetramethyluronium hexafluorophosphate
HBV	hepatitis B virus
HC	hydrocarbon
HCA	1,1,1,3,3,3-hexachloro-2-propanone; hexachloroacetone
HCA	human carbonic anhydrase
HCA	hydrochloric acid
HCCH	1,2,3,4,5,6-hexachlorocyclohexane; (incorrectly named benzene hexachloride)
HCCPD	hexachlorocyclopentadiene
HCFC-121	tetrachlorofluoroethane
HCFC-122	trichlorodifluoroethane
HCFC-123	trichlorotrifluoroethane
HCFC-124	chlorotetrafluoroethane
HCFC-131	trichlorofluoroethane
HCFC-132b	dichlorodifluoroethane
HCFC-133a	chlorotrifluoroethane
HCFC-141b	dichlorofluoroethane
HCFC-142b	chlorodifluoroethane
HCFC-21	dichlorofluoromethane
HCFC-22	chlorodifluoromethane

HCFC-221	hexachlorofluoropropane
HCFC-222	pentachlorodifluoropropane
HCFC-223	tetrachlorotrifluoropropane
HCFC-224	trichlorotetrafluoropropane
HCFC-225ca	dichloropentafluoropropane
HCFC-225cb	dichloropentafluoropropane
HCFC-226	chlorohexafluoropropane
HCFC-231	pentachlorofluoropropane
HCFC-232	tetrachlorodifluoropropane
HCFC-233	trichlorotrifluoropropane
HCFC-234	dichlorotetrafluoropropane
HCFC-235	chloropentafluoropropane
HCFC-241	tetrachlorofluoropropane
HCFC-242	trichlorodifluoropropane
HCFC-243	dichlorotrifluoropropane
HCFC-244	chlorotetrafluoropropane
HCFC-251	trichlorofluoropropane
HCFC-252	dichlorodifluoropropane
HCFC-253	chlorotrifluoropropane
HCFC-261	dichlorofluoropropane
HCFC-262	chlorodifluoropropane
HCFC-271	chlorofluoropropane
HCFC-31	chlorofluoromethane
HCFCs	hydrochlorofluorocarbons
HCG (hCG)	gonadotrop(h)in, human chorionic; human chorionic gonadotropin
HCH	hexachlorocyclohexane
HCMV	human cytomegalovirus
HCP	hexachlorophene
hCS	cholionic somatomammotropin
HCSH	homocysteine
HCSSCH	homocystine
HCTD	heptacyclo[$6.6.0^{2,6}.0^{3,13}.0^{4,11}.0^{5,9}.0^{10,14}$]tetradecane
HD	hexadecane
hd	1,5-hexadiene
HDAL	hexadecenal
HDAOS	*N*-(2-hydroxy-3-sulfopropyl)-3,5-dimethoxyaniline
HDCBS	2-hydroxy-3,5-dichlorobenzenesulfonic acid; 3,5-dichloro-2-hydroxybenzenesulfonic acid
7-11-HDDA	7,11-hexadecadien-1-yl acetate
HDI	hydroxyethylene dipeptide isostere
HDL	high density lipoprotein
HDMAC	*trans*-*N*,*N*-dimethyl-2-aminocyclohexanol
HDMAE	*N*,*N*-dimethyl-2-aminoethanol

HDMAP	*N*,*N*-dimethyl-1-amino-2-propanol
Hdmg	dimethylglyoximato(1⁻)
H_2dmg	dimethylglyoxime
HDODA	1,6-hexanediol diacrylate
HDPE	high-density polyethylene
HDQ	isobutyl 1,2-dihydro-2-isobutoxyquinoline-1-carboxylate
HEA	1-aziridineethanol; *N*-(2-hydroxyethyl)aziridine
Hea	ethanolamine
HECAMEG	methyl-6-*O*-(*N*-heptylcarbamoyl)-α-D-glucopyranoside
HEDTA	*N*-(2-hyroxyethyl)ethylenediamine-*N*,*N*′,*N*′-triacetic acid
Hedta	ethylenediaminetetraacetate(3-); ethylenediaminetetraacetato
H_2edta	ethylenediaminetetraacetate(2-); ethylenediaminetetraacetato
H_3edta	ethylenediaminetetraacetate(4-); ethylenediaminetetraacetato
H_4edta	ethylenediaminetetraacetic acid
HEEDTA	*N*-(2-hydroxyethyl)ethylenediaminetriacetate
HEEDTA	*N*-hydroxyethylethylenediaminetriacetic acid
HEEI	*N*-(2-hydroxyethyl)ethyleneimine
HEMA	2-hydroxyethyl methacrylate
HEOD	1,2,3,4,5,6,7,8,8-octachloro-3a,4,7,7a-tetrahydro-4,7-methanoindene
HEP	heparin
Hep	heptulose
HEPES (Hepes) (hepes)	4-(2-hydroxyethyl)-1-piperazineethanesulfonic acid; *N*-(2-hydroxyethyl)piperazine-*N*′-ethanesulfonic acid
HEPPS (Hepps) (hepps)	4-(2-hydroxyethyl)-1-piperazinepropanesulfonic acid; 4-(2-hydroxyethyl)piperazine-1-propanesulfonic acid; *N*-(2-hydroxyethyl)piperazine-*N*′-3-propanesulfonic acid
HEPPSO	4-(2-hydroxyethyl)piperazine-1-(2-hydroxypropane sulfonic acid)
HEPSO	*N*-hydroxyethylpiperazine-*N*′-2-hydroxypropanesulfinic acid
HEPY	2-(2-hydroxyethyl)pyridine
Het	heterocycle
HETCNQ	2-(2′-hydroxyethoxy)-7,7,8,8-tetracyano-*p*-quinodimethane
HETE	hydroxyeicosatetraenoic acid
5-HETE	5-hydroxy-6,8,11,14-eicosatetraenoic acid; 5-hydroxyeicosatetraenoic acid
8-HETE	8-hydroxy-5,9,11,14-eicosatetraenoic acid
9-HETE	9-hydroxy-5,7,11,14-eicosatetraenoic acid
11-HETE	11-hydroxy-5,8,12,14-eicosatetraenoic acid

12-HETE	12-hydroxy-eicosatetraenoic acid
15-HETE	15-hydroxy-eicosatetraenoic acid
HETP	tetraethylpyrophosphate
Hex	hexane
Hex	hexyl
HEX-BCH	hexachloronorbornadiene
HF	6-hydroxy-3-fluorone
HFA	(+)-1,2,3,4,4a,9a-hexahydro-4a-fluorenamine
HFA	hexafluoroacetone
HFAA	heptafluorobutyric anhydride
HFB	hexafluorobut-2-yne
HFBA	heptafluoro-*n*-butyric acid
HFBI	heptafluorobutyrylimidazole
hfc	3-(heptafluoropropylhydroxymethylene)-(+)-camphorato
HFC-134a	1,1,1,2-tetrafluoroethane
HFCS	high fructose corn syrup
HFF	human foreskin fibroblasts
HFIP	1,1,1,3,3,3-hexafluoro-2-propanol; hexafluoroisopropyl alcohol
HFP	1,1,1,3,3,3-hexafluoro-2-propanol
HFP	hexafluoropropene
HFTA	hexafluorothioacetone
HGF	glucagon
HGH (hGH)	pituitary growth hormone; human growth hormone
HGPRTase	hypoxanthine-guanine phosphorinbosyl transferase
HHPA	1,2-cyclohexanedicarboxylic anhydride
HHPA	hexahydrophthalic anhydride
HHSNNA	2-hydroxy-1-(2-hydroxy-4-sulfo-1-naphthylazo-3-naphthoic acid)
H_2ida	iminodiacetic acid
HIDC IODIDE	1,1,3,3,3′,3′-hexamethylindodicarbocyanine iodide
hIL-2	interleukin-2
Him	imidazole
HINAP	4-hydroxyisonitroso acetophenone
His	histidine
HISG	human immune serum globulin
HITC Iodide	1,1′,3,3,3′,3′-hexamethylindotricarbocyanine iodide
HK	β-hydroxyketone
HK	hexokinase
HLADH	horse liver alcohol dehydrogenase
HLE	horse liver esterase
HLE	human leukocyte elastase
HMAT	hexa[1-(2-methyl)aziridinyl]-1,3,5-triphosphatriazine
HMB	2-hydroxy-4-methoxybenzophenone

HMB	2-hydroxy-5-methoxybenzaldehyde
HMB	hexamethylbenzene
HMDB	hexamethylDewar benzene; 1,2,3,4,5,6-hexamethylbicyclo[2.2.0]hexa-2,5-diene
HMDS	1,1,1,3,3,3-hexamethyldisilazane; hexamethyldisilazane
HMDS	hexamethyldisilane
HMDS	hexamethyldisiloxane
HMDSO	hexamethyldisiloxane
HMF	5-(hydroxymethyl)-2-furaldehyde
HMG	gonadotropin
HMG CoA	3-hydroxy-3-methylglutaryl coenzyme A
HMGR	β-hydroxy-β-methylglutaryl coenzyme A reductase
HMI	hexamethyleneimine
HMN	2,2,4,4,6,8,8-heptamethylnonane
HMPA	hexamethylphosphoramide; hexamethylphosphoric triamide
HMPT	hexamethylphosphoramide, hexamethylphosphoric triamide
HMPT	hexamethylphosphorous triamide
HMPTA	hexamethylphosphoramide; hexamethylphosphoric triamide
HMTCANQ	2-(hydroxylmethyl)-11,11,12,12-tetracyanoanthraquinodimethane; 6-(hydroxymethyl)-9,9,10,10-tetracyanonaphthoquinodimethane
HMTT	3-hexadecanoyl-4-methoxycarbonyl-1,3-thiazolidine-2-thione
HMX	octahydro-1,3,5,7-tetranitro-1,3,5,7-tetrazocine
HN1	ethylbis(2-chloroethyl)amine
HN_2	mechlorethamine
$HNEA^+$	[1-(1-naphthyl)ethyl]ammonium cation
hnRNA	heterogenous nuclear ribonucleic acid
HNS	methyl 2-(1-methylethylidene)hydrazinecarbodithioate
H_3nta	nitrilotriacetic acid
HOAc	acetic acid
HOBB	*p*-(2-phenylethyl)phenol
2-HOBP	2-hydroxybiphenyl
3-HOBP	3-hydroxybiphenyl
4-HOBP	4-hydroxybiphenyl
HOBT (HOBt)	1-hydroxybenzotriazole; hydroxybenzotriazole
HOC	halogenated organic carbons
13(S)-HODE	13-(*S*)-hydroxyoctadeca-9-*cis*-11-*trans*-dienoic acid
HODPB	*p*-(4-phenylbutyl)phenol
homo Z	homobenzyloxycarbonyl
HON	2-amino-5-hydroxylevulinic acid

HONB	*N*-hydroxy-5-norbornene-2,3-dicarboximide; *N*-hydroxy-5-norbornene-2,3-dicarboxylic acid imide
HOP	2,5,7,10,11,14-hexaoxa[4.4.4]propellane
HOPG	highly oriented pyrolytic graphite
HOPip	*N*-hydroxypiperidine
HOS	hydroxyamine-*O*-sulfonic acid
HOSA	hydroxylamine-*O*-sulfonic acid
HOSu	*N*-hydroxysuccinimide
HO-T	enantiomer of Taber's chiral alcohol
HOTDO	2,5-diphenyl-4-hydroxy-3-thiophenone-1,1-dioxide
H_2ox	oxalic acid
HPC	6-(carboxymethyl)-2-hydroxy-2-pentadecyl-4,4-dimethylmorpholinium bromide
HPC	hemipalmitoylcarnitinium
$HPEA^+$	[1-phenylethyl]ammonium cation
HPETE	hydroperoxyeicosatetraenoic acid; monohydroperoxy-eicosatetraenoic acid
5-HPETE	5-hydroperoxy-6,8,11,14-eicosatetraenoic acid; 5-hydroperoxy-eicosatetraenoic acid
12-HPETE	12-hydroperoxy-5,8,10,14-eicosatetraenoic acid; 12-hydroperoxy-eicosatetraenoic acid
15-HPETE	15-hydroperoxy-5,8,11,13-eicosatetraenoic acid; 15-hydroperoxy-eicosatetraenoic acid
HPFH	hereditary persistance of fetal hemoglobin
HPG	*p*-hydroxyphenylglyoxal
HPMA	2-hydroxypropylmethacrylate
HPMPA	9-[3-hydroxy-2-(phosphonomethoxy)propyl]adenine
HPOA	hexafluoropropene diethylamine
13(S)-HPODE	13-(*S*)-hydroperoxyoctadeca-9-*cis*-11-*trans*-dienoic acid
HPP	allopurinol
HPPH	5-hydroxyphenyl-5-phenylhydantoin
HPrC	(2*S*,6*R*)-6-(carboxymethyl)-2-ethyl-2-hydroxy-4,4-dimethylmorpholinium chloride
HPrC	hemipropanoylcarnitinium
HPRtase	hypoxanthine phosphoribosyltransferase
Hpz	pyrazole
HQ	2-hydroxyquinolone
HQ	1,4-hydroxyquinone; hydroquinone
HR	isobutene-isoprene rubber
HRP	horseradish peroxidase
HSA	human serum albumin
HSA	hydroxylamine-*O*-sulfonic acid
HSAB	*N*-hydroxysuccinimidyl-4-azidobenzoate
HS-CoM Na	sodium 2-mercaptoethanesulfonate

HSDH	hydroxysteroid dehydrogenase
7β-HSDH	7β-hydroxysteroid dehydrogenase
20β-HSDH	3α,20β-hydroxysteroid dehydrogenase
β-HSDH	3β,17β-hydroxysteroid dehydrogenase
HS-HTP	(7-mercaptoheptanoyl)threonine phosphate
HSl	(*S*)-2-amino-4-butyrolactone hydrochloride
HSV	herpes simplex virus
HSV-1	herpes simplex virus type 1
5-HT	serotonin hydrochloride
HTA	*N*,*N*-diethylaminoethyl hexanoate
HTCl	hexadecyltrimethylammonium chloride
H_3tea	triethanolamine
HTMP	2,2,6,6-tetramethylpiperidine
HTP	4-hydroxythiophenol
5-HTP	5-hydroxytryptophan
HVA	homovanillic acid (4-hydroxy-3-methoxyphenylacetic acid)
HVB	hydrocortisone
Hyl	hydroxylysine
Hyl	hydroxylysyl
Hylv	α-hydroxyisovaleric acid
Hyp	hydroxyproline
Hyp	hydroxyprolyl, hydroxyproline
Hytra	2-hydroxy-1,2,2-triphenylethyl acetate
hZ	homobenzyloxycarbonyl
I	inhibitor
I	inosine
I	isoleucine
i	iso (as in *i*-Pr)
IAA	indol-3-yl acetic acid; indole acetic acid; 3-indolylacetic acid
I-acid	6-amino-1-naphthol-3-sulfonic acid
I-AEDANS	*N*-iodoacetyl-*N'*-(X-sulfo-1-naphthyl)ethylenediamine (X = 5: 1,5-I-AEDANS; X = 8: 1,8-I-AEDANS)
IAF	4-(iodoacetamido)fluorescein
5-IAF	5-iodoacetamidofluoroscein
6-IAF	6-iodoacetamidofluoroscein
IANBD	*N*-{{[2-(iodoacetoxy)ethyl]-*N*-methyl}amino}-7-nitrobenz-2-oxa-1,3-diazole
IBA	3-indolebutyric acid; indol-3-yl butyric acid
IBCF	isobutyl chloroformate
IBD	iodobenzene dichloride
IBF	isobenzofuran
IBIB	isobutyl isobutyrate
IBMX	3-isobutyl-1-methylxanthine

IBTMO	isobutyltrimethoxysilane
iBu	isobutyl
ICD	isocitric dehydrogenase
ICDH	isocitrate dehydrogenase
ICE	interleukin-1β converting enzyme
ICI	isophthaloyl chloride
IcP	inositol 1,2-cyclic phosphate
2^d-Icr	2-isocaranyl
Icr_2BH	di-2-isocaranylborane
ICSH	interstitial cell stimulating hormone
IDA	iminodiacetic acid
IDH	isocitrate dehydrogenase
IDL	intermediate density lipoprotein
IDP (5′-IDP)	inosine 5′-diphosphate
IDTr	3-(imidazol-1-ylmethyl)-4′,4″-dimethoxytriphenylmethyl
IDU (IdU)	5-iodo-2′-deoxyuridine; 2′-deoxy-5-iodouridine
IDUR	2′-deoxy-5-iodouridine
IF	intrinsic factor
IFN-γ	interferon-γ
IFP	glycerol
IF(x)	initiation factor (x)
Ig	immunoglobulin
IGF	insulin-like growth factor
IgG (igG)	immunoglobulin G
IH	immobilized histamine
IHP	inositol hexaphosphate
IIDQ	1-isobutoxycarbonyl-2-isobutoxy-1,2-dihydroquinoline; 2-isobutoxy-1-isobutoxycarbonyl-1,2-dihydroquinoline
IIR	isobutene-isoprene rubber
IL	interleukin
Ile	isoleucine
Ile	isoleucyl
Ileu (iLEU)	isoleucine
Im	imidazol-1-yl; 1-imidazolyl
Im	imidazole
Im_2CO	1,1′-carbonyldiimidazole; carbonyldimidazole
Imd	1-methyl-1,3-benzodiaza-2-yl
iMds	2,6-dimethoxy-4-methylbenzenesulfonyl
IMEO	imidazolinepropyltriethoxysilane
Imid (imid)	imidazole
IMP	inosine 5′-monophosphate; inosine 5′-phosphate; inosinic acid
5′-IMP	inosine-5′-monophosphate
5′-IMP	inosine-5′-monophosphoric acid

ImpN	imidizylphosphate nucleotide
INAH	isonicotinic acid hydrazide
INC	*N*-phenyl isopropyl carbamate
INDO 1	1-[2-amino-5-(6-carboxyindol-2-yl)-2-(2′-amino-5′-methylphenoxy]ethane-*N*,*N*,*N*′,*N*′-tetraacetic acid
INDO 1/AM	1-[2-amino-5-(6-carboxyindol-2-yl)-2-(2′-amino-5′-methylphenoxy]ethane-*N*,*N*,*N*′,*N*′-tetraacetic acid, pentaacetoxymethyl ester
INGAS	indium gallium arsonide
INH	isonicotinic acid hydrazide
Ino	inosine
Ins	inositol
$Ins(1,4)P_2$	inositol 1,4-bisphosphate
$InsP_5$	inositol pentakisphosphate
$InsP_6$	inositol hexakisphosphate
INT	2-(4-iodophenyl)-3-(4-nitrophenyl)-5-phenyltetrazolium chloride
INT	iodonitrotetrazolium chloride; *p*-iodonitrotetrazolium violet
IP	inositol 1-phosphate
I-5′-P	inosine 5′-monophosphate
Ip	isopropyl
IP_3	inositol 1,4,5-trisphosphate
IPA	isopropyl alcohol
Ipaoc	1-isopropylallyloxycarbonyl
IPC (Ipc)	isopinocampheyl
IPC	isopropyl *N*-phenylcarbamate; *N*-phenyl isopropyl carbamate
Ipc_2BCl	*B*-chlorodiisopinocampheylborane
$IPCBH_2$ ($IpcBH_2$)	monoisopinocampheylborane
IPC_2BH (Ipc_2BH) [$(Ipc)_2BH$]	diisopinocampheylborane
IPDI	isophorone diisocyanate; 3-isocyanatomethyl-3,5,5-trimethylcyclohexyl isocyanate
IPDMS	isopropyldimethylsilyl
IPF	9-isopropylfluorene
IPN	isophthalonitrile
IPOTMS	isopropenyloxytrimethylsilane
IPP	isopentenyl diphosphate; isopentenylpyrophosphate
iPr	isopropyl
IPTG	isopropyl β-D-thiogalactopyranoside; isopropyl β-D-thiogalactoside
IPTT	isopropyl-1,3-thiazolidine-2-thione
ISDN	isosorbide dinitrate

ITA	itaconic anhydride
ITP	inosine 5′-triphosphate; inosine triphosphate
IUDR	2′-deoxy-5-iodouridine
IZAA	5-chloroindazol-3-acetic acid ethyl ester
J Acid	6-amino-1-naphthol-3-sulfonic acid
JA	jasmonic acid
K	lysine
KAPA	potassium 3-aminopropylamide
K-ATPase	potassium pump
KBA	3-ketobutyraldehyde dimethyl acetal; acetylacetaldehyde dimethyl acetal
KBT	4-ketobenztriazine
KCTI	1-keto-3-carbomethoxy-1,2,3,4-tetrahydroisoquinoline
KDA	potassium diisopropylaminde
KDG	2-keto-3-deoxygluconate
KDO	2-keto-3-dioxyoctonate
KDO	3-deoxy-D-*manno*-2-octulosonic acid
KDO-8-P	3-deoxy-D-*manno*-2-octulosonic acid, 8-phosphate
KDPG	2-keto-3-deoxy-6-phosphogluconate
Kel-F	poly(chlorotrifluoroethylene)
α-KG	α-ketoglutarate dehydrogenase
KGA	ketoglutarate
α-KGA	α-ketoglutarate
KGDH	ketoglutarate dehydrogenase
KHMDS	potassium bis(trimethylsilyl)amide; potassium hexamethyldisilazide
K-IAO	potassium gluconate
KLH	keyhole limpet hemocyanin
KMH	1,2,3,6-tetrahydro-3,6-dioxopyridazine
L	leucine
L	ligand
L Acid	1-naphthol-5-sulfonic acid
LAAO	L-amino acid oxidase
Lac	lactose
LAC-EDA	lithium acetylide-ethylenediamine
LAD	lactic dehydrogenase
LAH	lithium aluminum hydride
LAK	lymphokine-activated killer (cells)
LAL	endotoxin test reagent
LAP	leucine aminopeptidase
LAS	linear alkylbenzene sulfonate
LBF	*Lactobacillus bulgaricus* factor
LBTMSA	lithium bis(trimethylsilyl)amide
LC-SPDP	succinimidyl 6[3-(2-pyridyldithio)propionamido]hexanoate

LCAT	lecithin-cholesterol acyl transferase; cholesterol acyl transferase
LD	lactic dehydrogenase
LDA	lithium diisopropylamide
LDAO	lauryldimethylamine oxide
LDAO	*N,N*-dimethyldodecylamine-*N*-oxide
LDBB	lithium 4,4′-di-*tert*-butylbiphenylide
LDH	lactate dehydrogenase
LDL	low density lipoprotein
LDMAN	lithium 1-(dimethylamino)naphthalenide
LDPE	low-density polyethylene
LDS	lipid disaccharide
Leu	leucyl
Leu (leu)	leucine
t-LeuPHOS	1-diphenylphosphino-2-dimethylamino-3,3-dimethylbutane
Lev	levulinoyl
LevS	levulinoyldithioacetal ester; 4,4-(ethylenedithio)pentanoyl
Lgf$_2$BH	dilongifolylborane
LH	luteinizing hormone
LHCP	light-harvesting chlorophast protein
LHMDS	lithium hexamethyldisilazane; lithium bis(trimethylsilyl)amide; lithium hexamethyldisilazide
LHRF (LH-RH)	luteinizing hormone releasing factor
LICA	lithium cyclohexyl(isopropyl)amide; lithium isopropylcyclohexylamide
LiDBB	lithium *p,p*′-di-*tert*-butylbiphenyl radical anion
LiDCHA	lithium dicyclohexylamide
LiHMDS	lithium hexamethyldisilazide
LiNaph	lithium naphthalenide
Lip(S_2)	*dl*-α-lipoic acid
LIS	3,5-diiodo-2-hydroxybenzoic acid lithium salt
LiTMP	lithium 2,2,6,6-tetramethylpiperidine
LLAT	lysolecthin: lecithin acyltransferase
LLD factor	vitamin B_{12}
LLF	fibrin
LLPACK	L-leucyl-leucyl-phenylalanyl-chloromethyl ketone
LM	lumazine; 2,4-dioxopteridine
α-Lnn	linolenic acid
LOS	lipidoligosaccharide
LP	lithiopinacolonate
LP-80	lipoprotein lipase 80
LPL	lipoprotein lipase
LPO	lauroyl peroxide

LPS	lipopolysaccharide; liposaccharide
LPTS	2,6-lutidinium *p*-toluenesulfonate
LR	Lawesson reagent
LSA	lithium *N*-benzyl-*N*-(trimethylsilyl)amide
LSD	lysergic acid diethylamide; lysergide
LSR	lanthanide shift reagent
LT	leukotriene (as in LTA_4, etc.)
LTA	lead tetraacetate
LTBH	lithium triethylborohydride
lTf	lactoferrin
LTH	luteotropic hormone; luteotropin
LTMAC	dodecyltrimethylammonium chloride
LTMP	lithium 2,2,6,6-tetramethylpiperidide
lut	lutidine
LVN	low viscosity nitrocellulose
LVP	lysine vasopressin
LYS (Lys)	lysine
Lys	lysyl
Lyso-PAF	1-hexadecyl-*sn*-glycero-3-phosphocholine
M	metal
M	methionine
M	morpholine
M	myosin
M Acid	5-amino-1-naphthol-3-sulfonic acid
β_2M	β-2-microglobulin
m	meta
2MlPT	2-methyl-1-propanethiol
mRNA	messenger RNA
MA	maleic anhydride
MA	methacrolein
MA	methyl *N*,*N*-diisopropylphosphoramidite
MA	methylamine
MAA	1-(-)-menthoxyacetic acid; menthoxyacetic acid
MAA	methanearsonic acid
MAA	methyl acetoacetate
MAADMA	methylaminoacetaldehyde dimethyl acetal
MAB	monoclonal antibodies
mAbs	monoclonal antibodies
MAC	methyl acrylate
MACE	*o*-chloroacetophenone
MAD	methyl aluminum bis(2,6-di-*tert*-butyl-4-methylphenoxide)
MADH	methylamine dehydrogenase
MADU	2′-deoxy-5-(methylamino)uridine
Mal	maleyl
Mal	maltose

MAM acetate	methylazoxymethyl acetate
Man	mannose
ManNAc	*N*-acetylmannosamine
MANPAP	1-methyl-4-[4-(4-aminonaphthylazo)phenylazo]pyridinium iodide
Man-Tan	dihydroxyacetone
MAO	monoamine oxidase
MAOA	2-methyl-3-*o*-tolyl-4(3*H*)-quinazolinone
MAP	methyl[(2,4-dinitrophenyl)amino]propanoate
MAPO	tris[1-(2-methyl)aziridinyl]phosphine oxide
MAPP	*N*-methyl-1-phenyl-2-(1-piperidinyl)ethylamine
MAPS	tris[1-(2-methyl)aziridinyl]phosphine sulfide
MAPTAC	methacrylamidopropyltrimethylammonium chloride
MAPTAM	bis-(2-amino-5-methylphenoxy)ethane-*N*,*N*,*N*′,*N*′-tetraacetic acid tetraacetoxymethyl ester
MAQ-Br	2-bromomethylanthraquinone
MASC	methylaluminum sesquichloride
MB	monobromobimane
p-MB	*p*-methylbiphenyl
Mb (mb)	myoglobin
MBA	2,2′-dichloro-*N*-methyldiethylamine
MBA	*N*,*N*′-methylenebisacrylamide
2-MBA	2-methylbutyrate
α-MBA	α-methylbenzylamine
MBA-Propionate	1-phenylethyl propionate
MBBA	4′-methoxybenzylidene-4-*n*-butylaniline
MBC	methylbenzylcarbamate
MBCP	*O*-[4-bromo-2,5-dichlorophenyl] *O*-methyl-phenyl phosphonothioate
MBD	4-(4-methoxybenzylamino)-7-nitrobenzofurazane
MBE	1-methyl-1-benzyloxyethyl
MBETA	*N*,*N*-diethylaminoethyl 4-methylbenzoate
MBF	2,3,3a,4,5,6,7,7a-octahydro-7,8,8-trimethyl-4,7-methanobenzofuran-2-yl
MBHA	4-methylbenzhydrylamine (resin)
MBHA-resin	4-methylbenzhydrylamine polymer bound
mbmp	6-methyl-2-(1-methylbenzimidazol-2-yl)pyridine
MbO_2	oxymyoglobin
MBOA	7-methoxy-1,4-benzoxazin-3-one
MBOCA	methylenebis(*o*-chloroaniline)
MBS	*m*-maleimidobenzoyl-*N*-hydroxysuccinimide ester
MBS (Mbs)	*p*-methoxybenzenesulfonyl
MBT	2-mercaptobenzothiazole
MBTFA	*N*-methyl-bis-trifluoroacetamide
MBTH	3-methyl-2-benzothiazolinone hydrazone

MBTS	2,2′-dithiobis(benzothiazole)
MC	magnesium chlorate
MC	monochlorobimane
3-MC	3-methylcholanthrene
MCA	chloroacetic acid; monochloroacetic acid
MCAA	chloroacetic acid; monochloroacetic acid
MCB	dibromobutane
MCBeTTB	1-methoxycarbonyl-3-benzyl-1,2,3,4-tetrahydro-1,3,5-triazino[1,2-*a*]benzimidazol
MCBS	monochloroborane-methyl sulfide
MCBuTTB	1-methoxycarbonyl-3-*n*-butyl-1,2,3,4-tetrahydro-1,3,5-triazino[1,2-*s*]benzimidazol
MCE	mixed cellulose ester
MCF	menthyl chloroformate
3,3-MCH	3-methyl-3-cyclohexen-1-one
MCLPXH	9-(*m*-chlorophenyl)xanthene
MCP	(4-chloro-*o*-toloxy)acetic acid
MCP	*meta*-cresol purple
MCP	metacyclophane
MCP	methylcyclopentane
MCP	methylenecyclopropane
MCPA	2-methyl-4-chlorophenoxyacetic acid; 4-chloro-2-methylphenoxyacetic acid; (4-chloro-*o*-toloxy)acetic acid
MCPB	4-(4-chloro-2-methylphenoxy)butanoic acid; 4-(2-methyl-4-chlorophenoxy)butyric acid
MCPBA (mCPBA)	3-chloroperoxybenzoic acid
MCPCA	2-methyl-4-chlorophenoxyaceto-*o*-chloroanilide
MCPDEA	*N*,*N*-di(2-hydroxyethyl)-*m*-chloroaniline
MCPP	2-(2-methyl-4-chlorophenoxy)propionic acid; 4-chloro-3-methylphenoxypropionic acid
MCT	mercury cadmium telluride
MDA	1,8-diamino-4-menthane
MDA	malondialdehyde
MDA	methylenedihydroanthracene
MDA	methylenedianiline
MDB	α-methyldeoxybenzoin
MDEA	*N*-methyldiethanolamine
M-DEA	4,4′-methylene-bis-(2,6-diethylaniline)
MDEB	*N*-dodecyl-*N*-methylephedrinium bromide; *N*-methyl-*N*-dodecylephedrinium bromide
MDEPAP	1-methyl-4-(4-diethylaminophenylazo)pyridinium lodide
MDH	malate dehydrogenase; malic dehydrogenase
MDI	di-*p*-phenylene isocyanate
MDMH	1-(hydroxymethyl)-5,5-dimethylhydantoin

MDN	malononitrile
MDNB	*m*-dinitrobenzene
MDP	methylenediphosphonic acid
MDP	muramyl dipeptide
MDP	*N*-acetylmuramyl-L-alanyl-D-isoglutamine
MDPF	2-methoxy-2,4-diphenyl-3(2*H*)-furanone
MDPP	menthyldiphenylphosphine
Mds	2,6-dimethyl-4-methoxybenzenesulfonyl
MDT	2-methyl-1,3-dithiane
ME (βME)	β-mercaptoethanol
Me	methyl
MEA	2-aminoethanethiol
MEA	2-methoxyethylamine
MEA	monoethanolamine
MEB	ethylenebisdithiocarbamic acid manganese salt
MeCbl	methylcobalamin
MeCCNU	1-(2-chloroethyl)-3-(4-*trans*-methylcyclohexyl)-1-nitrosourea
MED	2-butanone dioxolane
MeDBTh+	*S*-methyldibenzothiophenium ion
MEDNA	mitochondrial DNA
Me-DuPHOS	1,2-bis(2,5-methyl-1-phosphocyclopentyl)benzene
MEEC	membrane-enclosed catalyst
MEGA-8	*N*-octanoyl-*N*-methylglucamine
MEGA-9	*N*-nonanoyl-*N*-methylglucamine
MEGA-10	*N*-decanoyl-*N*-methylglucamine
MEHQ	hydroquinone monomethyl ether
MEI	2-morpholinoethyl isocyanide
2-Meim	2-methylimidazole
MEK	methyl ethyl ketone
MEKP	methyl ethyl ketone peroxide
MeLeu	*N*-methylleucine
MEM	(2-methoxyethoxy)methyl; methoxyethoxymethyl
MEMC	methoxyethylmercury chloride
MEMCl	β-methoxyethoxymethyl chloride
MEMO	3-(trimethoxysilyl)propyl methacrylate; 3-methacryloxy-propyltrimethoxysilane
Men	menthol
Men	menthyl
2MeODIOP	*trans*-4,5-bis[di(2-methoxyphenyl)phosphino]methyl-2,2-dimethyldioxolan
3MeODIOP	*trans*-4,5-bis[di(3-methoxyphenyl)phosphino]methyl-2,2-dimethyldioxolan
MeOH	methanol

MeOMOP	2-(diphenylphosphino)-2′-methoxy-1,1′-binaphthyl
1-MEO-PMS	1-methoxy-5-methylphenazinium methyl sulfate
ME^{ox}	bis(2-hydroxyethyl)disulfide
MeOZ	*p*-methoxybenzyloxycarbonyl
MEP	*O,O*-dimethyl *O*-(3-methyl-4-nitrophenyl) phosphorothioate
MEQ	9-*O*-(4′-methyl-2′-quinoyl) ether
MES (Mes)	2-morpholinoethanesulfonate; 2-(*N*-morpholino)-ethanesulfonic acid; 4-morpholinoethanesulfonic acid
MES	1,3,5-trimethylbenzene
Mes	mesylate
Mes (mes)	mesityl
mesal	*N*-methylsalicylaldimine
mesityl	2,4,6-trimethylphenyl
MESNA	2-mercaptoethanesulfonic acid
MET (Met)	methionine
Met	methionyl
Meth	2-mercaptoethanol
MetHb	methemoglobin
methyl-CAPP	methylcarbomyl-4-diphenylphosphino-2-diphenylphosphinopyrrolidine
Methyl-EDTA	propylenediamine-tetraacetic acid
Methyl-L-DOPA	3-(3,4-dihydroxphenyl)-2-methyl-L-alanine
MEZ	mezerein
MFA	*N*-methylformanalide
MG	monoglyceride
MG-Ch	methyl glycol chitosan
MH	1,2-dihydro-3,6-pyridazinedione
MH	maleic hydrazide
MH	morpholine
MHA	2-hydroxy-4-(methylthio)butyric acid
MHBI	methyl 4-hydroxybenzimidate hydrochloride
MHHPA	hexahydro-4-methylphthalic anhydride; methylhexahydrophthalic anhydride
MHPG sulfate	3-methoxy-4-hydroxyphenylethylene glycol sulfate; 4-hydroxy-3-methoxyphenylglycol sulfate, potassium salt; (3-methoxy-4-sulfonyloxyphenyl)glycol, potassium salt
MIA	*N*-methylisatoic anhydride
MIBC	methyl amyl alcohol
MIBK	methyl isobutyl ketone; 4-methyl-2-pentanone
MIC	methyl isocyanate
MIN	minocycline
MIPC	*O*-isopropylphenyl-methylcarbamate

MIPK	methyl isopropyl ketone; 3-methyl-2-butanone
MIT	3-(4,5-dimethyl-2-thiazolyl) 2,5-diphenyl-2*H*-tetrazolium bromide
MIT	monoiodotyranose
MIX	3-isobutyl-1-methylxanthine
MLCK	myosin light chain kinase
3MLF	3-methyllumiflavin
MMA	methyl methacrylate
MMAA	mono-*N*-methylacetoacetamide; *N*-methylacetoacetamide
MMC	methyl magnesium carbonate
MMCAA	2-chloro-*N*-methylacetoacetamide
MMD	2,5-bis(acetatemercurimethyl)-*p*-dioxane
MMH	methylmercuric hydroxide
MML	*Mucor meihei* lipase
MMO	methane monooxygenase
MMPP	monoperoxyphthalic acid, magnesium salt; magnesium monoperoxyphthalate; magnesium monoperphthalate
MMS	*dl*-methionine methylsulfonium chloride
MMS	methyl methanesulfonate
MMSE	methyl methylsilyl ether
α-MMT	α-methyl-*m*-tyrosine
MMTr (mmt)	4-monomethoxyltrityl; *p*-methoxyphenyldiphenylmethyl
MMTrCl	monomethoxytrityl chloride; 4-anisylchlorodiphenyl-methane
MMTS	methyl methylsulfinylmethyl sulfide
MNA	2-methyl-4-nitroaniline
MNA	methylnadic anhydride (methylnorbornene-2,3-dicarboxylic acid anhydride)
MNase	nuclease micrococcal
MNNG	1-methyl-3-nitro-1-nitrosoguanidine; *N*-methyl-*N*′-nitrosoguanidine
MNP	2-methyl-2-nitrosopropane
MNPT	*m*-nitro-*p*-toluidine
MNT-Cl	4-chloro-3-nitrobenzotrifluoride
MNU	*N*-methyl-*N*-nitrosourea
MO	methyl orange
MoAb	monoclonal antibody
MOC	methoxycarbonyl
MOM	methoxymethyl
MoOPD	oxodiperoxymolybdenum 1,3-dimethyl-3,4,5,6-tetrahydro-2-(1*H*)-pyrimidone complex
MoOPH	oxodiperoxymolybdenum hexamethylphosphoramide complex

MOPEG	3-methoxy-4-hydroxyphenylethylene glycol
MOPEG sulfate	4-hydroxy-3-methoxyphenylglycol sulfate potassium salt; (3-methoxy-4-sulfonyloxyphenyl)glycol potassium salt
MOPS	4-morpholinepropanesulfonic acid; 3-(*N*-morpholino)propanesulfonic acid
MOPSO	3-(*N*-morpholino)-2-hydroxypropanesulfonic acid
Morph-DAST (morph-DAST)	morpholinosulfur trifluoride
Morpho CDI	1-cyclohexyl-3-(2-morpholinoethyl)carbodiimide metho-*p*-toluenesulfonate
Moz	*p*-methoxybenzyloxycarbonyl
MOZ-ON	2-(4-methoxybenzloxycarbonyloxyimino)-2-phenylacetonitrile
MP	methyl propiolate
2-MP	2-methoxypropane
4,5-MP	9*H*-benzo[*def*]fluorene
6MP	6-methylpurine
6-MP (6MP)	6-mercaptopurine
7MP	7-methylpurine
9MP	9-methylpurine
MPA	1,2-propanediol monomethyl ether acetate
MPA	methylene penicillanic acid
MPB	3-phenoxybenzyl alcohol
MPCA	monoperoxycamphoric acid
MPD	2-methyl-2,4-pentanediol
MPE	methidium propyl-ethylenediaminetetraacetate
MPEMA	2-ethyl-2-(*p*-tolyl)malonamide
MPK	methyl propyl ketone; 2-pentanone
MPM	*p*-methoxybenzyl
MPM	*p*-methoxyphenylmethyl; *p*-methoxybenzyl
MPMA	phorbol, 4-*O*-methyl,12-myristate, 13-acetate
MPML	methoxy(phenylthio)methyl lithium
MPO	myeloperoxidase
MPP	methyl 3-phenylpropynoate
MPP	*N*-(4-nitrophenyl)prolinol
MPP	*O,O*-dimethyl *O*-(4-methylmercapto-3-methylphenyl) thiophosphate
MPPA	monoperoxyphthalic acid
MPPH	5-(4-methylphenyl)-5-phenylhydantoin
MPPM	*N*-methyl-4-diphenylphosphino-2-diphenylphosphinomethyl pyrrolidine
MPPP	methylphenyl-*n*-propylphosphine
MPS	3-mercapto-1-propane sulfonic acid
MPS	methyl phenyl sulfide; thioanisole

MPS	*N*-methylphenazonium methosulfate
Mps	*p*-methoxyphenylsulfonyl
MPT	methyl *p*-toluate
α-MPT	α-methyl-π-tyrosine
Mpt	dimethylthiophosphinyl
Mpt-Cl	methylphosphinothionyl chloride
MPTA	2-(2-methylphenoxy)triethylamine
MPTA	*N*,*N*-diethylaminoethyl 4-methylphenyl ether
MPTM	[2-(methylthio)phenyl]thiomethyl
MPTP	1-methyl-4-phenyl-1,2,3,6-tetrahydropyridine; *N*-methyl-4-phenyl-1,2,3,6-tetrahydropyridine
MQ	4-methyl-2-quinolone
MQ	monobromotrimethylammoniobimane
MR	methyl red
MRITC	methylrhodamine isothiocyanate; tetramethylrhodamine B isothiocyanate
mRNA	messenger RNA
MS (Ms)	mesyl; methanesulfonyl
MS-222	3-aminobenzoic acid ethyl ester
MSA	methanesulfonic acid
MSA	*N*-methyl-*N*-trimethylsilylacetamide
MSC	2-mesitylenesulfonyl chloride
MSC-ONSu	2-(methylsulfonyl)ethyl-*N*-succimidyl carbonate
MsCl	methanesulfonyl chloride
MSEO-PCl_2	2-(methylsulfonyl)ethyl dichlorophosphite
MSG	monosodium glutamate
MSH	2,4,6-trimethylbenzenesulfonyl hydrazide
MSH (α-MSH)	melanocyte-stimulating hormone, melanotropin; α-melanocyte stimulating hormone
MSHFBA	*N*-methyl-*N*-trimethylsilylheptafluorobutyramide
Msib	4-(methylsulfinyl)benzyl
MSMA	monosodium methanearsonate
MSNT	1-(2-mesitylenesulfonyl)-3-nitro-1*H*-1,2,4-triazole
MSO	4-methylanisole; *p*-cresyl methyl ether
MSOC	*N*-(2-methylsulfonyl)ethyloxycarbonyl
MST	1-(2-mesitylenesulfonyl)-1*H*-1,2,4-triazole
MSTFA	*N*-methyl-*N*-(trimethylsilyl)trifluoroacetamide
Msz	4-methylsulfinylbenzyloxycarbonyl
α-MT	α-methyltyrosine
MTA	5′-*S*-methyl-5′-thioadenosine
MTAD	1-methyl(1,2,4-triazoline-3,5-dione); *N*-methyl-1,2,4-triazoline-3,5-dione
MTB	methylthymol blue
Mtb	2,4,6-trimethoxybenzenesulfonyl

MTBD	7-methyl-1,5,7-triazabicyclo[4.4.0]dec-5-ene
MTBE	*tert*-butyl methyl ether
MTBSTFA	*N*-(*tert*-butyldimethylsilyl)-*N*-methyltrifluoroacetamide; *N*-methyl-*N*-(*tert*-butyldimethylsilyl)trifluoroacetamide
MTC	methyl isothiocyanate
MTCA	2-methylthiazolidine-4-carboxylic acid
MTD	*m*-toluenediamine
MTD	methyltetracyclododecene
MTDA	methyl trimethylsilyl dimethylketene acetal
MTDEA	*N*,*N*-di(2-hydroxyethyl)-*m*-toluidine; *m*-toluidine-*N*,*N*-diethanol
mtDNA	mitochondrial DNA
Mte	2,3,5,6-tetramethyl-4-methoxybenzenesulfonyl
MTES	methyltriethoxysilane
MTG	methyl β-D-thiogalactoside
MTH	methylthiohydantoin
MTHF	2-methyltetrahydrofuran
MTHP	4-methoxytetrahydropyranyl
MTHPA	methyltetrahydrophthalic anhydride
MTM	(methylthio)methyl ether
MTM	methylthiomethyl
MTMB	4-(methylthiomethoxy)butyryl
MTMC	4-(methylthio)-*m*-cresol
MTMC	*m*-tolyl-*N*-methylcarbamate
MTMECO	2-(methylthiomethoxy)ethoxycarbonyl
MTMP	*N*-bromomagnesium-2,2,6,6-tetramethylpiperidine
MTMS	methyltrimethoxysilane
MTMSE	methyl trimethylsilyl ether
MTMT	2-(methylthiomethoxymethyl)benzoyl
MTN	*m*-tolylnitrile
MTP	4-(methylthio)phenol
MTPA	α-methoxy-α-trifluoromethylphenylacetic acid
MTPA	α-methoxy-α-trifluoromethylphenylacetic acid chloride
MTPAA	α-methoxy-α-(trifluoromethyl)phenylacetic acid
MTPACl	α-methoxy-α-trifluoromethylphenylacetic acid chloride
Mtpc	4-methylthiophenoxycarbonyl
Mtr	2,3,6-trimethyl-4-methoxybenzenesulfonyl
mtRNA	mitochondrial ribonucleic acid
Mts	2,4,6-trimethylbenzenesulfonyl (mesitylenesulfonyl)
MTT	3-(4,5-dimethylthiazol-2-yl)-2,5-diphenyl-2*H*-tetrazolium bromide
MTT	thiazolyl blue
MTU	methylthiouracil
MTX	L-(+)-amethopterin

MTX	maitotoxin
MTX	methotrexate
4-MU	4-methylumbelliferone(β)
MUG	4-methylumbelliferyl-β-D-glucuronide
MUGB	4-methylumbelliferyl *p*-guanidinobenzoate
MUP	methylumbelliferyl phosphate
4-MUP	4-methylumbelliferone phosphate; 4-methylumbelliferylphosphate
Mur	muramic acid
MurNAc	*N*-acetylmuramic acid
MVA	mevalonic acid
MVK	methyl vinyl ketone
MVP	2-methyl-5-vinylpyridine
MXDA	*m*-xylylenediamine
M-Zyme	keratinase
N	asparagine
N	histidyl
N	(unspecified) nucleoside
n	normal
N. crassa	*Neurospora crassa*
2-NA	2-naphthaldehyde
NAA	1-naphthaleneacetic acid; 1-naphthylacetic acid
5-NAA	5-nitroanthranilic acid
NAAD	nicotinic acid adenine dinucleotide
[Na]ATPase	sodium ion activated ATPase
NABA	4-acetoxy-3-nitrobenzoic acid
NABS	4-acetoxy-3-nitrobenzenesulfonate
NAC	1-naphthyl *N*-methylcarbamate
NAD	α-nicotinamide adenine dinucleotide; nicotinamide adenine dinucleotide
β-NAD	β-nicotinamide adenine dinucleotide; β-nicotinamide adenine dinucleotide
$NAD(P)^+$	either NAD^+ or $NADP^+$
NAD(P)H	either NADH or NADPH
NAD^+	β-nicotinamide adenine dinucleotide; nicotinamide adenine dinucleotide (oxidized form)
Na-DAS	9,10-dimethoxyanthracene-2-sulfonic acid sodium salt
NADH	nicotinamide adenine dinucleotide (reduced form); nicotinamide adenine dinucleotide phosphate, reduced
NADIDE	β-nicotinamide adenine dinucleotide
NaDMAN	sodium (dimethylamino)naphthalenide
NADP	nicotinamide-adenine dinucleotide phosphate; triphosphopyridine nucleotide
β-NADP	β-nicotinamide adenine dinucleotide phosphoric acid

$NADP^+$	nicotinamide adenine dinucleotide phosphate (oxidized form)
NADPH	nicotinamide adenine dinucleotide phosphate (reduced form); reduced nictotinamide-adenine dinucleotide phosphate
NAG	*N*-acetylglucosamine
NaHMDS	sodium hexamethyldisilazide; sodium bis(trimethylsilyl)amide
NAI	1-acetylimidazole
NAM	9-maleimide-acridine; *N*-(9-acridinyl)maleimide
NAM	*N*-acetyl-L-methionine
NAMA	*N*-acetylmuramic acid
NAMS	*N*,*N*-bis(2-hydroxyethyl)-*p*-nitrosoaniline, dimethylsulfonate
NAN	*N*-acetylneuramic acid
NAN	sialic acids
NANA	*N*-acetylneuramic acid
NaNAP	sodium naphthalene
NaNp	sodium naphthalide
NAP	4-nitroaminophenol
Nap	naphthenyl
α-NAPAP	*N*α-(2-naphthalenesulfonylglycyl)-4-amidinophenylalaninepiperdide
NapEt	[α-(1-naphthyl)ethyl]ammonium perchlorate
Naph	naphthyl
NAPHOS	2,2′-bis(diphenylphosphinomethyl)-1,1′-binaphthyl
Naphthol AS-BO	2-hydroxy-3-naphthoic acid 1-naphthylamide
Naphthol AS-OL	2-hydroxy-3-naphthoic acid *O*-anisidide
NAS	*N*-acetoxysuccinimide
NB	*p*-nitrobenzyl
NB-Enantrane™	9-BBN-nopol benzyl ether adduct; lithium hydrido(9-BBN-nopol benzyl ether adduct)
NBA	3-nitrobenzyl alcohol
NBA	*N*-bromoacetamide
NBD (nbd)	norbornadiene
NBD CHLORIDE (NBD chloride) (NBD-Cl) (NBDCl)	7-chloro-4-nitrobenz-2-oxa-1,3-diazole; 4-chloro-7-nitrobenzofurazan; 4-chloro-7-nitrobenzo-2-oxa-1,3-diazole
NBD Taurine	*N*-(7-nitrobenz-2-oxa-1,3-diazol-4-yl)taurine
NBD-F	4-fluoro-7-nitrobenzofurazàn
NBDI	*N*,*N*′-diisopropyl-*O*-(4-nitrobenzyl)isourea; *O*-(4-nitrobenzyl)-*N*,*N*′-diisopropylisourea
NBDMO	3-bromo-4,4-dimethyl-2-oxazolidinone

NBHA	*O*-(4-nitrobenzyl)hydroxylamine
NBMPR	*S*-(4-nitrobenzyl)-6-thioinosine
Nbn	4-nitrobenzyl
NBR	acrylonitrile-butadiene rubber
NBS	*N*-bromosuccinimide
NBSac	*N*-bromosaccharin
NBSC	2-nitrobenzenesulfenyl chloride
p-NBSP	*p*-nitrobenzenesulfonyl peroxide
NBT	nitro blue tetrazolium
NBTGR	*S*-(4-nitrobenzyl)-6-thioguanosine
nBu	*n*-butyl
1,2-Nc	1,2-naphthalocyaninato
2,3-Nc	2,3-naphthalocyaninato
NCA	*N*-carboxy-α-amino acid anhydride; *N*-carboxy anhydride
NCA	*N*-chloroacetamide
NCDC	2-nitro-4-carboxyphenyl *N*,*N*-diphenylcarbamate
NCN	cyanonaphthalene
NCS	*N*-chlorosuccinimide
NCS	neocarzinostatin
ND	norbornadiene
NDC	nicotinium dichromate
NDEPAP	1-(4-nitrobenzyl)-4-(4-diethylaminophenylazo)pyridinium bromide
NDGA	4,4′-(2,3-dimethyltetramethylene)dipyrocatechol
NDGA	nordihydroguaiaretic acid
NDP	nucleoside 5′-diphosphate; nucleoside diphosphate
NDPP	4-(4-diethylaminophenylazo)-1-(4-nitrobenzyl)pyridinium bromide
NEI	1-(1-naphthyl)ethyl isocyanate
NEM	*N*-ethylmaleimide
NeoT	3,3′-(4,4′-biphenylene)bis(2,5-diphenyl)-2*H*-tetrazolium chloride; neotetrazolium chloride
NEP	1-ethyl-2-pyrrolidinone
NEPIS	*N*-ethyl-5-phenylisoxazolium-3′-sulfonate
NesMIC	(+)-(neomenthylsulfonyl)methyl isocyanide
NeuAc	*N*-acetylneuraminic acid
NeuNAc	*N*-acetylneuraminic acid
NFOBS	*N*-fluoro-*o*-benzenedisulfonimide
NFPy	*N*-fluoropyridinium pyridine heptafluorodiborate
NFZ	5-nitro-2-furfural semicarbazone
NGF	nerve growth factor
NHB	4-hydroxy-3-nitrobenzoic acid
NHC	natural hydrocarbons
NHMDS	sodium hexamethylsilazide

NHNP-E	*N*-(2-hydroxy-3-[1-naphthoxy]propyl)-ethylenediamine
NHP	*N*-hydroxyphthalimide
NHS	*N*-hydroxysuccinimide
NHS-AMCA	succinimidyl-7-amino-4-methylcoumarin-3-acetate
NHS-ASA	*N*-hydroxysuccinimidyl-4-azidosalicylic acid
NHS-Biotin	(+)-biotin *N*-hydroxysuccinimide ester
5-NIA	5-nitroisatoic anhydride
NIB	4-hydroxy-3-iodo-5-nitrobenzoic acid
NIP	4-hydroxy-3-iodo-5-nitrophenylacetic acid; 4-hydroxy-5-nitro-3-iodophenylacetic acid
NIP	2,4-dichlorophenyl 4′-nitrophenyl ether
NIS	*N*-iodosuccinimide
NITR	2-R-4,4,5,5-tetramethyl-4,5-dihydro-1*H*-imidazoyl-1-oxy 3-oxide (where R = ethyl, *n*-propyl or *iso*-propyl)
Nitro BT	nitro blue tetrazolium chloride
Nitron	4,5-dihydro-2,4-diphenyl-5-(phenylimino)-1*H*-1,2,4-triazolium hydroxide
Nitro-PAPS	2-(5-nitro-2-pyridylazo)-5-(*N*-propyl-*N*-sulfopropylamino) phenol
Nitroso-ESAP	3(*N*-ethyl-3-hydroxy-4-nitrosoanilino)propanesulfonic acid
Nitroso-PSAP	3(3-hydroxy-4-nitroso-*N*-propylanilino)propanesulfonic acid
Nitroso-R-salt	1-nitroso-2-naphthol-3,6-disulfonic acid
Nle	norleucine
Nle	norleucyl
NM	nitromethane
NMA	*N*-methyl-DL-aspartic acid
NMA	*N*-methylolacrylamide
N-MACR	9*H*,10-*N*-methylacridine
N-MATFA	*N*-methylanilinium trifluoroacetate
NMC	*N*-methylcarbazole
NMDA	*N*-methyl-D-aspartate
NMDA	*N*-methyl-D-aspartic acid
NMDPP	neomenthyldiphenylphosphine
NMED	*N*-methylethylenediamine
N-MFHA	*N*-methylfurohydroxamic acid
NMI	*N*-methylimidazole
NMM	*N*-methylmaleimide
NMM	*N*-methylmorpholine
NMMO	*N*-methylmorpholine-*N*-oxide
NMN	nicotinamide mononucleotide
β-NMN	β-nicotinamide mononucleotide
NMNH	nicotinamide mononucleotide, reduced form
NMO	*N*-methylmorpholine *N*-oxide

NMONS	3-methyl-4-methoxy-4′-nitrostilbene
NMP	1-methyl-2-pyrrolidinone; *N*-methyl-2-pyrrolidone; *N*-methylpyrrolidinone
NMP	*N*-methylphthalimide
NMP	nucleoside monophosphate
NMPP	neomenthyldiphenylphosphine
NMQ	*N*-methyl-2-quinolone
NMSO	4-methyl-2-nitroanisole
N-MSOC	*N*-(2-methanesulfonyl)ethoxycarbonyl
NMT	*N*-myristoyl transferase
N-m-t	*N*-*m*-tolylphthalamic acid
NMTT	1-methyl-1*H*-tetrazole 5-thiol
NNCD reagent	2-chloro-4-nitrobenzenediazonium naphthalene-2-sulfonate
NNOSBP®	2,4-dinitro-6-*sec*-butylphenol
NOBA	*m*-nitrobenzyl alcohol
Noc	4-nitrocinnimyloxycarbonyl
NODA	*n*-octyl *n*-decyl adipate
NODLN	nodularin
NORPHOS	2,3-bis(diphenylphosphino)bicyclo[2.2.1]heptene; bicyclo[2.2.1]hept5-en-2,3-bis(diphenylphosphine)
NP	bis(*m*-nitrophenyl)disulfide
NP	*p*-nitrophenyl
3-NP	3-nitropyridine
4-NP	4-nitropyridine
Np	naphthenyl
NPA	*N*-1-naphthylphthalamic acid
NPA	novel plasminogen activator
2NPANH	2-naphthylacetonitrile
NPCC	*O*-(4-nitrophenylphosphoryl)choline
NPD	*O*,*O*,*O*,*O*-tetrapropyl dithiopyrophosphate
p-NPDPP	*p*-nitrophenyl diphenyl phosphate
NPDTA	4-nitrophenyl dithioacetate
NPDTC	*O*-ethyl *S*-(4-nitrophenyl) dithiocarbonate; *p*-nitrophenyl *O*-ethyl dithiocarbonate
NPE	2-(4-nitrophenyl)ethyl
NPE	nonylphenol ethoxylates
NPE caged ATP	adenosine-5′-triphosphate P^3-[1-(2-nitrophenylether ester]
NPE caged GTP	guanosine-5′-triphosphate P^3-[1-(2-nitrophenylether ester]
NPGB	4-nitrophenyl-4-guanidinobenzoate
N_2ph	dinitrophenyl
NPM	*N*-phenylmaleimide
NPMI	*N*-phenylmaleimide
α-NPO	2-(1-naphthyl)-5-phenyloxazole

NPOE	2-nitrophenyl octyl ether
NPP	2-nitro-2-propenyl pivalate
NPP	*N*-(4-nitrophenyl)prolinol
NPP	nerolidyl diphosphate
NPPD	*trans,trans*-5-(4-nitrophenyl)-2,4-pentadienal
NPS	4-nitrothiophenoxide
NPS (Nps)	2-nitrophenylsulfenyl
NPS Reagent	2-nitrophenylsulfenyl chloride
NPSH	4-nitrothiophenol
NPSP (N-PSP)	*N*-(phenylseleno)phthalimide
NPTA	4-nitrophenyl thiolacetate
Npys	3-nitro-2-pyridinesulfenyl
NR	natural rubber
nRNA	nuclear ribonucleic acid
NSF	thiazyl fluoride
NSF_3	thiazyl trifluoride
NSG	*N*-substituted glycines
NSR	nitrile silicone rubber
NT	neotetrazolium chloride
NT	neurotensin
NTA	nitrilotriacetic acid
N-t-B	2-methyl-2-nitrosopropane
N-t-B	2-methyl-2-nitrosopropane dimer
NTC	3,3′-(4,4′-biphenylene)bis(2,5-diphenyl)-2*H*-tetrazolium chloride; neotetrazolium chloride
NTCB	2-nitro-5-thiocyanatobenzoic acid; 2-nitro-5-thiocyanobenzoic acid
NTMP	*N*-(bromomagnesio)-2,2,6,6,-tetramethylpiperidine
NTP	unspecified nucleoside 5′-triphosphate; nucleoside triphosphate
Nu	nucleophile
Nuc	nucleophile
NVA	*N*-methyl-*N*-vinylacetamide
NVOC	nitroveratryloxycarbonyl
NVOC-Cl	6-nitroveratryl chloroformate
NVP	*N*-vinyl-2-pyrrolidinone
NW Acid	1-naphthol-4-sulfonic acid
o	ortho
OA	oxyallyl
OAA	oxalacetic acid
OAA	oxaloacetate
OAc	acetate
OAG	1-oleoyl-2-acetyl-*sn*-glycerol; 2-acetyl-1-oleoylglycerol
OBN	*o*-nitrobiphenyl

Oc	1-bicyclo[2.2.2]octyl
OCAD	2-chlorobenzaldehyde
OCBA	2-chlorobenzoic acid
OCBC	2-chlorobenzyl chloride
OCBN	2-chlorobenzonitrile
OCCN	2-chlorobenzyl cyanide
OCDC	*o*-chlorodichlorotoluene
OCDD	octachlorodibenzo-*p*-dioxin
OCDF	octachlorodibenzofuran
OCOC	2-chlorobenzoyl chloride
OCPA	2-chlorophenylacetic acid
OCPT	2-chloro-4-aminotoluene; 3-chloro-4-methylaniline
OCT	2-chlorotoluene
OCT	ornithine carbamyl transferase
OCTC	*o*-chlorobenzotrichloride
OCTEO	octyltriethoxysilane
ODA	4,4′-oxydianiline; 4-aminophenyl ether
10-ODA	12-oxo-*trans*-10-dodecenoic acid
13-ODAL	13-octadecenal
ODCB	1,2-dichlorobenzene
ODH	octopine dehydrogenase
ODNB	*o*-dinitrobenzene
ODS	octadecylsilyl
OEC	octaethylchlorin
OEP	2,3,7,8,12,13,17,18-octaethylporphyrinate; octaethylporphyrinato
OEP	octaethylporphyrin
OG	octyl-β-D-glucopyranoside
OH-Cbl HCl	vitamin B_{12a}
2-OH-CTAB	(2-hydroxycetyl)trimethylammonium bromide
OM	octyl-β-D-maltopyranoside
OMEGA	octanoyl-*N*-methylglucamide
OMH-1	sodium diethyldihydroaluminate
OMI	*O*-methylisourea hydrogen sulfate; *O*-methylisourea sulfate
OMP	orotidine 5′-monophosphate
OMPA	octamethyl pyrophosphoramide
ONB	*o*-nitrobenzyl
ONB	*o*-nitrobiphenyl
ONPG	*o*-nitrophenyl β-D-galactopyranoside
OOBPD	*N*,*N*′-bis(4-octyloxybenzylidene)-*p*-phenylenediamine
OP	organophosphate
OPA	*o*-phthalaldehyde; phthaldialdehyde
OPD	1,2-phenylenediamine
OPG	oxypolygelatin

OPNB	*p*-nitrobenzoate
Orn	ornithine
Orn	ornithyl
OSC	2,3-oxidosqualene cyclases
OSGP	octyl-β-D-thioglucopyranoside
OST	oligosaccharyltransferase
OT	oxytocin
4-OT	4-oxalocrotonate tautomerase
OTB	*o*-toluidine boric acid
OTc	oxytetracycline
OTD	*o*-toluenediamine
OTf	ovotransferrin
OTG	octyl β-thioglucopyranoside
OTTM	oxatrimethylenemethane
ox	oxalate; oxalato
OXA	oxaloacetate
OXT	oxytocin
P (p)	phosphate
P	phosphate residues
P	polymer substituent
P	proline
p	para
P. notatum	*Penicillium notatum*
P. putida	*Pseudomonas putida*
P. roqueforti	*Penicillium roqueforti*
PA	3-*sn*-phosphatidic acid
PA	phosphatidic acid
PA	piperazine
PAA	phenylacetic acid
PAABA	4-acetamidobenzoic acid
PAB	*p*-aminobenzoate
PAB	*p*-aminobenzoic acid
PAB	polyclonal antibodies
PABA	4-aminobenzoic acid
PAC	polycyclic aromatic compounds
Pac	phenacyl
PACOPA	2-(phenylaminocarbonyloxy)propionic acid
PADA	poly(adipic anhydride)
PADA	potassium azodicarboxylate
PADA	pyridine-2-azo-*p*-dimethylaniline
PAF	1-hexadecyl-2-acetyl-*sn*-glycero-3-phosphocholine monohydrate
PAF	L-α-phosphatidylcholine, β-acetyl-γ-*O*-alkyl
PAF	platelet activating factor

PAH	4-aminohippuric acid
PAH	phenylalanine hydroxylase
PAH	piperazinium
PAH	polycyclic aromatic hydrocarbon
PAK	polycyclic aromatic ketones
PAL	phenylalanine ammonia lyase
P-Ald	phosphonoacetaldehyde
Paloc	3-(3-pyridyl)allyloxycarbonyl
PAM	2-pyridine aldoxime methiodide
PAM	peptidyl α-amidating monooxygenase
2-PAM	2-pyridinealdoxime methiodide
2-PAMCl	2-pyridinealdoxime methochloride
PAMS	poly(α-methylstyrene)
PAN	1-(2-pyridylazo)-2-naphthol
PAN	peroxyacetyl nitrate
PAO	plasma amine oxidase
PAP	adenosine 3′,5′-diphosphate
PAP	*O,O*-dimethyl *S*-α-(ethoxycarbonyl)benzyl phosphorothiolothioate
PAP	peroxidase-anti-peroxidase
PAPA	poly(azelaic anhydride)
PAPA	potassium 3-amino-propylamide
PAPI	polyisocyanate
PAPP	*p*-aminophenylethyl-*m*-trifluoromethylphenyl piperazine
PAPP	*p*-aminopropiophenone
PAPS	3′-phosphoadenosine-5′-phosphosulfate; adenosine 3′-phosphate-5′-phosphosulfate
PAR	4-(2-pyridylazo)resorcinol
PAS	4-aminosalicylic acid
PASAM	4-toluenesulfonamide
Pase	phosphatase
PB	polybutylene
PBA	*N*-benzyl-9-(2-tetrahydropyranyl)adenine
PBA	*p*-benzoquinone-2,3-dicarboxylic anhydride
PBAE	poly(biphenol A sebacate)
pbaOH	2-hydroxy-1,3-propylenebis(oximato) anion
PBB	polybrominated biphenyls
PBB	filixic acids
PBBO	2-(4-biphenylyl)-6-phenylbenzoxazole; 6-phenyl-2-(4-biphenylyl)benzoxazole
PBD	2-(4-biphenylyl)-5-phenyl-1,3,4-oxadiazole; 2-phenyl-5-(4-biphenylyl)-1,3,4-oxadiazole
PBD	polybutadiene
PBD	pyrrolo[1,4]benzodiazepine

PBG	porphobilinogen
PBI	*p*-benzoquinone-2,3-dicarboxylic imide
PBN	α-phenyl-*N*-*tert*-butylnitrone; *C*-phenyl-*N*-*tert*-butylnitrone; *N*-*tert*-butyl-α-phenylnitrone
pBNP	(+)-biotin 4-nitrophenyl ester
PBP	4-benzyloxyphenol
PBP	4,4′-(hydroxyphosphinylidene)bis-L-phenylalanine
PBP	filixic acids
m-PBPM	*m*-phenylenebis(phenylmethylene)
PBS	phosphate buffered saline
PBS	poly(butene-1-sulfone)
PBT	poly(butylene terephthalate)
PC	phosphatidylcholine
PC	plastocyanin
PC	polycarbonate
PC	propylene carbonate
PC	pterin-6-carboxylate
Pc	phthalocyaninato
Pc	phthalocyanine
PCA	5-amino-4-chloro-2-phenyl-3(2*H*)-pyridazinone
PCAD	4-chlorobenzaldehyde
PCB	phycocyanobilin
PCB	polychlorobiphenyl; polychlorinated biphenyl
PCBA	4-chlorobenzoic acid
PCBC	4-chlorobenzyl chloride
PCBN	4-chlorobenzonitrile
PCBTF	4-chlorobenzotrifluoride
PCC	pyridinium chlorochromate
PCCN	4-chlorobenzyl cyanide
PcCo	cobalt(II) phthalocyanine
PCDC	*p*-chlorodichlorotoluene
PCDD	polychlorinated dibenzo-*p*-dioxin
PCDF	polychlorinated dibenzofuran
PCIB	α-4-chlorophenoxyisobutyric acid
PCL	*Pseudomonas cepacia* lipase
PCMB	4-chloromercuribenzoic acid
pCMBS	4-(chloromercuri)benzenesulfonic acid
PCMP	cetylpyridinium tetrakis(diperoxomolybdo)phosphate
PCMX	4-chloro-3,5-dimethylphenol; *p*-chloro-*m*-xylenol; 4-chloro-3,5-dimethylbenzene
PCNB	pentachloronitrobenzene
PCOC	4-chlorobenzoyl chloride
PCONA	4-chloro-2-nitroaniline
PCOT	4-chloro-2-aminotoluene; 5-chloro-2-methylaniline

PCP	pentachlorophenol
PCP	pentachlorophenyl
PCP	phencyclidine
PCPA	4-chlorophenylacetic acid
1-PCPA	1-phenylcyclopropylamine
PCR	polymerase chain reaction
PCr	creatine phosphate
PCT	4-chlorotoluene
PCT	polychloroterphenyl
PCTA	poly(1,4-cyclohexanedimethylene terephthalic acid)
PCTC	4-chlorobenzotrichloride; *p*-chlorotrichlorotoluene
PCTFE	poly(chlorotrifluoroethylene)
PCUD	pentacyclo[5.4.0.0^{2,6}.0^{3,10}.0^{5,9}]undecane
PDA	phorbol 12,13-diacetate
PDA	potato dextrose sugar
PDANO	2,6-pyridinedicarboxylic acid *N*-oxide
PDBU	phorbol, 12,13-dibutyrate
PDBuAL	phorbol, 20-oxo-20-deoxy, 12,13-dibutyrate
PDBz	phorbol 12,13-dibenzoate
PDC	pyruvate decarboxylase
PDC	pyridinium dichromate
PDD	phorbol, 12,13-didecanoate
4αPDD (4α-PDD)	4α-phorbol, 12,13-didecanoate
PDDA	9-phenyl-10,10-dimethyl-9,10-dihydroanthracene
PDE	phosphodiesterase
PDEA	*N*-phenyldiethanolamine
PDGF	platelet derived growth factor
PDH	pyruvate dehydrogenase
PDI	protein disulfide-isomerase
PDK	1-phospho-2,3-diketo-5-*S*-methylthiopentose
$P450_{14}DM$	lanosterol 14α-dimethylase
PDNB	*p*-dinitrobenzene
PDP	pinacolone cyclic diperoxide
PDPBI	2,6-bis{[*N*-(1,2-diphenylethyl)imino]methyl}pyridine
PDPS	[4-(phenylthio)phenyl]diphenylsulfonium
PDQ	sodium (2-methyl-4-chlorophenoxy)butyrate
PDS	pentadienylsilane
PDT	2-phenyl-1,3-dithiane
PDT	3-(2-pyridyl)-5,6-diphenyl-1,2,4-triazine
PDT	pentadienyltin
PDT disulfonate	3-(2-pyridyl)-5,6-diphenyl-1,2,4-triazine-4′,4″-disulfonic acid
PDTA	propylenediamine tetraacetate
PDTA	phenyl dithioacetate

PDTC	phenyl *O*-ethyl dithiocarbonate
PDU®	1,1-dimethyl-3-phenylurea
PE	petroleum ether
PE	phosphatidylethanolamine
PE	poly(ethylene)
Pe	pentyl
PEA	*N*-(2-hydroxyethyl)aniline; *N*-phenylethanolamine; 2-anilinoethanol
PEA	polyesteramides
PEB	1-phenyl-2-ethylbenzenesulfonate
2-PEB	2-phenylethylbenzenesulfonate
PEBC	*S*-propyl butylethylthiocarbamate
PEC (4-PEC)	*S*-2-(4-pyridyl)ethyl-L-cysteine
PEEA	*N*-(2-hydroxyethyl)-*N*-ethylaniline; *N*-phenyl-*N*-ethylethanolamine
PEEK	poly ether ketone
PEG	polyethylene glycol
PEG 1	methoxypolyethylene glycol
PEG 1-Activated	methoxypolyethylene glycol—activated with cyanuric chloride
PEG-400	poly(ethylene glycol)400
PEI-Cellulose	polyethyleneimine-impregnated cellulose
PEMA	2-ethyl-2-phenylmalonamide
Pen	pentyl
Pent	pentenyl
Peoc	2-phosphonioethoxycarbonyl; 2-(triphenylphosphonio)ethoxycarbonyl
PEOC-Cl	2-(triphenylphosphinio)ethyl chloroformate chloride
c-PeOH	cyclopentanol
PEP	phosphoenolpyruvate
PEP	phosphoenolpyruvic acid
4-PEP	*S*-[2-(4-pyridyl)ethyl]-DL-penicillamine
PEP-3CHA	phosphoenolpyruvic acid tris(cyclohexylamine) salt
PEP-carboxylase	phosphoenolpyruvate carboxylase
PEP-K	phosphoenolpyruvic acid
PEPCK	phosphoenolpyruvate carboxykinase
PER	ammonium peroxydisulfate; ammonium persulfate
PET	poly(ethylene terephthalate)
Pet	2-(2′-pyridyl)ethyl
PETA	pentaerythritol triacrylate
PETFE	poly(ethylenetetrafluoroethylene)
PETN	pentaerythrityl tetranitrate
PFA	perfluoroalkoxy (resin)
PFAA	pentafluoropropionic anhydride

P-FAD	glucose oxidase
PFB	pentafluorobenzyl
PFBBr	pentafluorobenzyl bromide
PFBD	perfluorobutane-2,3-dione
PFBHA	*O*-(2,3,4,5,6-pentafluorobenzyl)hydroxylamine; (pentafluorobenzyl)hydroxylamine
PFC	pyridinium fluorochromate
PFK	perfluorokerosene
PFK	phosphofructokinase
PFL	*Pseudomonas fluorescens* lipase
PFP	pentafluorophenyl
PFPA	pentafluoropropionic anhydride
PFPA	perfluorophenyl azide
PFPI	pentafluoropropionylimidazole
PG	prostaglandin (as PGA_1,etc.)
PG	protective group
PG2(-8)	*p*-nitrophenyl-α-D-maltoside
PG#	*p*-nitrophenyl-α-D-malto(n)oside, where n = 3–8, i.e., tri through octa
2PG	2-phosphoglycerate
3PG	3-phosphoglycerate
6-PG	6-phosphogluconic acid
2-PG	D-2-phosphoglyceric acid
PGA	folic acid
6-PGDH	6-phosphogluconate dehydrogenase
PGE	phenyl glycidyl ether
PGI	phosphoglucose isomerase
PGI^2	prostacyclin
PGK	phosphoglycerate kinase
PGP	3-phosphoglyceroyl phosphate
Ph	phenyl
Ph-1,2-BF	11-phenyl-11[*H*]-benzo[*b*]fluorene
Ph-2,3-BP	11-phenyl-11[*H*]-benzo-[*b*]fluorene
Ph-3,4-BF	7-phenyl-7*H*-benzo[*c*]fluorene
PHA	phytohemagglutinin
PHAL	1,4-phthalazinediyl diether
PHB	poly-[(*R*)-3-hydroxybutyric acid]; polyhydroxybutyrate; poly(3-hydroxybutyric acid)
PHBF	phloroglucide
Phe	phenylalanine
Phe	phenylalanyl
PHELLANPHOS	*trans*-5,6-bis(diphenylphosphino)-8-isopropyl-2-methylbicyclo[2.2.2]oct-2-ene
phen	1,10-phenanthroline

Phenoc	4-methoxyphenacyloxycarbonyl
PHENPHOS	1-diphenylphosphino-2-dimethylamino-3-phenylpropane
Phenyl PAS	phenyl 4-aminosalicylate
Phenyl-CAPP	phenylcarbomyl-4-diphenylphosphino-2-diphenylphosphinopyrrolidine
Phenyl-MAPO	bis[1-(2-methyl)aziridinyl]phenylphosphine oxide
9-PhFl	9-phenylfluorene
PHGA	pteroylhexaglutamylglutamic acid
PhGLYPHOS	1-diphenylphosphino-2-dimethylamino-2-phenylpropane
pHMB	*p*-hydroxymercuribenzoate
PHN	9-*O*-(9′-phenanthryl) ether
PhO	phenanthrene 9,10-oxide
PHR	4β-phorbol; phorbol; 4β,9α,12β, 13α,20-pentahydroxytigilia-1,6-dien-3-one
4αPHR	4α,9α,12β,13α,20-pentahydroxytigilia-1,6-dien-3-one; 4α-phorbol; isophorbol
PhSFl	9-(phenylthio)fluorene
PHT	pyrrolidine hydrotribromide
Pht	phthalyl
Phth	phthaloyl
Phth	phthalyl
PI	phosphatidyl inositol
PI	piperidine
Pi	inorganic orthophosphate
Pi	inorganic phosphate
PIA	iodobenzene diacetate; phenyliodoso diacetate
PIC	phosphoinositidase C
pic	picolinic acid
PIDA	phenyliodine(III) diacetate
PIFA	phenyliodine(III) bis(trifluoroacetate)
PIFE	polytetrafluoroethylene; teflon
PIMP	2-pyridinal-3(iminomethyl)pinane
PIP	phosphatidylinositol 4-phosphate
pip	piperidine
PIP_2	phosphatidylinositol 4,5-bisphosphate
PIPES	1,4-piperazinebis(ethanesulfonic acid)
PIPSYL Chloride (Pipsyl chloride)	*p*-iodobenzenesulfonyl chloride
PITC	isothiocyanic acid phenyl ester; phenyl isothiocyanate
Piv	pivaloyl
Pixyl	9-(9-phenyl)xanthenyl
PK	protein kinase
PK	pyruvate kinase
PKC	protein kinase C

PLA	phospholipase
PLE	pig liver esterase
PLH	placental lactogenic hormone
PLP	pig liver esterase
PLP	pyridoxal 5′-phosphate; pyridoxal phosphate
PM	phenylmenthyl
PMA	phenylmercuric acetate
PMA	phorbol 12-myristate 13-acetate
PMA	4β,9α,12β,13α,20-pentahydroxytigilia-1,6-dien-3-one 12β-myristate, 13-acetate; phorbol, 12-myristate, 13-acetate
PMAC	phenylmercuric acetate
PMAP	1-phenyl-4-methylacetophenone
PMB	*p*-methoxybenzyl; *p*-methoxyphenylmethyl
PMBM	*p*-methoxybenzyloxymethyl
PMBOM	[(*p*-methoxybenzyl)oxy]methyl ether
Pmc	2,2,5,7,8-pentamethylchroman-6-sulfonyl
PMCC	(phenoxymethyl)chlorocarbene
PMCP	poly(methylene-1,3-cyclopentane)
PMDA	1,2,4,5-benzenetetracarboxylic anhydride; pyromellitic dianhydride
PMDBD	3,3,6,9,9-pentamethyl-2,10-diazabicyclo[4.4.0]dec-1-ene
PMDETA	pentamethyldiethylenetriamine
PMDTA	pentamethyldiethylenetriamine
Pme	pentamethylbenzenesulfonyl
PMEA	9-[(phosphonylmethoxy)ethyl]adenine; 9-[2-(phosphonomethoxy)ethyl]adenine
PMEA	*N*-(2-hydroxyethyl)-*N*-methylaniline; *N*-phenyl-*N*-methylethanolamine
PMEI	2-{[*N*-(1-methylethyl)imino]methyl}pyridine
8-PMen	8-phenylmenthyl
PMH	phenylmercuric hydroxide
PMHS	polymethylhydrosiloxane
PMI	3-phenyl-5-methylisoxazole
PMI-ACID	3-phenyl-5-methylisoxazole-4-carboxylic acid
PMMA	poly(methylmethacrylate)
PMN	polymorphonuclear
PMNL	polymorphonuclear leukocytes
PMP	1,2,2,6,6-pentamethylpiperidine
PMP	*O*,*O*-dimethyl *S*-(phthalimidomethyl)phosphorodithioate
PMP	*p*-methoxyphenyl
PMP	poly(methylpentene)
PMP	pyridoxamine 5′-phosphate; pyridoxamine phosphate
PMPXH	9-(*p*-methoxyphenyl)xanthene

PMS	*N*-methylphenazonium methosulfate
PMS	*p*-methylbenzylsulfonyl
PMS	phenazine methosulfate
PMSF	α-toluenesulfonyl fluoride; phenylmethanesulfonyl fluoride; phenylmethylsulfonyl fluoride
PMSG	pregnant mare serum gonadotrop(h)in
1-PMTR	1-phospho-5-*S*-methylthio-D-ribofuranoside
1-PMTRu	1-phospho-5-*S*-methylthioribulose
PMVE	poly(methyl vinyl ether)
PN	protease *N*
PN HCl	vitamin B_6 hydrochloride
Pn	phosphopantetheine
pN	phosphate nucleotide
pn	propylenediamine
PNA	peanut agglutinin
PNA	perinaphthenium
PNA	*p*-nitroaniline
PNA	polynuclear aromatic hydrocarbons
pNA	*p*-nitroanilide
PNASA	*p*-nitroaniline-*o*-sulfonic acid
PNBH	4-nitrobenzaldehyde hydrazone
PNBOH	*p*-nitrobenzoic acid
PNBS	pyridinium 3-nitrobenzenesulfonate
pNBSP	*p*-nitrobenzenesulfonyl peroxide
pNBz	*p*-nitrobenzoyl
PNEI	2-{[*N*-(1-naphthylethyl)imino]methyl}pyridine
PNMT	phenylethanolamine *N*-methyltransferase
PNO	pyridine 1-oxide
PNOT	2-methyl-4-nitroaniline
PNP	purine nucleoside phosphorylase
PNP-DTP	*p*-nitrophenyl-2-diazo-3,3,3-trifluoropropionate
PNPA (pNPA)	*p*-nitrophenyl acetate
PNPase	purine nucleoside phosphorylase
PNPD	*p*-nitrophenyl *n*-dodecanoate
PNPDPP	*p*-nitrophenyl diphenyl phosphate
PNPG	1-(4-nitrophenyl)glycerol; α-*p*-nitrophenylglycerine
pNPGB	4-nitrophenyl-4-guanidinobenzoate hydrochloride
PNPH	*p*-nitrophenyl *n*-hexanoate
PNPP	4-nitrophenyl phosphate
PNPP	*p*-nitrophenyl picolinate
pNPTA	*p*-nitrophenyl thioacetate
PO	polyolefin
POBN	α-(4-pyridyl-1-oxide)-*N*-*tert*-butylnitrone
POC	cyclopentyloxycarbonyl

Pol	polymerase
Polymin P	poly(ethylenimine)
POM	3-methyl-4-nitropyridine *N*-oxide
POM	4-pentenyloxymethyl; pivaloyloxymethyl
POM	chloromethyl pivalate
POM-Cl	chloromethyl pivalate
POMC	pro-opiomelanocortin
POPC	1-palmitoyl-2-oleoyl-*sn*-glycero-3-phosphocholine
POPG	1-palmitoyl-2-oleoyl-*sn*-glycero-3-phospho-*rac*-glycerol
POPOP	1,4-bis(5-phenyl-2-oxazolyl)benzene; 2,2′-*p*-phenylene-bis(5-phenyloxazole)
POPSO	piperazine-*N,N*′-bis(2-hydroxypropane-3-sulfonic acid)
Porofor BSH	benzenesulfonyl hydrazide
PP	poly(propylene)
PPA	polyphosphoric acid
PPB	1-phenyl-2-propyl benzenesulfonate
PPBI	2-{[*N*-(1-diphenylethyl)imino]methyl}pyridine
PPD	2,5-diphenyl-1,3,4-oxadiazole
PPDA	phenyl phosphorodiamidate
PPDP	4,4′-biphenol
PPDT	3-(4-phenyl-2-pyridyl)-5,6-diphenyl-1,2,4-triazine
PPDTS	2-(5,6-bis[4-sulfophenyl]-1,2,4-triazin-3-yl)-4-(4-sulfophenyl)pyridine
PPDTS	3-(4-phenyl-2-pyridyl)-5,6-diphenyl-1,2,4-triazinetrisulfonic acid
PPE	polyphosphate ester; ethyl *m*-phosphate
PPEI	2-pyridinalphenylethylimine; 2-[*N*-(1-phenylethyl)imino]methylpyridine
PPF	perchloro-9-phenylfluorenyl
PPFA	*N,N*-dimethyl-1-[1′,2-bis(diphenylphosphino)ferrocenyl] ethylamine
PPFA-IP	*N,N*-dimethyl-1-[1′,2-bis(diphenylphosphino)ferrocenyl]-2-methylpropylamine
PPFA-Ph	*N,N*-dimethyl-1-[1′,2-bis(diphenylphosphino)ferrocenyl] benzylamine
ppGpp	guanosine tetraphosphate
PP_i	inorganic pyrophosphate
PP_{ii}	inorganic pyrophosphate
PPL	pig pancrease lipase; porcine pancreatic lipase
PPM	4-diphenylphosphino-2-diphenylphosphinomethyl pyrrolidine
PPN	bis(triphenylphosphoranylidene)ammonium ion
PPN	bis(triphenylphosphoranylidene)iminium ion
ppN	diphosphate nucleotide

PPNCl	bis(triphenylphosphoranylidene)ammonium chloride
4-PPNO	4-phenylpyridine *N*-oxide
PPO	2,5-diphenyloxazole
PPO	poly(2,6-dimethyl-1,4-phenylene) ether
PPO	polyphenylene oxide
Ppoc	2-triphenylphosphonioisopropoxycarbonyl
PPP	presqualene pyrophosphate
pppGpp	guanosine pentaphosphate
PPPI	2-{[*N*-(1-phenylpropyl)imino]methyl}pyridine
PPPM	*N*-pivalyl-4-diphenylphosphino-2-diphenylphosphinomethyl pyrrolidine
pppN	triphosphate nucleotide
PPS	3-(1-pyridinio)-1-propanesulfonate
PPS	poly(phenylene sulfide)
PPSE	trimethylsilyl polyphosphate
PPST	2-(2-pyridyl)-5,6-bis(4-phenylsulfonic acid)-1,2,4-triazine
Ppt	(diphenylphosphino)thioyl; diphenylthiophosphinyl
PPTS	3-[5-(sulfophenyl)-2-pyridyl]-1,2,4-triazin-5-ylbenzenesulfonic acid, disodium salt
PPTS	pyridinium toluene-4-sulfonate
PpTS	phenyl *p*-tolyl sulfide
PPV	poly-(*p*-phenylenevinylene)
PQ	plastoquinone
PQQ	4,5-dihydro-4,5-dioxo-1*H*-pyrrolo[2,3-*f*]quinoline-2,7,9-tricarboxylic acid
PQQ	methyl 4,5-dihydro-4,5-dioxobenz[*g*]indole-2-carboxylate
PQQ	pyrroloquinoline quinone
PQQTME	trimethyl 4,5-dihydro-4,5-dioxo-1*H*-pyrrolo[2,3-*f*]quinoline-2,7,9-tricarboxylate
PR	phenol red
PR	progesteronic receptor
Pr	propyl
P-Rib-PP	5-phospho-D-ribose 1-diphosphate
Pro	proline
Pro	prolyl
Prop	*n*-propyl
PROPHOS	1,2-bis(diphenylphosphino)propane
PROXYL	3-[3-(2-iodoacetamido]-2,2,5,5-tetramethyl-1-pyrrolidinyloxy, free radical
PRPP	5-phosphorylribulose 1-pyrophosphate; phosphoribosylpyrophosphate
PS	polystyrene
P2S	2-pyridinealdoxime methyl methanesulfonate

PSAP	prostatic acid phosphatase
PS-Cl	2-pyridinesulfenyl chloride
ps-DNA	parallel strand DNA
PSDP	*p*-styryldiphenylphosphine
Psec	2-(phenylsulfonyl)ethoxycarbonyl
PSF	polysulfone
PSH	D-penicillamine thiol
PSL	*Pseudomonas* lipase
PSP	phenolsulfonphthalein
PSPA	poly(sebacic anhydride)
PSSeSP	bis(D-penicillamine)selenide
PSSP	penicillamine disulfide
PT	*p*-terphenyl
1-PT	1-propanethiol
PTA	factor XI; phosphotungstic acid
PTAA	3-phenyl-1,2,4-thiadiazol-5-yl-thioacetic acid
PTAD	4-phenyl-1,2,4-triazoline-3,5-dione
PTAH	phenyltrimethylammonium hydroxide
PTAP	phenyltrimethylammonium perbromide
PTBBA	4-*tert*-butylbenzoic acid
PTC	1-phenyl-2-thiourea
PTC	phase transfer catalysts
PTC	phenyl isothiocyanate
Ptdins	phosphatidylinsositol
PtdinsP$_2$	phosphatidylinsositol 4,5-bisphosphate
PteGlu	folic acid
PTFE	poly(tetrafluoroethylene)
PTH	parathyroid hormone; phenylthiohydantoin
PTH-Amino acids	amino acid phenylthiohydantoins
PTIO	2-phenyl-4,4,5,5-tetramethylimidazoline-3-oxide-1-oxyl
PTM	perchlorotriphenylmethyl
PTM	phenylthiomethyl
PTMH	perchlorotriphenylmethane
PTMO	*n*-propyltrimethoxysilane
PTMT	poly(tetramethylene terephthalate)
PTM	perchlorotriphenylmethyl radical
PTOC	2-thioxopyridinyloxycarbonyl
PTrMA	(+)-poly(triphenylmethyl)methacrylate
PTS	*p*-toluenesulfonic acid
PTSA	*p*-toluenesulfonic acid
PTSI	*p*-toluenesulfonyl isocyanate
PTT	phenyltrimethylammonium perbromide
PTX	palytoxin

PUCEB	*O*-(cellobiosylidenamino) *N*-phenylcarbamate
PUCIB	*O*-(*N*,*N'*-diacetylchitobiosylidenamino) *N*-phenylcarbamate
PUGLU	*O*-(D-glucopyranosylidenamino) *N*-phenylcarbamate
PUGNAC	*O*-(2-acetamido-2-deoxy-D-glucopyranosylidenamino) *N*-phenylcarbamate
Pv	pivaloyl
PVA	poly vinyl acetate
PVA	polyvinyl alcohol
PVAC (PVAc)	poly(vinyl acetate)
PVAL	poly(vinyl alcohol)
PVB	poly(vinyl butyrate)
PVC	polyvinyl chloride
PVDC	polyvinyl dichloride
PVDF	poly(vinyl difluoride); poly(vinylidene fluoride)
PVF	poly(vinyl fluoride)
PVI	polyvinyl isobutyl ether
PVK	pencillin V, potassium salt
PVM	polyvinyl methyl ether
PVP	polyvinylpyrrolidone
PVP-I	polyvinylpyrrolidone, iodine complex
PVPCC	pyridinium chlorochromate—polymer bound
PVPDC	poly(4-vinylpyridinium) dichromate; poly(vinylpyridinium dichromate); pyridinium dichromate—polymer bound
PVPHF	poly[4-vinylpyridinium poly(hydrogen fluoride]
PVPHP	pyridine hydrobromide perbromide—polymer bound
PVPP	polyvinylpolypyrrolidone
PVS	phenyl vinyl sulfoxide
PVSK	potassium polyvinyl sulfate
9-PX	9-phenylxanthene
PXH	9-phenylxanthene
Py (py)	pyridine
PyBOP	benzotriazol-1-yl-oxy-tripyrrolidinophosphonium hexafluorophosphate
Pyoc	2-(2'- and 4'-pyridyl)ethoxycarbonyl
4-PyPo	5-phenyl-2-(4-pyridyl)oxazole
Pyr (pyr)	pyridine
pyr	pyrazine
pyrr	pyrrolidine
pyz	pyrazine
pz	pyrazine
Q	2-quinolone
Q	coenzyme Q; ubiquinone
Q	glutamine

Q	quaternary ammonium salt
Q-4	coenzyme Q_4
Q-9	coenzyme Q_9
Q-10	coenzyme Q_{10}
QAC	quaternary ammonium compound
QAE	diethyl(2-hydroxypropyl)amino ethyl
QBS	quinine bisulfate
QC	quadricyclane
QD	quinodimethane
QDC	quinolinium dichromate
QPC	cuprolinic blue
QSNT	1-(8-quinolinesulfonyl)-3-nitro-1*H*-1,2,4-triazole
QT	quick tan; dihydroxyacetone
QUIBEC	*N*-benzylquininium chloride
Quin-2	2-(2-amino-5-methylphenoxy)methyl]-6-methoxy-8-aminoquinoline-*N*,*N*,*N'*,*N'*-tetraacetic acid
Quin-2/AM	2-(2-amino-5-methylphenoxy)methyl]-6-methoxy-8-aminoquinoline-*N*,*N*,*N'*,*N'*-tetraacetic acid tetraacetoxymethyl ester
R	alkyl
R	an organic group
R	arginine
R Acid	2-naphthol-3,6-disulfonic acid
R Salt	2-naphthol-3,6-disulfonic acid
Ra Ni	Raney nickel
RA_2	2,2',2"-trichlorotriethylamine
RAMA	rabbit muscle aldolase; D-fructose-1,6-bisphosphate aldolase
RAMEMP	(*R*)-1-amino-2-(2-methoxyethoxymethyl)pyrrolidine
RAMP	(*R*)-1-amino-2-(methoxymethyl)pyrrolidine
rANF	rat atrial natriuretic peptide
RB	Rose Bengal; rose bengal
RBD	Rose Bengal derivative
RBITC	rhodamine B isothiocyanate
RBP	retinol-binding protein
RBP	riboflavin binding protein
ψrd	pseudouridine
RDB	sodium dihydrobis(2-methoxyethoxy)aluminate
RDE	receptor destroying enzymes
RDPR	ribonucleoside diphosphate reductase
RDS	respiratory distress syndrome
RDX	hexahydro-1,3,5-trinitro-1,3,5-triazine
RE_2	restriction endonuclease site
Red-Al	sodium bis(2-methoxyethoxy)aluminum hydride

RF (x)	release factor (x)
R_f	perfluoroalkyl
RFLP	restriction fragment length polymorphism
RHMP	(*R*)-(−)-methyl 3-hydroxy-2-methylpropionate
Rib	ribose
RM	organometallic reagent
RNA	ribonucleic acid
RNase	ribonuclease
RNR	ribonucleotide reductase
RO-20-1724	4-(3-butoxy-4-methoxybenzyl)-2-imidazolidinone
R5P	ribose 5-phosphate
RPPRA	ribose-phosphate-phosphate-ribose-adenine
rRNA	ribosomal ribonucleic acid
RSH	1-dodecanethiol
RT	reverse transcriptase
RuBPCase	ribulose-bisphosphate carboxylase
RuDP	ribulose 1,5-diphosphate
Ru5P	ribulose 5-phosphate
S	serine
S	substrate
S ACID (S Acid)	8-amino-1-naphthol-5-sulfonic acid; 1-amino-8-naphthol-4-sulfonic acid
2S Acid	1-amino-8-naphthol-2,4-disulfonic acid
SA	sialic acid
SA	ubiquinones
SAA	succinic anhydride
SADH	*N*-dimethylaminosuccinamic acid
SADP	*N*-succinimidyl (4-azidophenyldithio)propionate
sadpen	bis(3-((2-hydroxybenzyl)imino)propyl)methylamine
SAED	sulfosuccinimidyl 2-(7-azido-4-methylcoumarin-3-acetamide)ethyl-1,3′-dithiopropionate
SAH	*S*-adenosylhomocysteine
SAIB	sucrose diacetate hexa-*iso*-butyrate
SAICAR	*N*-[(5-amino-1-ribofuranosylimidazo-4-yl)carbonyl] aspartic acid 5′-phosphate
SALADHP	*N*-salicylidene-2-amino-1,3-dihydroxypropane
SALEN (Salen) (salen)	*N*,*N*-bis(salicylidene)ethylenediamino; *N*,*N*-ethylenebis(salicylideneiminato); *N*,*N*′-bis(salicylideneamino)-1,2-diphenylethane
Salol	phenyl salicylate
salophen	*o*-phenylenebis(salicylideneiminato)
salpn	1,3-bis((2-hydroxybenzyl)imino)propane
SALPS	2-(salicylideneamino)phenyl disulfide

SAM	*S*-adenosylmethionine
SAM	*S*-(5′-adenosyl)-L-methionine chloride
SAMCA	succinimidyl-7-azido-4-methylcoumarin-3-acetate
SAMEMP	(*S*)-1-amino-2-(2-methoxyethoxymethyl)pyrrolidine
SAMP	(*S*)-1-amino-2-(methoxymethyl)pyrrolidine
SAN	styrene-acrylonitrile (copolymer)
SAND	sulfosuccinimidyl 2-(*m*-azido-*o*-nitrobenzamido)-ethyl-1,3′-dithiopropionate
SANPAH	succinimidyl 6-(4′-azido-2′-nitrophenylamino)hexanoate
Sar (sar)	sarcosyl
Sar	sarcosine; *N*-methylglycine
SAS	aluminum sodium sulfate
SASD	sulfosuccinimidyl 2-(*p*-azidosalicylamido)ethyl-1,3′-dithiopropionate
SATA	(*N*-succinimidyl-*S*-acetylthioacetate); *S*-acetylthioglycolic acid *N*-hydroxysuccinimide ester
SB	semibullvalenes
SBA	soybean agglutinin
SBAH	sodium dihydro-bis(2-methoxyethoxy)aluminate
SBD-F	7-fluorobenzofurazane-4-sulfonic acid ammonium salt
SBF-Chloride	4-chloro-7-sulfobenzofurazane ammonium salt
SBF-Cl	ammonium 4-chloro-7-sulfobenzofurazan
SBH	sodium borohydride
SBR	styrene-butadiene rubber
sBu	*sec*-butyl
SC	subtilisin carlsberg
SC-9	*N*-(6-phenylhexyl)-5-chloro-1-naphthalenesulfonamide
SC-10	*N*-(*n*-heptyl)-5-chloro-1-naphthalenesulfonamide
SCF	stem cell factor
Scm	*S*-carboxymethylsulfenyl
ScmCl	methoxycarbonylsulfenyl chloride
SCP	sodium cellulose phosphate
SDBP	*N*-hydroxysuccinimidyl 2,3-dibromopropionate
SDH	sorbitol dehydrogenase
SDH	succinic dehydrogenase
SDP	4,4′-sulfonyldiphenol
SDP	sedoheptulose 1,7-diphosphate
SDPDTPH	21-thia-5,20-diphenyl-10,15-di(*p*-tolyl)porphyrin
SDPP	*N*-succinimidyl diphenyl phosphate
SDS	sodium decyl sulfate
SDS	sodium dodecyl sulfate; sodium lauryl sulfate; sodium dodecylbenzenesulfonate
sec	secondary (as in sec-butyl, *sec*-Bu)

Selectride	tri-*sec*-butylborohydride
SEM	2-(trimethylsilyl)ethoxymethyl; trimethylsilylethoxymethyl; β(trimethylsilyl)ethoxymethyl
SEM-Cl	2-(trimethylsilyl)ethoxymethyl chloride
Ser	serine
Ser	seryl
SES	β-trimethylsilylethanesulfonyl
SEX	sodium ethyl xanthate
Sex-Ex	vitamin H_3
SFS	sodium formaldehyde sulfoxylate
SG	silica gel
SH	sulfhydryl
SHAM	salicylichydroxamic acid
SHb	sulfhemoglobin
SHPP	3-(4-hydroxyphenyl)propionic acid *N*-hydroxysuccinimide ester; *N*-succinimidyl 3-(4-hydroxyphenyl)propionate
Sia	1,2-dimethylpropyl
Sia_2BH	disiamylborane
SIAB	*N*-succinimidyl-(4-iodoacetyl)aminobenzoate
Silvex	2-(2,4,5-trichlorophenoxy)propionic acid
SIM	trimethylsilylimidazole
SITS	4-acetamido-4′-*iso*-thiocyanto-stilbene-2,2′-disulfonic acid
SKEWPHOS	2,4-bis(diphenylphosphino)pentane
SKTA	silyl ketene thioacetal
SLS	sodium dodecylsulfate; sodium lauryl sulfate
SMb	sulfmyoglobin
SMC	styrene-maleic acid (copolymer)
SMCC	succinimidyl 4-(*N*-maleimidomethylcyclohexane)-1-carboxylate
SMDC	sodium methyldithiocarbamate
SMEAH	sodium bis(2-methoxyethoxy)aluminum hydride
SMHP	(*S*)-(+)-methyl 3-hydroxy-2-methylpropionate
SMIFA	*syn*-2-methoxy-imino-2-(2-furyl)-acetic acid
SMOM	(phenyldimethylsilyl)methoxymethyl
SMP	(*S*)-2-(methoxymethyl)pyrrolidine
SMPB	succinimidyl 4-(*p*-maleimidophenyl)butyrate
SMPT	4-succinimidyloxycarbonyl-α-methyl-α-(2-pyridyldithio)-toluene
SNase	straphylococcal nuclease
SNG	solidified nitroglycerin
Snm	*S*-(*N*′-methyl-*N*′-phenylcarbamoyl)sulfenyl
SNOBS	sodium nonanoyloxybenzene sulfonate
SNP®	*O*,*O*-diethyl-*O*-*p*-nitrophenyl phosphorothioate
snRNA	small nuclear RNA

SO	somatostatin
SOC	synthetic organic compound
SOD	copper-zinc superoxide dismutase; superoxide dismutase
SP	substance P
SP	sulfopropyl
S3P	shikimate 3-phosphate
S7P	sedoheptulose 7-phosphate
SPA	super phosphoric acid
SPADNS	2-(4-sulfophenylazo)1,8-dihydroxy-3,6-naphthalenedisulfonic acid; 4,5-dihydroxy-3-(*p*-sulfophenylazo)-2,7-naphthalenedisulfonic acid
SPAN	Sorbitan
2-SPBN	*N-tert*-butyl-α-(2-sulfophenyl)nitrone
SPCA	factor VII
SPDP	3-(2-pyridyldithio)propionic acid *N*-hydroxysuccinimide ester; *N*-succinimidyl 3-(2-pyridyldithio)propionate
SPQ	6-methoxy-1(3-sulfopropyl) quinolinium; 6-methoxy-*N*-(3-sulfopropyl)-quinolinium
SR	synthetic rubber
SRS-A	slow reacting substance of anaphylaxis
SRSV	small round structured virus
SS	squalene synthetase
SS Acid	1-amino-8-naphthol-2,4-disulfonic acid, monosodium salt
SSB	single-strand binding proteins
SSP	1,2-distearoylpalmitin
STABASE	1,1,4,4-tetramethyldisilylazacyclopentane
STC	2,3-diphenyl-5-thienyl-(2)-tetrazolium chloride
sTf	serotransferrin
STH	pituitary growth hormone
STI	soybean trypsin inhibitor
stien	1,2-diamino-1,2-diphenylethane
STPP	sodium tripolyphosphate
STTPH	21-thia-5,10,15,20-tetra(*p*-tolyl)porphyrin
STX	saxitoxin
Styryl 7	2-[4-(4-dimethylaminophenyl)-1,3-butadienyl]-3-ethylbenzothiazolium *p*-toluenesulfonate
Su	succinimidyl
Suc	sucrose
SucNBr	*N*-bromosuccinimide
sulfo-GMBS	*N*-γ-maleimidobutyryloxysulfosuccinimide ester
Sulfo-HSAB	*N*-hydroxysulfosuccinimidyl-4-azidobenzoate
sulfo-MBS	*m*-maleimidobenzoyl-*N*-hydroxysulfosuccinimide ester
Sulfo-NHS	*N*-hydroxysulfosuccinimide
Sulfo-NHS-ASA	*N*-hydroxysulfosuccinimidyl-4-azidosalicylic acid

sulfo-NHS-Biotin	*N*-hydroxysulfosuccinimidobiotin
Sulfo-SAMCA	sulfo-succinimidyl 7-azido-4-methylcoumarin-3-acetate
SVP	snake venom phosphodiesterase
T	threonine
T	thymidine
T	transferase
t	tertiary (as in *t*-Bu)
T-3 (T_3)	3,3′,5-triiodothyronine
T_4	L-thyroxine; thyroxine
2,4,5-T	(2,4,5-trichlorophenoxy)acetic acid; 2,4,5-trichlorophenoxyacetic acid
TA	traumatic acid
TAAP	tetraalkylammonium perchlorate
TAC	triallyl cyanurate
TACDPI	5-(trimethylammonio)-methyl N^3-carbamoyl-1,2-dihydro-3*H*-pyrrolo[3,2-*e*]-indole-7-carboxylate
Tacm	trimethylacetamidomethyl
tacn	1,4,7-triazacyclononane
TAD	1,2,4-triazoline-3,5-diones
TAD	L-tyrosine decarboxylase apoenzyme
TAD	tiazofurin adenine dinucleotide
TADDOL	tetraaryldioxolane dimethanols
TADH	*Thermoanaerobium brockii* alcohol dehydrogenase
TAED	tetraacetylethylenediamine
TAES	tris(hydroxymethyl)methyl-2-aminoethanesulfonic acid
TAM	trialkylamine
TAMA	*N*-methylanilinium trifluoroacetate
TAME	α,*N*-tosyl-L-arginine methyl ester hydrochloride; *N*α-4-tosyl-L arginine methyl ester hydrochloride
TAMM	tetrakis(acetoxymercuri)methane
TAN	1-(2-thiazolylazo)-2-naphthol
TAO	4-(2-thiazolylazo)orcinol
TAO	5-methyl-4-[2-thiazolylazo]-1,3-benzenediol
TAP	pyrazino[2,3-*f*]quinoxaline
TAPA	α-(2,4,5,7-tetranitro-9-fluorenylideneaminoxy)propionic acid
TAPP	5,10,15,20-tetrakis[4-(trimethylammonio)phenyl]-21*H*,23*H*-porphine tetra-*p*-tosylate salt
TAPS	3-[tris(hydroxymethyl)methylamino]-1-propanesulfonic acid
TAPSO	3-[*N*-(tris(hydroxymethyl)methylamino]-2-hydroxypropanesulfonic acid
TAR	4-(2-thiazolylazo)resorcinol
TAS	tris(diethylamino)sulfonium

TASF (TAS-F)	tris(dimethylamino)sulfur (trimethylsilyl)difluoride; tris(dimethylamino)sulfonium difluorotrimethylsilicate
TB	thexylborane
TB	thymol blue
TBA	*tert*-butylamine
TBA	tetra-*n*-butylammonium; tetrabutylammonium
TBA	tribenzylamine
2,3,6-TBA®	2,3,6-trichlorobenzoic acid
TBAB	*tert*-butylamine borane
TBAB	tetrabutylammonium bromide
TBABF	tetrabutylammonium hydrogen difluoride
$TBABr_3$	tetrabutylammonium tribromide
TBAC	*tert*-butylacetyl chloride
TBACC	tetrabutylammonium chlorochromate
TBACN	tetrabutylammonium cyanide
TBADC	bis(tetrabutylammonium) dichromate
TBADH	*Thermoanaerobium brockii* dehydrogenase
TBAF	tetra-*n*-butylammonium fluoride; tetrabutylammonium fluoride
TBAF	tetra-*n*-butylammonium fluoroborate
TBAHS	tetrabutylammonium hydrogen sulfate
TBAI	tetrabutylammonium iodide
TBAP	tetra-*n*-butyl ammonium perruthenate
TBAP	tetra-*n*-butylammonium perchlorate; tetrabutylammonium perchlorate
TBAS	tetra-*n*-butylammonium succinimide
TBB	1-*tert*-butyl-1,3-butadiene
TBBA	*N,N'*-terephthalylidenebis(4-butylalanine)
TBC	4-*tert*-butylcatechol
TBCD	2,4,4,6-tetrabromocyclohexa-2,5-dienone
TBD	1,5,7-triazabicyclo[4.4.0]dec-5-ene
TBDA	thexylborane-*N,N*-diethylaniline
TBDMS	*t*-butyldimethylsilyl
TBDMSCN	*tert*-butyldimethylsilyl cyanide
TBDMSI	1-(*tert*-butyldimethylsilyl)imidazole
TBDMSIM	1-(*tert*-butyldimethylsilyl)imidazole
TBDPS	*t*-butyldiphenylsilyl
TBDS	tetra-*t*-butoxydisiloxane-1,3-diylidene
TBE	1,1,2,2-tetrabromoethane
TBH	1,2,3,4,5,6-hexachlorocyclohexane
TBHC	*tert*-butyl hypochlorite
TBHF	2,4,5,7-tetrabromo-6-hydroxy-3-fluorone
TBHP	*t*-butyl hydroperoxide
TBK	triethylammonium bicarbonate

TBMA	1-*tert*-butyl-3-methylallene
TBME	*tert*-butyl methyl ether
TBMPS	*t*-butylmethoxyphenylsilyl
TBMPSiBr	*tert*-butyl-methoxyphenylbromosilane
TBO	3-[(trimethylsilyl)oxy]-3-buten-2-one
TBOF	tributyl orthoformate
TBP	butyltriphenylphosphonium bromide; triphenylbutylphosphonium bromide
TBP	tetrabenzoporphyrin
TBP	tetrabenzoporphyrinato
TBP	tri-*n*-butyl phosphate
TBPE	tetrabromophenolphthalein ethyl ester
TBS	4-*tert*-butylphenyl salicylate
TBS	*t*-butyldimethylsilyl
TBSCl	*tert*-butyldimethylsilyl chloride
TBSE	*tert*-butyl silyl ether
TBSOP	*N*-*t*-Boc-2-(*tert*-butyldimethylsiloxy)pyrrole
TBT	tetrabutyl titanate
TBTD	tetrabutylthiuram disulfide
TBTH	tributyltin hydride
TBTO	bis(tributyltin)oxide
TBTr	4,4′,4″-tris(benzyloxy)triphenylmethyl
TBTU	*O*-(1*H*-benzotriazol-1-yl)-*N*,*N*,*N*′,*N*′-tetramethyluronium tetrafluoroborate
tBu (t-Bu)	*tert*-butyl
TBuAB	tetrabutylammonium bromide
TBUP	tri-*n*-butylphosphine
TC	2,3,4,5-tetraphenylcyclopentadienone
Tc	tetracycline
TC-NBT	thiocarbamyl nitro blue tetrazolium
TCA	trichloroacetate
TCA	trichloroacetic acid
TCA-O-PCP	trichloroacetic acid pentachlorophenyl ester
TCAc	trichloroacetyl
TCAP	cetrimonium pentachlorophenoxide
TCB	1,2,4-trichlorobenzene
TCB	trichlorobenzene (usually 1,3,5)
TCBA	2,3,6-trichlorobenzoic acid
TCBC	trichlorobenzyl chlororide
TCBOC chloride	2,2,2-trichloro-1,1-dimethylethyl chloroformate
TcBoc	1,1-dimethyl-2,2,2-trichloroethoxycarbonyl
TCBQ	2,3,5,6-tetrachlorobenzoquinone
TCC	3,3′,4′-trichlorocarbanilide

TCCD	2,3,7,8-tetrachlorobenzodioxin
TCDD	2,3,7,8-tetrachlorodibenzo-*p*-dioxin
TCDF	2,3,7,8-tetrachlorodibenzofuran
TCDI	1,1′-thiocarbonyldiimidazole
TCE	1,1,1-trichloroethylene
Tce	2,2,2-trichloroethyl
Tcec	β,β,β-trichloroethoxycarbonyl
TcecCl	β,β,β-trichloroethoxycarbonyl chloride; 2,2,2-trichloroethyl chloroformate
TCEP	tris(2-carboxyethyl)phosphine
TCHP	tricyclohexylphosphine
TCl	terephthaloyl chloride
TCNB	1,2,4,5-tetracarbonitrile
TCNB®	tetrachloronitrobenzene
TCNE	tetracyanoethylene
TCNEO	tetracyanoethylene oxide
TCNP	11,11,12,12-tetracyanopyreno-2,7-quinodimethane
TCNQ	7,7,8,8-tetracyanoquinodimethane
TCNQF4	tetrafluorotetracyanoquinodimethane
TCOB	*trans*-1,4-cyclohexanediyl-4,4′-bis(1-octyl benzoate)
TCP	trichlorophenol (usually 2,4,5 or 2,4,6)
TCP	tricresyl phosphate
TCP	trichloropropane
TCP	tritolyl phosphate
TCPO	bis(2,4,6-trichlorophenyl) oxalate
TCPP	tetrakis(4-carboxyphenyl)porphine
2,4,5-TCPPA	2-(2,4,5-trichlorophenoxy)propionic acid
TCPQ	tetracyanopentacenquinodimethane
TCR	T cell antigen receptor
Tcroc	2-(trifluoromethyl)-6-chromonylmethylenecarbonyl
Tcrom	2-(trifluoromethyl)-6-chromonylmethylene
TCTA	1,4,7-triazacyclononane-*N*,*N*′,*N*″-triacetate
TCTD	tetrachlorothiophene dioxide
2,3,6,5-TCTF	2,3,6-trichloro-5-(trifluoromethyl)pyridine
2,3,5,6,4-TCTF	2,3,5,6-tetrachloro-4-(trifluoromethyl)pyridine
TCTFP	1,1,2,2-tetrachloro-3,3,4,4-tetrafluorocyclobutane
TCTP	tetrachlorothiophene
TCU	tetracyclo[6.3.0.0^{4,11}.0^{5,9}]undecane
TDA	tetradecen-1-yl acetate
TDA-1	tris(dioxa-3,6-heptyl)amine; tris[2-(2-methoxyethoxy)-ethyl]amine
TDAB	1,3,5-tris(diphenylamino)benzene
TDBAC	tetradecyldimethylbenzylammonium chloride

TDBMB	1,3,5-tris[bis(biphenyl)methyl]benzene
TDBTU	*O*-(3,4-dihydro-4-oxo-1,2,3-benzotiazin-3-yl)-*N*,*N*,*N'*,*N'*-tetramethyluronium tetrafluoroborate
TDCPP	tetrakis(2,6-dichlorophenyl)porphyrinato
TDDA	tetradecen-1-yl acetate
TDE	1,1-dichloro-2,2-bis(*p*-chlorophenyl)ethane
TDE	triethylene glycol diglycidyl ether
p,p'-TDE	1,1-dichloro-2,2-bis(*p*-chlorophenyl)ethane
TDI	toluene-2,4-diisocyanate; tolylene diisocyanate
TDIAB	1,3,5-tris(diisopropylamino)benzene
TDO	thiophene dioxide
TDO-Carbonate	4,6-diphenylthieno[3,4-*d*]-1,3-dioxol-2-one 5,5-dioxide; Steglich's reagent
TDOL	tetradecen-1-ol
TDP	4,4'-thiodiphenol
TDP	ribosylthymidine 5'-diphosphate
TDS	thexyldimethylsilyl
TDSCl	thexyldimethylchlorosilane
TDTA	di-*p*-toluoyltartaric acid
TEA	tetraethylammonium
TEA	triethanolamine
TEA	triethylaluminum
TEA	triethylamine
TEAA	tetraethylammonium acetate
TEAB	tetraethylammonium borohydride
TEAB	tetraethylammonium bromide
TEAB	triethylamine borane
TEAB	triethylammonium bicarbonate
TEAE	triethylaminoethyl
TEAF	tetraethylammonium formate
TEAF	triethylammonium formate
TEAM	tetrahexylammonium tetrakis(diperoxomolybdo)phosphate
TEAP	tetraethylammonium perchlorate
TEAP	tetraethylammonium phosphate
TEAS	tetraethylammonium succinimide
TEBA	benzyltriethylammonium chloride
TEBAC	benzyltriethylammonium chloride
TEBACl	triethylbenzylammonium chloride
TED	1,4-diazabicyclo[2.2.2]octane
TEDP	tetraethyl dithionopyrophosphate
TEE	3,4-diethyl-3-hexene; tetraethylethylene
Teflon	poly(tetrafluoroethylene) resin
TEG	triethylene glycol
TEL	tetraethyllead

TEM	triethylenediamine; 1,4-diazabicyclo[2.2.2]octane
TEMED	*N,N,N',N'*-tetramethylethylenediamine
TEMPO	4-(2-iodoacetamido)
TEMPO	2,2,6,6-tetramethylpiperidinooxy, free radical
TEOA	triethanolamine
TEOC (Teoc)	[2-(trimethylsilyl)ethoxy]carbonyl; 2-(trimethylsilyl)ethoxycarbonyl
TEOC-ONp	2-(trimethylsilyl)ethyl *p*-nitrophenyl carbonate
TEOS	tetraethoxysilane
TEP	triethyl phosphate
TEPP	tetraethyl pyrophosphate
Ter	terpenyl
terpy	2,2':6',2"-terpyridine
tert	tertiary (as in *tert*-butyl; but *t*-Bu)
TES	2-[tris(hydroxymethyl)methylamino]-1-ethanesulfonic acid
TES	*N,N,N',N'*-tetraethylsulfamide
TES	*N*-tris(hydroxymethyl)methyl-2-aminoethanesulfonic acid
TES	triethylsilyl
TESCl	triethylchlorosilane
TESCN	triethylsilyl cyanide
TETA Resin	triethylenetetramine resin
TETD	tetraethylthiuram disulfide
TETM	tetraethylthiuram monosulfide
TETN	triethylamine
tetrakis	tetra(triphenylphosphine)palladium(0)
TEU	tetraethylurea
Tf	triflate; trifluoromethylsulfonyl; triflyl
TFA	trifluoroacetic acid
TFA	trifluoroacetyl
TFAA	trifluoroacetic anhydride
TFAI	trifluoroacetylimidazole
TFA-ME	methyl trifluoroacetate
TFAT	trifluoroacetyl triflate
tfc	3-(trifluoromethylhydroxymethylene)-(+)-camphorato; 3-(trifluoromethylhydroxymethylene)-*d*-camphorato
TFE	2,2,2-trifluoroethanol; trifluoroethanol
TFE	tetrafluroethylene
TFEB	trifluoroethyl butyrate
TFMC	3-(trifluoromethylhydroxymethylene)-*d*-camphorato
TfOH	trifluoromethanesulfonic acid
TFP	1,1,1-trifluoro-2-propanone
TFP	trifluoropropene
TFPA	trifluorophenyl azide
TFPB	tetrakis[3,5-bis(trifluoromethyl)phenyl]borate

m-TFPTAH	*m*-(trifluoromethyl)phenyltrimethylammonium hydroxide
m-TFPTAI	*m*-(trifluoromethyl)phenyltrimethylammonium iodide
TFPZ	trifluoroisopropenylzinc
TG	triacylglycerol; triglyceride
TGA	nitrilotriacetic acid
TH	tetrahydro
TH	thyrotropic hormone
TH	trihydro
Th	2-thiophenyl
Th	thianthrene
Th	thiazole
2-Th	2-thienyl
THA	9-amino-1,2,3,4-tetrahydroacridine
THA	5β-pregnane-3α,21-diol-11,20-dione; tetrahydro Kendall's Compound A
THA	tetrahydroaminoacridine
THAB	tetrahexylammonium benzoate
THAM (Tham)	tris(hydroxymethyl)aminomethane
1,3,5-THB	phoroglucinol
THBC	1,2,3,4-tetrahydro-β-carbolines
THC	tetrahydrocannabinol
THD (thd)	2,2,6,6-tetramethyl-3,5-heptanedionate; 2,2,6,6-tetramethyl-3,5-heptanedionato
ThDP	thiamin diphosphate
THE	tetrahydrocortisone
Thexyl (thexyl)	2-[2,3-dimethylbutyl]; 2,3-dimethyl-2-butyl; 1,1,2-trimethylpropyl
THF	tetrahydrofolic acid
THF	tetrahydrofuranyl
THF (thf)	tetrahydrofuran
THFA	tetrahydrofolic acid
THFA	tetrahydrofurfuryl alcohol
THFC-Eu	tris[3-(heptafluoropropylhydroxymethylene)-*d*-camphorato]europium(III)
THI	tetrahydro-1*H*-indole
THIP	4,5,6,7-tetrahydroisoxazolo[5,4-c]pyrimidin-3(2*H*)-one
THM	trihalomethanes
THN	1,2,3,4-tetrahydronaphthalene
ThO	thianthrene 5-oxide
THP	2-tetrahydropyranyl
THP	tetrahydropapaveroline
THP	tetrahydropyran
THP	tetrahydropyranyloxy
THPC	tetrakis(hydroxymethyl)phosphonium chloride

THPE	1,1,1-tris(4-hydroxyphenyl)ethane
THQ	tetrahydroxyquinone
Thr	threonine
Thr	threonyl
THTO	tetrahydrothiophene oxide
THU	tetrahydrouridine
Thy	thymine
THYM	tris(hydroxymethyl)methane
THYME	tetrakis(hydroxymethyl)ethylene
TI	trypsin inhibitor
TIBA (tiba)	triiodobenzoic acid (usually 2,3,5)
TIBA	triisobutylaluminum
2-TIC	tris(2imidazoyl) carbinol
4-TIC	tris[4(5)imidazoyl] carbinol
TIHF	2,4,5,7-tetraiodo-6-hydroxy-3-fluorone
TIM	triose phosphate isomerase
TIP	2,4,6-triisopropyl-1,3-dioxa-5-phosphacyclohexane
TIPB	triisopropylbenzene
TIPDS	1,3-(1,1,3,3-tetraisopropyldisiloxanylidene)
$TIPDSiCl_2$	1,3-dichloro-1,1,3,3-tetraisopropyldisiloxane
TIPH	sodium triisopropoxyhydroborate
TIPPS	iodophtalein sodium
TIPS	triisopropylsilyl
TIPSCl	1,3-dichloro-1,1,3,3-tetraisopropyldisiloxane
TIPSCl	triisopropylchlorosilane
TIQ	tetrahydroisoquinoline
TISCN	triisopropylsilyl cyanide
TLA	tris[(6-methyl-2-pyridyl)methyl]amine
TLCK	1-chloro-3-tosylamido-7-amino-2-heptanone; *N*α-*p*-tosyl-L-lysine chloromethyl ketone
TLTr	4,4′,4″-tris(levulinoyloxy)triphenylmethyl
TM	tetramethyl
TM	trimethyl
TMA	trimethylaluminum
TMA	trimethylamine
TMAB	tetramethylammonium borohydride
TMAB	trimethylamine borane
TMAC	trimellitic anhydride monoacid chloride
TMAEMC	2-trimethylammoniumethylmethacrylic chloride
TMAH	tetramethylammonium hydroxide
TMANO	trimethylamine *N*-oxide
TMAO	trimethylamine oxide
TMAP	tetramethylammonium perchlorate
TMAT	tetramethylammonium tribromide

TMAT	tris-2,4,6-[1-(2-methyl)aziridinyl]-1,3,5-triazine
TMB	1,2,4,5-tetrakis(methylene)benzene
TMB	3,3′,5,5′-tetramethylbenzidine
TMB	1,2,4,5-tetramethoxybenzene
TMB-4	1,1′-trimethylene-bis(4-formylpyridinium bromide)-dioxime; 1,1′-trimethylenebis[4-(hydroxyiminomethyl)-pyridinium bromide]
TMB-8	8-(diethylamino)octyl 3,4,5-trimethoxybenzoate; 3,4,5-trimethoxybenzoic acid 8-(diethylamino)octyl ester
TMBA	2,4,6-trimethylbenzoic acid
TMBA	3,4,5-trimethylbenzaldehyde
TMBCH	2,3,5,6-tetrakis(methylene)bicyclo[2.2.0]hexane
TMBS	trimethylbromosilane
TMBZ.PS	*N*-(3-sulfopropyl)-3,3′,5,5′-tetramethylbenzidine sodium salt
TMC	3,3,5-trimethylcyclohexanol
TMC-amine	3,3,5-trimethylcyclohexylamine
TMCS	trimethylchlorosilane
TMD	tetramethyl-1,2-dioxetane
TMDB	4,4,6-trimethyl-1,3,2-dioxaborinane
TMDS	1,1,3,3-tetramethyldisilazane
TME	tetramethyleneethane
TME	tetramethylethylene
TMED	1,1,2,2-tetramethylethylenediamine
TMEDA	*N*,*N*,*N*′,*N*′-tetramethylethylenediamine
TMG	methyl β-D-thiogalactoside
TMG	N^2,N^2,N^7-trimethylguanosine
TMG	tetramethylglucose
TMIS	iodotrimethylsilane
TML	tetramethyllead
TMM	trimethylenemethane
TMNO	trimethylamine *N*-oxide; trimethylamine oxide
TMO	trimethylamine *N*-oxide
Tmob	*S*-2,4,6-trimethoxybenzyl
TMOE	1,1,2-trimethoxyethane
TMP	4-*O*-(2,4,6-trimethylphenyl)
TMP	2,2,6,6-tetramethylpiperidine; tetramethylpiperidine
TMP	ribosylthymidine 5′-monophosphate; thymidine 5′-monophosphate
TMP	tetramesitylporphyrinato
TMP	triozsalen
2-TMP	(2-thiazolylmethylene)triphenylphosphorane
TMPA	tris(2-pyridylmethyl)amine
TMPE	1,1,1-tris(methoxyphenyl)ethane

TMPM	trimethoxyphenylmethyl
TMPP	tetra(*N*-methylpyridyl)porphyrin
TMPTA	2-ethyl-2-(hydroxymethyl)-1,3-propanediol trimethacrylate; trimethylolpropane triacrylate
TMPTMA	2-ethyl-2-(hydroxymethyl)-1,3-propanediol trimethacrylate; trimethylolpropane trimethacrylate
TMPyP	5,10,15,20-tetrakis(1-methyl-4-pyridinio)porphyrin tetra(toluene-4-sulfonate)
TMS	tetramethylsilane
TMS	trimethylsilyl
TMSA	*N*-(trimethylsilyl)acetamide
TMSacac	4-trimethylsiloxy-3-penten-2-one
TMSAN	trimethylsilylacetonitrile
TMS-BA	*N*,*O*-bis(trimethylsilyl)acetamide
TMSCl	trimethylsilyl chloride; chlorotrimethylsilane
TMSCN	trimethylsilyl cyanide
TMSDEA	*N*,*N*-diethyltrimethylsilylamine
TMSDMA	*N*,*N*-dimethyltrimethylsilylamine
Tmse	2-(trimethylsilyl)ethyl
TMSEC	2-(trimethylsilyl)ethoxycarbonyl
TMSI	1-(trimethylsilyl)imidazole
TMSI	trimethylsilyl iodide
TMSIM	1-(trimethylsilyl)imidazole
TMS-MVK	4-(trimethylsilyl)-3-buten-2-one
TMSO	3-trimethylsilyl-2-oxazolidnone
TMSOF	2-(trimethylsiloxy)furan
tmtacn	*N*,*N'*,*N''*-trimethyl-1,4,7-triazacyclononane
TMTAD	1,5,9-trimethyl-1,5,9-triazacyclododecane
TMTD	bis(dimethylthiocarbamoyl) disulfide; tetramethylthiuram disulfide
TMTM	tetramethylthiuram monosulfide
TMTP	tri-*m*-tolylphosphine
TMTr	tri-(*p*-methoxyphenyl)methyl
TMTSF	tetramethyltetraselenafulvalene
TMTTF	tetramethyltetrathiafulvalene
TMU	tetramethylurea
TMV	tobacco mosaic virus
tn	trimethylenediamine
TNA	2,4,6-trinitroanisole
TNB	5-thio-2-nitrobenzoic acid
1,2,4-TNB	1,2,4-trinitrobenzene
TNBA	tri-*n*-butylaluminum
TNBS	2,4,6-trinitrobenzene sulfonic acid; picrylsulfonic acid
TNBSA	2,4,6-trinitrobenzenesulfonic acid

TNBT	tetranitro blue tetrazolium
TNBT	titanium(IV) *n*-butoxide
TNBT Diformazan	1,1′-(3,3′-dimethoxy-4,4′-diphenylene)bis-3,5-di(*p*-nitrophenyl)diformazan
TNF	2,4,7-trinitro-9-fluorenone
TNF	tumor necrosis factor
TNM	tetranitromethane
2-TNO	2-thiazolyl carbonitrile *N*-oxide
TNP	trinitrophenol
TNPA	2,4,6-trinitrophenyl acetate
TNPA	tri-*n*-propylaluminum
TNPDTC	2,4,6-trinitrophenyl *O*-ethyl dithiocarbonate
TNPMC	2,4,6-trinitrophenyl methyl carbonate
TNPSH	2,4,6-trinitrobenzenethiol
TNPTA	2,4,6-trinitrophenyl thioacetate
TNS	2-*p*-toluidinylnaphthalene-6-sulfonate
2,6-TNS	6-(4-toluidino)-2-naphthalenesulfonic acid
TNT	2,4,6-trinitrotoluene
TNTU	*O*-(5-norbornene-2,3-dicarboximido)-*N*,*N*,*N*′,*N*′-tetramethyluronium tetrafluoroborate
TNV	tobacco mosaic virus
TOAF	tetraoctylammonium fluoride
TOBO	2-(*p*-tolyl)benzoxazole
TOC	total organic carbon
TOC	tri-*o*-carvacrotide
TOCP	triorthocresyl phosphate
T-OH	Taber's chiral alcohol
Tol (tol)	toluene
Tol (tol)	tolyl
Tolan	diphenylacetylene
tolbinap	2,2-bis(di-*p*-tolylphosphino)-1,1′-binaphthyl
TOMAC	trioctylmethylammonium chloride
TOP	tri-*n*-octylphosphine
TOPA	trihydroxyphenylalanine
TOPO	tri-*n*-octylphosphine oxide
TOPS	*N*-ethyl-*N*-sulfopropyl-*m*-toluidine
Tos	*p*-toluenesulfonyl; tosyl
Tosic Acid	*p*-toluenesulfonic acid
TosMIC	*p*-toluenesulfonylemethyl isocyanide; tosylmethyl isocyanide
Tosyl	*p*-toluenesulfonyl
TOT	tri-*o*-thymotide
TOTU	*O*-[(ethoxycarbonyl)cyanomethyleneamino]-*N*,*N*,*N*′,*N*′,tetramethyluronium tetrafluoroborate

TOX	tetradichloroxylene
TP	thymolphthalein
T-5′-P	thymidine 5′-monophosphate
2,4,5-TP	2-(2,4,5-trichlorophenoxy)propanoic acid
TPA	4β,9α,12β,13α,20-pentahydroxytigilia-1,6-dien-3-one 12β-myristate, 13-acetate; phorbol, 12-myristate, 13-acetate
TPA	tetraphthalic acid
TPA	tissue plasminogen activator
TPA	tris(2-pyridylmethyl)amine
TPA	12-*O*-tetradecanoylphorbol-13-acetate
TPAP	tetra-*n*-propylammonium perruthenate; tetrapropylammonium perruthenate
TPB	1,1,4,4-tetraphenyl-1,3-butadiene
TPC	*N*-(trifluoroacetyl)prolyl chloride
TPC	thymolphthalein complexone
TPCD	tetraphenylcyolopentadienone
TPCK	L-1-*p*-tosylamino-2-phenylethyl chloromethyl ketone; *N*-tosyl-L-phenylalanine chloromethyl ketone
TPCP	triphenylcyclopenium ion
TPDMDS	1,3-dimethyl-1,1,3,3-tetraphenyldisilazane
TPE	1,1,1-triphenylethane
TPE	tetraphenylethylene
TPE	thermoplastic elastomers
TPEN (tpen)	*N,N,N′,N′*-tetrakis(2-pyridylmethyl)ethylenediamine
TP2H	1,3,3-triphenylpropene
TPM	triphenylmethane
tpma	tris(2-pyridylmethyl)amine
TPN	β-nicotinamide adenine dinucleotide phosphoric acid
TPNH	nicotinamide adenine dinucleotide phosphate—reduced form
TPO	thermoplastic polyolefins
TPP	1,1,3-triphenylpropene
TPP	5,10,15,20-tetraphenyl-21*H*,23*H*-porphine; *meso*-tetraphenylporphine; tetraphenylporphyrin
TPP	cocarboxylase tetrahydrate
TPP	tetraphenylporphyrinato
TPP	thiamine pyrophosphate
TPP	triphenyl phosphate
TPP	triphenylphosphine
TPPS	tetraphenylporphine tetrasulfonic acid
TPS	2,4,6-triisopropylbenzenesulfonyl
TPS	2,4,6-triisopropylbenzenesulfonyl chloride
TPS	triphenylsilyl

TPS	triphenylsulfonium chloride
TPSA	triphenylsilylamine
TPSCl	triphenylchlorosilane
TPSH	2,4,6-triisopropylbenzenesulfono hydrazide
TPSI	1-(2,4,6-triisopropylbenzenesulfonyl)imidazole
TPSNI	1-(2,4,6-triisopropylbenzenesulfonyl)-3-nitro-1*H*-1,2,4-triazole
TPST	1-(2,4,6-triisopropylbenzenesulfonyl)-1*H*-1,2,4-triazole
TPTA	triphenyltin acetate
TPTH	triphenyltin hydroxide
TpTM	tri-*p*-tolylmethane
tptn	tetrakis(2-pyridylmethyl)-1,3-propylenediamine
TPTOH	triphenyltin hydroxide
TPTU	*O*-(1,2-dihydro-2-oxo-1-pyridyl)-*N*,*N*,*N'*,*N'*-tetramethyluronium tetrafluoroborate
TPTZ	2,4,6-tris(2'-pyridyl)-*s*-triazine
TPTZ	triphenyltetrazolium chloride
TPU	thermoplastic polyurethanes
Tr	triphenylmethyl; trityl
TRAM	1,3,5-tris(aminomethyl)benzene
tren	2,2',2"-triaminotriethylamine; tris(2-aminoethyl)amine
Tresyl Chloride	2,2,2-trifluoroethanesulfonyl chloride
TRH	thyrotropin releasing hormone; thyroxin releasing hormone
TRI	4-methyl-2,6,7-trithiabicyclo[2.2.2]octane
TRIAMO	triaminosilane
TRICINE (Tricine) (tricine)	*N*-[tris(hydroxymethyl)methyl]glycine; 2-*N*-glycine-2-hydroxymethylpropane-1,3-diol
Triflate (Tf)	trifluoromethanesulfonyl
TRIPHOS	tris(diphenylphosphinomethyl)ethane
TRIS (Tris)	tris(hydroxymethyl)aminomethane; 2-amino-2-hydroxymethylpropane-1,3-diol
TRITC	tetramethylrhodamine B isothiocyanate; tetramethylrhodamine-5-(and-6)-isothiocyanate
TritonB	benzyltrimethylammonium hydroxide
Tritylone	9-(9-phenyl-10-oxo)anthryl
tRNA	transfer ribonucleic acid
TROC (Troc)	2,2,2-trichloroethoxycarbonyl
Troc	trichlorethoxycarbonyl
Trolox®	6-hydroxy-2,5,7,8-tetramethylchroman-2-carboxylic acid
Trp	tryptophan
Trp	tryptophyl
TRPGDA	tripropylene glycol diacrylate
Trt	triphenylmethyl; trityl
TS	*trans*-stilbene

Ts	tosyl; *p*-toluenesulfonyl
TSA (p-TSA)	*p*-toluenesulfonic acid
TSA-F	tris(dimethylamino)sulfur (trimethylsilyl)difluoride
TSAG	antimony sodium gluconate
Tse	2-(*p*-toluenesulfonyl)ethyl
TSeF	tetraselenafulvenes
TSH	thyroid-stimulating hormone; thyrotrop(h)in; thyrotropic hormone
TSIM	*N*-trimethylsilylimidazole
TSNI	1-(*p*-toluenesulfonyl)-4-nitroimidazole
TsOH	tosic acid; *p*-toluenesulfonic acid
T-Solvent	2,3(tetrahydrofurfuryloxy)tetrahydropyran
TSP	3-(trimethylsilyl)propionate
TSP	3-(trimethylsilyl)propionic acid
TSP	tribasic sodium phosphate; trisodium phosphate
TSPP	tetrasodium pyrophosphate; sodium pyrophosphate
2-TST	2-(trimethylsilyl)thiazole
TSTU	*O*-(*N*-succinimidyl)-*N*,*N*,*N'*,*N'*-tetramethyluronium tetrafluoroborate
TT2	2,3,5-triphenyltetrazolium
TTA	2-thenoyltrifluoroacetone
TTA	trimethylsilyl trichloroacetate
TTC	2,3,5-triphenyl-2*H*-tetrazolium chloride
TTD	bis(diethylthiocarbamoyl) disulfide
TTEGDA	tetraethyleneglycol diacrylate
TTEPP	tetrakis(1,3,5-triethylphenyl)porphyrinato
TTF	tetrathiafulvalene
TTFA	thallium(III) trifluoroacetate
TTFA	trityl trifluoroacetate
TTFD	thiamine tetrahydrofurfuryl disulfide
TTH	thyrotrop(h)in
TTHA	triethylenetetramine-*N*,*N*,*N'*,*N''*,*N'''*,*N'''*-hexaacetic acid
TTiPP	tetrakis(1,3,5-triisopropylphenyl)porphyrinato
TTMAPP	5,10,15,20-tetrakis(4-trimethylammoniophenyl)porphyrin tetra(toluene-4-sulfonate)
TTMSS	tris(trimethylsilyl)silane
TTN	thallium(III) nitrate
TTOC	3-hydroxy-4-methylthiazole-2(3*H*)-thione
TTP	ribosylthymidine 5'-triphosphate; thymidine triphosphate
TTP	tetra-*p*-tolylporphyrinato
t-tpchxn	tetrakis(2-pyridylmethyl)-*trans*-cyclohexane-1,2-diamine
TTQ	tryptophylquinone
TTS	tris(4-tolyl)sulfonium
TTTMSS	tris(trimethylsilyl)silane

TTX	tetrodotoxin
TTZ	2,3,5-triphenyltetrazolium chloride
tu	thiourea
TWEEN	polyoxyethylenesorbitan
TX	thromboxane (as in TXA_2, etc.)
Tyr	tyrosine
Tyr	tyrosyl
Tyr-OH	tyrosine
Tyr-OMe	tyrosine methyl ester
tz	tetrazine
U	uracil
U	uridine
U-5′-P	uridine-5′-monophosphate
UC	uncompetitive
UDCS	ursodeoxycholic acid
UDMH	1,1-dimethylhydrazine; *unsym*-dimethylhydrazine
UDP (5′-UDP)	uridine 5′-diphosphate; uridine diphosphate
UDP-Gal	uridine 5′-diphosphogalactose
UDP-GalNAc	uridine 5′-diphospho-*N*-acetylgalactosamine
UDP-Glc	uridine 5′-diphosphoglucose
UDP-GlcDH	uridine 5′-diphosphoglucose dehydrogenase
UDP-GlcNAc	uridine 5′-diphospho-*N*-acetylglucosamine
UDP-GlcUA	uridine 5′-diphosphoglucuronic acid
UDP-Xyl	uridine 5′-diphosphoxylose
UDPAG	uridine-5′-diphospho-*N*-acetylglucosamine
UDPG	uridine-5′-diphosphoglucose
UDPGA	uridine-5′-diphosphoglucose ammonium salt
UDPGA	uridine-5′-diphosphoglucuronic acid
UDPGDH	uridine-5′-diphosphoglucose dehydrogenase
UDPGE	uridine diphosphategalactose 4′-epimerase
UHMWPE	ultrahigh molecular weight polyethylene
UMP	5′-uridylic acid
UMP (5′-UMP)	uridine 5′-monophosphate; uridine monophosphate; uridine 5′-phosphate
UNCA	urethane-protected amino acid *N*-carboxyanhydride
UQ	coenzyme Q
ur	urea
Ura	uracil
Urd	uridine
UTP	uridine 5′-triphosphate; uridine triphosphate
V	valine
Val	valine
Val	valyl
ValPHOS	1-diphenylphosphino-2-dimethylamino-3-methylbutane

VCM	vinyl chloride monomer
VFA	volatile fatty acid
VIP	vasoactive intestinal peptide; vasoactive intestinal polypeptide
VLB	vinblastine sulfate
VLDL	very low density lipoprotein
VMA	DL-4-hydroxy-3-methoxymandelic acid
VMA	vanillylmandelic acid
Voc	vinyloxycarbonyl
VRC	vanadyl ribonucleoside complexes
VSG	variable surface glycoprotein
VTC	trichlorovinylsilane; vinyltrichlorosilane
VTEO	vinyltriethoxysilane
VTMO	vinylmethoxysilane
VTMOEO	vinyltris(2-methoxyethoxy)silane
VUL	vinylogous urethane lactone
VX	*O*-ethyl *S*-[2-(diisopropylamino)ethyl] methylphosphonothioate)
W	tryptophan
W-5	*N*-(6-aminohexyl)-1-naphthalenesulfonamide
W-7	*N*-(6-aminohexyl)-5-chloro-1-naphthalenesulfonamide, hydrochloride
W-12	*N*-(4-aminobutyl)-2-naphthalenesulfonamide
W-13	*N*-(4-aminobutyl)-5-chloro-2-naphthalenesulfonamide
WSC	water-soluble carbodiimide
X	xanthosine
XAN	xanthene
Xao	xanthosine
XDP	xanthose 5′-diphosphate
X-Gal (X-gal)	5-bromo-4-chloro-3-indolyl-β-D-galactopyranoside
XMP	xanthosine 5′-monophosphate; xanthosine 5′-phosphate
XOD	xanthine oxidase
X5P	xylulose 5-phosphate
XPD	xanthosine 5′-diphosphate
XTP	xanthosine 5′-triphosphate
Xy	xylene
Xyl	xylose
Y	tyrosine
YAC	yeast artificial chromosomes
YADH	yeast alcohol dehydrogenase
Z	benzyloxycarbonyl
ZDBC	zinc dibutyldithiocarbamate
ZDEC	zinc diethyldithiocarbamate
ZDMC	zinc dimethyldithiocarbamate

ZMA	zinc metaarsenite
Z_2O	dibenzyl dicarbonate
Z-ONSu	*N*-(benzyloxycarbonyloxy)succinimide
ZPCK	*N*-carbobenzyloxy-L-phenylalanyl chloromethyl ketone
ZPD	5-aminoimidazole-4-carboxamide-1-β-D-ribofuranosyl-5′-diphosphate
ZPO	zinc peroxide
ZPS	3-(benzothiazol-2-ylthiol)-1-propanesulfonic acid
ZTP	5-aminoimidazole-4-carboxamide-1-β-D-ribofuranosyl-5′-triphosphate

Chapter 2

Physical Acronyms

This chapter contains a list of physical acronyms, abbreviations, and symbols used in many chemical fields. Although the list is fairly comprehensive, no attempt has been made to list all techniques, abbreviations, and symbols; only those likely to be found in the modern organic chemical literature have been included. A few words are needed to explain the conventions used in the list.

Any numbers, parentheses, and hyphens have not been considered for the purposes of alphabetizing when part of the name. The alphabetization of Greek symbols has caused some problems. The convention used here follows the ASCII code from the symbol font used by Apple® and they follow the corresponding Roman letter.[†] Periods have been omitted for abbreviations. Capitalized entries will always precede lower case entries. Letters with subscripts have been alphabetized as single letters. When there is more than one way of representing the same technique, and the acronyms only vary by letter case, alternatives are given in parentheses as in, for example, NMR (nmr).

Note, that in many cases, one acronym can refer to a number of techniques or physical constants. Also, many techniques may be combined such as in COSY-COSY, NOESY-COSY etc.; these may be elucidated via the parent acronyms. The techniques and physical constants have been listed in alphabetical order; it is for the reader to decide which one is correct in that specific context. In some cases, an acronym may have been used incorrectly in the literature. Many acronyms have more than definition consequently, care should be taken to ensure that interpretation of the acronym is correct.

† Macintosh computers were used to prepare the entries, hence this convention was followed.

A	absorption; absorbance
A	amperes
A (*A*)	atomic weight; mass number; nucleon number
A	optical density
a	absorptivity
a	antisymmetric
α	angle of optical rotation
α	degree of dissociation
α	alpha radiation
α	relative retention
α	specific rotation
α_0	Bohr radius
AA	atomic absorption
AAS	atomic absorption spectrophotometry
A/B	amphoteric
AC (ac)	alternating current
ACE	affinity capillary electrophoresis
ACS	American Chemical Society grade reagent
A/D	analog to digital
ADC	analog to digital converter
ADC	automatic developing chamber
ADRF	adiabatic demagnetization in the rotating frame
AE	attachment energy
ae	acid equivalent
AED	atomic emission detection
AEM	automated electron microscopy
AES	atomic emission spectroscopy
AES	Auger electron spectroscopy
AFC	advanced flow control
AFM	atomic force microscope
AFS	atomic fluorescence spectroscopy
AIPA	automated individual particle analysis
AISEFT	abundant isotope signal elimination
AJCP	adiabatic *J*-cross polarization
AL	acceptable level
ALARA	as low as reasonably achievable
ALC	application limiting constituent
ALD	approximate lethal dose
ALR	action leakage rate
alt	alternating (as in poly(A-*alt*-B))
ALU	arithmetic-logic unit
amu	atomic mass unit
AO (ao)	atomic orbital
AOD	argon-oxygen decarbonization

AOES	actinometric optical emission spectroscopy
AOTS	acousto-optically tuned scanning
APCI	atmospheric pressure chemical ionization
API	atmospheric pressure ionization
APS	appearance potential spectroscopy
aps	average particle size
APT	attached proton test
AQ	aquisition time
aq	aqueous
AQTX	aquatic toxicity
AR	analytical reagent
A_r	relative atomic mass; atomic weight
arb unit	arbitrary unit
ARC	accelerating rate calorimetry
ARPES	angle-resolved photoelectron spectroscopy
ARRF	adiabatic remagnetization
AS	absorption spectroscopy
as (*as*)	asymmetric
ASA	atomic sphere approximation
ASCII	American standard code for information exchange
ASE	aromatic stabilization energies
ASIS	aromatic solvent induced shift
ASV	anodic stripping voltammetry
ASVR	automated small volume recirculator
asym	asymmetric
AT	acquisition time
AT	argentometric titration
at	atomic
at wt	atomic weight
ATC	automatic temperature compensation
atm	atmosphere
atom %	atom percent
ATR	attenuated total reflectance
ATR-FTIR	attenuated total reflectance-Fourier transform infrared spectroscopy
ATR-IR	attenuated total reflectance-infrared spectroscopy
AU	absorbance units
au	atomic units
AUFS	absorbance units full scale
AVSS	automated voltametric sampling system
AWLN	Advanced Wiswesser Line Notation—computerized notation for chemical structures
b	block
b	broad

b	molality
B_{eff}	effective magnetic field
B_0	static magnetic field
B_1	transverse magnetic field
B_2	additional transverse magnetic field
β^-	beta radiation
β^+	positron
bar	unit of pressure
BB	broad band (decoupling)
BCD	binary coded decimal
BD	Brownian dynamics
BDE	bond dissociation energy
BEBO	bond energy-bond order
BEI	backscattered electron image
BET	back-electron-transfer
BET	Brunauer-Emmett-Teller (adsorption isotherm)
BFS	beam-foil spectroscopy
BIRD	bilinear rotational decoupling operator
BL	bioluminescence
BM	Böhr magneton
BMR	base metabolic rate
BN	bond number
BNCT	boron neutron capture theory
BO	Born-Oppenheimer
BOA	bond orbital approximation
BOD	biological oxygen demand; biochemical oxygen demand
BP (bp)	boiling point
bp	base pair
Bq	Becquerel
BR	biradical
br	broad
BRN	Beilstein registry number
BSB	background-subtraction process
BSO	benzene soluble organics
BSSE	basis set superposition error
Btu (BTU)	British thermal unit
BUN	blood urea nitrogen
C	Celsius; centigrade
C	combustible
C	competitive
C	coulomb
C	crystal phase of a liquid
C (c)	specific heat capacity
c	centered

c	concentration
c	speed of light
C_m	molar heat capacity
C_v	heat capacity at constant volume
CA	collisional activation
CAD	collision activated decomposition; collisionally activated dissociation
CAD	computer-assisted design
Cal	kilocalorie
cal	calorie
CAM	carbon absorption method
CAM	computer-assisted make up
CAMD	computer-assisted molecular design
CAMELSPIN	cross-relaxation appropriate for mini-molecules emulated by locked spins
CAMM	computer aided molecular modeling
CAMP	continuous air monitoring program
CAR	carcinogenic; carcinogenic effects
CARC	carcinogenic
CARS	coherent anti-Stokes Raman spectroscopy
CAT	catalytic transfer hydrogenation
CAT	computed axial tomography
CAT	computer averaging of transients
cat	catalyst
CB (cB) (cb)	conjugate base
CC	column chromatography
CC	coupled-cluster
CCC	counter-current chromatography
CCI	carbocationoid intermediate
CD	circular dichroism
cd	candela
cd	current density
CDA	chiral derivatizing agents
CE	capillary electrophoresis
CE	Cotton effects
CE-FAST	contrast enhanced Fourier aquired steady state
CEC	cation-exchange capacity
CEIE	conformational equilibrium isotope experiments
CEM	continuous emissions monitoring
CEPA	coupled electron pair approximation
CER	carbon dioxide exchange rate
CFA	continuous flow analysis
CFE	continuous flow enthalpimetry
CFFAB	continuous flow fast atom bombardment

CFSE (cfse)	crystal field stabilization energy
CFT	critical floculation temperature
CGC	capillary gas chromatography
CGC	complexation gas chromatography
CGC-MS	capillary gas chromatography-mass spectroscopy
CGE	capillary gel electrophoresis
CGM	conjugated gradient minimization
CH (CHN)	elemental analysis
ch	cholesteric phase of a liquid crystal
CHEF	chelation-enhanced fluorescence
CHEMFET	chemically modified field effect transistor
CHEQ	chelation-enhanced quenching
CHF	coupled Hartree-Fock
CHORTLE	carbon-hydrogen correlation from one dimensional polarization-transfer spectra by least squares analysis
CHPLC	chiral high performance liquid chromatography
CI	chemical ionization
CI	color index
CI	configuration interaction
Ci	Curie
CID	collision induced dissociation; collision induced decomposition
CIDEP	chemically induced dynamic electron polarization
CIDMS	collisionally induced dissociation mass spectrometry
CIDNP	chemically induced dynamic nuclear polarization
CIDNP-COSY	photochemical excitation-correlation spectroscopy
CIDNP-NOESY	photochemical excitation-nuclear Overhauser spectroscopy
CIEEL	chemically initiated electron exchange chemiluminescence
CIEF	capillary isoelectric focusing
CI-MIKES	chemical ionization-metastable ion kinetic energy spectrometry
CIMS	chemical ionization mass spectrometry
CINCH-C	calibrate indirect carbon hydrogen correlation spectra-correlation by long range coupling
CIP	contact ion pair
CIR	cylindrical internal reflectance
CIRP	contact ion radical pairs
CISD	single and double excitations
CISD	single- and double-substituted configuration interaction method; configuration interaction wave functions
CL	cathodoluminescence
CL	chemiluminescence
CLA	complete line shape analysis
CLC	cross-linked enzyme crystals

CLEC	cross-linked enzyme crystals
CLS	characteristic loss spectroscopy
cls	complete line shape
CLSM	confocal laser scanning microscope
CLSR	chiral lanthanide shift reagent
CM	corrosive material
CMA	cylindrical mirror analysis
CMC (cmc)	critical micelle concentration
CMD	count median diameter
CMO	canonical molecular orbital
CMR (cmr)	^{13}C nuclear magnetic resonance
CMS	cytoplasmic male sterility
CN	coordination number
CNDO	complete neglect of differential overlap
CNS	central nervous system
co	copoly
COCONOSY	combined correlated and nuclear Overhauser enhancement spectroscopy
COD	chemical oxygen demand
coef (coeff)	coefficient
COLOC	correlation by long range couplings
COLOC-S	correlation by long range coupling-sensitive
compd	compound
CONJ	conjugate
const	constant
COR	corrosive
COSY	correlation spectroscopy
COSYLR	long range correlation spectroscopy
COSYRCT	relayed coherence transfer correlation spectroscopy
CP (cp)	chemically pure
CP	cross polarization
Cp (C_p) (c_p)	heat capacity; specific heat capacity
cp	candlepower
cP	centipoise
CPC	chemical-protective clothing
cpd	compound
cpd	contact potential difference
CPE	controlled-potential electrolysis
CPF	carcinogenic potency factor
CPG	chromatopyrography
CPHF	coupled perturbed Hartree-Fock
CPK	Corey-Pauling-Koltun (models)
CPL	circular polarization of luminescence
cpm	counts per minute

CPMAS (CP-MAS) (CP/MAS)	cross-polarization-magic angle spinning
CPS	controlled power system
cps	counts per second
cps	cycles per second
CPU	central processing unit
CR	charge resonance
CRA	chiral resolving agent
CRAMPS	combined rotation and multiple pulse sectroscopy
CRSE	conventional ring strain energy
CRTA	controlled rate thermal analysis
CRU	constitutional repeating unit
CSA	chemical shift agent
CSA	chemical shift anisotropy; chemical shield anisotropy
CSCM	carbon-proton shift correlation spectroscopy; correlation spectroscopy for carbon multiplicities; chemical shift correlation map
CSCMLR	long-range correlation spectroscopy for carbon multiplicities
CSD	Cambridge Structural Database
CSI	chemical shift imaging
CSIN	Chemical Substances Information Network
CSP	chiral stationary phase
CSRS	coherent Stokes Raman spectrometry
CT	charge-transfer
CTC	charge-transfer complex
CTE	coefficient of thermal expansion
CTEM	conventional transmission electron microscope
CUPID	continuous probability distribution of rotamers
CUT	content uniformity testing
CV (C_v)	calibrated volume
CV	cyclic voltammetry
CVD	chemical vapor deposition
CVE	cluster valence electrons
CW	constant width
CW	continuous wave
CYCLOPS	cyclically ordered phase sequence
CZE	capillary zone electrophoresis
D	bond dissociation energy
D	configurational
D	Debye
D	deuterium
D	diffusion coefficient
D	electric displacement
D	symmetry group

1-D (1D)	one dimensional
2-D (2D)	two dimensional
3-D (3D)	three dimensional
d	density; relative density
d	deuteron
d_n	deuterium (where n = number of atoms per molecule)
d	diameter
d	diffuse
d	distance
d	doublet
d	spacing
d	thickness
Δ	change
δ	scale delta position
Dq	crystal field splittings
D_T	thermal diffusion coefficient
Dt	crystal field splittings
D/A	digital to analog
D/A	donor/acceptor
DAC	digital to analog converter
daf	dry ash free
DANTE	delays alternating with nutations for tailored excitation
DAS	dynamic-angle spinning
DBD	dielectric barrier discharge
DBE	double bond equivalent
DBSP	double-bond stabilizing perameters
DC (d-c) (dc)	direct-current
DCCA	drying control chemical agent
DCCC	droplet counter current chromatography
DCI	desorption chemical ionization
DCI	direct chemical ionization
DCP	direct current polarography
DD	dipole-dipole
DDE	dynamic data exchange
DDTA	derivative differential thermal analysis
DEF	deformation energy
DEFT	driven-equilibrium Fourier transform spectroscopy
deliq	deliquescent
denat	denaturated; denatured
DEPT	distortionless enhancement polarization transfer
DEPT-GL	distortionless enhancement polarization transfer-grand lux
DETA	dielectric thermal analysis
df	degrees of freedom
DFI	direct fluid interface

DGEOM	distance geometry
DH°	bond dissociation energies
DHPLC	dynamic high performance liquid chromatography
DIDA	direct isotopic dilution analysis
DIE	direct injection enthalpimetry
DIFNOE	difference nuclear Overhauser spectroscopy
dil	dilute
dim	dimensions
DISCO	differences and sums in correlation spectroscopy
DISP	disproportionation equilibrium
disp	disposal procedures
2DJ	two dimensional *J*-resolved experiment
DKAM	double known-addition method
DLE	delayed light emission
DLS	dynamic light scattering
DM	diatomic molecules
DMA	direct memory access
DMA	dynamic mechanical analysis
DME (dme)	dropping mercury electrode
DMM	digital multimeter
2D-NMR	two dimensional nuclear magnetic resonance
3D-NMR	three dimensional nuclear magnetic resonance
DMR (dmr)	^{2}H nuclear magnetic resonance
DMS	dynamic mechanical spectrometer
DMTA	dynamic mechanical thermal analysis
DNMR	deuterium nuclear magnetic resonance
DNMR (dnmr)	dynamic nuclear magnetic resonance
DNOE	dynamic nuclear Overhauser effect
DOAS	differential optical absorption spectroscopy
DOC	dissolved organic carbon
DOPE	double phase encoding
DOPS	direct optical position sensors
DOUBTFUL	double quantum transition to find unresolved lines
DP (dp)	degree of polymerization
DP	differential pulse
DPE	deprotonation energies
dpm	disintegrations per minute
dps	disintegrations per second
DQ	double quantum
DQCRCT	proton double quantum relayed conference transfer
DQF	double quantum filtered
DQF-COSY (DQFCOSY)	double quantum filtered *J*-correlated spectroscopy
DR	diffuse reflectance

DRE	Dewar resonance energies
DRESS	depth resolved surface-coil spectroscopy
DRF	detector response factor
DRIFT	diffuse reflectance Fourier transform infrared
DRIFTS	diffuse reflectance infrared Fourier transform spectroscopy
DRS	dielectric relaxation spectroscopy
DSAC	dry state adsorption conditions
DSB	desolvation barrier
DSC	differential scanning calorimetry
DSIMS	dynamic secondary ion mass spectrometry
DSPT	double selective population transfer
DTA	differential thermal analysis
DTC	depolarization thermocurrent
DTC	differential thermal calorimetry
DVM	digital voltmeter
DW	distilled water
DW (*DW*)	dwell time
DXRD	dynamic x-ray diffraction
dyn	dyne
DZ	double zeta
DZP	double zeta polarization
E	electric field strength
E	electrical potential
E(s)	electrode potential
E	energy
E	energy of activation
E	envelope
E	specific extinction coefficient
e	electric
e (e^-)	electron
e	elementary charge
ε	emittance
ε	epsilon position
ε	molar absorptivity
ε	molar extinction coefficient; molar decadic absorption coefficient
ε	permittivity
E_F	Fermi energy
E_i	ionization energy
E_k	kinetic energy
E_{MF} (emf)	electromotive force
ε_o	permittivity
E_p	potential energy
ε_r	relative permittivity

e_s^-	solvated electron
EAG	electroantennogram
EAN	effective atomic number
EB	electron bombardment
EC	electrochemical detection
EC	electron capture
ec	emulsifiable concentrate
EC_{50}	effective concentration to cause a response in 50% of the individuals
EC_{100}	effective concentration to cause a response in 100% of the individuals
ECD	electron capture detection
ECD	electronic circular dichroism
ECL (ecl)	electrochemiluminescence
E COSY(E-COSY)	exclusive correlation spectroscopy
ECP	electronic conducting polymers
ECP	effective core potentials
ED	effective dose
ED	electron diffraction
ED_{50} (ED50)	effective dose in 50% of test subjects
ED_{100} (ED100)	effective dose in 100% of test subjects
EDA	electron donor-acceptor
EDA	energy decomposition analysis
EDS	electron donating substituent
EDS	energy-dispersive spectroscopy
EDX	energy-dispersive x-ray detector; energy-dispersive x-ray spectroscopy
ee	enantiomeric excess
EEM	emission-excitation matrix
EFF	empirical force field
eff	efficiency
EFFF	electrical field-flow fractionation
EFG	electric field gradients
EFISH	electric-field-induced second-harmonic generation
EFP	electric field pulse
EGA	evolved gas analysis
EGD	evolved gas detection
EGR	evolved gas recirculation
EH	extended Hückel
EI	electron impact; electron-bombardment ionization
EIA	enzyme immunoassay
EI-HRMS	high resolution-electron impact mass spectrometry
EI-LRMS	low resolution-electron impact mass spectrometry
EIMS	electron impact mass spectrometry

EIS	electrical impedance spectroscopy
EL	electroluminescence
ELCD	electrolytic conductivity detector
ELD	energy level diagram
ELDOR	electron double resonance
ELISA	enzyme-linked immunosorbent assay
ELS	electrophoretic light scattering
ELSA	enzyme-linked immunosorbent assay
ELSD	evaporative light-scattering detector
EM	electromagnetic (radiation)
EM	electron microscopy
EMI	electromagnetic interference
EMIRS	electrochemically modulated infrared spectroscopy
EMIT	enzyme-multiplied immunoassay technique
EMR	electromagnetic radiation
EMS	electron momentum spectroscopy
emu	electromagnetic unit
ENDOR	electron-nuclear double resonance
ENDOR	external nuclear double resonance
enzym	enzymatic assay
EOF	electroosmotic flow
EPA	effective proton affinity
EPC	electronic pressure control
EPC	enantiomerically pure compound
EPP	electronic pressure programming
EPR (epr)	electron paramagnetic resonance
EPT	exclusive polarization transfer
EPXMA	electron probe x-ray microanalysis
eQ	quadruple moment
equil	equilibrium
equiv	equivalent
ER	electroreflectance
er	enantiomer ratio
erf	error function
erfc	error function complement
ES	electrospray
ESA	electron spin alignment
ESBO	edge-sharing bioctahedral
ESCA	electron spectroscopy for chemical analysis (also known as x-ray photoelectron spectroscopy -XPS)
ESD	electron-stimulated desorption
ESDIAD	electron-stimulated desorption ion angular distribution
ESE	electron spin echo
ESE-ESR	electron spin echo detected electron spin resonance

ESEEM	electron spin echo envelope modulation
ESES	electron spin echo spectroscopy
ESI	electrospray interface
ESI	electrospray ionization
ESI	element specific image
ESMS	electrospray mass spectrometry
ESP	elimination of solvation procedure
ESP	exchangeable sodium percentage
ESR (esr)	electron spin resonance
esu	electrostatic unit
ET	electron transfer
ETAAS	electrothermal atomization atomic absorption spectrometry
ETE	electron-transfer equilibrium
ETM	emission-time matrix
ETV	electrothermal vaporization
EU (eu)	enzyme unit
eV	electron volt
EWS	electron withdrawing substituent
EXAFS	extended x-ray absorption fine structure
exp	exponential
EXSY	2-D exchange spectroscopy
F (f)	farad
F	Fermi
F	force
F	formal
F	free energy
F	Helmholtz function
F	hyperfine quantum number
F	rotational term
f	activity coefficient (mole fraction basis)
f	fine
f	focal length
f	frequency
f	function (e.g. f(x))
Φ	magnetic flux
Φ	potential energy
ϕ	electric potential
ϕ	heat flow rate
ϕ	osmotic coefficient
ϕ	photochemical yield; quantum yield
ϕ	volume fraction
F_m	magnetomotive force
Φ_o	magnetic flux quantum
f_1	first frequency domain in 2D nuclear magnetic resonance

f_2	second frequency domain in 2D nuclear magnetic resonance
FA	flame absorption
FAAS	flame atomic absorption spectroscopy
FAAS	flameless atomic absorption spectroscopy
FAB	fast atom bombardment
FABMS (FAB MS)	fast atomic bombardment mass spectrometry
FABRMS	fast atom bombardment rapid multistream sampler
FACM	friable asbestos-containing material
FACT	fully automated calibration technology
FAES	flame atomic emission spectroscopy
FAFS	flame atomic fluorescence spectroscopy
FANES	furnace atomization nonthermal excitation spectroscopy
FAPES	furnace atomization plasma excitation spectroscopy
FARCE	freezing at reactive centers of enzymes
far-IR	far infrared
FAS	flame absorption spectroscopy
Fc	fragment crystallizable
FCC	fluid catalytic cracking
fcc(fccub)	face centered cubic
FD	field desorption
Fd	fragment domain
FDCD	fluorescence detected circular dichroism
FEM	flame emission spectroscopy
FES	flame emissive spectrometry
FET	field-effect transistor
FETM	fluorescence-excitation time matrix
FFEM	freeze-fracture electron microscopy
FFF	field-flow fractionation
FFM	friction force microscope
FFS	flame fluorescence spectroscopy
FFT	fast Fourier transform
FHT	Fisher-Hirschfelder-Taylor space filling models
FI	field ionization
FI	flow injection
FIA	flow injection analysis
FIA-EC	flow injection analysis-electrochemical detection
FID	flame ionization detection
FID (fid)	free induction decay
FIK	field ionization kinetics
FIM	friable insulation material
FIMS	field ionization mass spectrometry
FIR	far infrared
FIR	fast inversion recovery

FIR-VRT	far-infrared vibration-rotation-tunneling
FISP	fast imaging with steady precession
FL	flammable liquid
FL	function test by luminescence spectroscopy
fl	fluid
FLASH	fast low angle shot
FLC	ferroelectric liquid crystal
Flpt	flash point
FM	frequency modulation
FMIR	frustrated multiple internal reflectance
FMO	frontier molecular orbital
fnp	fusion point
FOCSY	foldover-corrected spectroscopy
FOMA	fast optical multichannel analyzer
FOPPA	first-order polarization propagator approach
FOS	fluorine opsin shift
FP (Fp)	flash point
fp	freezing point
FPC	fixed partial charge
FPD	flame photometric detection
FPRF	flash-photolysis resonance fluorescence
FPT	finite perturbation theory
Fr	Franklin
Fr	Froude number
FRA	full range analyzer
FRP	free radical pair
FS	flammable solid
FSBO	face-sharing bioctahedral
FSE	fast spin echo
FSE	formal steric enthalpy
FSGO	floating spherical Gaussian orbital
FT	Fourier transform; Fourier transformation
FTICR (FT-ICR)	Fourier transform ion cyclotron resonance; Fourier transform ion cyclotron resonance mass spectrometry
FTIR (FT-IR) (FT/IR) (FT IR)	Fourier transform infrared
FTIR	frustrated total internal reflectance
FTIRRAS	Fourier transform infrared reflection absorption spectroscopy
FTMS (FT-MS) (FT/MS)	Fourier transform mass spectrometry
FTNMR (FT-NMR)	Fourier transform nuclear magnetic resonance
FT-RAIRS	Fourier transform reflection-absorption infrared spectroscopy

FTS	Fourier transform spectroscopy
FUCOUP	long range heteronuclear correlation using the fully coupled technique
FVP	flash vacuum pyrolysis
FVT	flash vacuum thermolysis
FW (fw)	formula weight
fwhm	full width at half maximum
G	conductance
G	Gauss
G	Gibbs free energy
G (g)	gravitational constant
G	vibrational term
G	weight
g	g-factor
Γ	ionic strength
Γ	surface concentration
γ	activity coefficient (molality basis)
γ	gamma radiation
γ	magnetogyric ratio
γ	surface tension
γ	wavelength
g_c	proportionality constant
g_o	standard gravity
Ga	Galileo number
GARP	globally optimized alternating phase rectangular pulse
GASP	gated spin tickling
GASPE	gated spin echo
g-atom	gram atom
GC	gas chromatography
GC-CI	chemical ionization-gas chromatography
GCCIMS	chemical ionization-gas chromatography-mass spectrometry
GC-EI	electron impact-gas chromatography
GCEIMS	electron impact-gas chromatography-mass spectrometry
GC-FID (GC/FID)	gas chromatography-flame ionization detection
GC-FTIR (GC/FTIR)	gas chromatography-Fourier transform infrared
GCGC	glass capillary-gas chromatography
GC-IR	gas chromatography-infrared
GC-MS	gas chromatography-mass spectrometry
GDC	gas displacement chromatography
GE	gel electrophoresis
GED	gas-phase electron deffraction
gem	geminal

GFAAS (GF-AAS)	graphite furnace-atomic absorption spectrometry
GFC	gas frontal chromatography
GFCCP-AES	graphite furnace capacitively coupled plasma-atomic emission spectrometry
gfw	gram formula weight
Gi	Gilbert
GIAO	gauge-invariant atomic orbital
GIR	grazing incidence reflectance
GLC	gas-liquid chromatography
GPC	gel permeation chromatography
GPC-LALLS	gel permeation chromatography-low angle laser light scattering
GROPE	generalized compensation for resonance offset
GRP	geminate radical pair
GS	ground state
GSC	gas-solid chromatography
GSD	geometric standard deviation
GTO	Gaussian-type orbital
GTP	group transfer polymerization
GVB	generalized valence bond
H	Boltzmann function
H	enthalpy
H	Hamiltonian function
H	height equivalent of a theoretical plate
H	Henry
$\mathbf{H}$	magnetic field strength
h	coefficient of heat transfer
h	Planck constant
η	viscosity
H_m	change in heat of melting
H_t	heat of transition
H_v	heat of vaporization
HAT	hydrogen atom transfer
HB	hydrogen bonding
HCCOSY	proton-carbon heteronuclear shift correlation spectroscopy
HCP	hollow-cathode plume
HC-RELAY	heteronuclear relayed coherence transfer experiment
HD	high density
HD	hydrolysis degree
HDC	hydrodynamic chromatography
HDCOSY	broadband proton decoupled correlated spectroscopy
HDPE	high efficiency particulate absolute
HETCOR	heteronuclear correlation

HETLOC	heteronuclear long-range couplings with heteronuclei in abundance
HETP	height equivalent of a theoretical plate
HF	Hartree-Fock
hfs	hyperfine splitting
hfs	hyperfine structure
hfsc	hyperfine splitting constant
HH	head to head (polymer)
HID	helium ionization detector
H&L	heavy and light
HLB	hydrophile-lipophile balance number
HLD	higher limits of detection
HMBC	heteronuclear multiple bond connectivity
hmde	hanging mercury drop electrode
HMO	Hückel molecular orbital
HMQC	heteronuclear multiple quantum coherence
HNDQC	heteronuclear double quantum coherence
HNZQC	heteronuclear zero quantum coherence
HOE	heteronuclear Overhauser effect
HOESY	heteronuclear 2D nuclear Overhauser spectroscopy
HOHAHA	homonuclear Hartman-Hahn spectroscopy
HOM2DJ	homonuclear (^{1}H) two dimensional *J*-resolved spectrum
HOMO	highest occupied molecular orbital
HPAE-PAD	high-performance anion exchange chromatography-pulsed amperometric detection
HPCE	high-performance capillary electrophoresis
HPIC	high performance immunoaffinity chromatography
HPLC	high performance liquid chromatography; formerly high pressure liquid chromatography
HPPLC	high-pressure planar liquid chromatography
HPRC	high-performance radial chromatography
HPSEC	high-performance size exclusion chromatography
HPTLC	high performance thin layer chromatography
HQI	hit quality index
HREELS	high-resolution electron energy loss spectrometric
HREIMS	high resolution electron impact mass spectrometry
HRFABMS	high resolution fast atom bombardment mass spectrometry
HRMS	high resolution mass spectrometry
HSAB	hard and soft acids and bases
HSC	heteronuclear shift correlation
HSC	hyperfine splitting constant
HSCCC	high-speed countercurrent chromatography
HSES	hydrostatic equilibrium system

HSL	heteronuclear spin locking
HSP	heat shock proteins
HSP	homospoil pulse
HSQC	heteronuclear single quantum correlation
HST	high-sensitivity trap
HT	head to tail (polymer)
HT	high temperature
HT	hypothermally treated
HT/DXRD	high-temperature dynamic x-ray diffraction
HTMD	high-temperature molecular dynamics
HTP	high-temperature and pressure
HTSC	high-temperature superconductor
HTT	high-temperature treatment
HT-TGA	high-temperature thermal gravimetric analysis
HVIO	high volume industrial organics
HVL	half-value layer
HW	hazardous waste
hygr	hygroscopic
Hz	hertz
I	electric current
I	inhibitor
I	intermittent
I	ionic strength
I	ionization energy
I	isotropic phase of a liquid crystal
I	moment of inertia
I	nuclear spin quantum number
I	radiant intensity
i_d	diffusion current
IA	international angstrom
IC	ion chromatography
IC	integrated circuit
ic	intracerebrally
ICA	*in vivo* biological activity assays
ICD	induced circular dichroism
IC-ES	ion chromatography-eluent suppression
ICLC	industrial continuous liquid chromatography
ICP	inductively coupled plasma
ICP-AES	inductively coupled plasma-atomic emission spectroscopy
ICP-MS	inductively coupled plasma-mass spectrometry
ICR	ion cyclotron resonance
ics	internal chemical shift
ICT	International critical tables
ICT	intramolecular charge transfer

ID	ineffective dose
ID	inhibitory dose
id	inside diameter
ID_{50}(ID50)	ineffective dose in 50% of the test subjects
IDAS	instrument data acquisition system
IDF	inverse diffusion flame
IE	ionization energy
IEC	ion exchange capacity
IEC	ion exchange chromatography
IEF	isoelectric focusing
IEP	immunoelectrophoresis
IFA	immunofluorescent assay
IFSOFT	irradiated fused silica open tubular
IGC	inverse gas chromatography
IGLO	individual gauge for localized molecular orbitals
IHB	intramolecular hydrogen bonding
IIDA	inverse isotopic dilution analysis
IKE	ion kinetic energy
IKES	ion kinetic energy spectroscopy
IMFP	inelastic mean free path
INAA	instrumental neutron activation analysis
INADEQUATE	incredible natural abundance double quantum transfer experiment
INAPT	insensitive nuclei assigned by polarization transfer
INDO	intermediate neglect of differential overlap internucleus; incomplete neglect of differential overlap
INDOR	internal nuclear double resonance; internucleus (nucleus-nucleus) double resonance
INDOSCF	intermediate-neglect of differential overlap self-consistent field
INEPT	insensitive nuclei enhanced by polarization transfer
INFERNO	irradiation of narrow-frequency envelopes by repeated nutation
INO	iterative natural orbital
INSIPID	inadequate sensitivity improvement by proton indirect detection
INT	interaction energy
I/O	input/output
IP	ionization potential
ip	intraperitoneal(ly)
IPC	ion-pair chromatography
IPTC	inverse phase transfer catalysis
IR	infrared
IR	internal reflectance

IRC	infrared reconstructed chromatograph
IRC	intrinsic reaction coordinate
IRD	infrared detection
IRDO	intermediate retention of differential overlap
IRE	internal reflectance element
IRMP	infrared multiple photon dissociation
IRP	internal reflection photolysis
IRR	irritant effects (systemic)
IRS	internal reflection spectroscopy
IRS	inverse Raman scattering
ISC(isc)	intersystem crossing
ISCA	ionization spectroscopy for chemical analysis
ISE	ion-selective electrode
ISFET	ion sensitive field effect transistor
i-spin	isotopic spin
ISS	ion-scattering spectroscopy
ITC	isothermal titration calorimeter
ITMS	ion trap-mass spectrometry
ITP	isotachophoresis
IU	International unit
iv	intravenous(ly)
J (*J*)	coupling constant
J	joule
J (*j*)	total angular momentum quantum number
j	electric current density
j	pressure drop correction factor
ϑ	Bragg angle
JCP	*J*-cross polarization
JFET	junction field-effect transistor
JMSE	*J*-modulated spin-echo
JR-COSY	*J*-resolved correlation spectroscopy
J_x	flux of a quantity x
K	electrolytic conductivity
K	equilibrium constant
K	isothermal compressibility
K	Kelvin
K	kinetic energy
k	Boltzmann constant
K	dissociation constant
k	kinetic
k	rate constant, rate coefficient
k	thermal conductivity
κ	conductivity
k_{cat}	turnover number

K_i (K_I)	inhibition constant; inhibitor constant
K_M(K_m)	Michaelis-Menton constant
K_o	overall rate constant; number of molecules of substrate transformed per second per mole of enzyme—turnover number
kbp	kilobase pair
KC	kinetic control
KE	kinetic energy
KIE	kinetic isotope effect
KT	complexometric titration
KVP	kilovolt power
kX	crystallographic unit
L	Lambert
L	large
L (L) (*l*)	levorotary configuration
L	ligand
L (*l*)	orbital angular momentum quantum number
L	self-inductance
I	electric current
l	length
Λ	molar conductivity of an electrolyte
λ	decay constant
λ	molar conductivity of an ion
λ	thermal conductivity
λ	wavelength
λ_C	Compton wavelength
λ_{max}	maximum wavelength in nm
λ_{min}	minimum wavelength in nm
LALLS	low-angle laser light scattering
LARIS	laser-ablative resonant ionization spectroscopy
LASER	light amplification by stimulated emission of radiation
LAT	light-absorbing transients
LC	lethal concentration
LC	liquid chromatography
LC	liquid cystal
LC_{100}	a lethal concentration which kills 100% of the individuals
LC_{50}	a lethal concentration which kills 50% of the individuals
LCAO	linear combination of atomic orbitals
LCD	liquid crystal display
LC-EC (LC/EC)	liquid chromatography-electrochemical detection; liquid chromatography-electrochemistry
LCGTO-MCP-DF	linear combination of Gaussian type orbitals-model core potential-density functional formalism
LCICD	liquid crystal induced circular dichroism

LC-IR	liquid chromatography-infrared
lcm	least common multiple
LC-MS	liquid chromatography-mass spectrometry
LCST	lowest critical solution temperature
LCVAO	linear combination of virtual atomic orbitals
LD	laser desorption
LD	lethal dose
LD	linear dichroism
LD_{50} (LD50)	the lethal dose which kills 50% of a group of test animals
LD_{100} (LD100)	the lethal dose which kills 100% of a group of test animals
LDA	linear diode array
LDFT	laser desorption Fourier transform
LDF-TMS	laser desorption Fourier transform mass spectrometry
LDR	linear dynamic range
LDV	low-dead-volume
LE	locally excited
LED	light emitting diode
LEED	low energy electron diffraction
LEED/ESDIAD	low energy electron diffraction-electron stimulated desorption ion angular distribution
LEI	laser enhanced ionization
LEIS	laser enhanced ionization spectrometry
LEL	lower explosive limit
LEMF	local effective mole fraction
LF	ligand field
LFER	linear free-energy relationship
LFP	laser flash photolysis
LIESST	light-induced excited-state stabilization
LIF	laser-induced fluorescence
LIL	laboratory interface language
lim	limit
LIMA	laser ionization mass analysis
LIMS	laboratory information management system
LIMS	laser ionization mass spectrometry
LINUP	laser-induced nuclear polarization
LIS	lanthanide-induced shift
Lk	linking number
LLC	liquid-liquid chromatography
LLD	lower limits of detection
LM	light microscopy
LM	linear molecules
lm	lumen
LMCT	ligand-to-metal charge-transfer
LMO	localized molecular orbital

LMO-VCD	localized molecular orbital method
LMTO	linear muffin-tin orbital
LNDO	local neglect of differential overlap
LOD	limit of detection
LOD	loss on drying
LORG	localized orbital-local origin
Lp	Lorentz-polarization
LPC	liquid phase conditions
LPDA	linear photodiode array
LPSIRS	linear potential-sweep infrared reflectance spectroscopy
LR	long range (spectroscopy)
LRCOSY	long range correlated spectroscopy
LRE	local resonance energy
LREIMS	low resolution electron impact mass spectrometry
LREPE	local resonance per electron
LRET	long-range electron transfer
LRHMBC	long-range heteronuclear multiple quantum chemical shift correlation
LRS	laser Raman spectroscopy
LSC	liquid-solid chromatography
LSI	large-scale integration
LSIMS	liquid secondary ion mass spectrometry
LSR	lanthanide shift reagent
LT	low temperature
LT-HPLC	low temperature-high performance liquid chromatography
LUMO	lowest unoccupied molecular orbital
LUO	laboratory unit operation
LVDT	linear variable differential transformer
LW	long wave
lx	lux
M	magnetic quantum number
M	magnetization; magnetic moment
[M]	molecule rotation
M	moment of force
[M]	monomer concentration
m	mass
m	medium
m	moderately strong
m	multiplet
μ	chemical potential; electrochemical potential
μ	Joule-Thomson coefficient
μ	magnetic moment of particle
μ	permeability
μ	reduced mass

μ	viscosity
M_r	relative molecular mass
M_w	weight-averaged molecular weight
M_z	z-average molecular weight
m_a	atomic mass
m_d	deuteron mass
m_e	electron mass
M_I	nuclear spin quantum component
m_I	nuclear magnetic quantum number
$m_J (m_j)$	total angular momentum component
$m_L (m_l)$	orbital quantum number component
m_N	nuclear mass
m_n	neutron mass
m_p	proton mass
$m_S (m_s)$	electron spin quantum component; electron magnetic quantum number
$\mu_B (\mu B)$	Bohr magneton
μ_e	magnetic moment of electron
μ_ν	magnetic moment of neutron
μ_o	permeability
μ_p	magnetic moment of proton
μ_r	relative permeability
MALD	matrix-assisted laser desorption
MALDI	matrix-assisted laser desorption-ionization
MALD-TOF	matrix-assisted laser desorption-time of flight
MAS	magic-angle spinning
MASS	magic angle sample spinning
MCD	magnetic circular dichroism
MCL	maximum contaminant level
mcl	monoclinic
MCS	micro calorimetry system
MCXD	magnetic circular x-ray dichroism
MD	metered dose
MD	molar rotation
MD	molecular dynamics
MDL	method detection limits
MDSC	modulated differential scanning calorimetry
m/e	mass to charge ratio
MECC	micellar electrokinetic chromatography
MECI	monoexcited configuration interaction
MED	mean effective dose
MEEC	membrane-enclosed enzymatic catalysis
MEK	micellar electrokinetic chromatography
MEKC	micellar electrokinetic chromatography

MEM	maximum entropy method
MEP	minimum energy pathway
MEP	molecular electrostatic potentials
MERP	minimum energy reaction pathway
meq	milliequivalent
MF	mean fluorescence
MF	molecular fragment
mf	mole fraction
MFC	multifunction controller
MFP	magnetic field perturbation
MHD	magnetohydrodynamics
MHE	multiple headspace extraction
mho	conductivity
MIC	minimum growth inhibitory concentration
MIE	magnetic isotope effect
MIKES	mass-analyzed ion kinetic energy spectrometry
MINDO	modified intermediate neglect of differential overlap
MIR	medium infrared (mid-infrared)
MIR	multiple internal reflectance
MIRS	multiple internal reflectance spectroscopy
MLCT	metal-to-ligand charge-transfer
MLD	minimum lethal dose
MLM	multidimensional luminescence measurement
MLR	multiple linear regression
MM	molecular mechanics
MN (M_n)	number-average molecular weight
MO	molecular orbital
MODEPT	modified distortionless enhancement polarization transfer
MOLE	molecular optical laser examiner
MOR	magneto-optical rotation spectroscopy
MORE	microwave oven-induced reaction
MOSFET	metal-oxide semiconductor field effect transistor
Mp	peak maximum; molecular mass peak at maximum
MP2	second order Møller-Plesset theory
MPI	multiphoton ionization
MPLC	medium pressure liquid chromatography
MQC	multiple quantum coherence
MQF	multiple quantum filtration
MQF-RCOSY	multiple quantum filtration single relayed coherence transfer experiment
MR	molecular refraction
MRCI	multi-reference configuration interaction
MRD-CI	multi-reference single and double excitation configuration-interaction

MRES	mass resolved excitation spectrum
MRI	magnetic resonance imaging
MS	mass spectrometry; mass spectrum
m-s	moderate to strong
MSD	mass selective detection; mass selective detector
MSI	medium-scale integration
MS-MS	tandem mass spectroscopy
MS-RTP	mass spectroscopy-room temperature phosphorimetry
MTD	maximum tolerated dose
MTD	mean therapeutic dose
mtfe	mercury thin-film electrode
MTS	mechanical test system
mu (m_u)	atomic mass constant; unified atomic mass unit
MUPI-MS	multiphoton-ionization-mass spectrometry
MVS	multiple-variable storage
MW	molecular weight
MWD	molecular weight distribution
Mx	Maxwell
m/z (m/z)	mass to charge ratio
N	isotopic label
N	nematic phase of a liquid crystal
N	neutron number
N	Newton
N (n)	normal; normality
N	number of entities, individuals or molecules
N	transference number
n	amount of substance
n	number concentration; number density of entities
n	overall order of reaction
n	principal quantum number
n	refractive index
ν	frequency
ν	kinematic viscosity
ν	wavenumber
N_a	Avogadro constant
NAA	neutron activation analysis
NAE	non-additivity effects
NAT	non-aqueous titration
NAZ	normal analytical zone
NBMO	non-bonding molecular orbitals
NBO	natural bond orbital
N/C	nitrogen to carbon ratio
NC	non-competitive
NCI	negative chemical ionization

NDDO	neglect of diatomic differential overlap
NEER	non-equilibration of excited rotamers
NEMO	non-empirical molecular orbital
NEXAFS	near edge x-ray absorption fine structure
NGI	neighboring group interaction
NHC	natural hydrocarbons
NHE	normal hydrogen electrode
NICI	negative ion chemical ionization
NIMBY	not in my back yard
NIR	near infrared
NIRA	near infrared reflectance analysis
NIRS	near infrared reflectance spectroscopy
NLO	nonlinear optical (responses)
nm	nuclear magneton
NMR (nmr)	nuclear magnetic resonance
NMRS	neutralization-reionization-mass spectrometry
NO	natural orbital
NOCOR	neglect of core orbitals
NOE (nOe)	nuclear Overhauser effect
NOED	nuclear Overhauser effect difference
NOEDS	nuclear Overhauser effect difference spectroscopy
NOESY	nuclear Overhauser enhancement and exchange spectroscopy
NONBO	nonorthogonal natural bond orbitals obtained prior to interatomic orthogonalization
NORD	noise off-resonance decoupling
NPD	nitrogen phosphorous detection
NP-LC	normal-phase-liquid chromatography; normal-phase high-performance liquid chromatography
NP-HPLC	normal-phase high-performance liquid chromatography
NPR	net protein retention
NQI	nuclear quadrupole interaction
NQR	nuclear quadrupole resonance
NR	not regulated
NRS	nuclear reaction spectrometry
NRTL	non-random two-liquid
NSGC	non steady-state gas chromatography
NT	non-aqueous titration
NTP	normal temperature and pressure
Nu	nucleophile
O/C	oxygen to carbon ratio
OD	optical density
OD	outer diameter
ODD	ouchterlony double diffusion

ODMR	optically detected magnetic resonance
ODP	ozone-depletion potentials
ODU	optical density units
Oe	oersted
OEC	oxygen evolving complex
OES	optical emission spectrometry
OFDR	off-frequency decoupling resonance
OITB	orbital interactions through bonds
OITS	orbital interactions through space
OM	organic matter
OM	oxidizing material
OOP	out of plane
OPLC	over-pressure layer chromatography
ORD	optical rotatory dispersion
ORF	open reading frames
osM (osm)	osmolar
OSWV	Osteryoung square-wave voltammetry
ot	oven temperature
OTC	open tubular column
OTE	optically transparent electrode
OTR	open tubular reactor
OTTLE	optically transparent thin layer electrode
Ox	oxidized
OXY	oxidizer
P	dielectric polarization
P (***p***)	electric dipole moment
P	optical purity
P	poise
P	poison
P	polymer support
P	power
P	probability
[P]	product
p	density
p	momentum
p	negative logarithm (e.g. pH)
p	pressure
p	resistivity
Π	osmotic pressure
π	pros (near) in nuclear magnetic resonance measurements
P_B	break pressure
P_e	electric dipole moment
P_R	retention polarity
PA	proton affinity

PACVD	plasma assisted chemical vapor deposition
PAGE	polyacrylamide gel electrophoresis
PAL	photoaffinity labeling
PAPR	powered air purifying respirator
PAS	photoacoustic spectrometry
PAS	principal axis system
PB	particle beam
PBR	packed bed reactors
PBR	plant biochemical regulators
PC	paper chromatography
PCD	plasma coupled device
pcH	measure of hydrogen concentration
PCILO	perturbed configuration interaction with localized orbitals
PCJCP	phase corrected *J*-cross polarization
PCR	principle components regression
PD	plasma desorption
pd	potential difference
PDA	photodiode array
PDI	polydispersity index
PDMS	plasma-desorption mass spectrometry
PDT	photodynamic therapy
PE	photoelectron
PE	potential energy
P E COSY	primitive exclusive correlation spectroscopy
PEL	permissible exposure limit
PeP	proton-electron-proton
PEPICO	photoelectron photoion coincidence
PES	photoelectron spectroscopy
PES	potential energy surface
PESIS	photoelectric spectroscopy of the inner shell
PESOS	photoelectric spectroscopy of the outer shell
PET	photochemically induced electron transfer
PET	positron emission tomography
PETM	phosphorescence-excitation time matrix
PFBC	pressurized fluidized-bed combustion
PG	polarography
phar	pharmaceutical
PHC	principle hazardous constituent
PHPMS	pulsed high pressure mass spectrometry
PI	photoionization
PICSY	pure in-phase correlation spectroscopy
PID	photometric ionization detection; photoionization detector
PIDS	polarization intensity differential scattering
PIXE	particle-induced x-ray emission

PIXE	photon-induced x-ray emission
PIXE	proton-induced x-ray emission
PL	isoelectric point
PL	photoluminescence
PLC	preparative layer chromatography
PLD	pulsed laser ablation and deposition
PLM	polarized light microscope
PLM	principles of least motion
PLOT	porous-layer open tubular
PLS	partial least squares
PM	permanent magnet
PM	photomultiplier
PM	polyatomic molecules
PMF	potential mean force
PMO	perturbational molecular orbital
PMR (pmr)	^{1}H nuclear magnetic resonance
PMR	^{31}P nuclear magnetic resonance
PMR	polymerization of monomeric reactants
PMS	pressure mass spectrometry
PNDO	partial neglect of differential overlap
POE	point of exposure
POEMS	plasma optical emission mass spectrometry
POHC	principal organic hazardous constituent
POI	point of interception
POM	polarizing optical microscopy
POMMIE	phase oscillations to maximize editing
por	porosity
PPB (ppb)	parts per billion
PPC	personal protective clothing
PPE	personal protective equipment
PPI/NICI	pulsed positive ion-negative ion chemical ionization
PPM (ppm)	parts per million
PPS	photophoretic spectroscopy
PR	progesterone receptor
PRDDO	partial retention of diatomic differential overlap
PREP	population redistribution for enhancement with proton decoupling
PRFT	partially relaxed Fourier transform
PRINCE	programmable injector for capillary electrophoresis
PRT	platinum resistance thermometer
PSD	particle size distributions
PSE	passive sampling element
PSE	pulsed-sonication extraction
PSEPT	polyhedral skeletal electron pair theory

PSI (psi)	pounds per square inch
psig	pounds per square inch gauge
PTC	phase-transfer catalysis
PVD	physical vapor deposition
PXD	powder x-ray diffraction
PZT	piezoelectric transducer
Q	canonical ensemble partition function
Q	electric charge
Q (q)	heat
Q (θ)	quadrupole moment
Q	reaction energy
q	partition function
q	quartet (in spectra)
θ	Celsius temperature
θ	characteristic temperature
q_m	mass flow rate
q_v	volume flow rate
QCISD	quadratic configuration interaction, single and double
QCOSY	double quantum correlation spectroscopy
QELS	quasielastic light scattering
QELS-SEF	quasielastic light scattering-sinusoidal electric field
QET	quasi-equilibrium theory
QM	quantum mechanics
QMS	quadrupole mass spectrometer
QSAR	quantitative structure-activity relationships
QUANTA	molecular modeling user interface
R	degree Réaumur
R	molar gas constant
R	resistance
R	Roentgen
r	position vector
ρ	charge density
ρ	mass; density
R_f	retardation factor
R_m	molar refraction
RA	radioactive
RAD	radiation absorbed dose
RAIR	reflection-asorption infrared
RAIRS	reflection-absorption infrared spectroscopy
RAM	random access memory
RARE	rapid aquisition with relaxation enhancement
RARP	radical anion-radical pair
RBA	relative binding affinity
RBS	Rutherford backscattering spectroscopy

RC	reaction coordinate
RC	reference cell
RCF	relative centrifugal force
RCOSY	double relayed coherence transfer experiment; single relayed coherence transfer experiment
RCRA	resource conservation and recovery act
RCT	relayed coherence transfer
RE	resonance energies
Red	reduced
REDOR	rotational-echo double resonance
REL	recommended exposure limit
RELAY	single relayed coherence transfer experiment
RELAY-COSY	single relayed coherence transfer experiment-correlation spectroscopy
RELAYH	proton-single relayed coherence transfer experiment
REM	roentgen equivalent man
REMPI	resonance enhanced multiphoton ionization
REPE	resonance energy per π-electron
RF (rf)	radio frequency
RF	rapid frequency
RFLP	restriction fragment length polymorphism
RGA	residual gas analyzer
RH	relative humidity
RHF	restricted Hartree-Fock
RI	refractive index
RIA	radioimmunoassay
RID	radioimmunodiffusion
RIMS	resonance ionization-mass spectroscopy
RINMR	rapid injection
RIS	resonance ionization spectroscopy; resonant ionization spectroscopy
RIS	rotational isomeric state
RJCP	refocused *J*-cross polarization
rMD	restrained molecular dynamics
RMS	rapid multistream sampler
rms	root-mean-square
RMSD	root-mean-square deviation
RNAA	radiochemical neutron activation analysis
ROA	Raman optical activity
ROE	rotational frame nuclear Overhauser effect
ROESY	rotational frame nuclear Overhauser spectroscopy
ROESY-TOCSY	rotational frame nuclear Overhauser—total correlation spectroscopy
ROHF	restricted Hartree-Fock open-shell

ROI	residue on ignition
ROM	read-only memory
ROTO	rotational frame nuclear Overhauser—total correlation spectroscopy
RP	radical pair
RP	reverse phase
RPE	rotating platinum electrode
RP-HPLC	reverse phase-high-performance liquid chromatography
RPLC	reverse phase-liquid chromatography
RP-TLC	reverse phase-thin layer chromatography
RQ	reportable quantity
RQ	respiratory quotient
RR	resonance Raman
RRDE (rrde)	rotated ring-disk electrode
RRE	ring resonance energy
RRS	resonant Raman spectroscopy
RSD	relative standard deviation
RSE	radical stabilization energies
RSM	rapid scanning spectrophotometry
RT	redox titration
RTP	room temperature phosphorescence
RTP	room temperature phosphorimetry
RVB	resonant valence bond
RVI	remote visual inspection
Ry	Rydberg
$R\infty$	Rydberg constant
S (S)	electron spin quantum number
S	entropy
S	Siemens (conductance)
S	smetic phase of a liquid crystal
S ([S])	substrate
S	Svedberg unit
s	path; length of arc
s	sedimentation coefficient
s	single bond
s	singlet
s (σ)	standard deviation (analytical)
s	strong
s	symmetrical
s (σ)	symmetry number
σ (r)	collision (reaction) cross section
σ	Stefan-Boltzmann constant
σ	surface tension
σ	wavenumber

SA	self-assembled
SA	specific activity
SAED	selected area electron diffraction
SALS	small angle light scattering
SAM	scanning Auger microprobe
SAM	self-assembled monolayers; self assembling organic monolayers
SANS	small angle neutron scattering
SAR	structure-activity relationships
SAXD	small angle x-ray diffraction
SAXS	small angle x-ray scattering
SC	sample cell
sc	synclinal
sc	subcutaneously
SCBAF	self-contained breathing apparatus with full facepiece
SCC	size-consistency correction
SCD	segmented-array charge-coupled-device detector
SCE	standard calomel electrode; saturated calomel electrode
SCF	self-consistent field
SCF-MO	self-consistent field-molecular orbital
SCR	selective catalytic reduction
SCRF	self-consistent reaction field
SCSE	single conformational strain energy
SD	standard deviation
SDEPT	selective distortionless enhancement polarization transfer
SdFFF	sedimentation field-flow fractionation
SDL	symmetrical double-labeling method
SE	spectroscopic ellipsometry
SE	stabilization energy
SE	standard error
SE	steric energy
SEC	size-exclusion chromatography; steric-exclusion chromatography
SEC	standard error of calibration
SEC-LALLS	size exclusion chromatography-low-angle laser light scattering
SECM	scanning electron microscopy
SECS	simulation and evaluation of chemical synthesis
SECS	single-electron capacitance spectroscopy
SECSY	spin-echo correlated spectroscopy
SEDS	spin-echo difference spectroscopy
SEF	sinusoidal electric field
SEFT	spin-echo Fourier transform
SEI	secondary electron image

SEM	scanning electron microscope
SEM	standard error of the mean
SEMINA	amalgamation of subspectral editing using a multiple quantum trap and incredible natural abundance double quantum transfer experiment
SEMUT	subspectral editing using a multiple quantum trap
SEMUT GL	subspectral editing by a multiple-quantum trap grand lux
SEP	standard error of prediction
SEPT	selective insensitive nuclei enhanced by polarization transfer
SERS	surface enhanced Raman scattering
SESET	semi-selective excitation for polarization transfer
SESETACR	semi-selective excitation for polarization transfer and assignment of carbon resonances
SET	single electron transfer
SET	solution electron transfer
SFC	supercritical fluid chromatography
SFM	scanning force microscopy
SFORD	single frequency on- or off-resonance decoupling
SFR	spin-flip Raman
SFS	sum-frequency spectroscopy
SH	Sherwood number
sh	sharp
sh	shoulder
SHACV	second harmonic alternating current voltammetric
SHC	shape and Hamiltonian consistent
SHE	standard hydrogen electrode
SHECOR	selective heteronuclear correlation
SHG	second harmonic generation
SHOMO	second highest occupied molecular orbital
SI	International System (of units)
SI	secondary ion
si	stereochemical descriptor
sicc	dry
SIC-LSD	self interaction corrected-local spin density approximation
SIFT	selected ion flow tube
SIM	selected ion monitoring
SIM	single ion monitoring
SIMPLE	secondary isotope multiplets of partially labeled entities
SIMS	secondary ion mass spectrometry
SIMS	static secondary ion mass spectrometry
SINEPT	soft selective insensitive nuclei enhanced by polarization transfer
SIR	selected ion recording

SIT	silicon intensified target
SLAP	sign-labeled polarization transfer
SLPTC (SL-PTC)	solid-liquid phase transfer catalysis
SLR	spin-lattice relaxation
SLS	static light scattering
SMB	simulated moving bed
smde	static mercury drop electrode
SN	saponification number
S/N (S:N)	signal to noise ratio
SNCR	selective catalytic reduction
SNIF-NMR	site-specific natural isotope fractionation by nuclear magnetic resonance
SNIFTIRS	subtractively normalized interfacial Fourier transform infrared spectroscopy
SNO	second-order non-linear optical
SNO	semiempirical natural orbital
SNR	signal to noise ratio
SOMO	singly occupied molecular orbital
SOP	standard operating procedures
SP	square planar
sp	specific
SPE	single pulse experiment
SPE	solid-phase extraction
SPEC	specific
SPI	selective population inversion
SPI	spin population inversion
SPLITT	split-flow thin
SPM	scanning probe microscopy
SPME	solid-phase microextraction
SPOTS	spin-polarization torsional spectroscopy
SPP	short path pyrolysis
SPPS	solid phase peptide synthesis
SPR	stroposcopic pulse radiolysis
SPRA	side port reflectance accessory
SPT	selective population transfer
SPT	spin population transfer
SPY	square pyramidal
SQM	scaled quantum mechanical
SQW	square wave
SQUID	superconducting quantum interference device
sr	steradian
SRET	scanning reference electrode technique
SRG	stimulated Raman gain
SRP	signal recognition particle

SSB	single-strand binding protein
SSD	steady-state distribution
SSET	singlet-singlet energy transfer
SSI	small-scale integration
SSIMS	static secondary ion mass spectrometry
SSIRP	solvent separated ion radical pairs
SSR	segmented steam reactors
St	Stokes
STEL	short-term exposure limit
STEM	scanning transmission electron microscopy
STM	scanning tunneling microscope
STM	symmetric top molecules
STO	Slater-type orbital
STO-3G	Slater-type orbitals, three Gaussian
STP	standard temperature and pressure
SUCZESS	successive zero quantum single quantum coherences for spin correlation
SUP	standard unit of processing
Sv (sv)	Sievert
SVL	single vibrational level
SW	short wave
SW	spectral width
swg	standard wire gauge
SWR	standing-wave ratio
SXAPS	soft x-ray appearance potential spectroscopy
synth	synthetic
SYS	systemic effects
T	acidimetric titration
T	kinetic energy
T	tautomeric
T	Tesla
T	thermodynamic temperature
T	torque
T	total term
t	temperature
t	transport number
t	triplet
t	triton
τ	relaxation time
τ	tele
τ	transmittance
$\mathrm{T_B}$	break temperature
T_c	Curie temperature
T_M	melting temperature

T_N	Neel temperature
t_R	retention time
t_1	evolution time
t_1 noise	noise that appears in 2D nuclear magnetic resonance parallel to the ω_1 axis
$t_{1/2}$	half-life
t_2	detection time
TA	thermal analysis
TA-MS	thermal analysis-mass spectrometry
TANGO	testing for adjacent nuclei with a gyration operator
TBP	trigonal bipyramidal
TC	target concentration, toxic concentration
TC	thermodynamic control
T/C	treated *vs.* cured
TCC	tag closed cup
TCD	thermal conductivity detection
TCM	temperature control module
TCP	thermodynamic cycle perturbation
TCP	transmission control protocol
TCRI	toxic chemical release inventory
TD	toxic dose
Td (T_d)	thermal decomposition temperature
TDDI	three-dimensional digital imaging
TDS	total dissolved solids
TE	thermal expansion
TEA	transversely excited atmospheric
TED	total energy distribution
TEDOR	transferred-echo double resonance
TEM	transmission electron microscope
TEM	trial and error method
TEM	tunneling electron microscope
TER	teratogenic effects
TFX	toxic effects
TG	thermogravimetry
Tg (T_g)	glass transition temperatures
TGA (tga)	thermogravimetric analysis
TG/DTA	thermogravimetry-differential thermal analysis
TGL	temperature gradient lamp
TIC	total ion chromatogram
TIC	total ion current
TID	time interval digitizer
TIP	temperature-independent paramagnetism
TISAB	total ionic strength adjustment buffer
TISE	take it somewhere else

TL	triboluminescence
TLC	thin-layer chromatography
TL_m	median tolerated limit
TLV	threshold limit value
TM	temperature midrange
TMA	thermal-mechanical analysis
TnL	tunnel luminescence
TOC	total organic carbon
TOCSY	total correlation spectroscopy
TOCSY-ROESY	total correlation-rotational frame nuclear Overhauser spectroscopy
TOD	total oxygen demand
TOF	time-of-flight
TOF-MS	time-of-flight mass spectrometry
TOF-SIMS	time-of-flight secondary ion mass spectrometry
TON	turnover number
TORO	total correlation-rotational frame nuclear Overhauser spectroscopy
TP	triple point
TPD	temperature-programmed desorption; thermal programmed desorption spectroscopy
TPD/R	temperature-programmed desorption/reaction
TPE	thermoplastic elastomers
TPO	thermoplastic polyolefins
TPO	transaction processing option
TPPI	time proportional phase incrementation
TPR	temperature-programmed reaction
TPU	thermoplastic polyurethanes
TQF	triple quantum filtered
TQF-COSY	triple quantum filtered correlation spectroscopy
Tr	trace
TRC	total response chromatography
TRI	toxic release inventory
TS	thermospray
TS	transition state; transition structures
TSC	thermal stimulated current
TSP	thermospray
TSP	total suspended particulate matter
TSQ	triple stage quadrupole
TSR	thermal scanning rheometer
TT	tail to tail
U	electric potential difference
U	internal energy
U	uniformly labeled

u	atomic mass constant
u	electric mobility
u	unified atomic mass unit
u (υ)	velocity
UC	uncompetitive
UEL	upper explosive limit
UFL	upper flammability limit
UHF	unrestricted Hartree-Fock
uhf	ultrahigh frequency
UHP	ultrahigh purity
UHV	ultrahigh vacuum
Um	magnetic potential difference
UME	ultramicroelectrode
UPS	ultraviolet photoelectron spectroscopy
UPT	universal polarization transfer
UTM	universal transverse mercator
UV (uv)	ultraviolet; ultra-violet spectroscopy
UVCD	vacuum ultraviolet circular dichroism
UV-PES	ultraviolet photoelectron spectroscopy
UVRR	ultraviolet resonance Raman
UV-VIS (UV-vis)	ultraviolet-visible spectroscopy
V(ς)	electric potential
V	maximum velocity of an enzyme catalyzed reaction (saturated with substrate)
V	potential energy
v	rate of reaction
v	specific volume
v	stoichiometric coefficient
v	velocity; velocity of an enzyme catalyzed reaction
v	very
v	vibrational quantum number
v_B	rate of concentration change of substance B (through chemical reaction)
V_m	mobile-phase retention volume
V_m	molar volume of an ideal gas
V_{max}	maximum velocity
V_o	initial velocity
va	volt-ampere
var	variable
VAS	variable-angle spinning
VASS	variable angle sample spinning
VB	valence bond
VCD	vibrational circular dichroism
VCT	vibronic coupling theory

VDU	video display unit
VE	valence electron
VESCF	variable electronegativity self-consistent field
VLE	vapor-liquid equilibrium
VLPR	very low pressure reactor
VLSI	very large-scale integration
VOA	vibrational optical activity
VOC	volatile organic compounds
VP (vp)	vapor pressure
VPC (vpc)	vapor phase chromatography; vapor pressure chromatography
VPO	vapor pressure osmometry
vs	very strong
VSCC	valence shell of charge concentration
VSD	virtually safe dose
VSIP	valence state ionization potential
VSWR	voltage standing wave ratio
VTAT	variable temperature adsorption trap
VUV	vacuum ultraviolet
VVk	van Vleck
vw	very weak
VZPE	vibrational zero-point energies
W	radiant energy
W (w)	Watt
W (w)	weight
W (w)	work
w	mass fraction
w	weak
Ω	microcanonical ensemble partition function
Ω (ω)	solid angle
ω	angular velocity
ω	circular frequency
ω_L	Larmor frequency
ω_1	first frequency domain in 2D nuclear magnetic resonance
ω_2	second frequency domain in 2D nuclear magnetic resonance
WALTZ	wideband, alternating phase, low power technique for zero residual splitting
WATR	water attenuation by transverse relaxation
WAXD	wide angle x-ray diffraction
WAXS	wide angle x-ray scattering
Wb	Weber
WCOT	wall-coated open tubular
WDX	wavelength-dispersive x-ray spectroscopy

WEFT	water exclusion Fourier transform
WGK	German water hazard classification
WLN	Wiswesser Line Notation—computerized notation for chemical structures
w-m	weak to moderate
WOC	water oxidation catalyst
WT	wild type
WWOT	whisker-walled open tubular
X	magnetic susceptibility
X	reactance
x	mole fraction
Ξ	Grand canonical ensemble partition function
ξ	extent of reaction
ξ	rate of conversion
X_e	electric susceptibility
XANES	x-ray absorption near-edge structure
XC	exchange-correlation
XCORFE	x-nucleus correlation with fixed evolution time
XHCORR	direct carbon-proton correlation spectroscopy
X,H-COSY	inverse heteronuclear correlation spectroscopy
XPS	x-ray photoelectron (photoemission) spectroscopy (also known as electron spectroscopy for chemical analysis-ESCA)
XRD	x-ray diffraction
XRF	x-ray fluorescence
y	activity coefficient (concentration basis)
$Z(Z)$	proton number; atomic number
Z	canonical ensemble partition function
Z	charge number of an ion; net charge
Z	impedance
z	partition function
Z_{AB}	collision frequency factor
z-COSY	z-filtered correlation spectroscopy
ZDO	zero differential overlap
ZECSY	zero quantum echo correlated spectroscopy
ZFS (zfs)	zero-field splitting
zfsc	zero-field splitting constant
ZPE	zero-point energy
ZPVE	zero-point vibrational energy
ZQ	zero quantum
ZQC	zero quantum coherence
ZQCOSY	zero quantum correlated spectroscopy
ZRL	zero risk level

Chapter 3

Glossary of Organic Stereochemical and Synthetic Terms

The contents of this chapter provide a guideline to some of the common stereochemical and retrosynthetic terms[1] encountered by an organic chemist. The explanations are not intended to be definitions. Abbreviations and acronyms for the terms are given in parentheses. The list is not intended to cover all stereochemical terms as some are only intended for use in very specific situations.[2] Where appropriate, references have been given for further elaboration or precise definitions.

α position	Position next to a functional group.
A strain	Allylic strain.
a synthon	An electron deficient, reactive substrate produced by a retrosynthetic disconnection (transform).[3]
A values	Gibbs free energy term that is used to estimate the steric bulk of groups.
Absolute configuration	Precise orientation of ligands or groups around a chiral center (usually designated by the Cahn-Ingold-Prelog priority rules)
Achiral	Center, axis or plane which possesses some element of symmetry.
Actuate	To make a retrosynthetic transformation possible in the same way that a structural subunit can activate a molecule for a chemical reaction (see Control Group).
Ancillary keying groups	Structural subunits which provide keying of a particular transform over and above those associated with the synthon itself.
Ancat	Descriptor to define relative stereochemistry. When a molecule is written in the zig-zag form, stereocenters on the opposite side (up or down) are ancat.[4]
Anti	On the opposite side. This can be used in the context of isomers about a bond with hindered rotation, substituents on a ring, addition and elimination reactions, and a descriptor for the aldol reaction when the zig-zag form is used to depict the product.[5]
Anticlinal	The groups are at 120° to one another when viewed down the carbon-carbon bond.
Antiperiplanar (ap)	The groups are on opposite sides of a carbon-carbon bond yet all of the atoms are planar (e.g. *trans* elimination).
Antipode	The mirror image, or opposite of another molecule.
Antithetic analysis	Synonymous with retrosynthetic analysis.
APD	Appendage disconnection.
Appendage	Structural subunit which is bonded to a ring or functional group origin.
Asymmetric	Possessing no symmetry.
Asymmetric induction	Effect produced by a chiral substituent or reagent which enhances the reaction at one face of a substrate over the other.
Asymmetric synthesis	A process or reaction that acts on a prochiral substrate or moiety to produce unequal amounts of stereoisomers.
Atropisomers	Isomers that can be separated but are interchanged by rotation about a single bond.
Axial	A group that is (close to) perpendicular to the molecular plane.

β position	Position two removed from a functional group.
Boat	Ring conformation that resembles the shape of a boat. For cyclohexane this is: 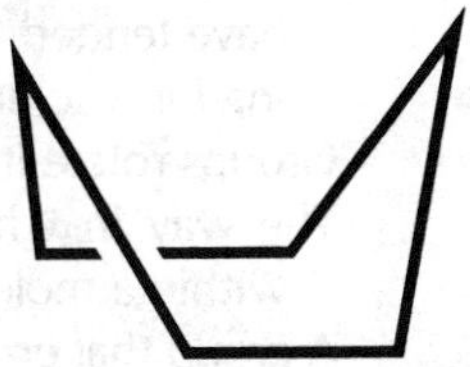
Blocking group (BG)	A group that stops reaction at a specific site.
Bowsprit	Synomym for equatorial.
B-strain	Back strain (in crowded molecules).
Cahn-Ingold-Prelog (CIP)	See priority rules.
Carbogen	An organic molecule.
Chair	Ring conformation that resembles the shape of a chair. For cyclohexane this is: 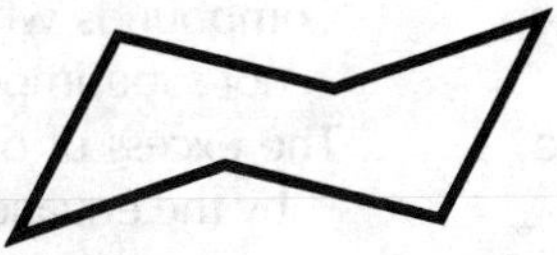
Chemoselective (cs)	A reagent that is selective for one type of functional group within a molecule.
Chemzymes	Catalytic chemical agents that act like enzymes.
Chiral	A center, axis, or plane which does not possess symmetry.
Chiral auxiliary	A chiral unit which is attached to a substrate to enable or enhance stereoselection. The auxiliary is removed after the reaction.
Chiral controller	Synonymous with chiral auxiliary.
Chiral pool	A collection of chiral starting materials, often natural products, that are readily available (see chiron).
Chiron	An optically active starting material.
Chirotopic	A group or atom whose local environment is chiral.
Cis-isomer	Where two alkyl, or other groups are on the same side of a bond with restricted rotation, or ring system.
Clearable stereocenter	A stereocenter which can be removed retrosynthetically by application of a transform with stereochemical control.
Cocyclic bonds	Endo bonds that are within the same ring.
Configuration	The way in which atoms are related to one another in three-dimensional space.

Conformer	Conformational isomer; isomers arising from rotation about a specific bond.
Conglomerate	A racemate where molecules of the same configuration have tended to crystallise on themselves, thus, allowing for resolution. (c.f. racemate)
Conrotatory	Groups rotate in the same direction.
Constitution	The way in which atoms are bonded to one another within a molecule.
Control group	A group that enables a reaction to be performed. A control group can be either an integral part of the substrate or an auxiliary unit.
Core functional groups	A member of the most simple class of functional groups, which are the most versatile, such as carbonyl, hydroxyl, alkene, alkyne, and amine.
d synthon	An electron rich, reactive substrate produced by a retrosynthetic disconnection (transform).[3]
Dextrorotatory	The enantiomer which rotates the plane of polarized light in a clockwise manner.
Diastereoisomers	Compounds which have the same constitution, yet are not superimposable mirror images.
Diastereomeric excess (de)	The excess of one diastereoisomer over another. Given by the equation:

$$de = \left| \frac{[RR] - [SR]}{[RR] + [SR]} x100 \right|$$

Diastereoselective (ds)	A reaction that leads to the formation of one diastereoisomer in preference to all others.
Diastereospecific	A reaction that produces only one diastereoisomer to the total exclusion of all others.
Diastereotopic	Ligands or face which, if substituted, would give rise to the formation of diastereoisomers.
Disconnection	See transform.
Disrotatory	Groups rotate in opposite directions.
Distal	The site of equivalent reactivity about a functional group which is furthest removed from a reference point contained within the molecule.
Distomer	The undesired enantiomer (with respect to biological effects).
Double asymmetric induction	The effect produced when the facial selectivity of two reacting, chiral substrates work in harmony to produce a larger asymmetric induction than either single reactant would achieve.

Eclipsed	The dihedral angle between two groups is 0° (see syn-periplanar).
E-isomer	The *trans*-isomer. See priority rules.
Enantiomeric excess (ee)	The excess of one enantiomer over another. Given by the equation: $$ee = \left\| \frac{[R]-[S]}{[R]+[S]} x100 \right\|$$
Enantiomers	A pair of molecules that have the same constitution yet are nonsuperimposable mirror images.
Enantioselective (es)	A reaction that leads to the formation of one enantiomer in preference to the other.
Enantiospecific	A reaction that produces only one enantiomer to the total exclusion of the other.
Enantiotopic	Substituents or faces, which, if they were changed for a different ligand, or underwent reaction to form an sp^3 center, would give rise to enantiomers.
Endo bond	Bond within a ring.
Endo (isomer)	One with the bond within (inside) the ring system.
ent (isomer)	One where the stereocenters are reversed.
Envelope ring	See perimeter ring.
epi (isomer)	Inversion of the normal configuration.
Epimerization	The interconversion of a stereocenter through a deprotonation-reprotonation procedure.
Equatorial (e)	A group that is (approximately) in the molecular plane.
Equivalent of reactive intermediate	See synthetic equivalent.
Erythro	The isomer which when written in the Fischer projection form can be related to erythrose. Heathcock suggested a reversal of this nomenclature system in the "aldol" notation.[6]
Eutomer	The desired enantiomer (usually in a biological sense).
Ex-target tree (EXTGT Tree)	A branching tree structure formed by retrosynthetic analysis of a target molecule.
Exendo bond	A bond which is directly attached to a ring yet within another ring.
Exo bond	A bond which is directly attached to a ring.
Exo (isomer)	One with the bond directed away from (outside of) the ring system.
α face	The bottom face.
β face	The top face.
FGA	Functional group addition.

FGI	Functional group interchange.
FGR	Functional group removal.
Fischer projection	A method of depicting stereochemistry: 
Flagstaff	A synonym for axial.
F-strain	Front strain (in crowded molecules).
Functional group origin	The atom to which a functional group is attached.
Functional group equivalent	A functional group from which the target functional group can be obtained.
Functional group based strategy	The use of functional groups to guide the retrosynthetic analysis.
γ position	Position three times removed from a functional group.
Gauche	Groups have a dihedral angle of 60° between them.
Geminal (*gem*)	On the same carbon atom.
Haworth	A method to visualize carbohydrates.
Heterotopic	Ligands or faces which can be distinguished by an asymmetric reagent.
Homochiral	A molecule that contains only chiral centers of the same chirality (it does not necessarily mean optically pure)[7].
Homotopic	Ligands or faces which are indistinguishable from one another.
Ipso	As in *ipso*-substitution, reaction at the same center.
Levorotatory	The enantiomer which rotates the plane of polarized light in a counterclockwise manner.
Like (*lk*)	The priorities of two reagents that form a new bond are similar, as in *R*,*Re* or *R*,*R*.[8]
Matched pair	Reagents whose chirality work in harmony to provide double asymmetric induction.
Mechanism control of stereochemistry	Stereochemical control in a reaction or transform as a result of mechanistic factors rather than structure alone.
Mechanistic transform	A transform involving a sequence of reactive intermediates generated in a stepwise fashion which lead to a stable precursor structure.
Meso compound	A compound that contains both chiral centers and a plane of symmetry.
Mis-matched pair	Reagents whose chirality are not working together to afford double asymmetric induction.

Molecular complexity — A measure of the combined effects of molecular size, functional groups, constitution, stereochemistry, and reactivity which cause synthetic difficulties.

Multistrategic analysis — Use of two or more retrosynthetic strategies at the same time.

Newman projection — A method to visual relative stereochemistry about a specific bond:

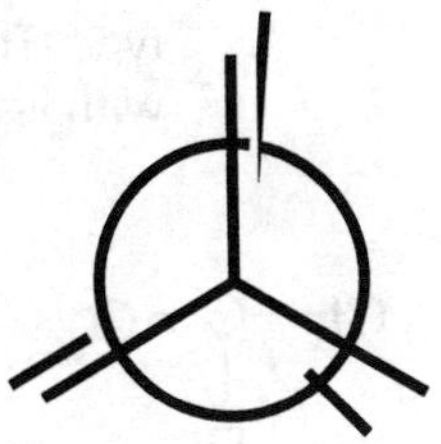

Nomenclature — An accepted naming system.

Offexendo bond — A bond connected to an exendo bond.

Optical purity — A variation on enantiomeric excess, but optical purity(P) is derived from optical rotation measurements, and expressed by the equation:

$$P = \frac{[\alpha]}{[\alpha]_{max}} x100$$

Partial retron — An incomplete retron for a particular transform.

Perimeter ring — A ring which can be constructed from the atoms of two smaller rings that are fused or bridged to one another.

Positional selectivity — Synonymous with regioselectivity.

Preserve bond — A bond which should not be broken retrosynthetically.

Preserve ring — A ring which should be preserved for strategic reasons during retrosynthetic analysis.

Primary ring — A ring within a structure which cannot be described as a perimeter ring.

Priority rules — Cahn-Ingold-Prelog Priority Sequence Rules

Tetrahedral Compounds

1. Determine the priority of the four ligands about the center
   a) Atoms with the higher atomic number have precedence.
   b) For atoms of the same atomic number, but different atomic mass, the higher mass has the higher precedence.

Priority rules, cont'd.

c) Atoms attached to the asymmetric center are sequenced first. If two, or more atoms have the same priority, then each of the atoms attached to these must be compared. Repeat this procedure moving out from the atom under consideration, until a priority is assigned.

d) Multiple bonded atoms are treated as four coordinate by the addition of replica atoms of the same type. The replica atoms are treated as bare atoms with ligands of zero priority. Eg.

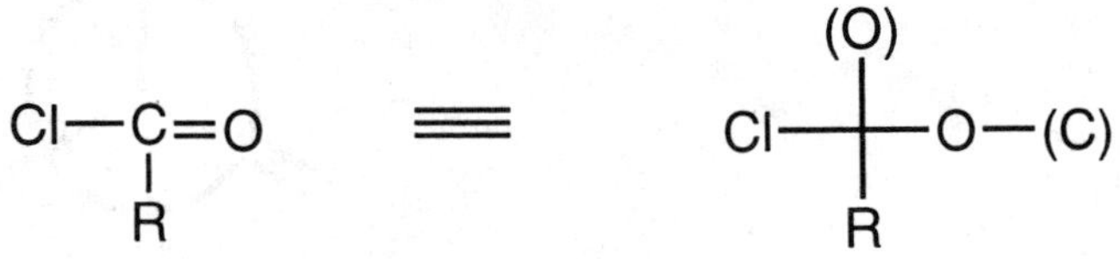

2. The asymmetric atom is viewed with the ligand of lowest priority pointing away from the viewer. If the direction of rotation in moving from the ligand of highest priority to the next highest, to third highest is clockwise, the configuration is designated R. A counterclockwise rotation is designated S.

Compounds With More Than One Chiral Center

1. Each of the centers are sequenced separately.
2. A further rule is necessary; a ligand with the R-configuration has priority over a constitutionally identical ligand of the S-priority.

Molecules With Axial Chirality

Allenes provide examples of such molecules.

1. View the molecules down the chiral axis—it does not matter from which end.
2. Ligands are assigned priorities as above, except the ligands nearer the viewer take priority.
3. Assign configuration as before.

Diastereisomers From Restricted Rotation About Multiple Bonds

1. Assign priorities to the two ligands for each atom of the multiple bond. Lone pairs have a lower priority than H.
2. If the ligands of highest priority are on the same side, the configuration is Z, on the opposite side, E.

Prochiral Centers

1. Apply the sequence rules to the ligands.
2. Ligands in decreasing order, with the other group remote, give pro-R (Re) if the direction is clockwise; pro-S (Si) arises from an counterclockwise sequence.

Prochiral Faces

1. As for prochiral centers, except the face is viewed from above.

Prochiral	A center, axis, or plane which when acted upon by a chiral agent could lead to the production of stereoisomers.
Protecting group (PG)	Group used to protect a functional group from reaction during a specific transformation.
Proximal	The site of equivalent reactivity about a functional group which is closest to a reference point contained within the molecule.
Pseudo-axial	Group that is (close to) perpendicular to the molecular plane.
Pseudo-equatorial	Group that is (approximately) in the molecular plane.
Racemate (rac)	Equal mixture of two enantiomers—see also racemic compound
Racemic compound	A compound that crystallizes with equal amounts of each enantiomer in a unit cell. These crystals cannot be resolved (c.f. Conglomerate).
Racemic modification	See racemate.
Re-face	The face of a planar center (eg sp^2 carbon) where the order of priorities of the substituents are viewed as being clockwise.
Reagent	The compound which brings about a chemical transformation.
Regioselective (rs)	A reaction which occurs at only one of several possible reaction sites.
Resolution	The separation of two enantiomers by physical means through the formation of diastereoisomers.
Retron	The minimal substructure element in a target structure which keys a transform (retrosynthetic disconnection) to generate a synthetic precursor. See synthon.
Retrosynthetic	The opposite of an experimental synthetic step.
Retrosynthetic analysis	The analysis of a synthesis through the disconnection of bonds to provide simpler structures which ultimately leads to available starting materials.

Retrosynthetic disconnection — See transform

RGD — Ring disconnection.

S Goal — A structure (substructure) which corresponds to a potential intermediate and can be used for retrosynthetic guidance.

Sawhorse — Representation method for relative stereochemistry about a specific bond:

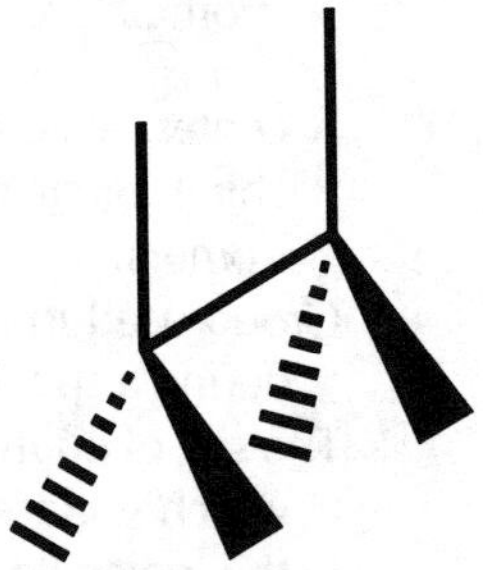

Scalemic — A mixture of enantiomers enriched in one isomer. Often used for natural products.

Selectivity — A reaction or process that forms certain products more predominantly.

Sequence rules — See priority rules.

Si face — The face of a planar center (eg sp^2 carbon) where the order of priorities of the substituents are viewed as being counterclockwise.

SM goal — A starting material which can be used for retrosynthetic guidance.

SS goal — A retrosynthetic intermediate generated from the TM by use of a control group.

Staggered — Where two groups have a dihedral angle of 60° between them; sometimes used to denote a dihedral angle of 180° (see synclinal and antiperiplanar)

Stereogenic — A center which gives rise to a stereoisomers if any two ligands were transposed.

Stereoselective (ss) — A reaction which leads to the formation of one stereoisomer in preference to all others.

Stereospecific — A reaction which produces only one stereoisomer to the total exclusion of all others.

Structure goal strategy — The use of a particular structure to guide retrosynthetic analysis.

Syn — On the same side. This can be used in the context of isomers about a bond with hindered rotation, substituents on a ring, addition and elimination reactions,

	and a descriptor for the aldol reaction when the zig-zag form is used to depict the product.[5]
Syncat	A descriptor to define relative stereochemistry. When a molecule is written in the zig-zag form, stereocenters on the same side (up or down) are syncat.[4]
Synclinal	The groups are on the same side of a carbon—carbon bond, but are at a 60° angle. (See gauche)
Synperiplanar (sp)	All of the groups lie in one plane, while the two groups being considered are on the same side of a carbon—carbon bond.
Synthetic equivalent	A molecule which reacts in a manner equivalent to a synthon or reactive intermediate.
Synthon	The structure produced by a retrosynthetic disconnection.
Target molecule (TGT or TM)	The molecule whose synthesis is under examination in the retrosynthetic analysis.
Threo	The isomer which when written in the Fischer projection form can be related to threose. Heathcock suggested a reversal of this nomenclature system in the "aldol" notation.[6]
Topological strategy	The use of particular bonds eligible for disconnection.
Torsonal asymmetry	Asymmetry arising from hindered rotation about a sigma bond where free rotation does not allow interconversion of enantiomers.
Transform based strategy	A strategic guide in which a transform guides the overall strategy.
Transform	The reverse of a synthetic reaction which results in the generation of synthons. See retrosynthetic disconnection.
Trans-Isomer	Where two alkyl, or other groups are on the opposite side of a bond with restricted rotation, or ring system.
Umpolung synthon	A synthon in which the natural polarity of a functional group has been reversed.[3]
Unlike (*ul*)	The priorities of two reagents that form a new bond are not similar, as in *S*,*Re* or *S*,*R*.[8]
Vicinal (*vic*)	Next to one another as in 1,2-disubstituted.
Z-Isomer	The *cis*-isomer. See priority rules.

References:

1. Corey, E. J.; Cheng, X.-M. *The Logic of Chemical Synthesis*; Wiley: New York, 1989.
2. Dodziuk, H.; Mirowicz, M. *Tetrahedron: Asymmetry* **1990**, 1, 171 and references cited therein.

3. Seebach, D. *Angew. Chem. Int. Ed. Engl.* **1979**, 18, 239.
4. Carey, F. A.; Kuehne, M. E. *J. Org. Chem.* **1982**, *47*, 3811.
5. Masamune, S.; Ali, S. A.; Snitman, D. L.; Garvey, D. S. *Angew. Chem. Int. Ed. Engl.* **1980**, *19*, 557.
6. Heathcock, C. H.; Buse, C. T.; Kleschick, W. A.; Pirrung, M. C.; Sohn, J. E.; Lampe, J. *J. Org. Chem.* **1980**, *45*, 1066.
7. Masamune, S.; Choy, W.; Petersen, J. S.; Sita, L. R. *Angew. Chem. Int. Ed. Engl.* **1985**, *24*, 1.
8. Seebach, D.; Prelog, V. *Angew. Chem. Int. Ed. Engl.* **1982**, *21*, 654.

Chapter 4

Named Reactions

This chapter contains a list of acronyms for named reactions and a list of named reactions. For the purposes of this chapter, a Named Reaction is derived from a person or is commonly associated with a particular transformation. Named Processes and Methods have been excluded. Named reagents may be accessed in Chapter 5.

The focus of this chapter is to let the reader quickly identify the transformation associated with the reaction. Should additional information be required references have been included. Where possible, reviews with leading references have been included rather than the original papers. Further information on many named reactions may be accessed from the following texts: March, *Advanced Organic Chemistry* 3rd Ed., Wiley, New York, 1985.; *Comprehensive Organic Chemistry*, Barton D. and Ollis, W. D., Pergamon, Oxford, 1979, and; *Merck Index* 10th Ed., Windholz, M. ed., Merck, Rathway, NJ, 1983.

Authors have not been included solely on space limitations. For many reactions subheadings have been included in the references which deal with a particular aspect of the reaction.

Many named reactions have been given alternate names, these are listed either in brackets or in notes below the scheme. We have attempted to remove any confusion and resorted to what appears to be the more acceptable name. Readers are then referred to the "real" named reaction.

Part 1 Acronyms of Named Reactions

AD	Asymmetric Dihydroxylation Reaction
ADH	Asymmetric Dihydroxylation Reaction
ADMET	Acylic Diene Metathesis
AE	Asymmetric Epoxidation
AE	Addition-Elimination
AFO	Algar-Flynn-Oyamada Reaction
ANRORC	Addition of the Nucleophile, Ring Opening and Ring Closure
BZ	Belousov-Zhabotinsky Reaction
CACR	Carbanion-Accelerated Claisen Rearrangement
CCPA	Cationic-Cyclopentannelation Reaction
DPM	Di-π-Methane Rearrangement
EA	Elimination-Addition
HDA	Homo-Diels-Alder Reaction
HEW	Horner-Emmons-Wittig
IBC	Intramolecular Benzyne Cycloaddition
IBIE	Intramolecular Base-Induced Elimination
IMDA	Intramolecular Diels-Alder Reaction
INHC	Intramolecular Nitrile oxide-olefin-Heterocycle Cycloaddition
INOC	Intramolecular Nitrile Oxide Cycloaddition
INOC	Intramolecular Nitrile oxide-Olefin Cycloaddition
IOOC	Intramolecular Oxime Olefin Cycloaddition
ISMS	Intramolecular Silyl Modified Sakurai Reaction
LACDAC	Lewis Acid Catalyzed Diels-Alder Cycloaddition Reaction
MIMIRC	Michael-Michael Ring Closure
MIRC	Michael-Induced Ring Closure
MORE	Microwave-Induced Organic Reaction Enhancement Chemistry
ODPM	Oxadi-π-Methane Reaction
PCR	Polymerase Chain Reaction
ROMP	Ring Opening Metathesis Polymerization
ROP	Ring Opening Polymerization
SAE	Sharpless Asymmetric Epoxidation
SCOOPY	α-Substitution plus Carbonyl Olefination via β-Oxido Phosphorous Ylides
SiCAC	Silylcarbocyclization Reaction
SMS	Silyl Modified Sakurai Reaction
SPAC	Sulfoxide Piperidine and Carbonyl
TARA	Tandem Addition Rearrangement
VNS	Vicarious Nucleophilic Substitution
W-H-W-E	Wittig-Horner-Wadsworth-Emmons Reaction
WGS	Water Gas Shift Reaction
WGSR	Water Gas Shift Reaction
WK	Wolff-Kishner Reduction

Part 2 Named Reactions

Abramovitch-Shapiro Synthesis (Process)

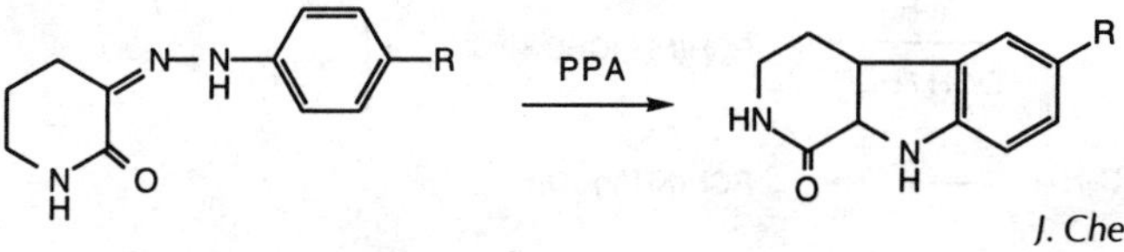

J. Chem. Soc. **1956**, 4589.

Acetoacetic Ester Condensation

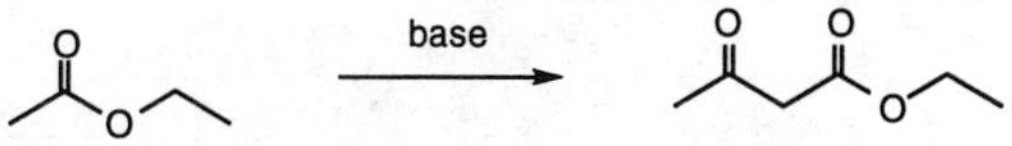

Org. Reactions **1942**, *1*, 266.

Acetoacetic Ester Synthesis

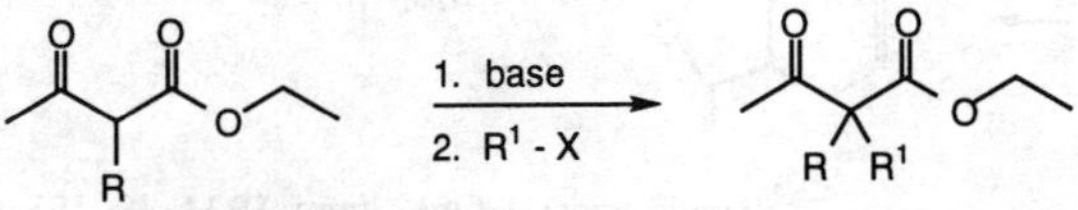

Comp. Org. Chem. **2**, 606–9.
Org. Reactions **1942**, *1*, 266.

Acyloin Condensation

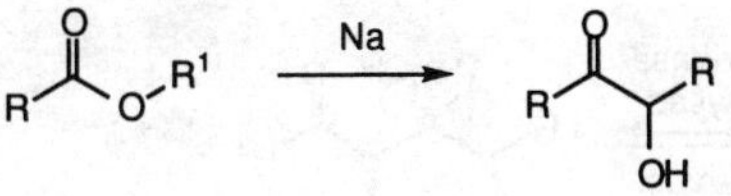

Org. Reactions **1976**, *23*, 259.
Synthesis **1971**, 236.
Chem. Rev. **1964**, *64*, 573.
Org. Reactions **1948**, *4*, 256.

Note: The esters may be on the same chain to yield cyclic α-hydroxyketones (Acyloin Cyclization).

Acyloin Rearrangement

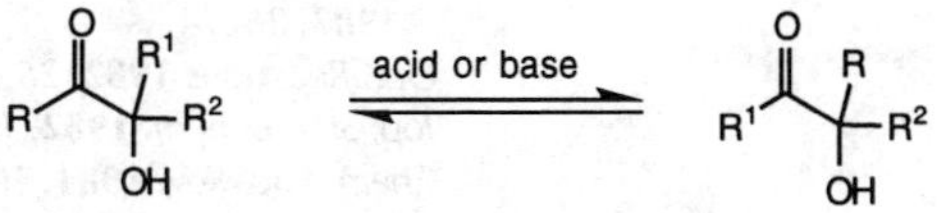

Comp. Org. Chem. **1**, 1090.

Acyloxylation Reaction (See Kharasch-Sosnovsky Reaction)

Adamantane Rearrangement

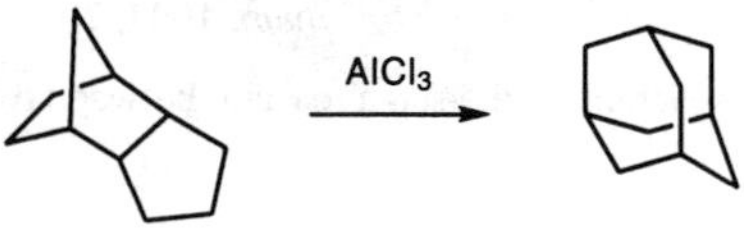

Chem. Soc. Rev. **1974**, *3*, 479.
Org. Synth. **1973**, *Coll. Vol. 5*, 16.

Adkins-Peterson Reaction

$$CH_3OH \xrightarrow[\text{metal oxide}]{\text{air} / \Delta} CH_2O$$

J. Am. Chem. Soc. **1931**, *53*, 1512.

Akabori Amino Acid Reactions

Reactions of amino acids:

$$RCH(NH_2)CO_2H + \text{sugar} \longrightarrow RCHO + CO_2 + NH_3$$

$$\underset{(R^1 = H \text{ or alkyl})}{RCH(NH_2)CO_2R^1} \xrightarrow[\text{EtOH / HCl}]{\text{Na-Hg}} RCH(NH_2)CHO \cdot HCl$$

$$ArCHO + RCH(NH_2)CO_2H \xrightarrow{\Delta} RCH(NHAr)CO_2H$$

Merck ONR-2.
J. Chem. Soc. **1956**, 307.

Note: The second reaction is also known as the Akabori Reduction.

Alder (Ene) Reaction

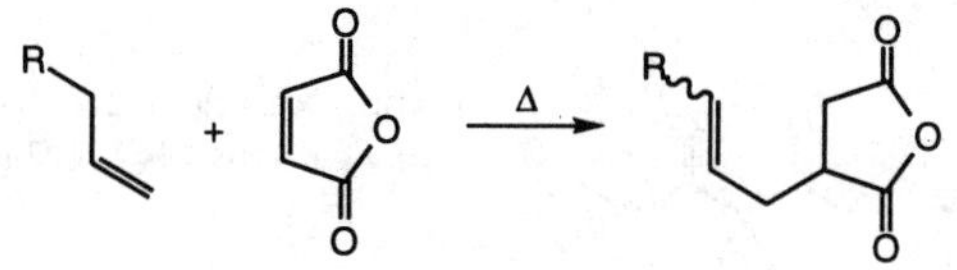

Acc. Chem. Res. **1990**, *23*, 34.
J. Organometal. Chem. **1975**, *101*, 187.

Lewis acids: *J. Org. Chem.* **1974**, *39*, 255.

Aldol Condensation (Reaction)

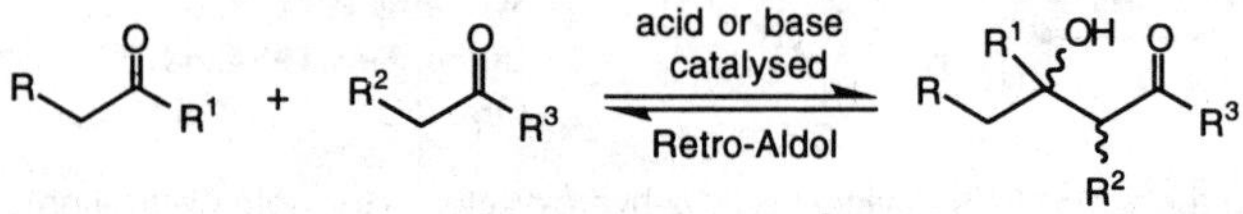

Pure & Appl. Chem. **1988**, *60*, 1.
Angew. Chem. Int. Ed. Engl. **1987**, *26*, 24.
Org. Reactions **1982**, *28*, 203.
Top. Stereochem. **1982**, *13*, 1.
Chem. Soc. Rev. **1981**, *10*, 83.
Org. Reactions **1968**, *16*, 1.

Asymmetric Aldol: *Angew. Chem. Int. Ed. Engl.* **1987**, *26*, 24.

Directed Aldol Condensation: *Org. Reactions* **1982**, *28*, 203.
J. Chem. Soc., Chem. Commun. **1982**, 291.

Note: The Crossed Aldol Condensation (Claisen-Schmidt) refers to a reaction between two different aldehydes.

Algar-Flynn-Oyamada Reaction

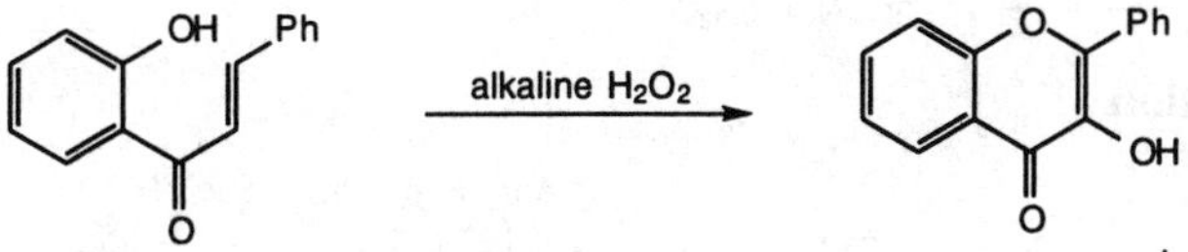

J. Am. Chem. Soc. **1948**, *70*, 1686.

Allan-Robinson Reaction

Tetrahedron **1964**, *20*, 2977.

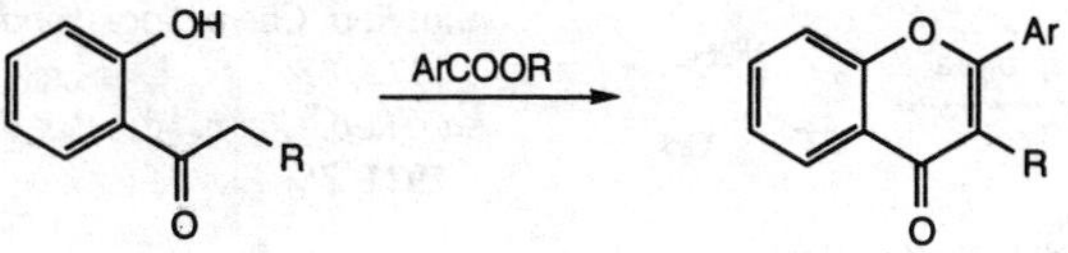

Allylic Rearrangement

$$RCHXCH{=}CH_2 \xrightarrow{Nu^-} RCH{=}CHCH_2Nu + X^-$$

Chem. Rev. **1956**, *56*, 753.

Enzyme Catalyzed: *Chem. Rev.* **1990**, *90*, 1203.

Note: Under S_N1 conditions a mixture of products results, whereas under S_N2 conditions one product results.

Aluminum Alkoxide Reduction (see Meerwein-Ponndorf-Verley Reduction)

Amadori Rearrangement

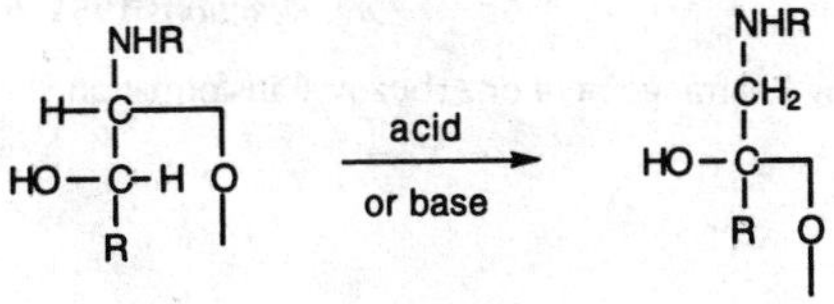

Merck ONR-3.
Adv. Carbohydr. Chem. **1955**, *10*, 169.

Amino-Claisen Rearrangement

Org. Reactions **1975**, *22*, 44.

Anderson Synthesis (Method)

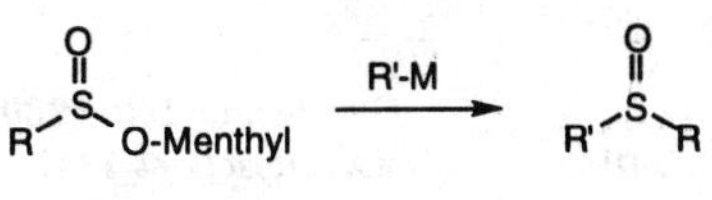

Int. J. Sulfur Chem. **1971**, 6, 69.
J. Am. Chem. Soc. **1964**, *88*, 5637.
Tetrahedron Lett. **1963**, 93.

Andrussov Oxidation

$$2NH_3 + 2CH_4 + 3O_2 \longrightarrow 2HCN + 6H_2O$$

Angew. Chem. **1935**, *48*, 593.

Angeli (Aldehyde) Reaction

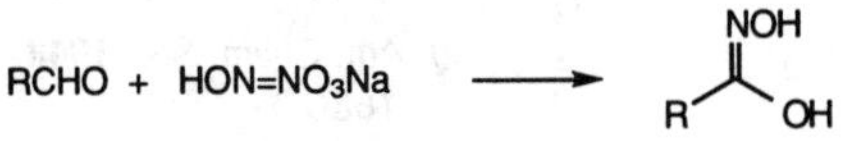

Ann. Rep. Chem. Soc. (London) **1923**, 87.
Ann. Rep. Chem. Soc. (London) **1921**, 76.

Angeli-Rimini (Aldehyde) Reaction

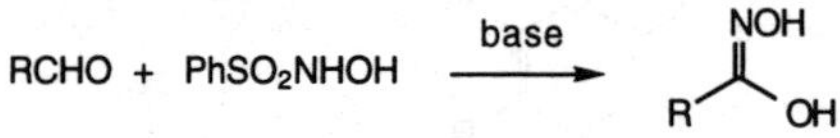

Ann. Rep. Chem. Soc. (London) **1923**, 87.
Ann. Rep. Chem. Soc. (London) **1921**, 76.

Note: Also known as the Rimini Aldehyde Reaction.

Arbusov-Wittig (see Wittig Reaction)

Arbusow Reaction (see Arbuzov Reaction)

Arbuzov Reaction (Rearrangement)

$$(RO)_3P + R^1CH_2X \longrightarrow (RO)_2P(O)CH_2R^1 + RX$$

$X = Cl, Br, I$

Chem. Rev. **1984**, *84*, 577.
Chem. Rev. **1981**, *81*, 415.
Z. Chem. **1974**, *14*, 41.
Pure & Appl. Chem. **1964**, 9, 307.
Org. Reactions **1951**, 6, 276.

Note: Also known as the Michaelis-Arbuzov Rearrangement or Arbuzov Transformation.

Arens-van Dorp Synthesis

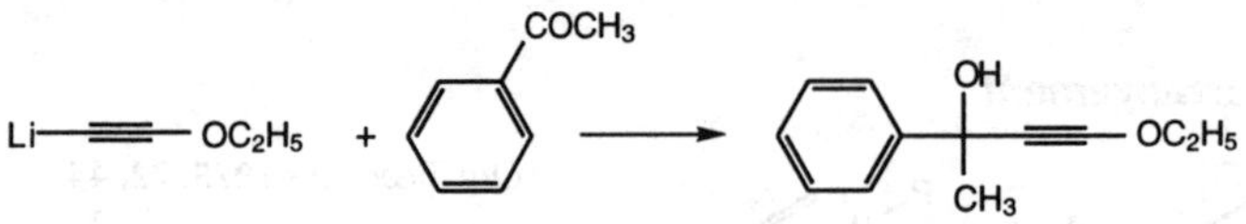

Helv. Chim. Acta **1956**, *39*, 259.

Arndt-Eistert Synthesis (Reaction)

$$RCO_2H \xrightarrow[\text{2. } 2CH_2N_2 \quad \text{3. } R^1OH]{\text{1. } SOCl_2} RCH_2CO_2R^1$$

Tetrahedron Lett. **1980**, 4461.
Org. Reactions **1942**, *1*, 38.

Atherton-Todd Reaction

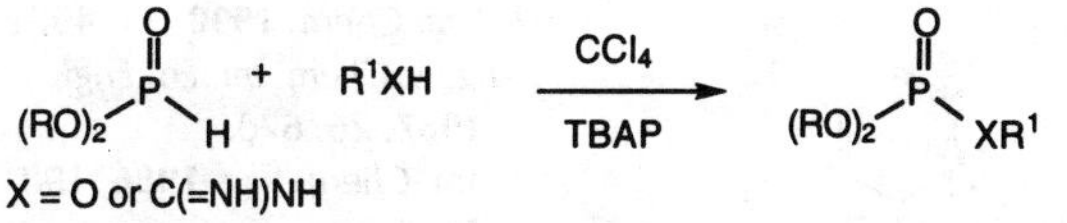

Synthesis **1985**, 971.
Synthesis **1976**, 305.

Auwers Synthesis

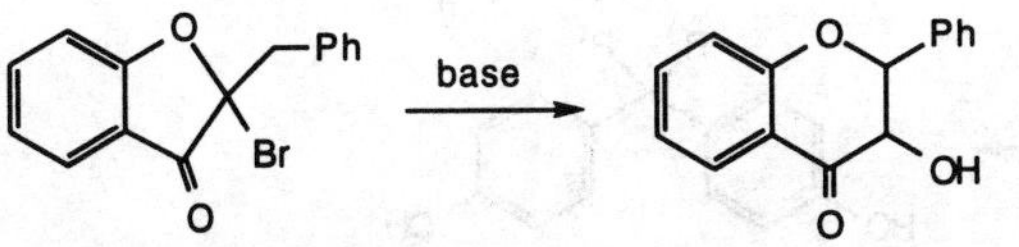

J. Chem. Soc. **1940**, 817.

Aza-Claisen Rearrangement

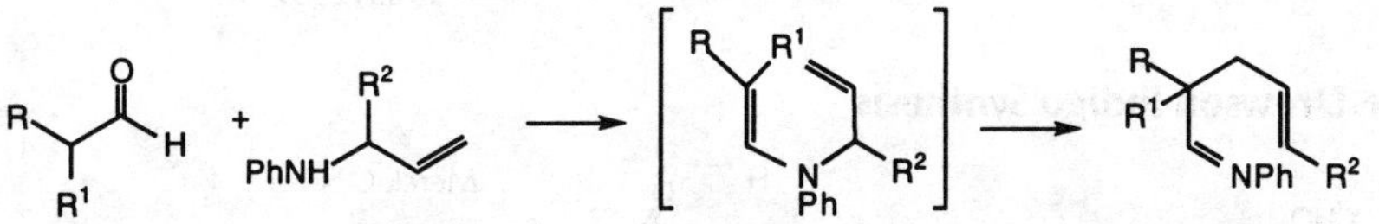

Synthesis **1988**, 362.
Tetrahedron Lett. **1978**, 4337.

Aza-Cope Rearrangement

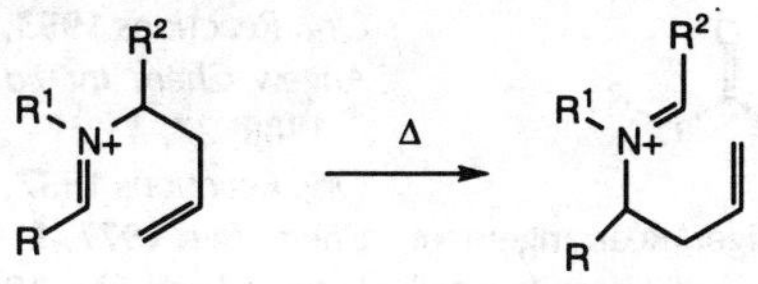

Angew. Chem. Int. Ed. Engl. **1987**, *26*, 670.
J. Am. Chem. Soc. **1986**, *108*, 2400.
J. Am. Chem. Soc. **1983**, *105*, 6622.
Org. Reactions **1975**, *22*, 42.

Aza-Ene Reaction

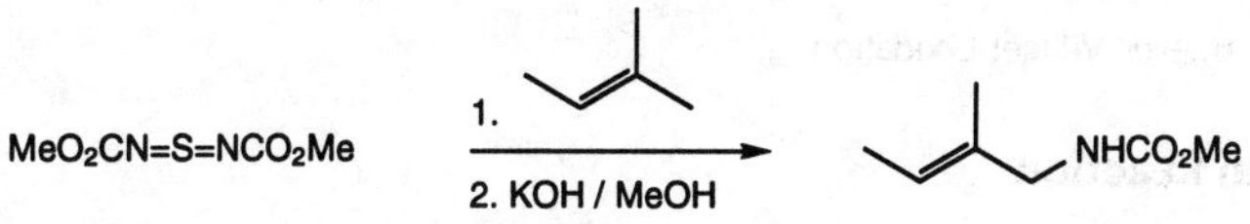

Org. Synth. **1987**, *65*, 159.

Aza-Wittig Reaction

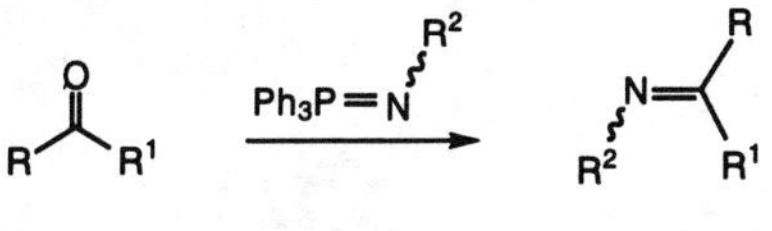

Russ. Chem. Rev. **1991**, *60*, 146.
Tetrahedron **1989**, *45*, 4263.
J. Chem. Soc., Chem. Commun. **1982**, 1224.

Note: Also see the Wittig Reaction.

Azo-Cope Rearrangement

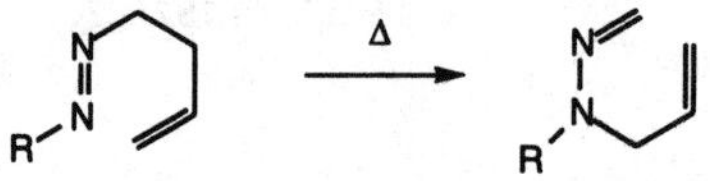

J. Org. Chem. **1990**, *55*, 4981.
Angew. Chem. Int. Ed. Engl. **1987**, *26*, 670.
J. Am. Chem. Soc. **1986**, *108*, 2400.

Baeyer Reaction

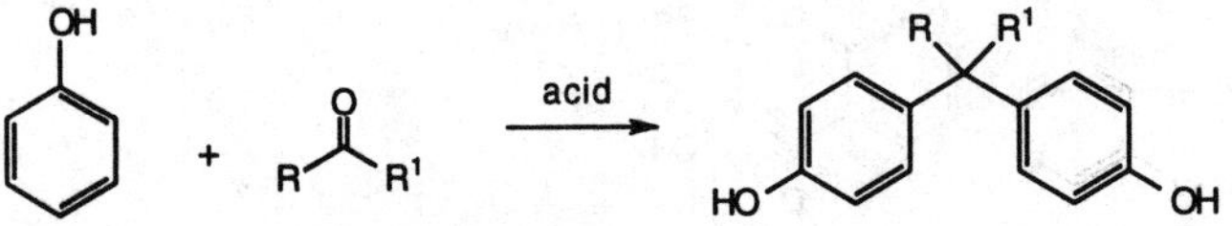

Angew. Chem. Int. Ed. Engl. **1963**, *2*, 373.

Baeyer-Drewson Indigo Synthesis

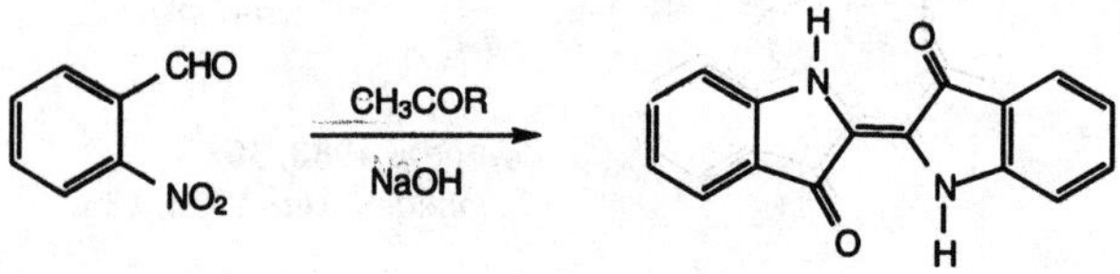

Merck ONR-5.

Baeyer-Villiger Rearrangement (Reaction)

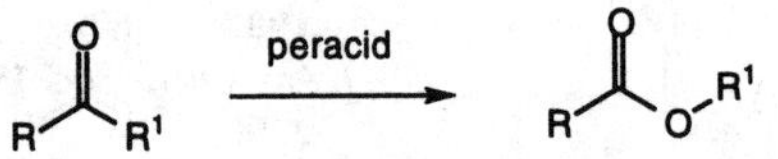

Org. Reactions **1993**, *43*, 251.
Angew. Chem. Int. Ed. Engl. **1988**, *27*, 333.
Org. Reactions **1957**, *9*, 73.
Retro-'Baeyer-Villiger' Rearrangement: *Chem. Ber.* **1977**, *111*, 1228.
Silicon Directed: *J. Am. Chem. Soc.* **1980**, *102*, 6894.
Tin Directed: *J. Am. Chem. Soc.* **1990**, *112*, 6729.

Note: Also known as the Baeyer-Villiger Oxidation.

Baker Venkataraman Reaction

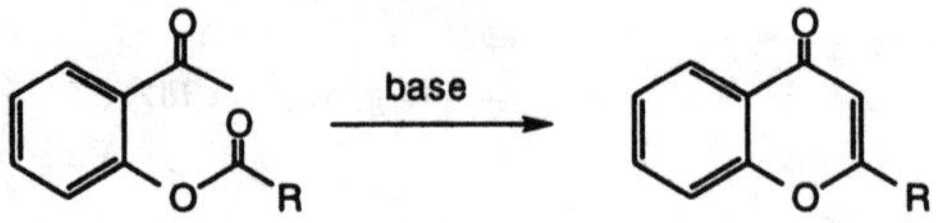

Synthesis **1985**, 697.

Baker-Venkataraman Rearrangement (Transformation)

base

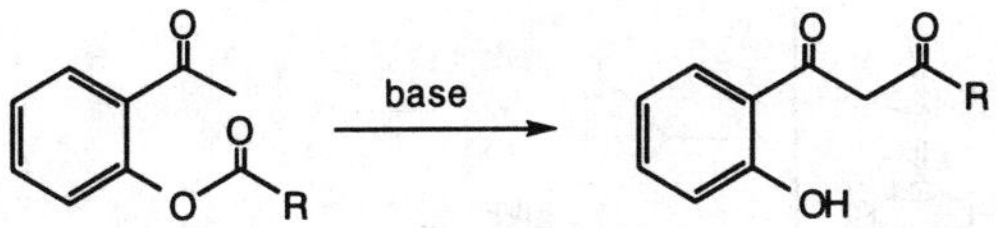

Comp. Org. Chem. **4**, 678.

Bally-Scholl Synthesis

HOCH$_2$CH(OH)CH$_2$OH

H$_2$SO$_4$

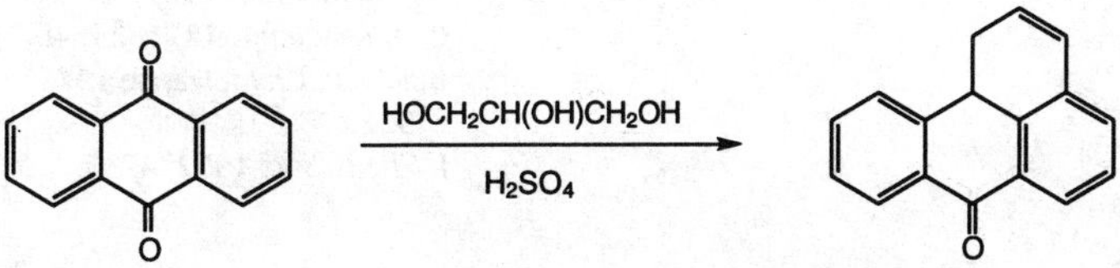

Merck ONR-6
J. Chem. Soc. **1938**, 401.

Note: Also known as Bally's Reaction.

Balz-Schiemann Reaction

ArNH$_2$ $\xrightarrow[\text{2. }\Delta]{\text{1. HNO}_2\text{ - HBF}_4}$ ArF

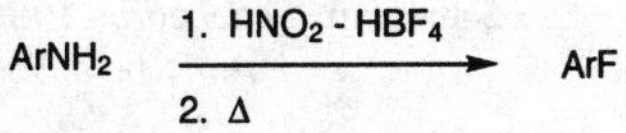

Chem. Ber. **1990**, *123*, 1673.
Adv. Fluorine Chem. **1965**, *4*, 1.
Org. Reactions **1949**, *5*, 193.

Note: Also known as the Schiemann Reaction.

Bamberger Reaction

PhCOCl

NaOH

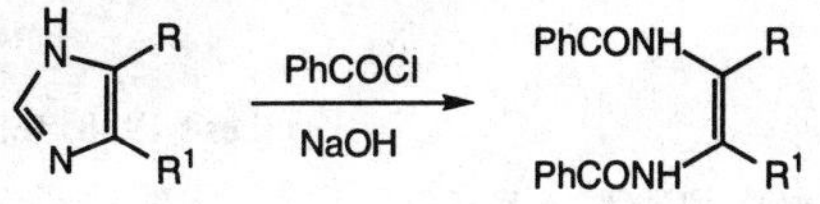

J. Org. Chem. **1990**, *55*, 3370.

Bamberger Rearrangement

acid

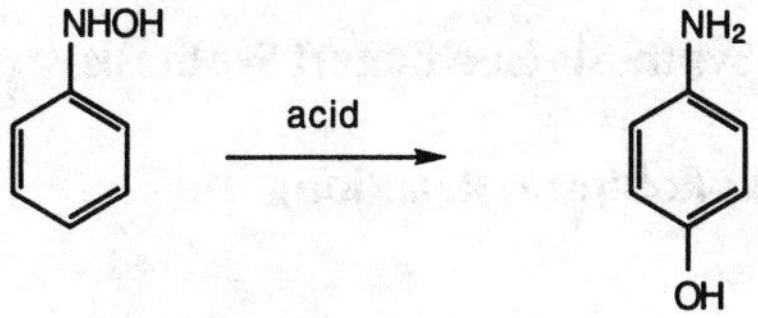

Comp. Org. Chem. **2**, 177; 200.

Bamberger Triazine Synthesis

Δ / acid

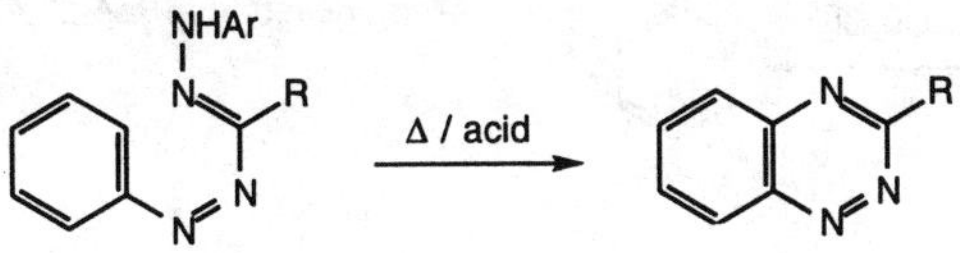

J. Chem. Soc. **1955**, 2326.

Bamford-Stevens Reaction

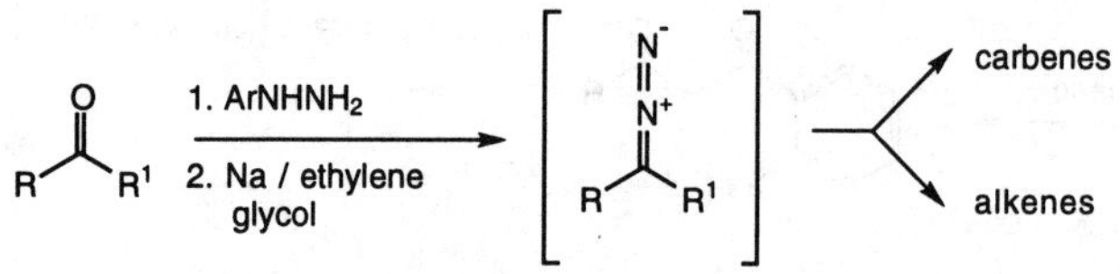

Synthesis **1982**, 418.
Org. Reactions. **1976**, *23*, 407.
Bull. Soc. Chim. France **1975**, 922.
J. Chem. Soc. **1952**, 4735.

Barbier Reaction

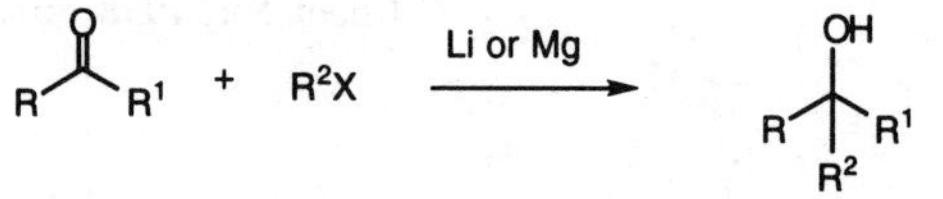

Can. J. Chem. **1987**, *65*, 229.
J. Am. Chem. Soc. **1982**, *104*, 3481.
J. Am. Chem. Soc. **1980**, *102*, 7926.
Synthesis **1977**, 18.
Samarium: *Tetrahedron* **1986**, *42*, 6573.
J. Am. Chem. Soc. **1980**, *102*, 2693.
Ultrasonic: *Synthesis* **1990**, 805.

Barbier-Grignard Reaction (see Barbier Reaction)

Barbier-Wieland Degradation

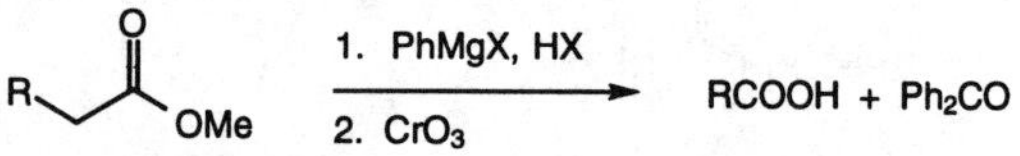

Synthesis **1978**, 468.

Note: Also known as the Barbier-Wieland procedure and Barbier-Locquin Degradation.

Bardhan-Sengupta Phenanthrene Synthesis (see Bogert Synthesis)

Bargellini Reaction (see Kostanecki-Robinson Reaction)

Bart Reaction

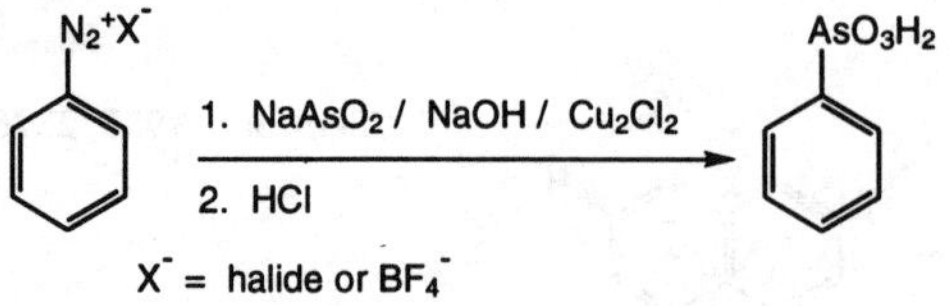

Org. Synth. **1955**, *3*, 665.
Org. Reactions **1944**, *2*, 415.

Barton Decarboxylation Reaction

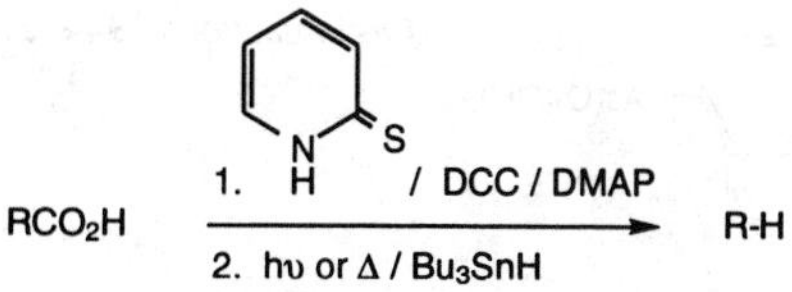

J. Org. Chem. **1991**, *56*, 2193.
J. Org. Chem. **1990**, *55*, 376.
Tetrahedron **1985**, *41*, 3901.

Barton Deoxygenation

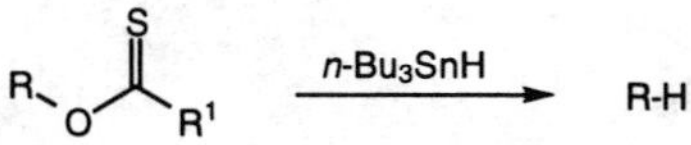

Can. J. Chem. **1988**, 66, 1258.
Tetrahedron **1983**, *39*, 2609.
J. Chem. Soc., Perkin Trans. 1 **1975**, 1574.

Barton Fragmentation

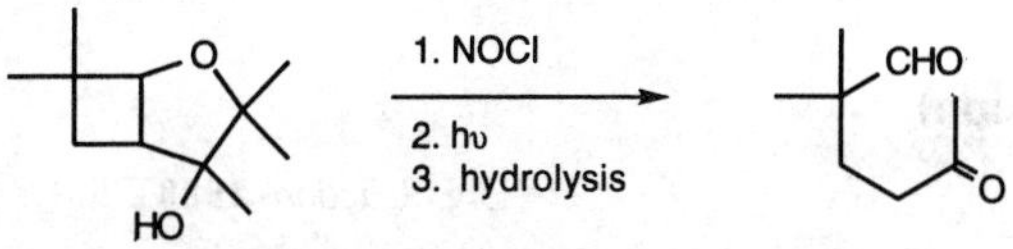

J. Am. Chem. Soc. **1982**, *104*, 4990.

Barton Olefination Reaction

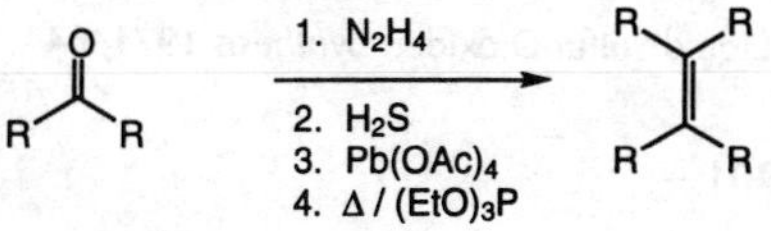

J. Chem. Soc., Perkin Trans. 1 **1974**, 1794.
J. Chem. Soc., Perkin Trans. 1 **1972**, 305.

Barton Reaction

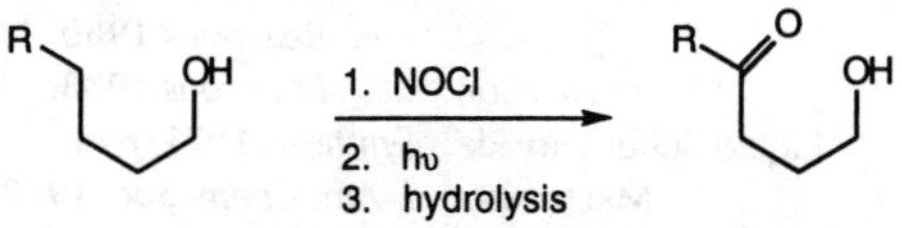

Tetrahedron **1971**, *27*, 291.
Adv. Free-Radical Chem. **1969**, *3*, 83.
Angew. Chem. Int. Ed. Engl. **1964**, *3*, 525.

Mechanism: *J. Chem. Soc., Perkin Trans. 1* **1979**, 1159.

Bashkirov Reaction

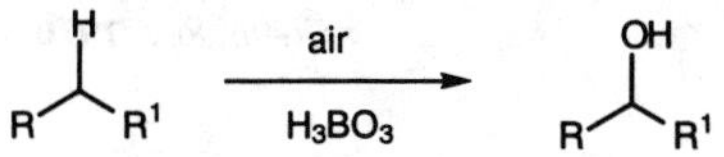

Comp. Org. Chem. **1**, 588.

Baudisch Reaction

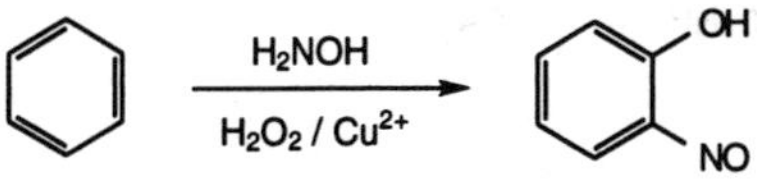

Merck ONR-8.
J. Org. Chem. **1967**, *52*, 2516.

Béchamp Reaction

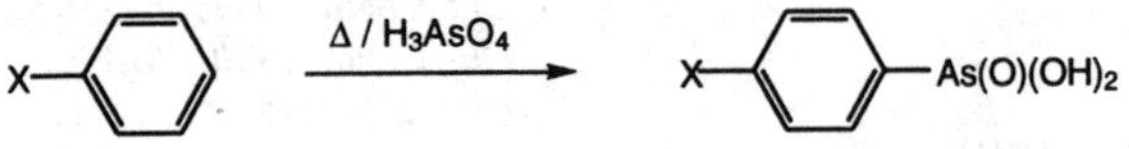

X—⟨benzene ring⟩ $\xrightarrow{\Delta / H_3AsO_4}$ X—⟨benzene ring⟩—As(O)(OH)$_2$

Org. Reactions **1944**, *2*, 428.

X = OH, NH$_2$

Béchamp Reduction

ArNO$_2$ $\xrightarrow{Fe / HCl}$ ArNH$_2$

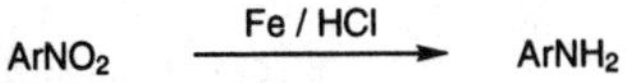

Bull. Chem. Soc. Jpn. **1956**, *29*, 194.

Beckmann Elimination-Addition Reaction

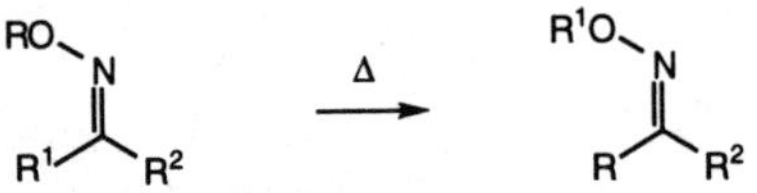

$\xrightarrow{\Delta}$

Org. Reactions **1988**, *35*, 1.

Beckmann Fragmentation (Fission)

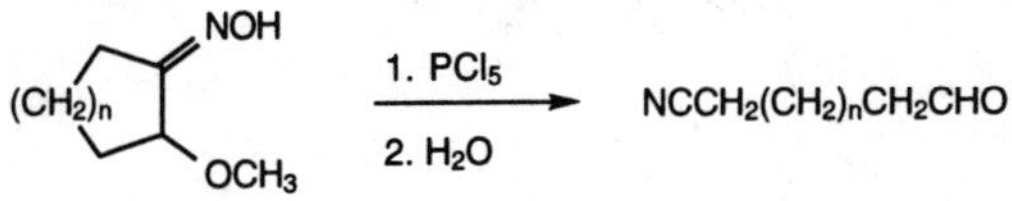

$\xrightarrow[\text{2. H}_2\text{O}]{\text{1. PCl}_5}$ NCCH$_2$(CH$_2$)$_n$CH$_2$CHO

Org. Reactions **1988**, *35*, 1.

Liquid Sulfur Dioxide: *Synthesis* **1971**, 644.

Beckmann Rearrangement (Reaction)

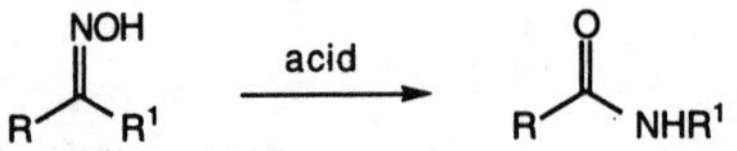

$\xrightarrow{acid}$

Org. Reactions **1988**, *35*, 1.
Tetrahedron **1980**, *36*, 1279.
Tetrahedron **1971**, *27*, 1317.
Org. Reactions **1960**, *11*, 1.
Cyclization: *Org. Reactions* **1988**, *35*, 1.
Liquid Sulfur Dioxide: *Synthesis* **1971**, 644.
Mechanism: *J. Am. Chem. Soc.* **1959**, *81*, 612.
J. Org. Chem. **1955**, *20*, 488.
Nitrone-Beckmann: *J. Chem. Soc., Perkin Trans. 2* **1975**, 1764.
J. Chem. Soc. (D) **1971**, 945.
Photo Beckmann: *Pure & Appl. Chem.* **1977**, *49*, 305.
Chem. Rev. **1970**, *70*, 23.

Beecham Reaction

ArSO$_2$NHC(RR^1)COCl $\xrightarrow[\text{2. H}_2\text{O}]{\text{1. NaOH}}$ ArSO$_2$NH$_2$ +

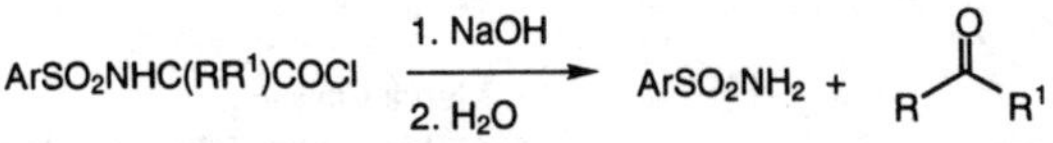

Org. Reactions **1962**, *12*, 165.

Behrend Rearrangement

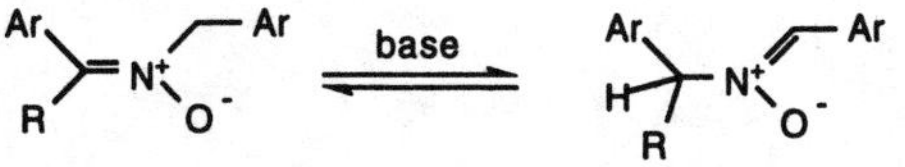

Comp. Org. Chem. **2**, 506.

Bénary Reaction

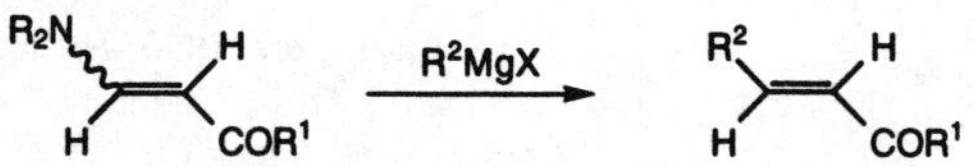

Merck ONR-9.
Bull. Chem. Soc. France. **1965**, 2328.

Benkeser Reduction

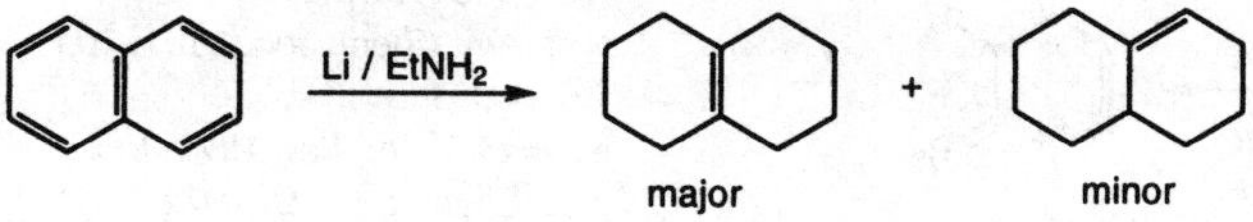

Synthesis **1972**, 391.

Benzidine Rearrangement

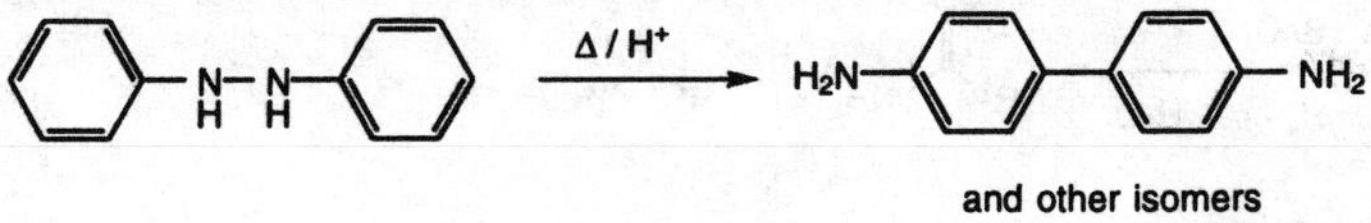

Comp. Org. Chem. **1**, 337; **2**, 230.

Benzilic Acid Rearrangement

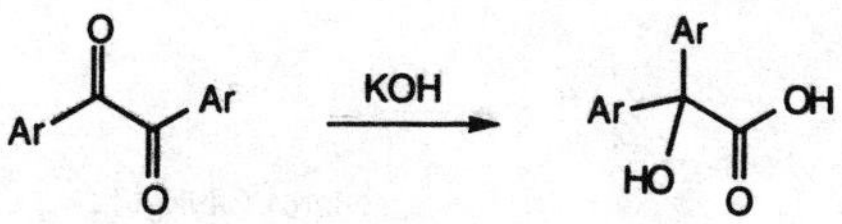

Quart. Rev. **1960**, *14*, 221.

Note: Also known as the Benzil-Benzilic Acid Rearrangement.

Benzoin Condensation

$ArCHO + Ar^1CHO \xrightarrow{CN^-} ArCH(OH)COAr^1$

Org. Reactions **1976**, *23*, 296.
Org. Reactions **1948**, *4*, 272.
Catalysts: *J. Chem. Soc., Chem. Commun.* **1982**, 580.

Benzylic Acid Rearrangement (see Benzilic Acid Rearrangement)

Bergmann Azlactone Peptide Synthesis

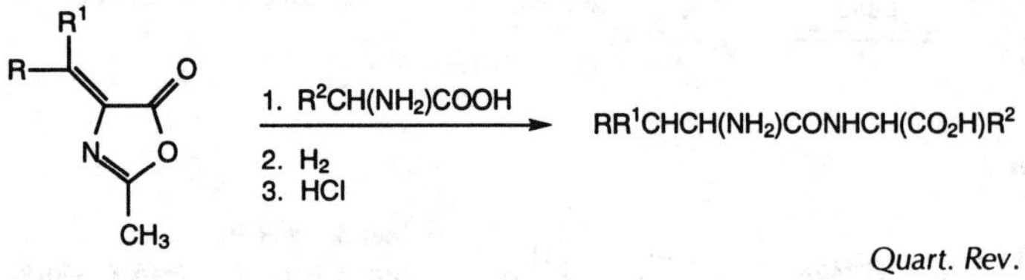

Quart. Rev. **1956**, *10*, 235.

Bergman Cyclization

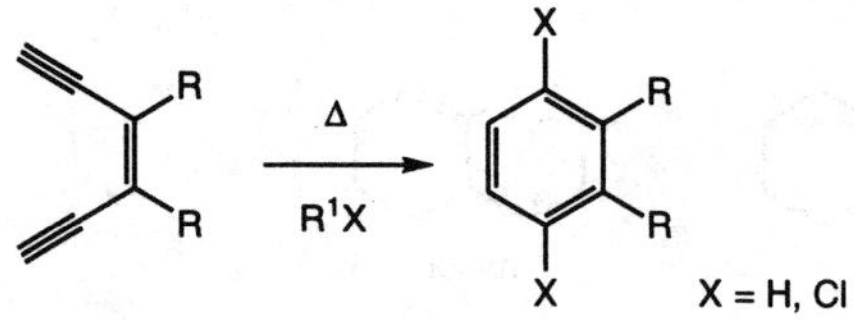

Angew. Chem. Int. Ed. Engl. **1991**, *30*, 1387.
J. Am. Chem. Soc. **1981**, *103*, 4091.
Acc. Chem. Res. **1973**, 6, 25.
J. Chem. Soc. (D) **1971**, 1516.

Bergmann Degradation

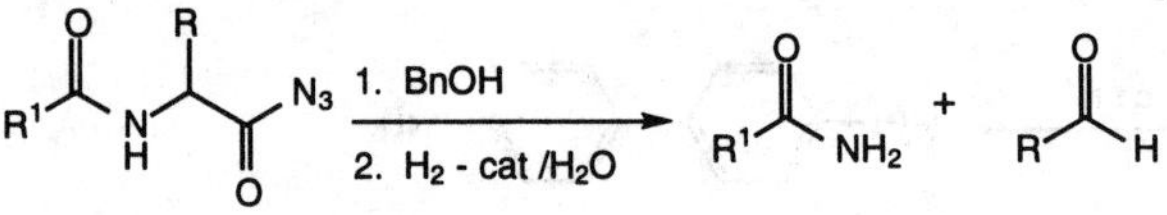

Merck ONR-10.

Bernet-Vasella Reaction

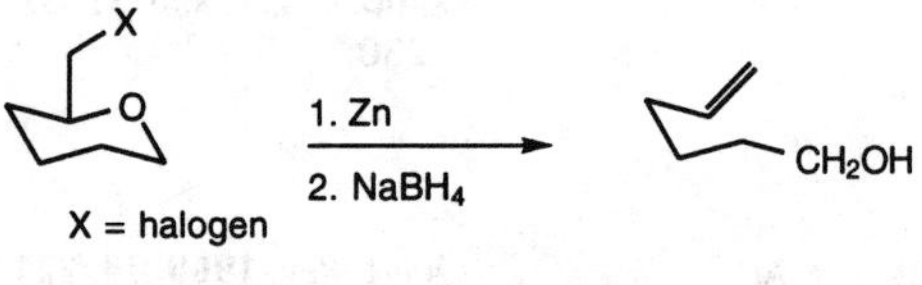

Can. J. Chem. **1984**, *62*, 2728.

Bernthsen Acridine Synthesis

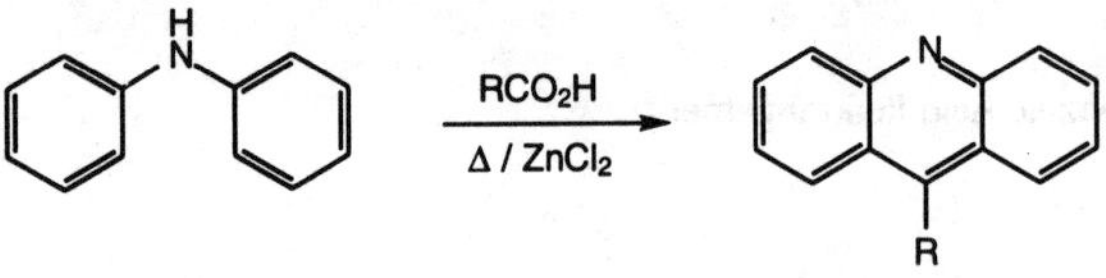

Merck ONR-11.
J. Org. Chem. **1962**, *27*, 2658.

Betti Reaction

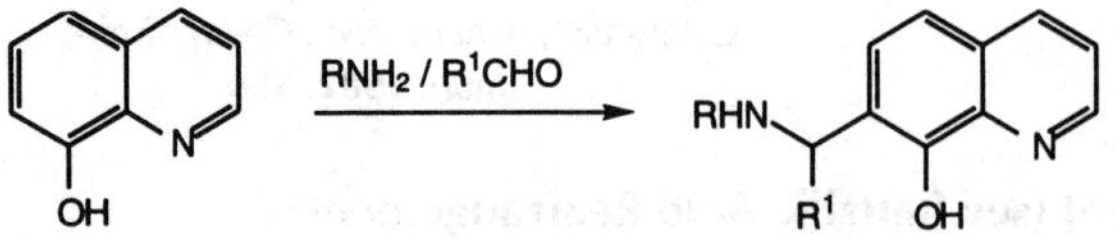

Chem. Rev. **1956**, *56*, 286.
J. Org. Chem. **1954**, *19*, 907.

Beyer Synthesis (Beyer Method for Quinolines)

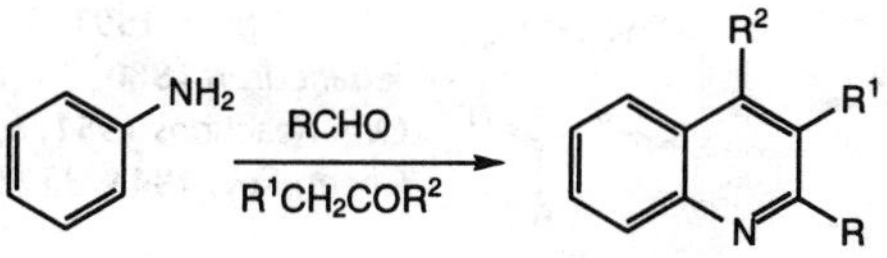

Comp. Org. Chem. **4**, 161.

Note: The Beyer Synthesis is also refered to as the Doebner Reaction.

Biginelli Pyrimidine Synthesis

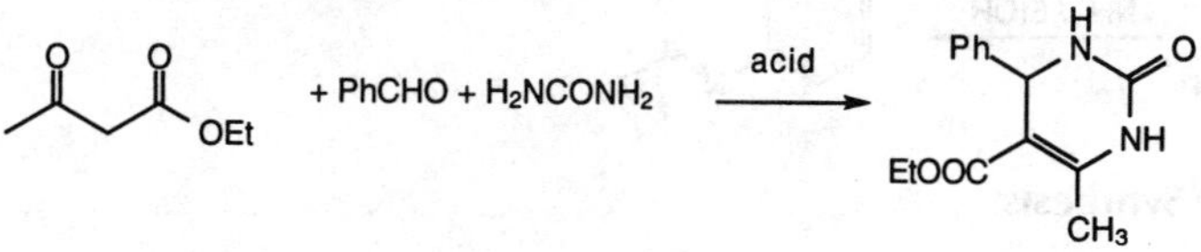

Org. Reactions **1965**, *14*, 88.

Birch Reduction

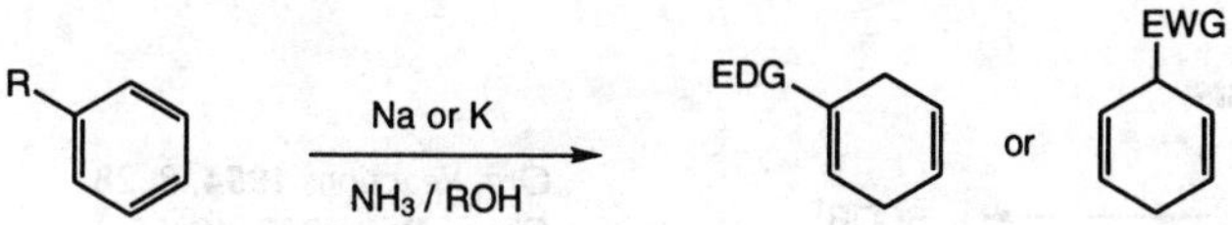

Tetrahedron **1989**, *45*, 1579.
Acc. Chem. Res. **1982**, *15*, 245.
Org. Reactions **1976**, *23*, 1.
Synthesis **1972**, 391.
Quart. Rev. **1950**, *4*, 69.
Alkenes: *Org. Reactions* **1976**, *23*, 1.
Alkynes: *J. Chem. Soc., Chem. Commun.* **1968**, 634.
J. Am. Chem. Soc. **1963**, *85*, 622.
Aromatics: *Org. Reactions* **1992**, *42*, 1.
Heterocycles: *Chem. Rev.* **1969**, *69*, 785.
Regioselectivity: *J. Am. Chem. Soc.* **1990**, *112*, 1280.

Bischler (Indole) Synthesis

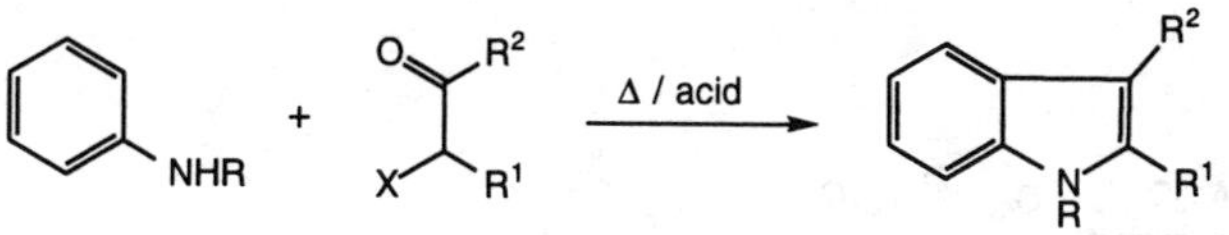

X = halogen or hydroxy

Comp. Org. Chem. **4**, 461.

Note: Also known as the Bischler-Möhlau Indole Synthesis.

Bischler-Napieralski Reaction

Dehydrating agent

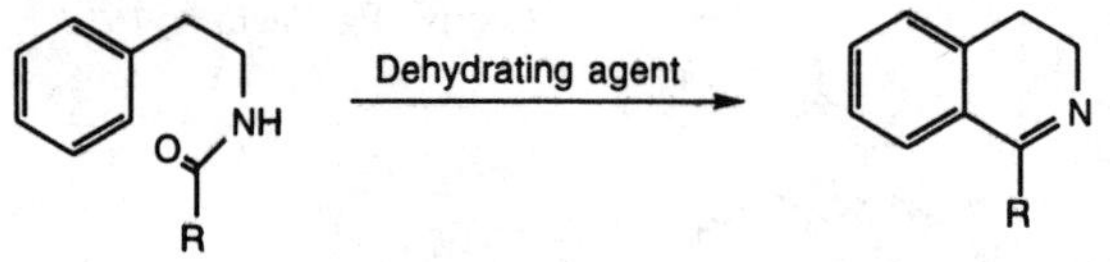

J. Org. Chem. **1991**, 56, 6034.
Tetrahedron **1980**, 36, 1279.
Org. Reactions **1951**, 6, 74.
Chem. Rev. **1944**, 35, 218.

Bischler Quinoxaline Synthesis

NH_3 / EtOH

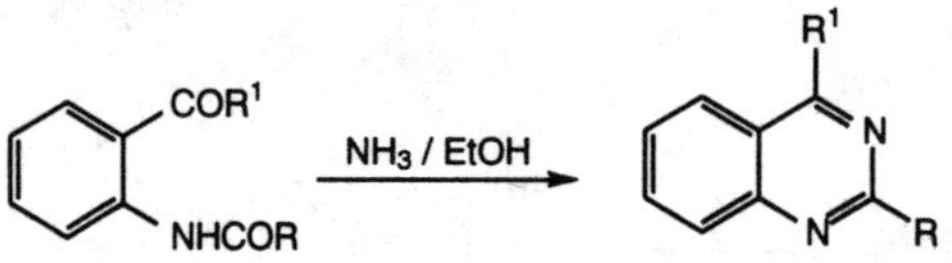

Comp. Org. Chem. **4**, 126.

Bischler Triazine Synthesis

acid

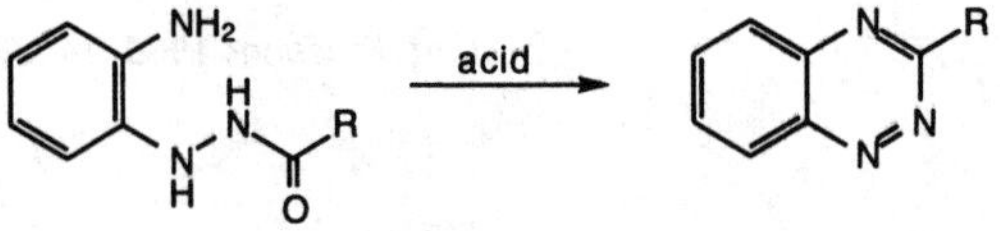

J. Chem. Soc. **1955**, 2326.

Blaise Ketone Synthesis

$$RZnCl + R^1COCl \longrightarrow RCOR^1$$

Org. Reactions **1954**, 8, 28.
Chem. Rev. **1947**, 40, 15.

Blaise-Maire Reaction

$$AcOCH_2CHRCOCl \xrightarrow[\text{2. } H_2SO_4]{\text{1. } R^1ZnCl} CH_2=CR\text{-}COR^1$$

Merck ONR-12.

Blaise Reaction

$$RR^1CBrCO_2Et \xrightarrow[\substack{\text{2. } R^2CN \\ \text{3. } H_2O}]{\text{1. Zn}} R^2COCRR^1CO_2Et$$

J. Org. Chem. **1983**, 48, 3833.

Blanc (Chloromethylation) Reaction

$$ArH + CH_2O \xrightarrow[ZnCl_2]{CH_2O \text{ / } HCl} ArCH_2Cl$$

Org. Reactions **1942**, 1, 63.

Blanc Reaction

$$HO_2CCH_2(CH_2)_xCO_2H \xrightarrow{Ac_2O}$$

$x = 1, 2$

when x = >2 cyclic
ketones result

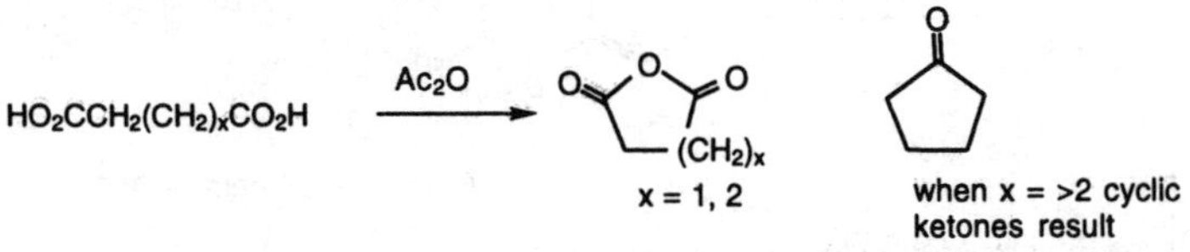

Merck ONR-13.

Bodroux-Chichibabin Aldehyde Synthesis

$$HC(OR)_3 \xrightarrow[\text{2. } H_3O^+]{\text{1. } R^1MgX} R^1CHO$$

Org. Synth. **1955**, *3*, 701.
J. Org. Chem. **1941**, *6*, 437.

Note: Also known as the Bodroux-Tschitschibabin Aldehyde Synthesis.

Bodroux Reaction

$$RCO_2Et \xrightarrow{XMgNHR^1} RCONHR^1$$

Comp. Org. Chem. **2**, 977.

Boekelheide-Weinstock Rearrangement

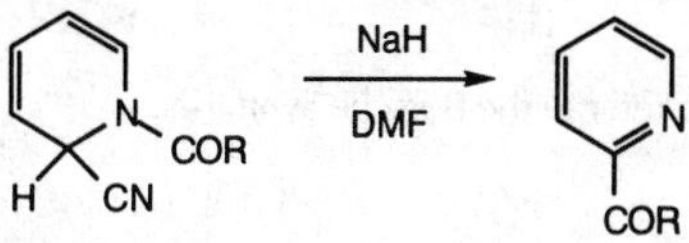

Arch. Pharm. **1973**, *306*, 648.

Bogert (-Cook) Synthesis

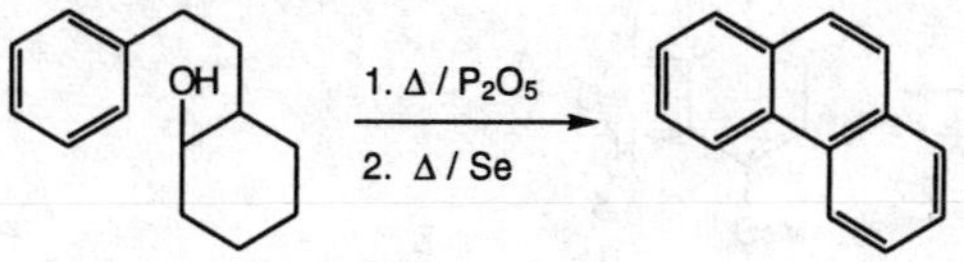

Chem. Rev. **1976**, *76*, 527.
J. Am. Chem. Soc. **1951**, *73*, 317.
Chem. Rev. **1941**, *29*, 536.

Note: Also known as the Bardhan-Sengupta Phenanthrene Synthesis.

Bohn-Schmidt Reaction

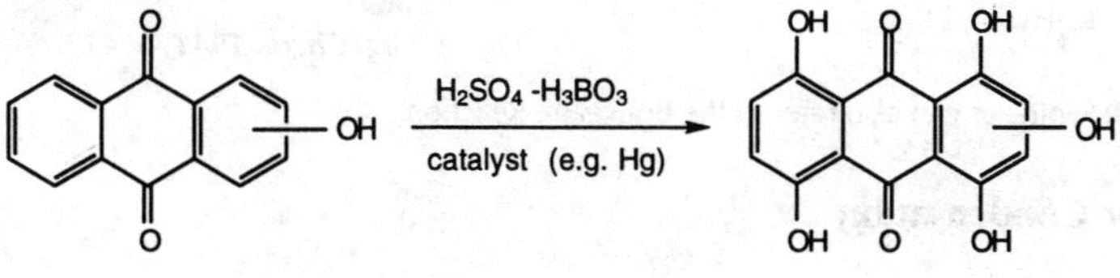

Helv. Chim. Acta **1965**, *48*, 119.
Chem. Rev. **1929**, *6*, 168.

Boord (Olefin) Synthesis

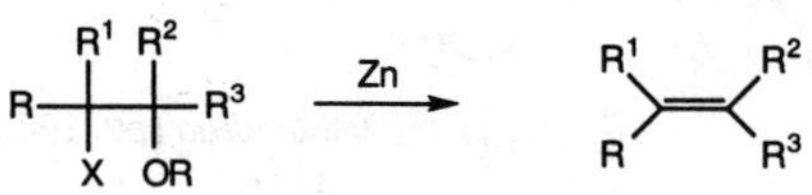

Quart. Rev. **1952**, *6*, 131.
J. Am. Chem. Soc. **1951**, *73*, 3329.

Note: Also known as the Boord Reaction.

Borodine (-Hunsdiecker) Reaction (see Hunsdiecker Reaction)

Borsche (Cinnoline) Synthesis

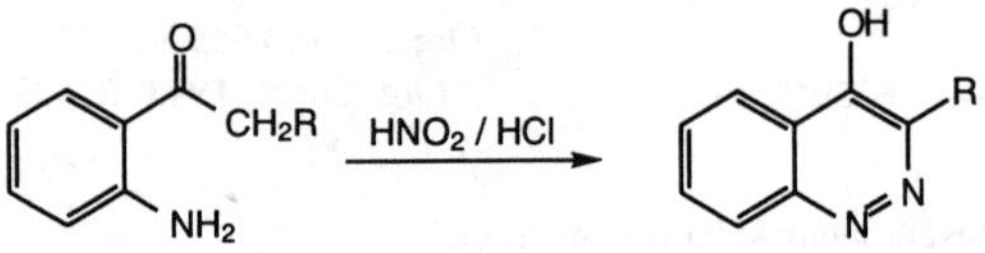

Comp. Org. Chem. **4**, 123.
Org. Reactions **1959**, 10, 28.

Note: The Borsche Synthesis can also refer to the Borsche-Drechel Cyclization.

Borsche-Drechsel Cyclization (Synthesis)

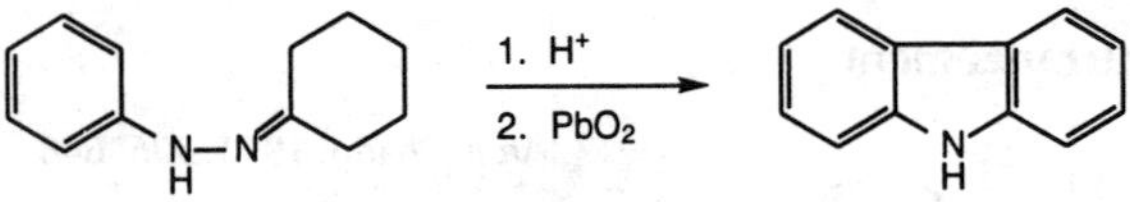

Chem. Rev. **1947**, 40, 359.

1. H⁺
2. PbO₂

Note: The Borsche-Drechel Cyclization can also refer to the Borsche Synthesis.

Boulton-Katritzky Rearrangement

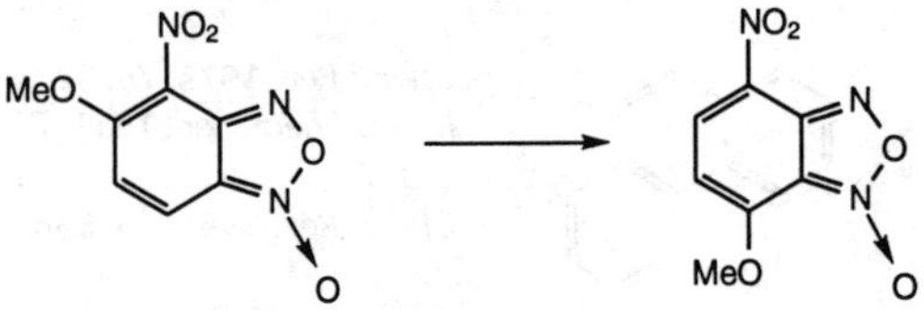

Can. J. Chem. **1990**, 68, 2239.

Bouveault (Aldehyde) Synthesis

$$R_2NCHO \xrightarrow[\text{2. } H_3O^+]{\text{1. } R^1MgX} R^1CHO + HC(R^1)_2NR_2$$

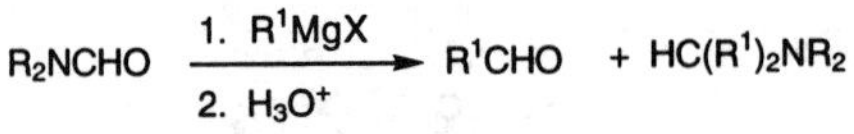

Chem. Ber. **1958**, 91, 1867
J. Am. Chem. Soc. **1953**, 75, 3697.
J. Org. Chem. **1941**, 6, 437.

Note: The Bouveault Synthesis can also refer to the Bouveault Reaction.

Bouveault-Blanc Condensation

$$RCOOR^1 \xrightarrow[\text{Me}_3\text{SiCl}]{\text{Na}}$$

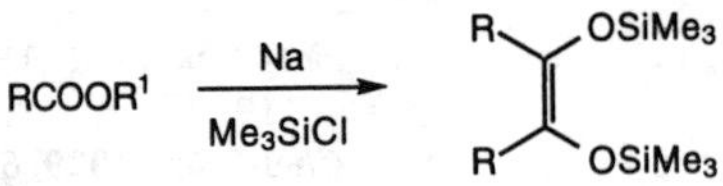

Synthesis **1971**, 236

Note: Also known as the Rühlmann Reaction.

Bouveault-Blanc Reduction

$$R \overset{O}{\underset{}{\mathord{\text{—}}}} O\text{—}R^1 \xrightarrow[\text{Toluene}]{\text{Na / EtOH}} R \overset{O}{\underset{}{\mathord{\text{—}}}} OH$$

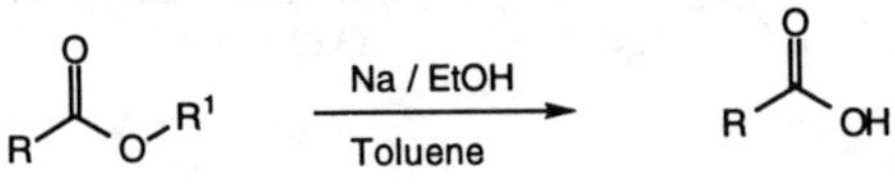

Tetrahedron Lett. **1982**, 23, 3397.
Synthesis **1972**, 409

Note: Also known as the Bouveault-Blanc Procedure.

Bouveault Reaction

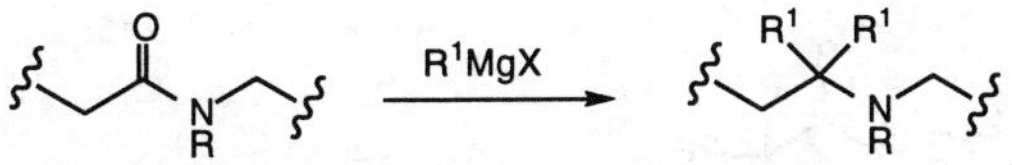

Comp. Org. Chem. **2**, 16.

Note: The Bouveault Reaction can also refer to the Bouveault Synthesis.

Boyland-Sims Oxidation

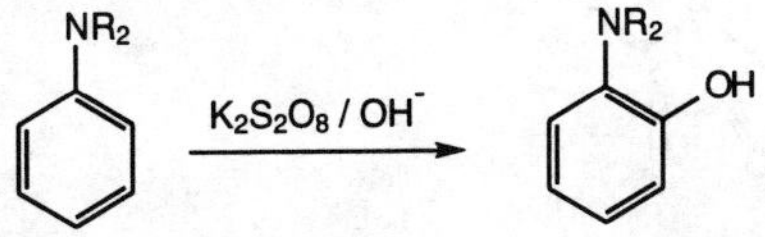

Org. Reactions **1988**, *35*, 421.

Note: For ortho disubstituted amines, para substitution results.

Bradsher Reaction (Cyclization)

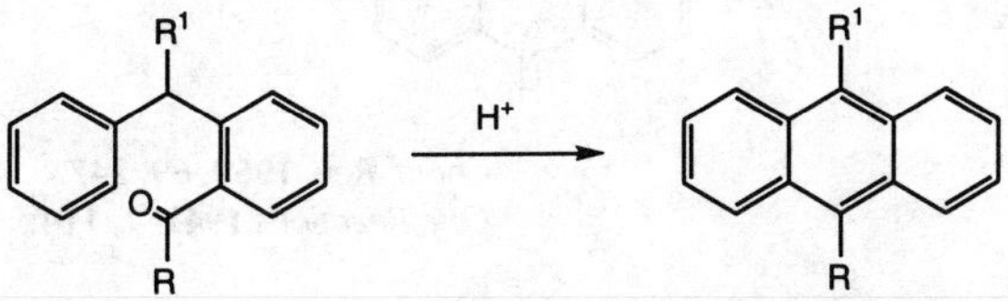

Synthesis **1970**, 655.

Note: The use of O, S, Se in place of CHR^1 is also known as the Bradsher Cyclization.

Braun Reaction (see von Braun Reaction)

Brown-Walker Synthesis

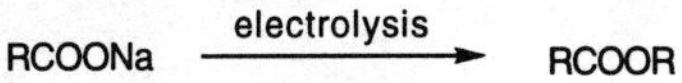

Quart. Rev. **1952**, 6, 384.

Bredereck Imidazole Synthesis

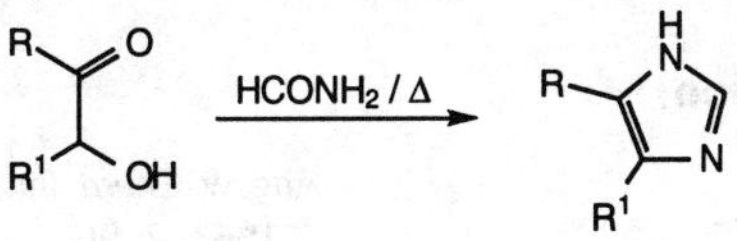

Angew. Chem. **1958**, *71*, 753.

Brook Rearrangement

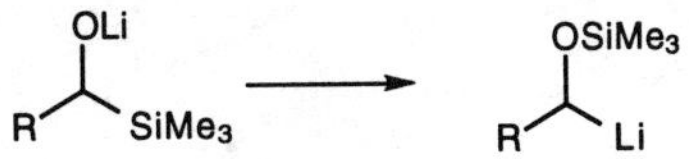

Organometal. **1984**, *3*, 1317.
Acc. Chem. Res. **1974**, *7*, 77.

Bruylants Reaction

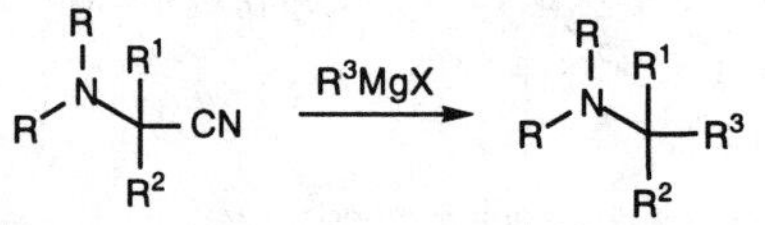

Synthesis **1985**, 745.

Bucherer-Bergs Reaction

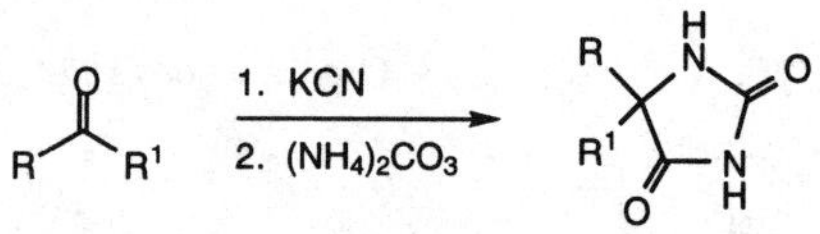

Chem. Rev. **1950**, 46, 422.

Bucherer Carbazole Synthesis

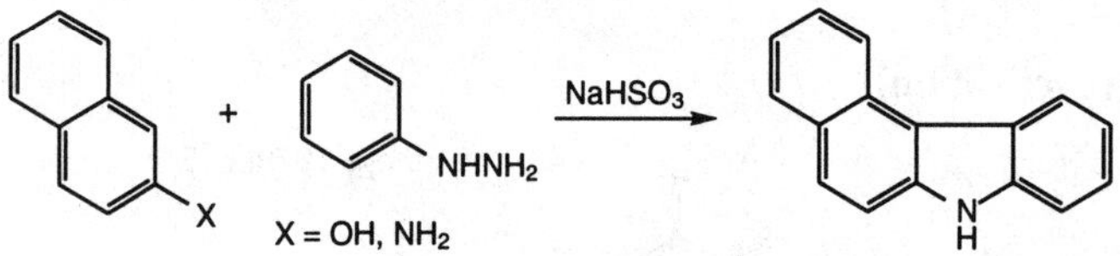

Chem. Rev. **1969**, 69, 247.
Org. Reactions **1942**, *1*, 114.

Buchner-Curtius-Schlotterbeck Reaction

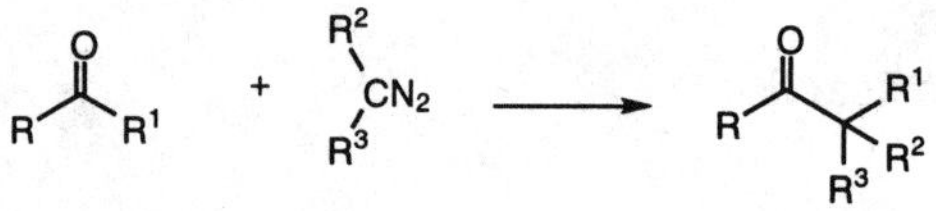

Org. Reactions **1954**, *8*, 364.

Bucherer Reaction

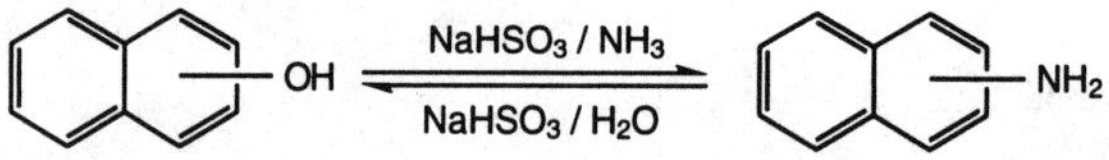

Angew. Chem. Int. Ed. Engl. **1967**, 6, 307.
Org. Reactions **1942**, *1*, 105.

Buchner Ring Enlargement (Reaction)

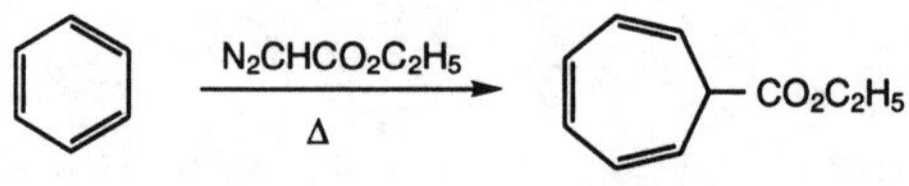

Angew. Chem. Int. Ed. Engl. **1963**, *2*, 96.
J. Am. Chem. Soc. **1963**, *85*, 1199.

Burgstahler Rearrangement

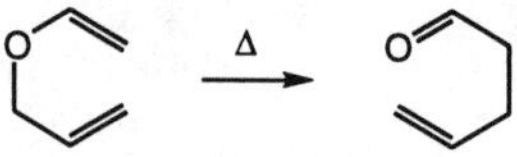

J. Am. Chem. Soc. **1961**, *83*, 198.

Butlerow Reaction

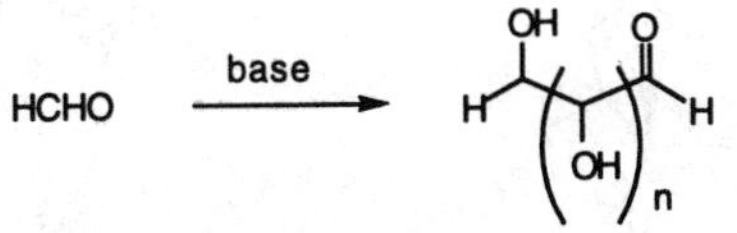

Org. Reactions **1968**, *16*, 13.
Angew. Chem. Int. Ed. Engl. **1962**, *1*, 79.

Cadiot-Chodkiewicz Coupling (Reaction)

$$R{-}C{\equiv}C{-}H + Br{-}C{\equiv}C{-}R^1 \xrightarrow{Cu^{2+}/\text{ amine}} R{-}C{\equiv}C{-}C{\equiv}C{-}R^1$$

Synth. Commun. **1986**, *16*, 847.
Angew. Chem. Int. Ed. Engl. **1976**, *15*, 764.
Adv. Org. Chem. **1963**, *4*, 225.

Note: Also known as the Cadiot-Chodkiewicz Procedure.

Cagliotí (Magi) Reaction

$$RR^1C{=}NNHTs \xrightarrow[\text{2. } H_2O]{\text{1. } LiAlH_4} RR^1CH_2$$

Org. Reactions **1976**, *23*, 430.

Camps Quinoline Synthesis

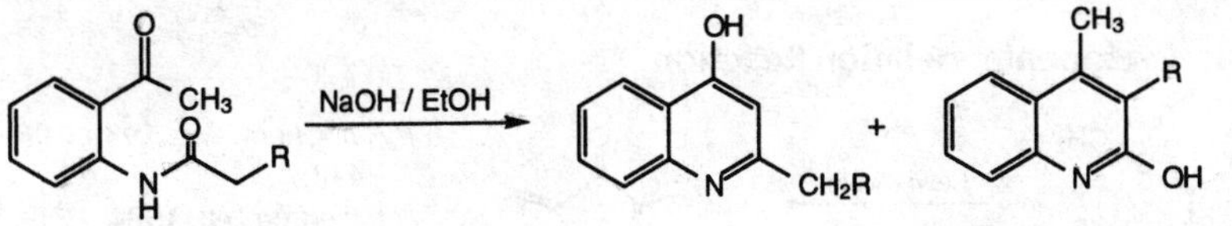

Org. Reactions **1982**, *28*, 52.
Chem. Rev. **1942**, *30*, 127.

Cannizzaro Reaction

$$2\ RCHO \xrightarrow{NaOH} RCOONa + RCH_2OH$$

Org. Reactions **1959**, *10*, 210.

Crossed Cannizzarro: *Org. Reactions* **1944**, *2*, 94.
Intramolecular: *Org. Reactions* **1959**, *10*, 210.

Note: 1. Aldehydes must have no α-hydrogens.
2. The Crossed Cannizzarro Reaction involves the oxidation and reduction of two different aldehydes.

Carbylamine Reaction (see Hofmann Isonitrile Synthesis)

Cargill Rearrangement

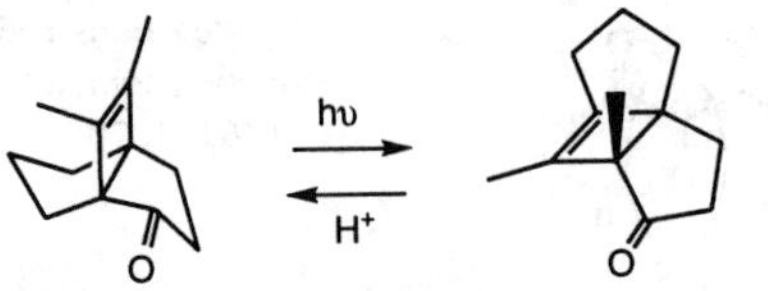

Acc. Chem. Res. **1974**, *7*, 106.

Carroll Reaction

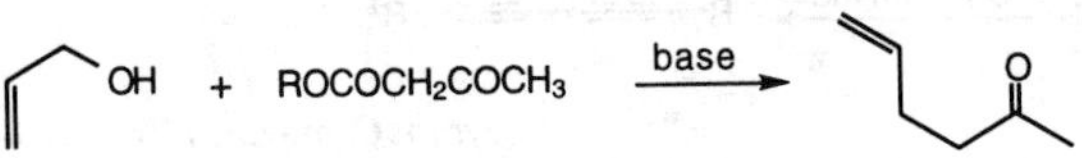

J. Chem. Soc. **1940**, 704; 1266.

Carroll Rearrangement

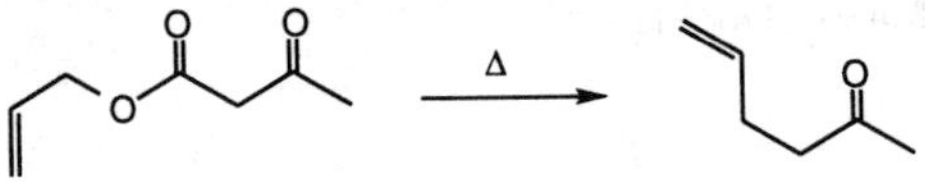

Org. Synth. **1990**, *68*, 210.
J. Am. Chem. Soc. **1980**, *102*, 862.
J. Chem. Soc. **1940**, 704, 1266.

Note: Also known as the Kimel-Cope Rearrangement.

Castro-Stephens Coupling

ArX + Cu—C≡C—R ⟶ Ar—C≡C—R

Comp. Org. Chem. **1**, 196.

Cationic-Cyclopentannelation Reaction

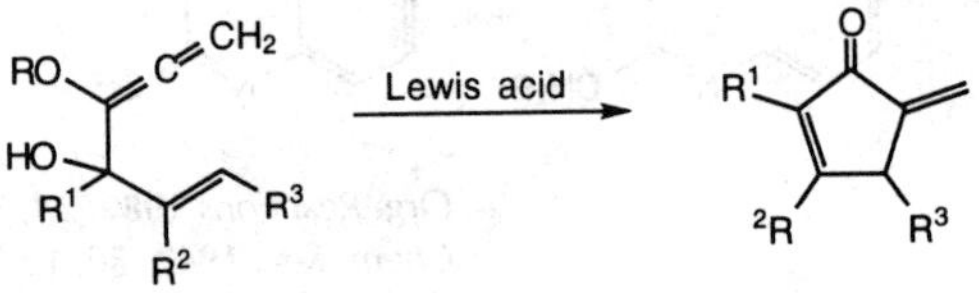

J. Am. Chem. Soc. **1986**, *108*, 3438.
Tetrahedron Lett. **1984**, 1539.

Chapman Rearrangement

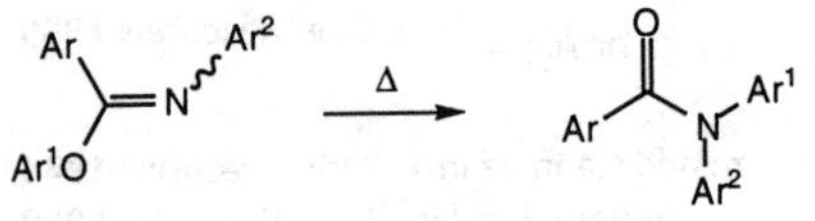

Org. Reactions **1965**, *14*, 1.

Chichibabin Pyridine Synthesis

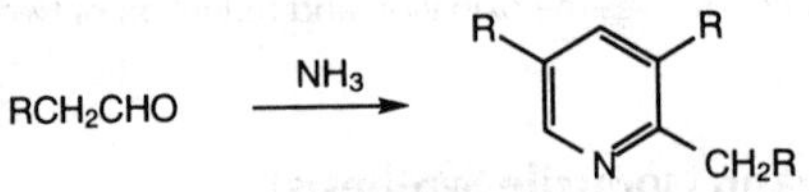

J. Am. Chem. Soc. **1956**, *78*, 3477.
Chem. Rev. **1940**, *26*, 301.

Note: Also known as the Tschitschibabin Reaction (see below).

Chichibabin Reaction

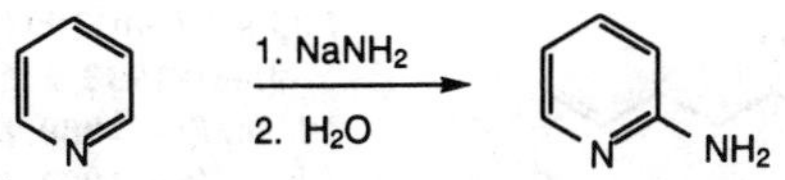

Russ. Chem. Rev. **1978**, *47*, 1042.
J. Am. Chem. Soc. **1956**, *78*, 3477.

Note: Also known as the Tschitschibabin Amination or Tschitschibabin Reaction.

Chloromethylation Reaction (see Blanc Reaction)

Chugaev Reaction

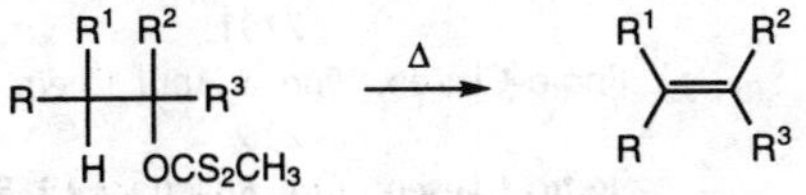

Synthesis **1972**, 107.
Org. Reactions **1962**, *12*, 57.

Note: Also known as the Tschugaeff Method or Tschugaeff Reaction.

Ciamician-Dennstedt Rearrangement

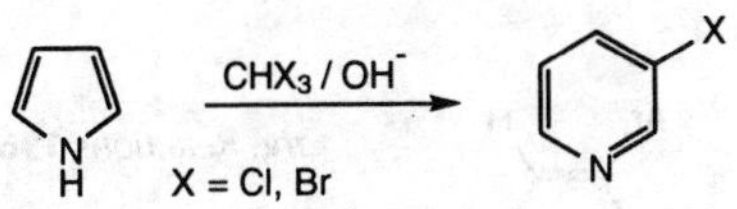

Merck ONR-18.

Cine Rearrangement

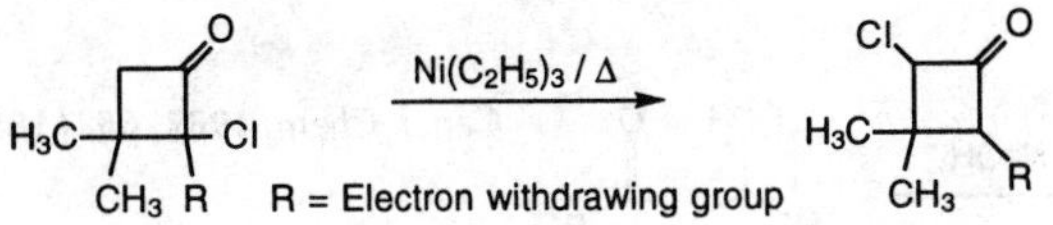

J. Am. Chem. Soc. **1979**, *101*, 5853.

Claisen Condensation

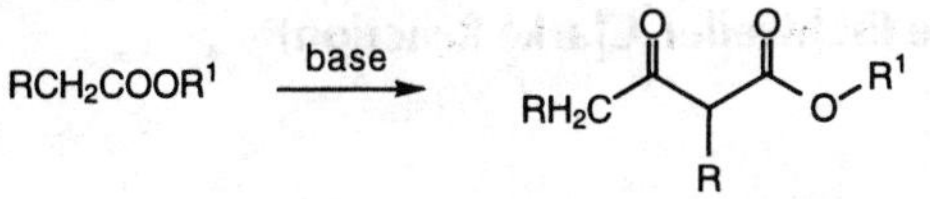

Pure & Appl. Chem. **1989**, *61*, 303.
Org. Synth. **1940**, *1*, 252.

Biosynthesis: *Pure & Appl. Chem.* **1989**. *61*, 303.

Claisen Rearrangement

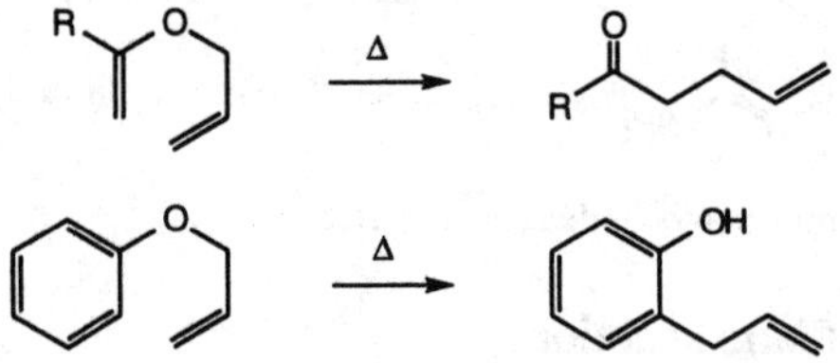

J. Org. Chem. **1991**, *56*, 650.
Synthesis **1989**, 71.
Chem. Rev. **1988**, *88*, 1423.
Chem. Rev. **1984**, *84*, 205.
Acc. Chem. Res. **1977**, *10*, 227.
Synthesis **1977**, 589.
Org. Reactions **1975**, *22*, 1.
Progr. Phys. Org. Chem. **1971**, *8*, 154.
Catalysts: *Synthesis* **1977**, 589.
Chiral: *J. Am. Chem. Soc.* **1990**, *112*, 7791.
Photo-Claisen: *Pure & Appl. Chem.* **1977**, *49*, 223.
Retro-Claisen: *Org. Reactions* **1975**, *22*, 29.

Note: 1. Also known as the Claisen Cope Rearrangement.
2. See also the Amino-Claisen, Aza-Claisen, Ireland-Claisen, Orthoester-Claisen, Para-Claisen and Ynamine Claisen Rearrangements.

Claisen-Schmidt Condensation

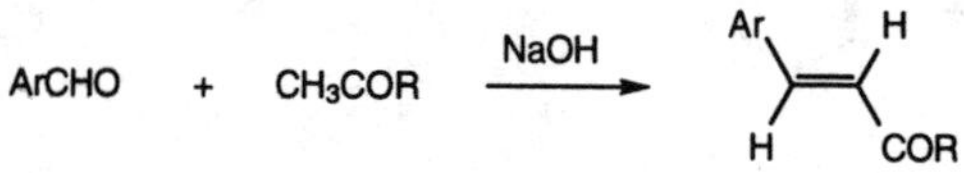

Org. Reactions **1968**, *16*, 1.

Claisen-Schmidt Reaction

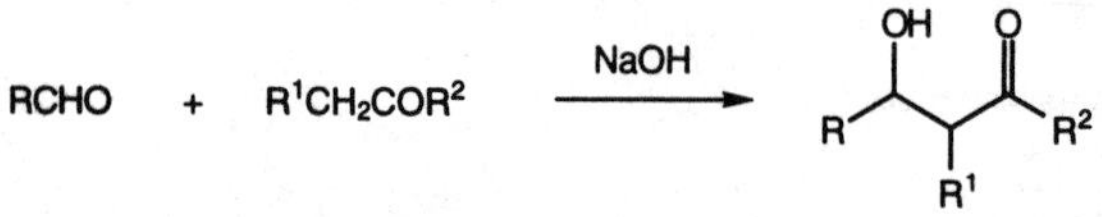

Can. J. Chem. **1987**, *65*, 1165.

Clarke-Eschweiler Reaction (see Eschweiler-Clarke Reaction)

Clauson-Kass Synthesis

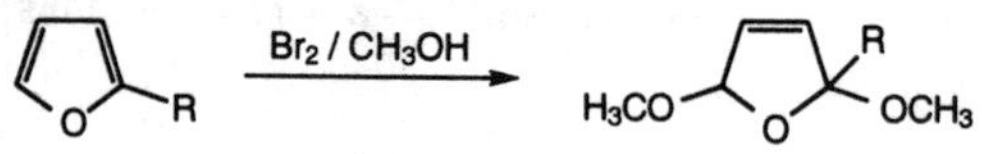

Acta Chem. Scand. **1955**, *9*, 17.
Acta Chem. Scand. **1952**, *6*, 667.

Clemmensen Reduction

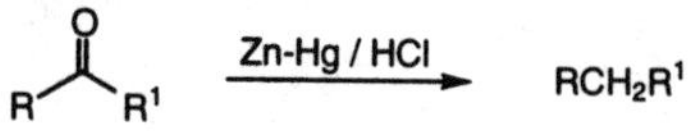

J. Org. Chem. **1991**, *56*, 4269.
Helv. Chim. Acta **1976**, *59*, 962.
Org. Reactions **1975**, *22*, 401.
Angew. Chem. **1959**, *71*, 726.

Collins Oxidation

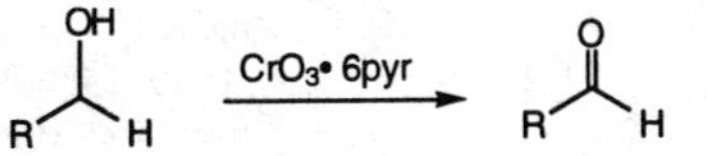

March 1057.

Collman's Reaction

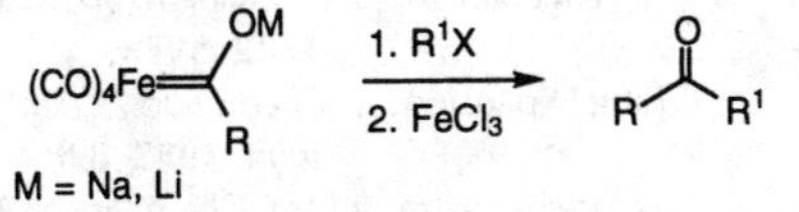

J. Am. Chem. Soc. **1983**, *105*, 4099.
Acc. Chem. Res. **1975**, *8*, 342.

Colonge-Mukherji Reaction

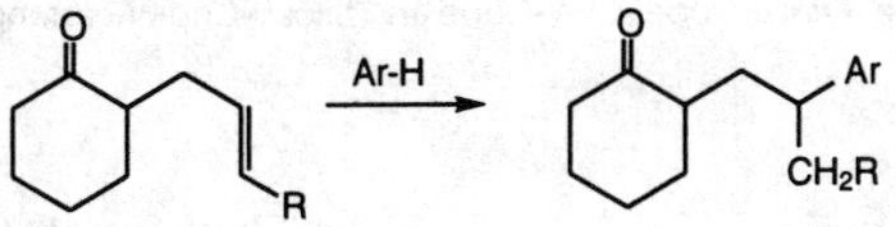

Chem. Rev. **1976**, *76*, 529.
J. Org. Chem. **1952**, *17*, 1202.

Combes (Quinoline) Synthesis

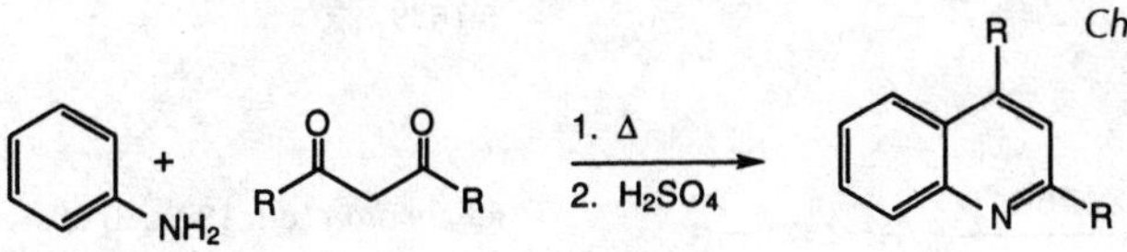

Chem. Rev. **1944**, *35*, 156.

Conrad-Limpach Reaction (Synthesis)

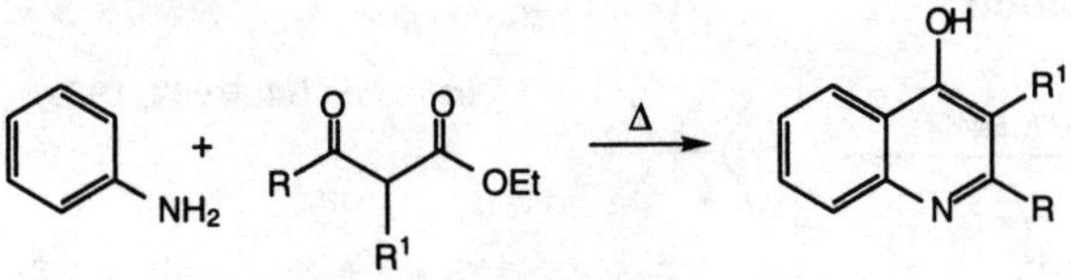

Org. Reactions **1982**, *28*, 52.

Cook-Weiss Reaction (see Weiss Reaction)

Cope (Elimination) Reaction

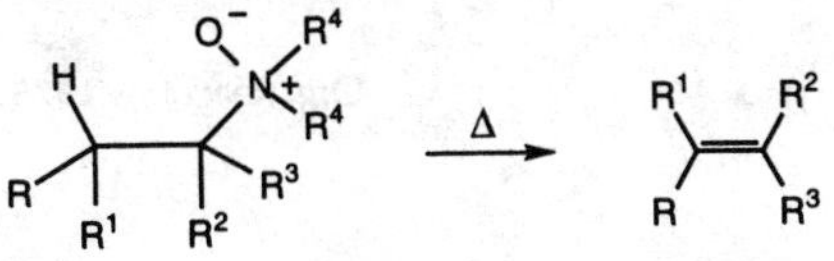

Pure & Appl. Chem. **1990**, *62*, 1531.
Org. Reactions **1960**, *11*, 361.

Cope Rearrangement

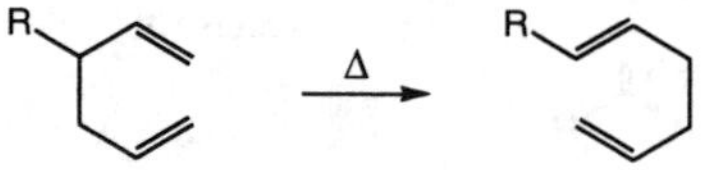

Angew. Chem. Int. Ed. Engl. **1990**, *29*, 609.
Chem. Rev. **1984**, *84*, 205.
Org. Reactions **1975**, *22*, 51.

Hetero-Cope Rearrangement: *Synthesis* **1989**, 71.
Mechanism: *J. Am. Chem. Soc.* **1990**, *112*, 1732; 5973.
Ring Expansion: *J. Chem. Soc., Chem. Commun.* **1982**, 828.
Stereochemistry: *J. Am. Chem. Soc.* **1990**, *112*, 5973.

Note: See also the Aza-Cope, Azo-Cope, Diaza-Cope, Oxy-Cope and Siloxy-Cope Rearrangements.

Corey Condensation

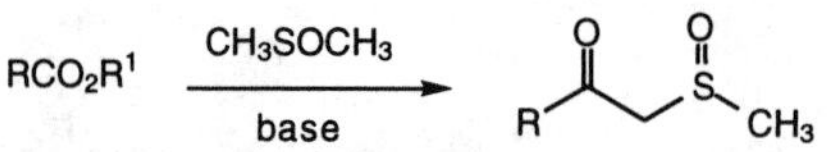

J. Am. Chem. Soc. **1966**, *88*, 5498.
J. Am. Chem. Soc. **1964**, *86*, 1639.

Corey-Fuchs Reaction

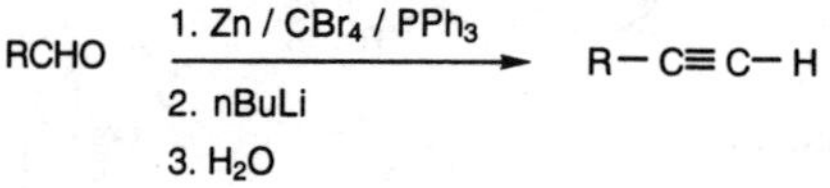

Tetrahedron Lett. **1972**, 3769.

Corey-Hopkins Deoxygenation

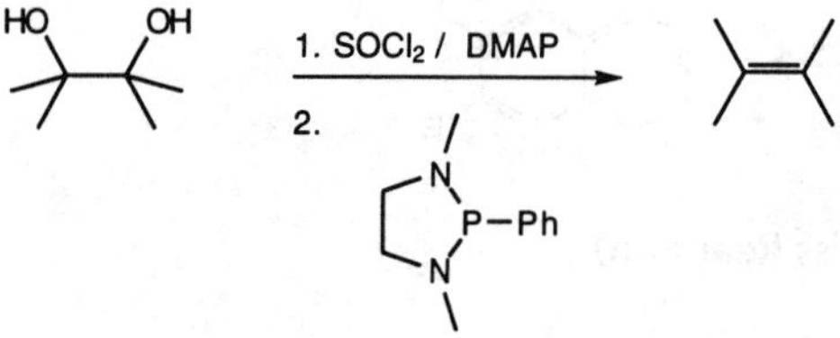

Tetrahedron Lett. **1982**, 1979.

Corey House Synthesis

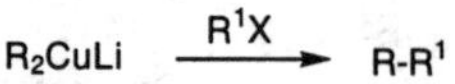

Org. Reactions **1975**, *22*, 253.

Corey-Kim Oxidation

Tetrahedron, **1978**, 1651.

Corey-Winter (Olefination) Reaction

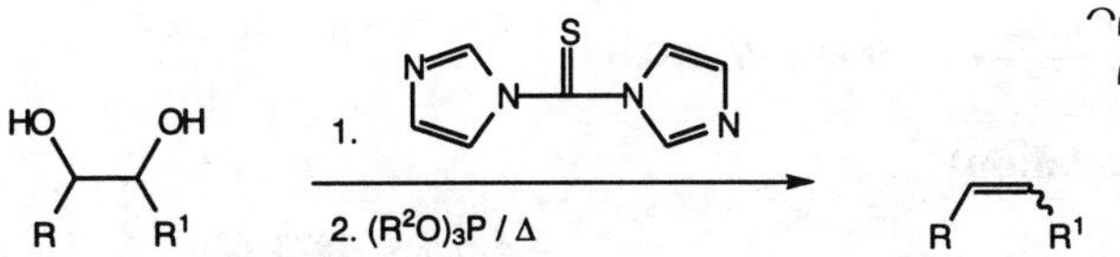

Org. *Reactions* **1984**, *30*, 457.
Org. *Reactions* **1976**, *23*, 263.

Note: Also known as the Corey-Winter Alkene Synthesis.

Cornforth Rearrangement

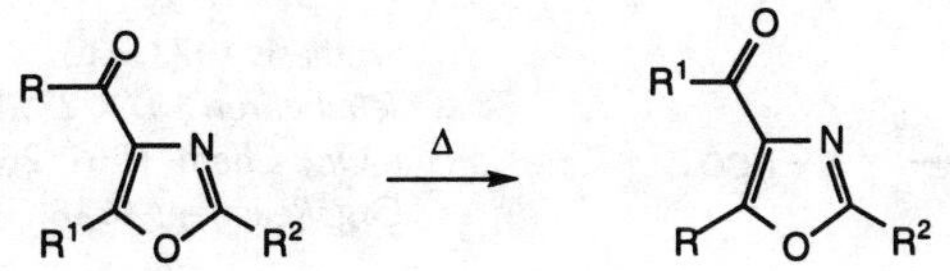

Comp. Org. Chem. **4**, 972.

Criegee Decomposition

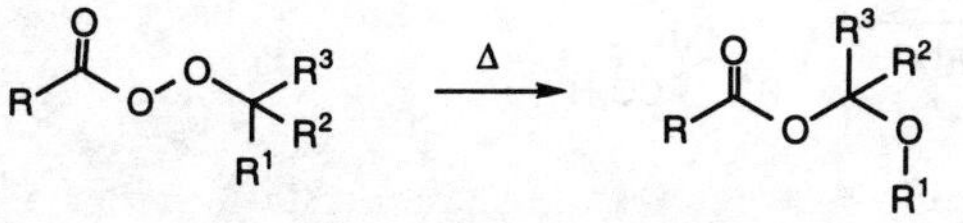

Comp. Org. Chem. **2**, 1114.

Note: Also called the Criegee Reaction.

Criegee Reaction

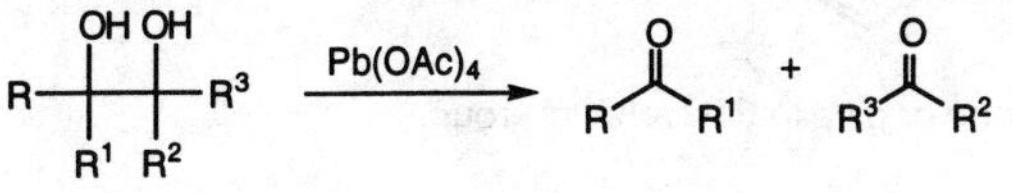

Merck ONR-21.

Note: Also known as the Criegee Glycol Cleavage.

Cristol-Firth-Hunsdiecker Reaction

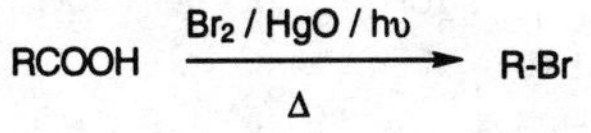

J. Org. Chem. **1979**, *44*, 3405.

Note: Also see the Hunsdiecker Reaction and Simonini Reaction.

Crossed Aldol Condensation (see Aldol Condensation)

Crossed Cannizzarro Reaction (see Cannizzarro Reaction)

Crum Brown-Walker Reaction

$$2\ RO_2C(CH_2)_nCO_2^- \xrightarrow{\text{electrolysis}} RO_2C(CH_2)_{2n}CO_2R$$

Org. Synth. **1981**, 60, 4.

Curtius Reaction (Degradation)

$$RCOCl \xrightarrow[\text{2. } \Delta / H_2O]{\text{1. NaN}_3} RNH_2$$

Synthesis **1973**, 678.
J. Org. Chem. **1961**, 26, 3511.
Org. Reactions **1946**, 3, 337.

Curtius Rearrangement (Reaction)

$$RCON_3 \xrightarrow{\Delta} R-NCO$$

Synthesis **1973**, 413.
Tetrahedron **1971**, 27, 1317.
J. Org. Chem. **1961**, 26, 3511.
Org. Reactions **1946**, 3, 337.

Cyanoacetic Ester Synthesis

$$EtO-\overset{O}{\overset{\|}{C}}-CH_2CN \xrightarrow[\text{3. } H_3O^+ / \Delta]{\substack{\text{1. base / RX} \\ \text{2. base / R}^1X}} R-\overset{R^1}{\underset{}{CH}}-CO_2H$$

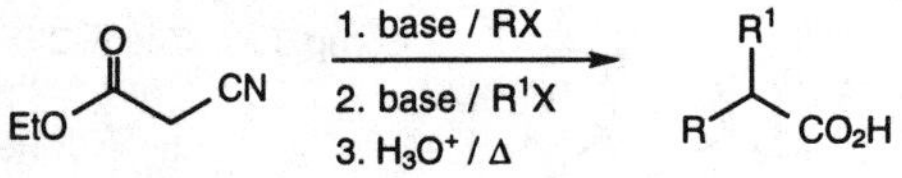

March 413.

Dakin Reaction (Oxidation)

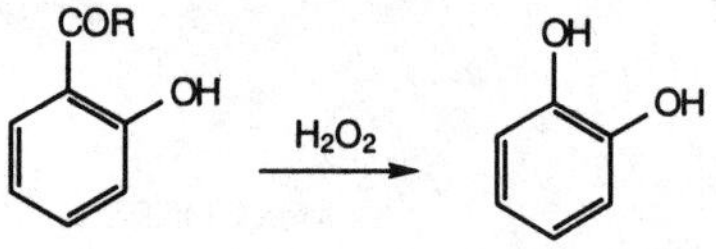

Quart. Rev. **1967**, 21, 454.
Org. Reactions **1957**, 9, 85.

Note: The hydroxy group must be ortho or para to the carbonyl group.

Dakin-West Reaction

$$H_2N-\overset{R}{\underset{}{CH}}-\overset{O}{\overset{\|}{C}}-OH \xrightarrow{Ac_2O / pyr} H_3COCHN-\overset{R}{\underset{}{CH}}-\overset{O}{\overset{\|}{C}}-CH_3$$

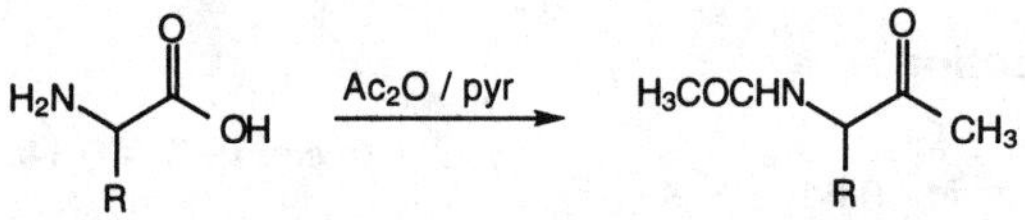

Chem. Soc. Rev. **1988**, 17, 91.
Org. Reactions **1962**, 12, 180.
J. Chem. Soc. **1957**, 4427.

Dalton Reaction

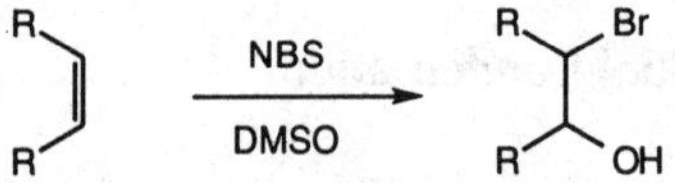

J. Chem. Soc., Chem. Commun. **1966**, 591.

Darapsky Degradation (Procedure)

$$NC-\overset{R}{\underset{}{CH}}-\overset{O}{\overset{\|}{C}}-OEt \xrightarrow[\substack{\text{2. HNO}_2 \\ \text{3. } \Delta / EtOH \\ \text{4. } H_3O^+}]{\text{1. N}_2H_4} NC-\overset{R}{\underset{}{CH}}-NH_2$$

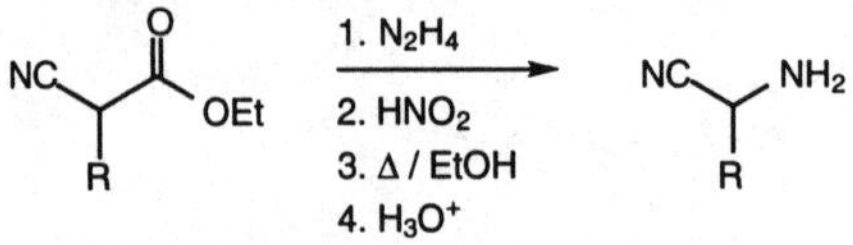

Comp. Org. Chem. 5, 199.

Darzens Condensation (Reaction)

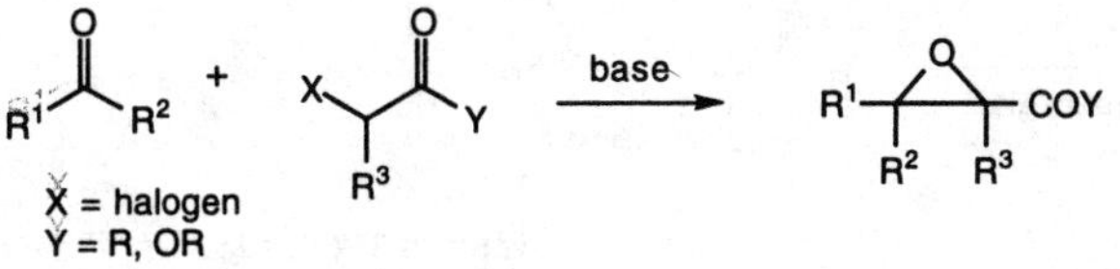

J. Chem. Soc., Chem. Commun. **1988**, 1515.
Can. J. Chem. **1969**, *47*, 2875.
J. Am. Chem. Soc. **1963**, *85*, 955.
Chem. Rev. **1955**, *55*, 283.
Org. Reactions **1949**, *5*, 413.

Note: Also known as the Darzens Procedure, Darzens-Claisen Reaction or Darzens Glycidic Ester Condensation.

Darzens-Reaction

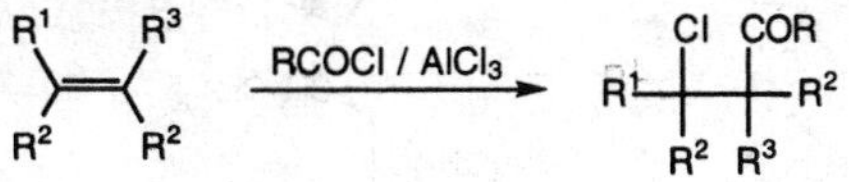

J. Org. Chem. **1961**, *26*, 4232.

Note: 1. Also known as the Darzens-Nenitzescu Ketone Synthesis.
2. The formation of an acid chloride with thionyl chloride is known as Darzens Procedure.

Darzens Synthesis of Tetralin Derivatives

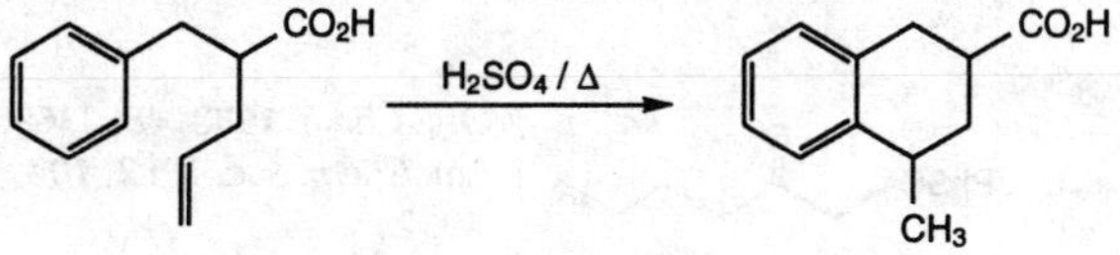

Chem. Rev. **1941**, *29*, 536.

Dauben-Dietsche Rearrangement

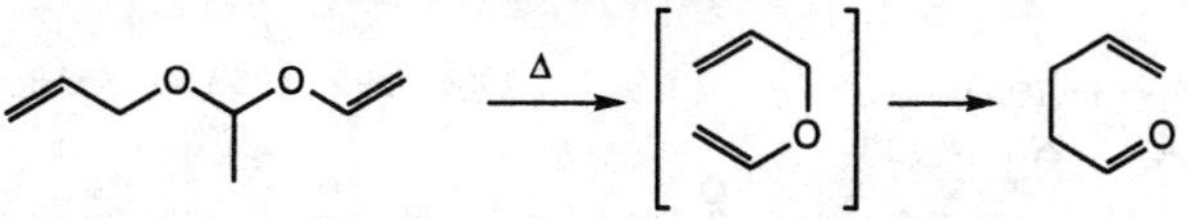

J. Org. Chem. **1972**, *37*, 1212.

Dauben Reaction

$$C_7H_8 + (C_6H_5)_3C^+X^- \longrightarrow C_7H_7^+X^- + (C_6H_5)_3CH$$

J. Org. Chem. **1960**, *25*, 1442.
J. Am. Chem. Soc. **1957**, *79*, 4557.

Delépine Reaction

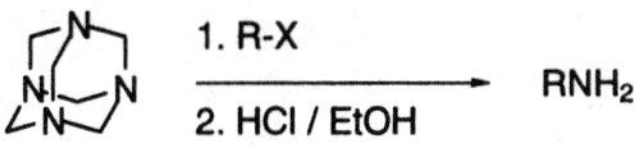

Org. Reactions **1954**, *8*, 204.

De Mayo Reaction

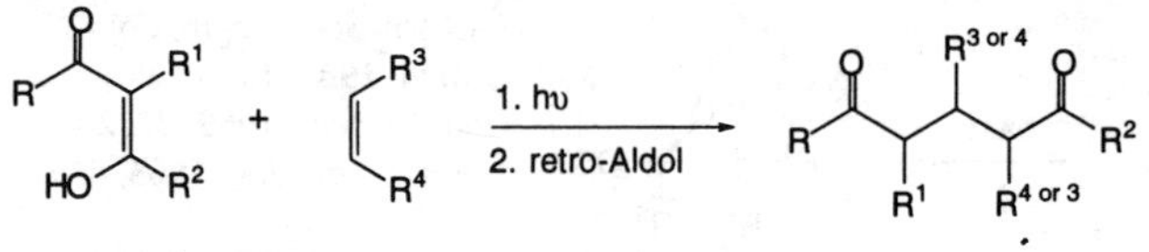

Synthesis **1989**, 145.
Chem. Rev. **1988**, *88*, 1453.

Demjanov Rearrangement (Reaction)

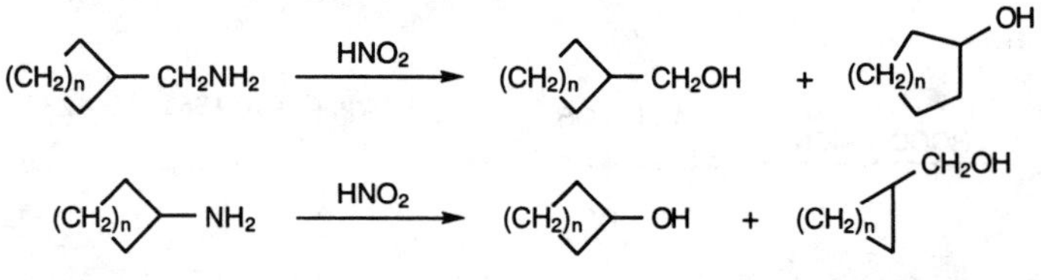

Org. Reactions **1960**, *11*, 157.
Chem. Rev. **1952**, *51*, 156.

Note: Also known as the Demyanov Rearrangement or Demyanov Ring Expansion.

Denmark Rearrangement

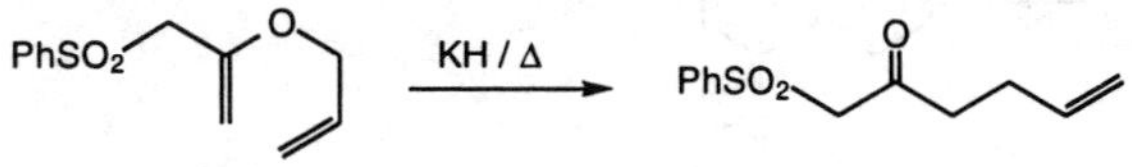

J. Org. Chem. **1983**, *48*, 3369.
J. Am. Chem. Soc. **1982**, *104*, 4972.

Dess-Martin Oxidation

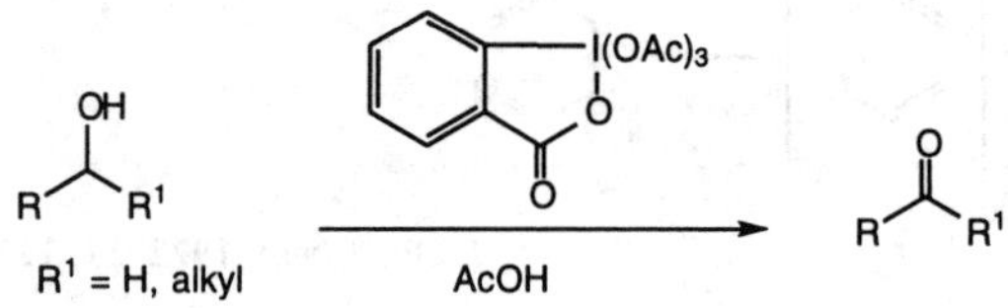

J. Org. Chem. **1983**, *48*, 4155.

Di-π-Methane Rearrangement (see Zimmermann Rearrangement)

Diaza-Cope Rearrangement

Org. Reactions **1975**, *22*, 65.

Diazo Reaction

$ArN_2X + Ar^1H \longrightarrow Ar\text{-}Ar^1$

Org. Reactions **1944**, *2*, 224.

Dieckmann Condensation (Reaction)

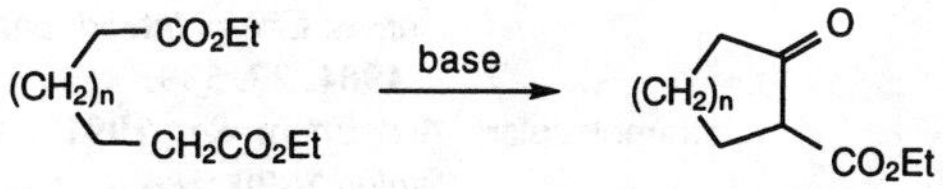

Tetrahedron **1971**, *27*, 1363.
Org. Reactions **1967**, *15*, 1.
Org. Reactions **1951**, 6, 449.

Retro-Dieckmann: *Synthesis* **1979**, 518.

Note: Also known as the Dieckmann Cyclization.

Diels-Alder Reaction

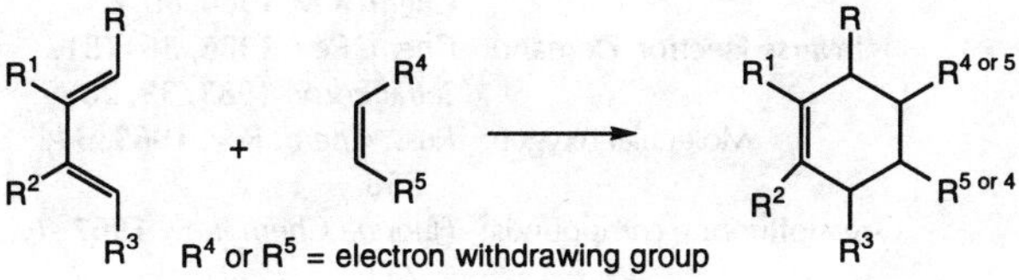

Angew. Chem. Int. Ed. Engl. **1984**, *23*, 876.
Angew. Chem. Int. Ed. Engl. **1980**, *19*, 779.
Adv. Alicyclic Chem. **1968**, 2, 1.
Angew. Chem. Int. Ed. Engl. **1967**, 6, 16.
Angew. Chem. Int. Ed. Engl. **1966**, 5, 211.
Org. Reactions **1949**, 5, 136.
Org. Reactions **1948**, *4*, 1.

Asymmetric: *Synthesis* **1991**, 1.
Tetrahedron: Asymmetry **1991**, 2, 1173.
Angew. Chem. Int. Ed. Engl. **1984**, *23*, 876.

Asymmetric Catalysts: *Chem. Rev.* **1992**, 92, 1007.
Synthesis **1991**, 1.

Catalytic Methods: *Chem. Rev.* **1993**, *93*, 741.
Synthesis **1991**, 1.

Hetero-Diels Alder Reaction: *Heterocycl. Chem.* **1990**, 47.
Acc. Chem. Res. **1988**, *21*, 313.
Angew. Chem. Int. Ed. Engl. **1987**, 26, 15.
Acc. Chem. Res. **1986**, *19*, 250.
Chem. Rev. **1986**, *86*, 781.
Org. Reactions **1984**, *32*, 1.
Synthesis **1984**, 181.
Angew. Chem. Int. Ed. Engl. **1976**, *15*, 639.

Hetero Diels Alder Reaction-Natural products: *Acc. Chem. Res.* **1986**, *19*, 250.

High pressures-Reaction mechanisms: *Acc. Chem. Res.* **1974**, *7*, 251

Homo Diels-Alder Reaction: *J. Org. Chem.* **1992**, *57*, 8.

(Cont'd.)

(Diels-Alder Reaction, cont'd.)

Angew. Chem. Int. Ed. Engl. **1986**, *25*, 1.
Angew. Chem. Int. Ed. Engl. **1984**, *23*, 539.

Intramolecular: *Acc. Chem. Res.* **1991**, *24*, 229.
Synlett **1990**, 186.
Chem. Soc. Rev. **1987**, *16*, 187.
Acc. Chem. Res. **1985**, *18*, 16.
Org. Reactions **1984**, *32*, 1.
Pure & Appl. Chem. **1981**, *53*, 1181.
Chem. Rev. **1980**, *80*, 63.

Inverse Electron Demand: *Chem. Rev.* **1986**, *86*, 781.
Tetrahedron **1983**, *39*, 2869.

Molecular oxygen: *Russ. Chem. Rev.* **1965**, *34*, 558.

Organofluorine compounds: *Fluorine Chem. Rev.* **1967**, *1*, 253.

Polymerization: *J. Org. Chem.* **1962**, *27*, 3088.

Reaction mechanisms: *Angew. Chem. Int. Ed. Engl.* **1980**, *19*, 779.
Adv. Alicyclic Chem., **1968**, *2*, 1.
Angew. Chem. Int. Ed. Engl. **1967**, *6*, 16.

Retro-Diels Alder Reaction: *Synthesis* **1987**, 270.
Tetrahedron **1978**, *34*, 19.
Progr. Phys. Org. Chem. **1971**, *8*, 201.
Chem. Rev. **1968**, *68*, 415.

Stereochemistry: *Chem. Rev.* **1961**, *61*, 537.

Siloxy dienes: *Acc. Chem. Res.* **1981**, *14*, 400.

Tetracyanoethylene: *Synthesis* **1987**, 749.

Theoretical aspects: *Pure & Appl. Chem.* **1983**, *55*, 237.

Diels-Reese Reaction

PhNHNHPh + RO$_2$C≡CO$_2$R $\xrightarrow{\text{CH}_3\text{OH} / \Delta}$

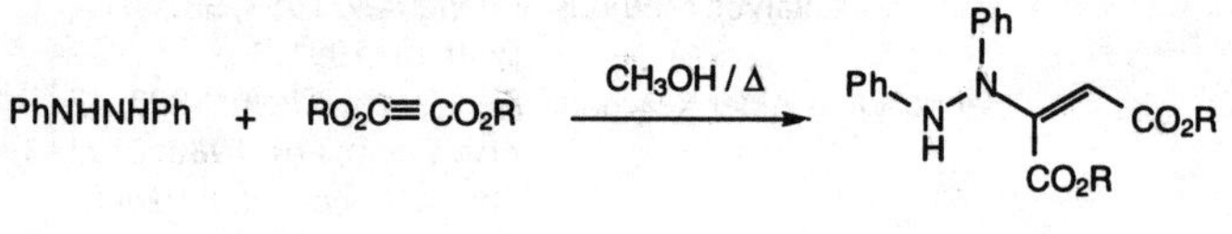

J. Am. Chem. Soc. **1956**, *78*, 2225.

Dienol-Benzene Rearrangement

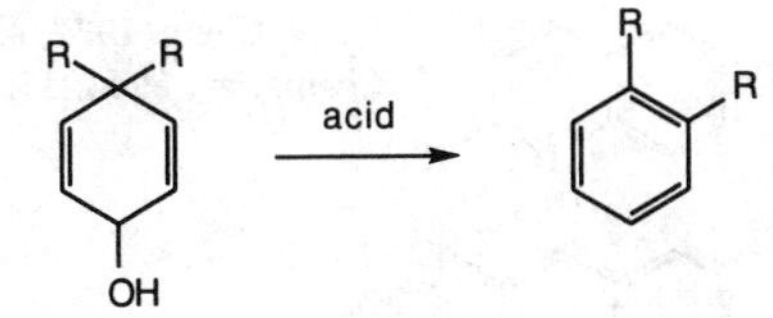

Angew. Chem. **1956**, *68*, 618.

Dienone-Phenol Rearrangement

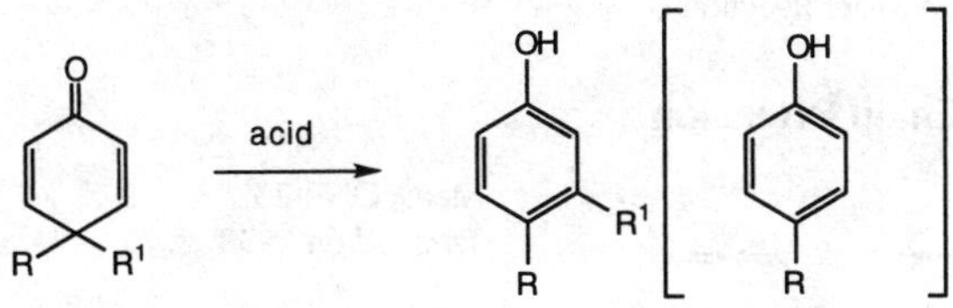

If R^1 is a leaving group

Comp. Org. Chem. **1**, 726.

Dimroth Rearrangement

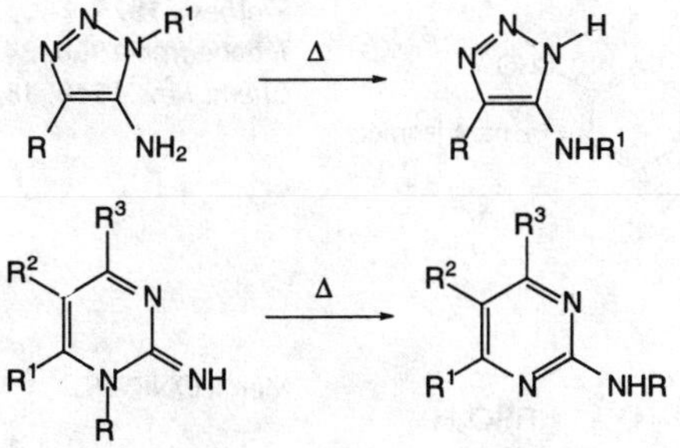

Synthesis **1988**, 851.
Heterocycl. Chem. **1974**, *16*, 33.
Tetrahedron **1972**, *28*, 535.
Z. Chem. **1969**, *9*, 241.

Directed Aldol Condensation (see Aldol Condensation)

Doebner Condensation

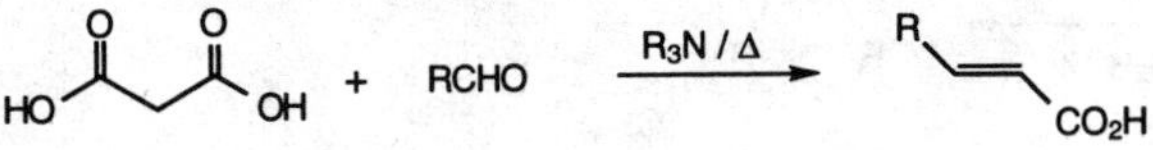

Chem. Rev. **1944**, *35*, 156.

Note: Also known as the Doebner Reaction.

Doebner Reaction (Synthesis)

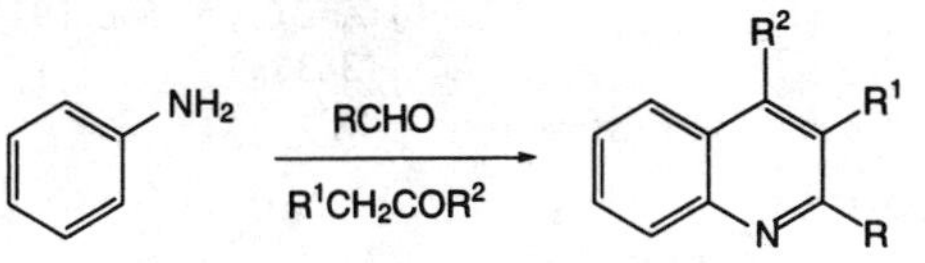

Comp. Org. Chem. **4**, 161.
Chem. Rev. **1944**, *35*, 153.

Note: 1. Also referred to as the Beyer Synthesis.
2. The Doebner Condensation is also refered to as the Doebner Reaction.

Doebner-von Miller Synthesis

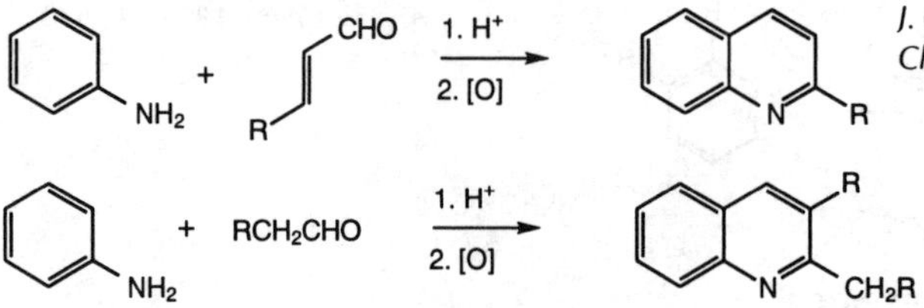

J. Org. Chem. **1977**, *42*, 911.
Chem. Rev. **1944**, *35*, 153.

Note: Also known as the Doebner-von Miller Reaction

Doering-LaFlamme Carbon Chain Extension

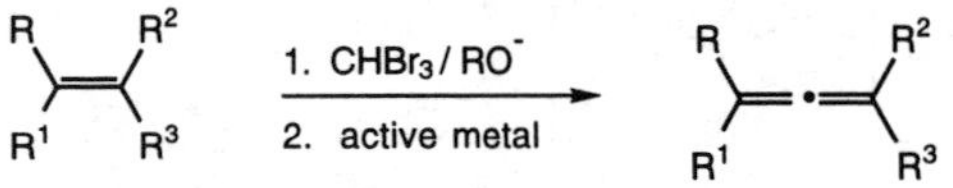

Merck ONR-25.
Tetrahedron **1958**, *2*, 75.

Duff Reaction

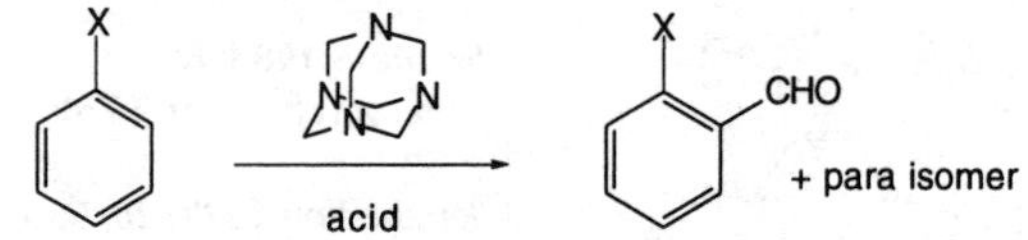

X = OH, NR_2

Synthesis **1979**, 166.
Tetrahedron **1968**, *24*, 5001.
Chem. Rev. **1946**, *38*, 227.

Dutt-Wormall Reaction

$$ArN_2Cl \xrightarrow[2.\ OH^-]{1.\ RSO_2NH_2} ArN_3 + RSO_2H$$

Merck ONR-26.

Eastwood Reaction (Deoxygenation)

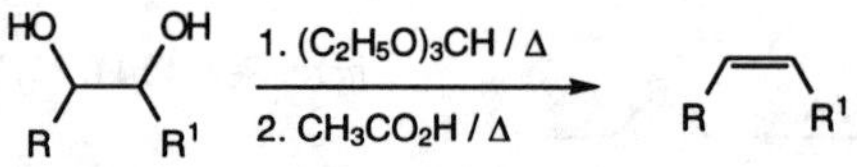

Org. Reactions **1984**, *30*, 480.

Edman Degradation

$$PhNCS + H_2NCHRCONHCHR^1COR^2 \xrightarrow{H^+} H_2NCHR^1COR^2$$

J. Chem. Soc. **1956**, 3689.
J. Am. Chem. Soc. **1953**, *75*, 3638.

Eglinton Reaction (Coupling)

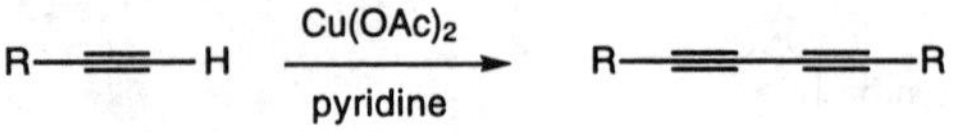

Comp. Org. Chem. **1**, 204.

Ehrlich-Sachs Reaction

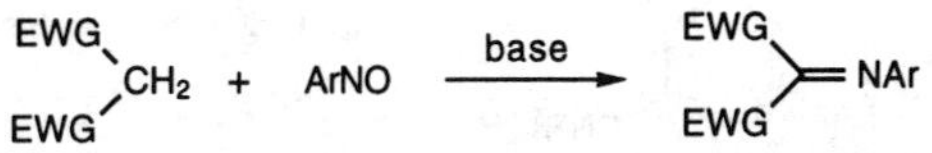

J. Chem. Soc. **1961**, 1631.

Einhorn-Brunner Reaction

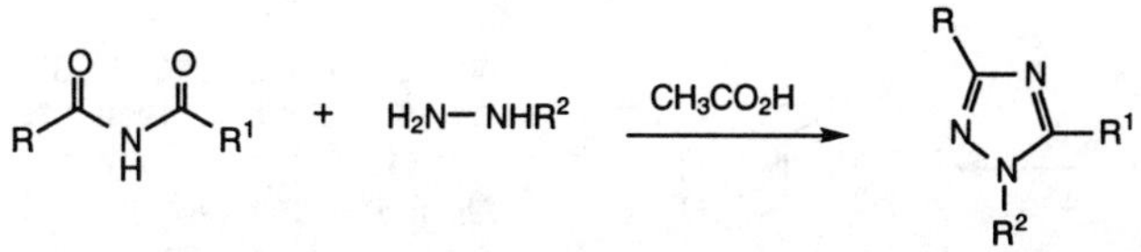

Chem. Rev. **1961**, *61*, 103.

Elbs (Oxidation) Reaction

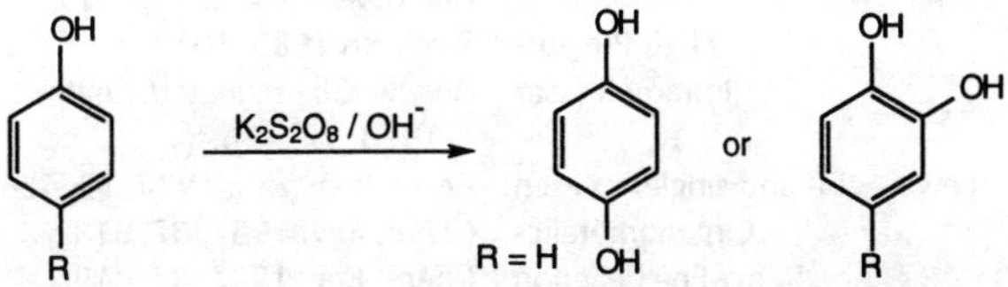

ɉ. Reactions **1988**, *35*, 421.
art. Rev. **1967**, *21*, 453.
ɜm. Rev. **1951**, *49*, 91.

Note: 1. Also known as the Elbs Persulfate Oxidation.
2. See also the Elbs Reaction.

Elbs Reaction

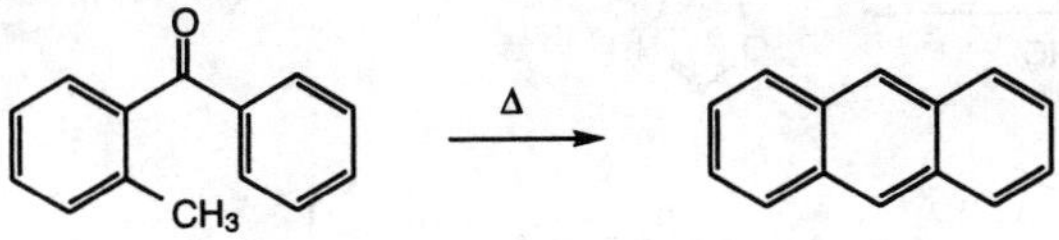

Org. Reactions **1942**, *1*, 129.

Note: See also Elbs (Oxidation) Reaction.

Eltekoff Reaction

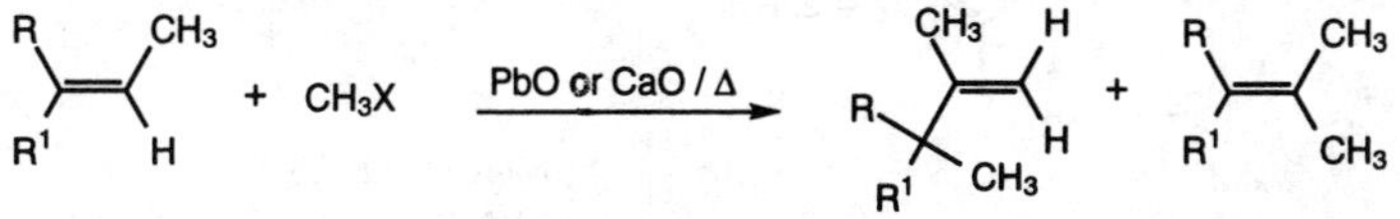

Merck ONR-27.

Emdé Reduction (Degradation)

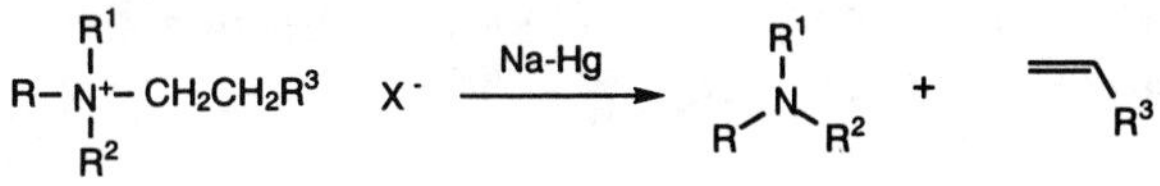

Org. Reactions **1953**, *7*, 143.

Emmert Reaction

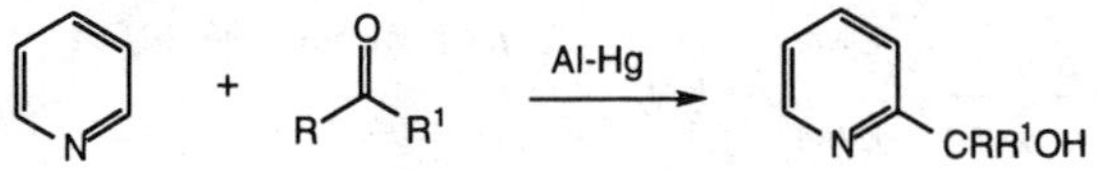

Comp. Org. Chem. **4**, 62.

Ene Reaction

Y = C, Se   Z = C, O

Acc. Chem. Res. **1990**, *23*, 34.
Pure & Appl. Chem. **1990**, *62*, 1941.
Chem. Rev. **1987**, *87*, 319.
Angew. Chem. Int. Ed. Engl. **1984**, *23*, 876.
Pure & Appl. Chem. **1981**, *53*, 1181.
Org. Reactions **1973**, *20*, 141.

Asymmetric: *Chem. Rev.* **1992**, *92*, 1021.
High Pressure: *Synthesis* **1985**, 1019.
Intramolecular: *Angew. Chem. Int. Ed. Engl.* **1978**, *17*, 476.
Lewis acids and singlet oxygen: *Acc. Chem. Res.* **1980**, *13*, 426.
Organometalics: *Chem. Rev.* **1987**, *87*, 319.
Retro-Ene Reaction: *Chem. Rev.* **1987**, *87*, 319.

Erlenmeyer Azlactone Synthesis

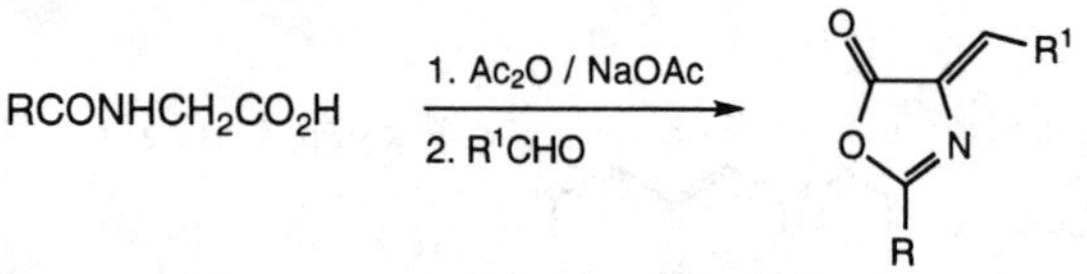

Org. Reactions **1946**, *3*, 198.

Erlenmeyer-Plöchl Azlactone and Amino Acid Synthesis

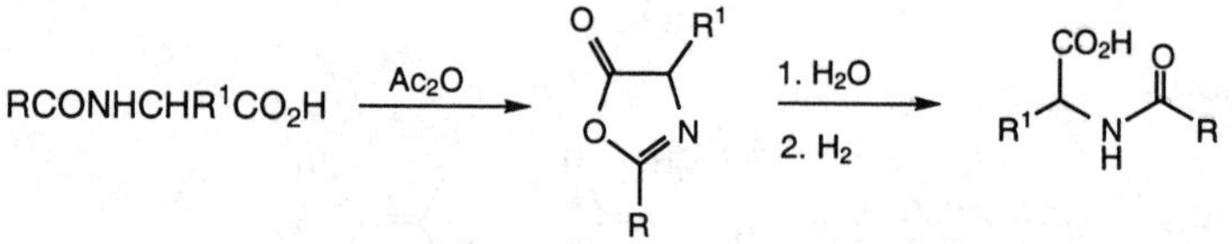

Merck ONR-28.
Org. Reactions **1946**, *3*, 214.

Eschenmoser Fragmentation (Rearrangement)

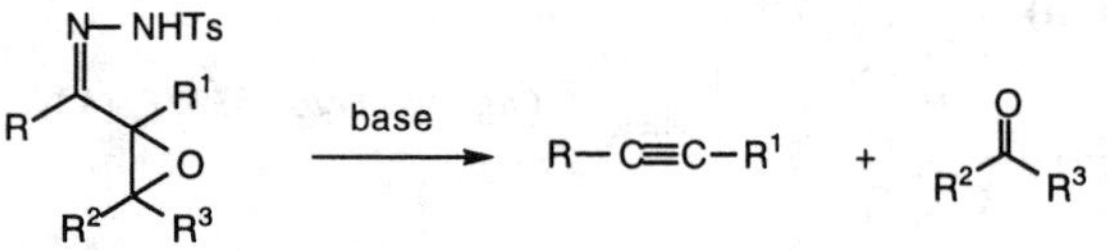

Synthesis **1981**, 276.
Helv. Chim. Acta **1967**, *50*, 708.

Eschenmoser Rearrangement

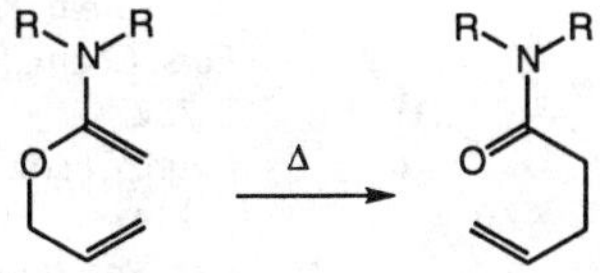

Tetrahedron **1978**, *34*, 127.
Helv. Chim. Acta **1969**, *52*, 1030.

Note: Also see the Meerwein-Eschenmoser Reaarrangement.

Eschweiler-Clarke Reaction (Methylation)

$$RR^1NH \xrightarrow{H_2CO\ /\ HCO_2H} RR^1NCH_3$$

Org. Reactions **1949**, *5*, 301.
Cyclization: *Chem. Rev.* **1975**, *75*, 389.

Note: Also known as the Eschweiler-Clark Procedure.

Étard Reaction

$$ArCH_3 \xrightarrow{CrO_2Cl_2} ArCHO$$

Chem. Rev. **1958**, *58*, 1.
Chem. Rev. **1946**, *38*, 237.

Evans Asymmetric Aldol Protocol (Reaction)

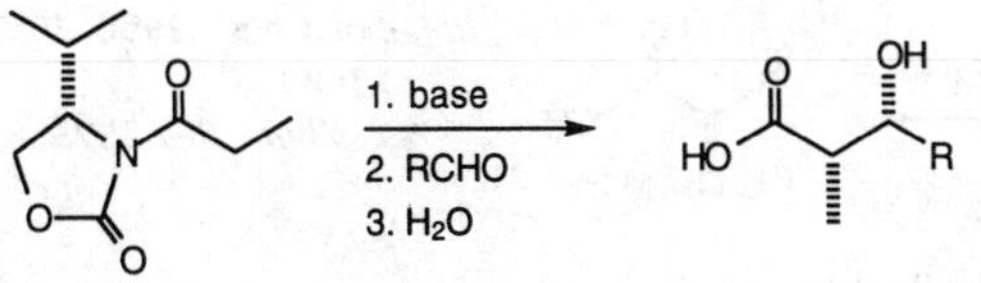

J. Org. Chem. **1991**, *56*, 5747.
Aldrichimica Acta **1982**, *15*, 23.
Top. Stereochem. **1982**, *13*, 1.
J. Am. Chem. Soc. **1981**, *103*, 2127.

Evans (-Mislow) Rearrangement

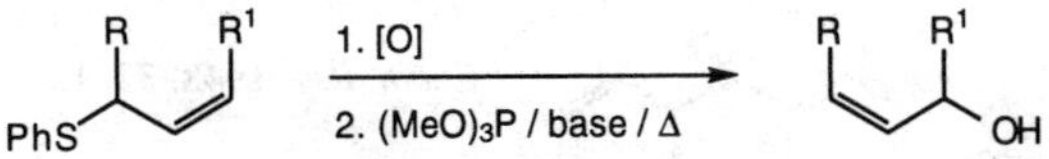

Acc. Chem. Res. **1974**, *7*, 147.
J. Am. Chem. Soc. **1968**, *90*, 4869.

Note: The formation of rearranged sulfoxides from allyl alcohols is also known as the Evans Rearrangement.

Favorskii Reaction

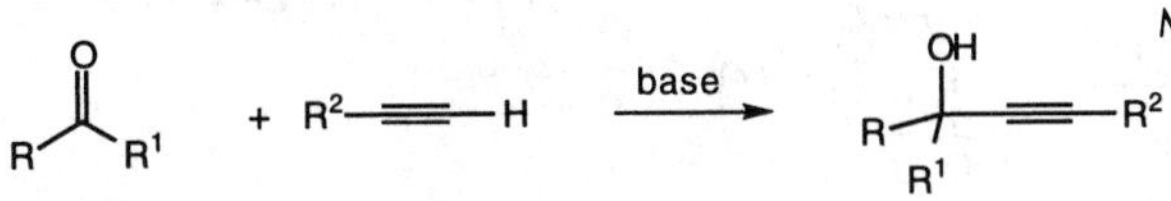

March 838.

Note: Also known as the Favorskii-Babayan Synthesis.

Favorskii Rearrangement

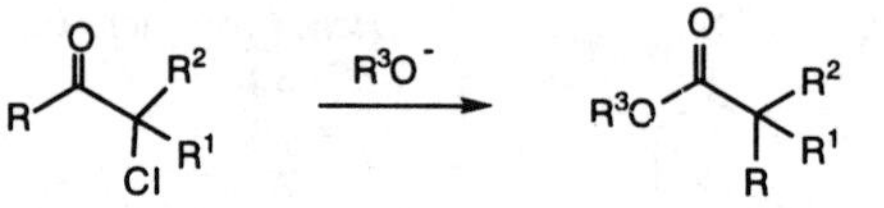

Acc. Chem. Res. **1970**, *3*, 286.
Russ. Chem. Rev. **1970**, *39*, 732.
J. Am. Chem. Soc. **1963**, *85*, 234.
Org. Reactions **1960**, *11*, 261.

Note: Also known as the Favorsky Rearrangement.

Mechanism: *J. Org. Chem.* **1991**, *56*, 6773.
Acc. Chem. Res. **1970**, *3*, 281.

Feist-Bénary Synthesis

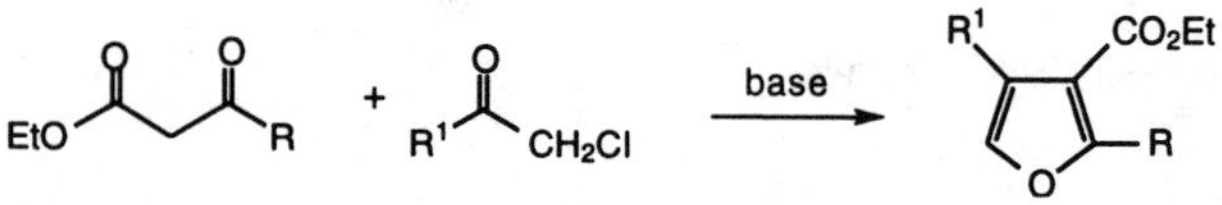

Comp. Org. Chem. **4**, 715–6.

Note: Also known as the Feist-Bénary Furan Synthesis.

Fenton Reaction

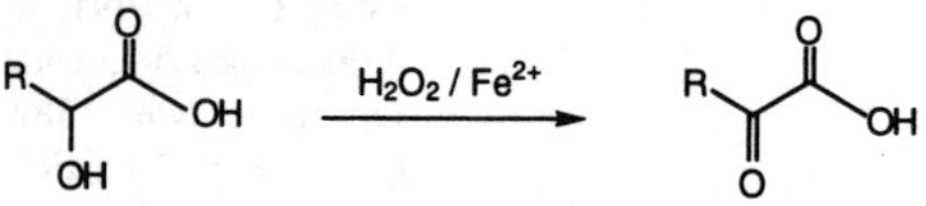

J. Biol. Chem. **1990**, *265*, 13589.
Acc. Chem. Res. **1975**, *8*, 125.

Note: Also known as the Haber-Weiss Reaction.

Ferrario Reaction

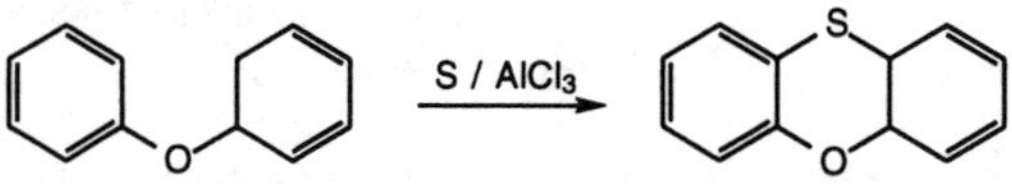

Chem. Rev. **1943**, *32*, 173.

Ferrier Rearrangement

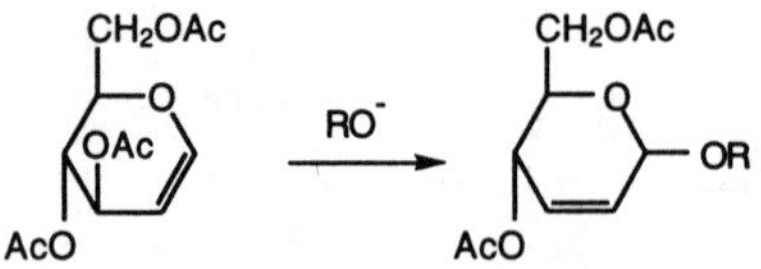

J. Chem. Soc. (C) **1969**, 570.
J. Chem. Soc. **1964**, 5443.

Ficini-Barara Rearrangement (see Ynamine-Claisen Rearrangement)

Fiesselman (Thiophene) Synthesis

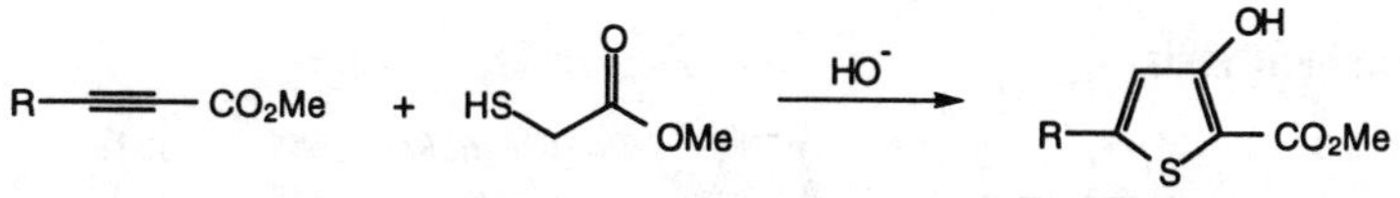

Comp. Org. Chem. **4**, 830.

Finkelstein Reaction (Substitution)

Chem. Lett. **1978**, 1435.
Bull. Chem. Soc., Jpn. **1976**, *49*, 1989.
J. Chem. Soc., Perkin Trans. 1 **1976**, 416.

Fischer Cyanohydrin Synthesis (see Kiliani-Fischer Synthesis)

Fischer Esterification

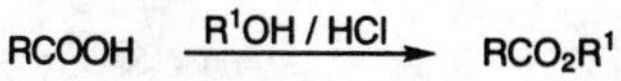

Fieser & Fieser 'Advanced Organic Chemistry', Reinhold, New York, **1961**, 371.

Note: Also known as the Fischer-Speier Esterification.

Fischer Glycosidation

Chem. Ind. (London) **1974**, 727.
Chem. Rev. **1969**, 69, 407.

Fischer-Hepp Reaction (Rearrangement)

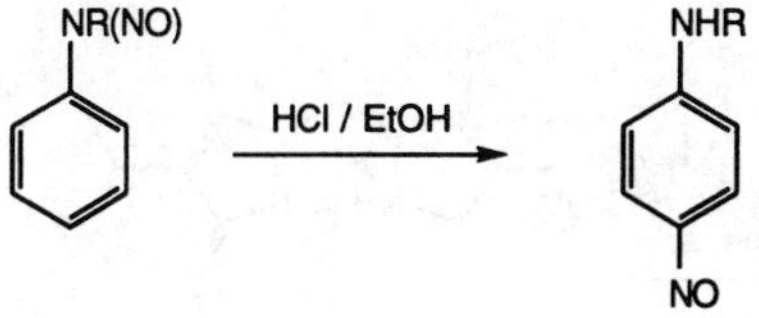

Tetrahedron **1975**, *31*, 1343.

Fischer Indole Synthesis

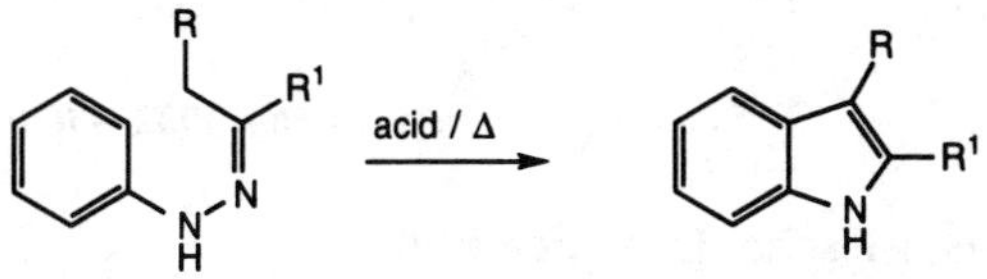

Acc. Chem. Res. **1981**, *14*, 275.
Russ. Chem. Rev. **1974**, *43*, 115.
Chem. Rev. **1969**, 69, 227.
Chem. Ind. (London) **1967**, 547.
Chem. Rev. **1963**, 63, 373.
Org. Reactions **1959**, *10*, 153.

Note: Also known as the Fischer Reaction, Fischer Synthesis or Fischer Indolization.

Fischer-Kiliani Synthesis (see Kiliani-Fischer Synthesis)

Fischer Oxazole Synthesis

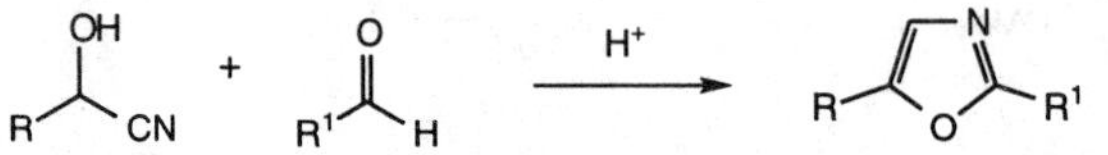

Chem. Rev. **1975**, *75*, 391.
Chem. Rev. **1945**, *37*, 410.

Fischer Peptide Synthesis

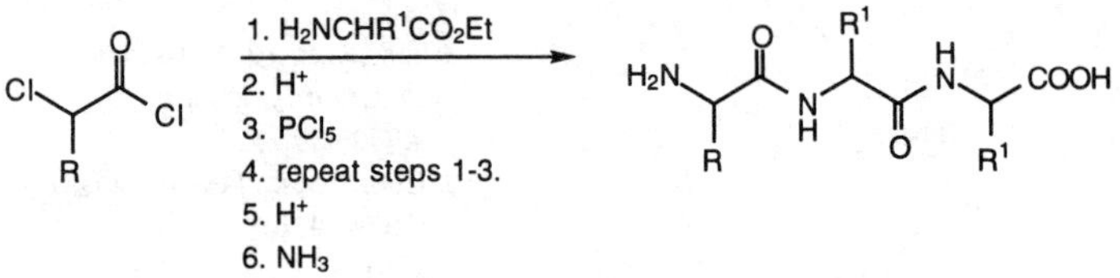

Merck ONR-30.

Note: Also known as the Fischer Polypeptide Synthesis.

Fischer Phenylhydrazine Synthesis

$$ArN_2X \xrightarrow[\text{2. HCl}]{\text{1. NaHSO}_3} ArNHNH_2 \cdot HCl$$

Merck ONR-31.
Org. Synth. **1932**, *Coll. Vol. 1*, 442.

Fischer Phenylhydrazone and Osazone Reaction

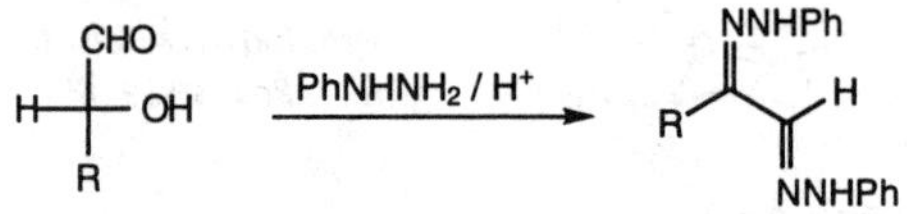

Merck ONR-31.

Fischer Porphyrin Synthesis

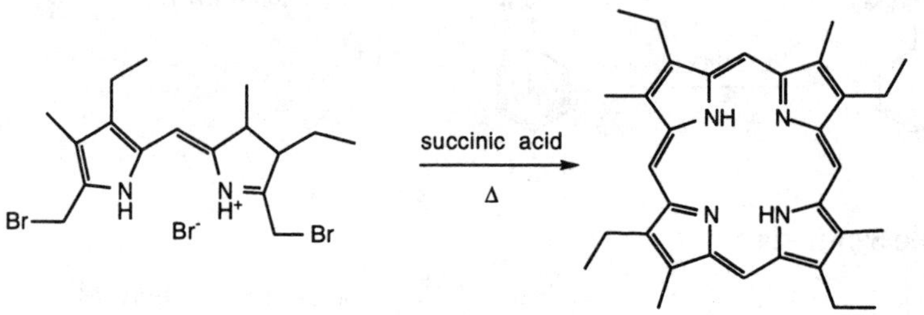

Liebigs Ann. **1932**, *496*, 1.

Fischer-Speier Esterification (see Fischer Esterification)

Fischer-Tropsch Synthesis (Reaction)

$$CO + H_2 \xrightarrow{\text{metal catalyst}} \text{alkanes and alkenes}$$

Acc. Chem. Res. **1984**, *17*, 103.
Angew. Chem. Int. Ed. Engl. **1982**, *21*, 117.
Chem. Rev. **1981**, *81*, 447.
Angew. Chem. Int. Ed. Engl. **1976**, *15*, 136.

Note: Also known as the Fischer-Tropsch Process or Fischer-Tropsch Gasoline Synthesis.

Fittig Reaction

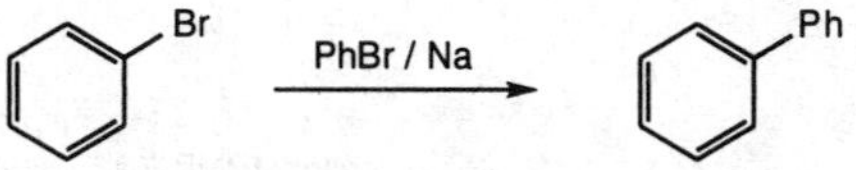

From 'Organic Chemistry' by Whitmore **1937**, D. van Nostrand, New York 830.

Note: In the Fittig Reaction both reactants are aryl (also see the Wurtz Reaction and Wurtz-Fittig Reaction).

Flood Reaction

$$(R_3Si)_2O \xrightarrow{H_2SO_4 / NH_4Cl} R_3SiCl$$

J. Am. Chem. Soc. **1933**, *55*, 1735.

Formazan Reaction

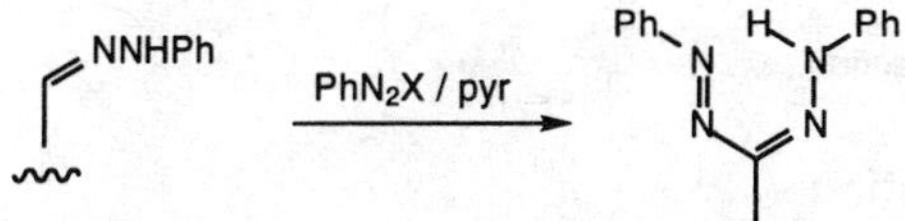

(carbohydrate)

Chem. Ber. **1960**, *93*, 1684.
Adv. Carbohydrat. Chem. **1958**, *13*, 105.

Formose Reaction

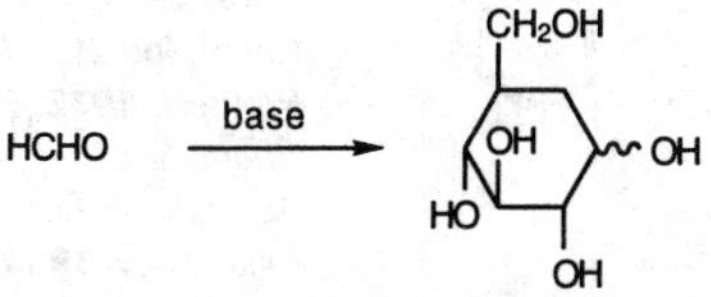

J. Am. Chem. Soc. **1978**, *100*, 1309.
Bull. Chem. Soc., Jpn. **1977**, *50*, 1527; 2138.

Forster Reaction

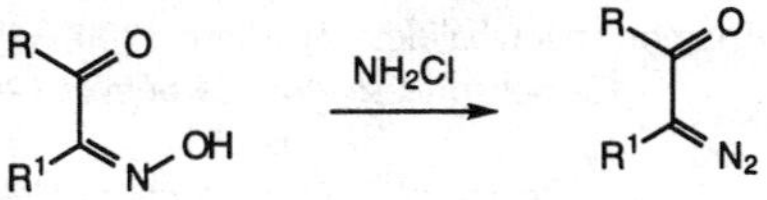

Chem. Rev. **1973**, *73*, 283.

Note: The synthesis of Schiff's bases is sometimes referred to as the Forster Reaction or Forster-Decker Method.

Franchimont Reaction

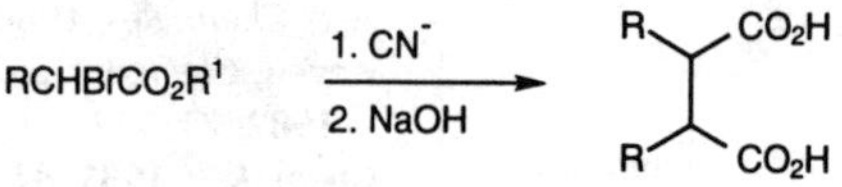

Merck ONR-32.

Frankland-Duppa Reaction

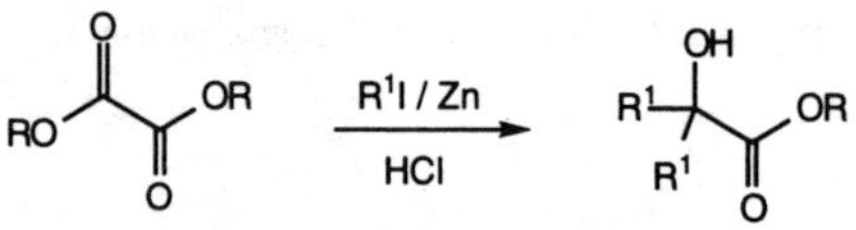

Merck ONR-32.

Frankland Synthesis (Method)

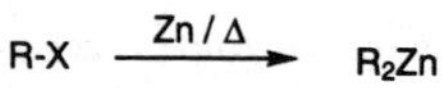

Merck ONR-32.
Org. Synth. **1943**, *Coll. Vol. 2*, 184.

Freund Reaction (Method)

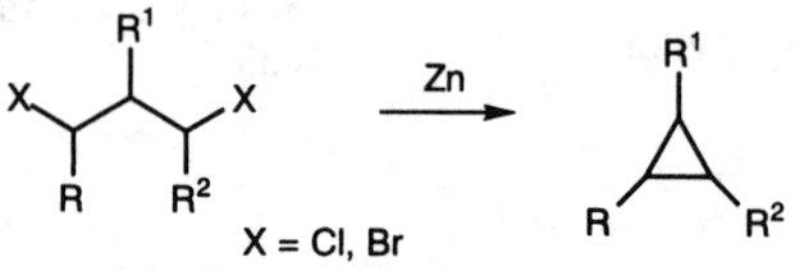

Org. Reactions **1949**, *5*, 328.
J. Am. Chem. Soc. **1948**, *70*, 946.
J. Am. Chem. Soc. **1946**, *68*, 2513.

Note: Also known as the Gustavson Reaction.

Friedel-Crafts (Acylation) Reaction

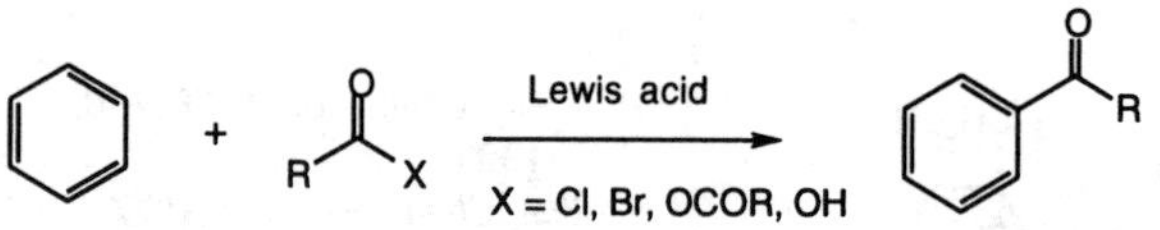

Acc. Chem. Res. **1980**, *13*, 330.
Synthesis **1980**, 345.
Angew. Chem. Int. Ed. Engl. **1974**, *13*, 1.
Chem. Ind. (London) **1974**, *727*.
Synthesis **1972**, 533.
Chem. Soc. Rev. **1972**, *1*, 73.
Acc. Chem. Res. **1971**, *4*, 240.
Quart. Rev. **1954**, *8*, 355.

Acid anhydrides: *Org. Reactions* **1949**, *5*, 229.
Alkenes: *Chem. Soc. Rev.* **1972**, *1*, 73.
Acetylation: *Chem. Ind. (London)* **1974**, *727*.
Antimony pentahalides: *Synthesis* **1980**, 345.
Carbohydrates: *Adv. Carbohydr. Chem.* **1951**, 6, 251.
Cyclizations: *Org. Reactions* **1944**, *2*, 130.
Dicarboxylic acid anhydrides; Phosphazenes: *Org. Reactions* **1949**, *5*, 229.
Reaction mechanisms: *Acc. Chem. Res.* **1971**, *4*, 240.

Friedel-Crafts (Alkylation) Reaction

$$Ar + RX \xrightarrow{\text{Lewis acid}} Ar\text{-}R$$

X = Cl, Br, I, OH   R = alkyl, alkene

Z. Chem. **1975**, *15*, 135.
Org. Reactions **1946**, *3*, 1.

Friedel-Crafts (Arylation) Reaction (see Scholl Reaction)

Friedel-Crafts-Karrier Synthesis

$$Ar + BrCN \xrightarrow{AlCl_3} Ar\text{-}CN$$

Chem. Rev. **1948**, *42*, 220.

Friedländer (Quinoline) Synthesis

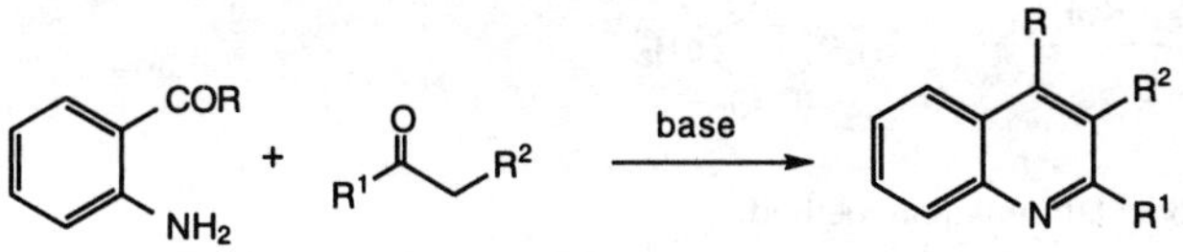

J. Org. Chem., **1991**, *56*, 7288.
Org. Reactions **1982**, *28*, 37.
Chem. Rev. **1944**, *35*, 151.

Note: Also known as the Friedlaender Synthesis.

Fries Rearrangement (Reaction)

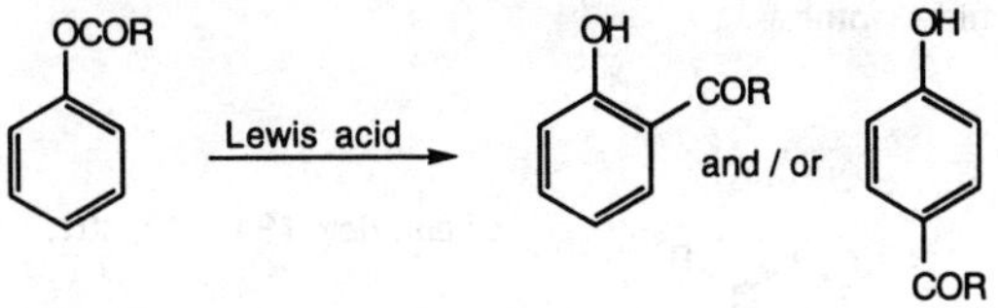

Bull. Soc. Chim., France **1974**, 983.
Tetrahedron **1970**, *26*, 973; 1001.
Org. Reactions **1942**, *1*, 342.
Chem. Rev. **1940**, *27*, 413.

Note: Also see the Photo-Fries Rearrangement.

Fritsch-Buttenberg-Wiechell Rearrangement

Org. Reactions **1979**, *26*, 42.

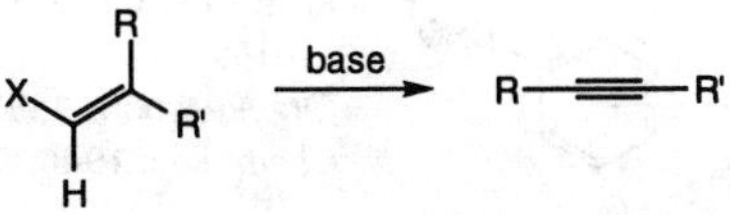

Note: Also known as the Fritsch-Wiechell Rearrangement.

Fujimoto-Belleau reaction

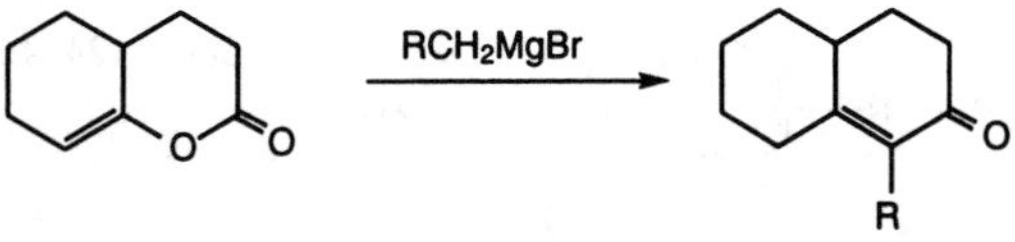

Synthesis **1969**, 49.

Gabriel-Colman Rearrangemant

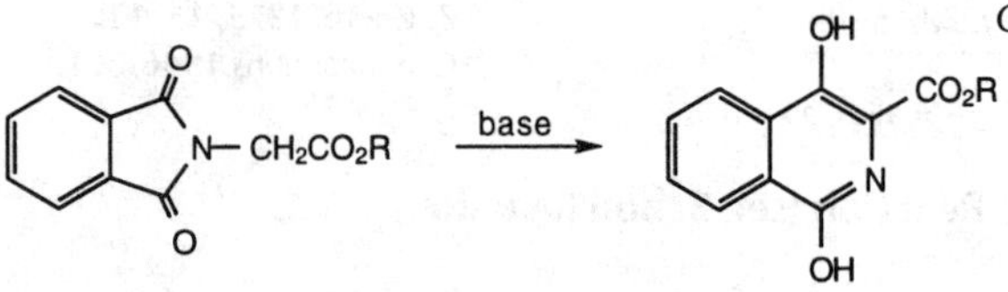

Chem. Rev. **1950**, *47*, 284.

Note: Also known as the Gabriel Isoquinoline Synthesis.

Gabriel-Marckwald Ethylenimine Synthesis

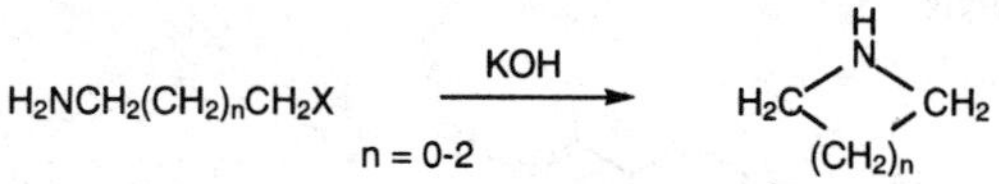

Merck ONR-34.

Note: Also known as the Gabriel Ethylenimine Method.

Gabriel Synthesis (Reaction)

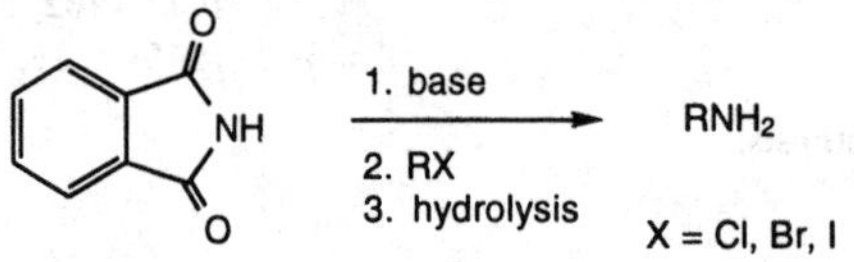

Acc. Chem. Res. **1991**, *24*, 285.
Synthesis **1981**, 472.
Angew. Chem. Int. Ed. Engl. **1968**, *7*, 919.

Note: Also known as Gabriel's Phthalimide Synthesis.

Gastaldi (Pyrazine) Synthesis

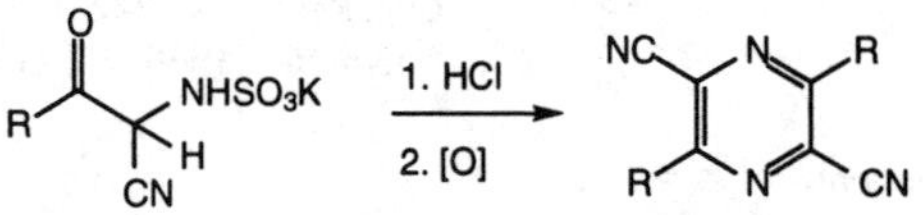

Chem. Rev. **1947**, *40*, 301.

Gattermann (Aldehyde) Reaction

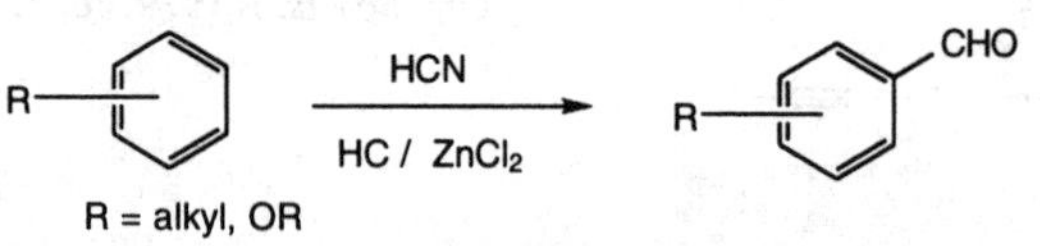

Russ. Chem. Rev. **1962**, *31*, 615.
Org. Reactions **1957**, *9*, 37.
Chem. Rev. **1946**, *38*, 227.

Note: Also known as the Gattermann Formylation or Gattermann Aldehyde Synthesis.

Gattermann-Adams Formylation

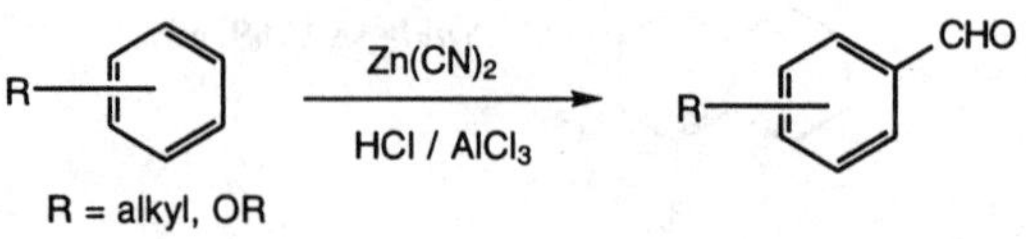

J. Am. Chem. Soc. **1924**, *46*, 1518.

Gattermann Amide Synthesis

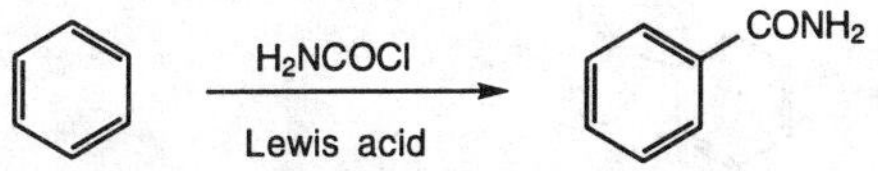

March, 491.

Gattermann-Cantzler Reaction

$$ArN_2X + CNO^- \xrightarrow{Cu} ArNCO$$

Quart. Rev. **1952**, 6, 363.

Gattermann-Koch Reaction

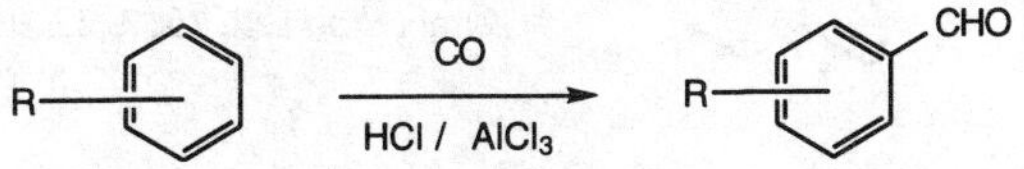

R = simple alkyl substituents

Org. Reactions **1949**, *5*, 290.
Chem. Rev. **1946**, *38*, 227.

Note: Also known as the Gattermann Carbon Monoxide Synthesis of Aldehydes.

Gattermann Reaction

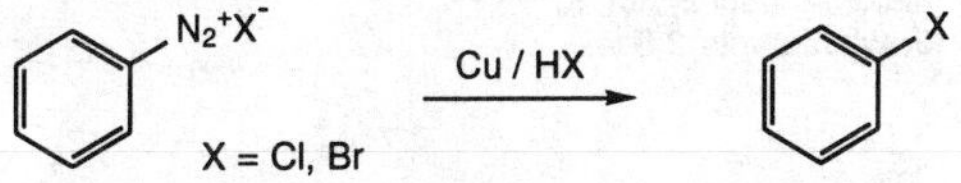

Quart. Rev. **1952**, 6, 358.

Gattermann-Sandmeyer Reaction

$$ArNH_2 \xrightarrow[\text{2. Cu powder / }\Delta]{\text{1. aq. HBr / NaNO}_2} ArBr$$

Fieser & Fieser 'Advanced Organic Chemistry', Reinhold, New York, **1961**, 729.

Gattermann-Skita Synthesis

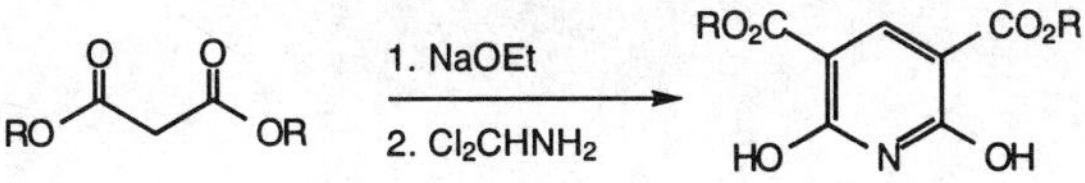

Chem. Ber. **1916**, *49*, 494.

Gewald Reaction

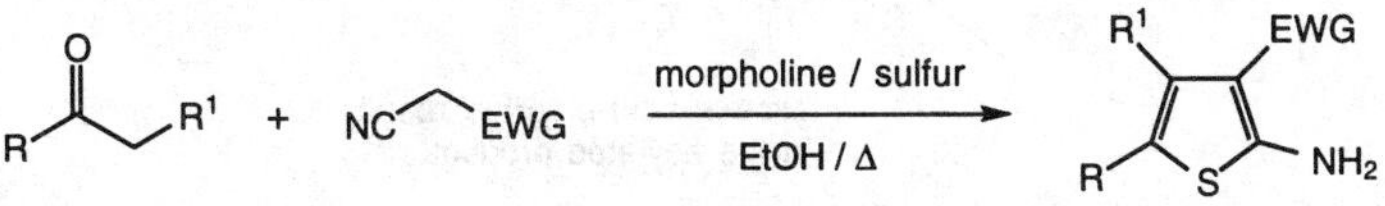

Chem. Ber. **1966**, 99, 94.
Chem. Ber. **1965**, *98*, 3571.

Gibbs Phthalic Anhydride Process

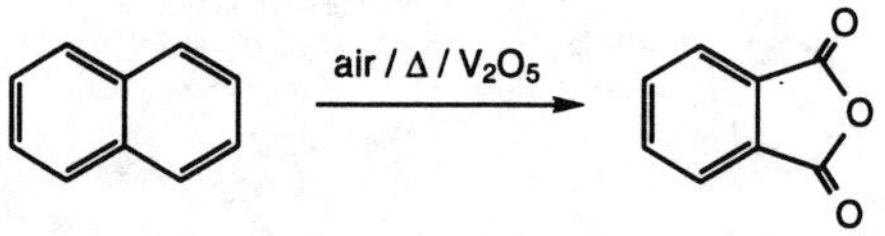

Merck ONR-35

Note: Also known as the Gibb-Wohl Oxidation.

Giese Reaction

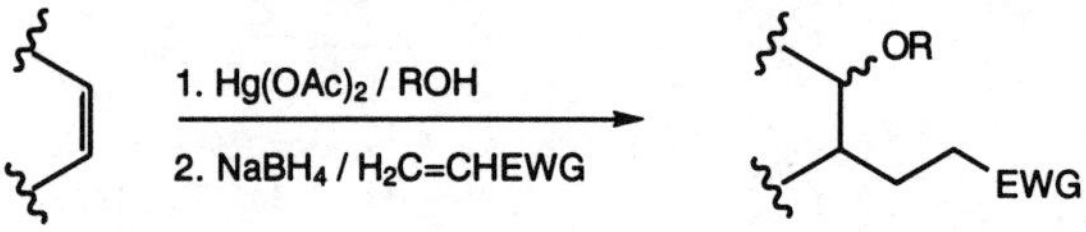

Tetrahedron Lett. **1984**, *25*, 2743.
Tetrahedron Lett. **1982**, *23*, 931.

Gilman Reaction

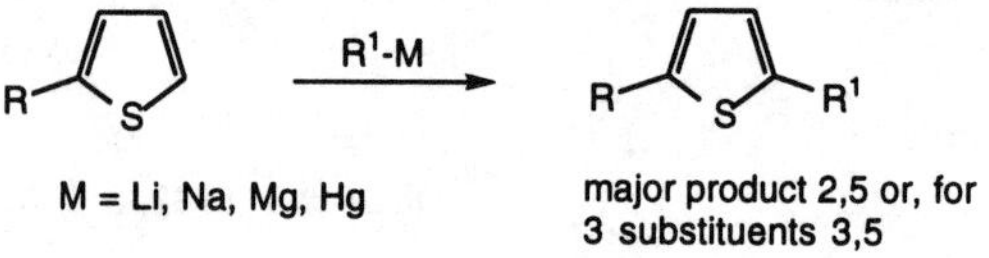

Comp. Org. Chem. **4**, 808.

Glaser (Coupling) Reaction

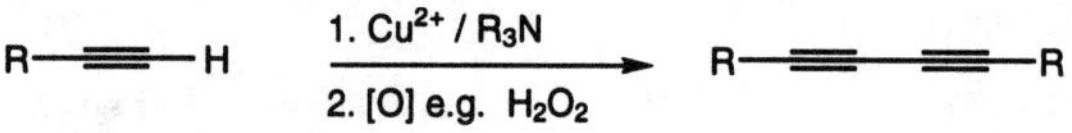

Adv. Org. Chem. **1963**, *4*, 225.
Chem. Rev. **1957**, *57*, 215.

Note: Also known as the Glaser Oxidative Coupling.

Glycidic Ester Condensation (see Darzens Condensation)

Gogte Synthesis

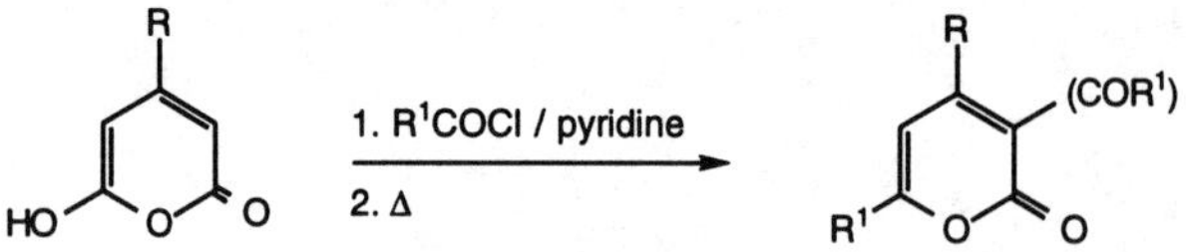

excess acid chloride results
in the acylated product

Merck ONR-36.

Goldberg Reaction

$$\text{ArNHAc} \xrightarrow{\text{Ar}^1\text{X/ CuI / K}_2\text{CO}_3} \text{ArAr}^1\text{NAc}$$

March 591.
Org. Reactions **1965**, *14*, 19.

Gomberg-Bachmann-Hey Reaction

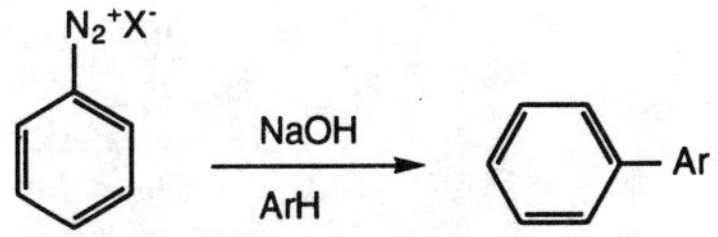

Adv. Free-Radical Chem. **1966**, *2*, 47.
Chem. Rev. **1957**, *57*, 77.

Note: Also known as the Gomberg Reaction, Gomberg-Hey Reaction or Gomberg-Bachmann Reaction

Gomberg-Bachmann Pinacol Synthesis

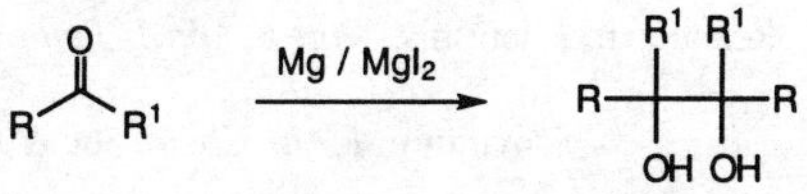

March 1110.

Gomberg Free Radical Reaction

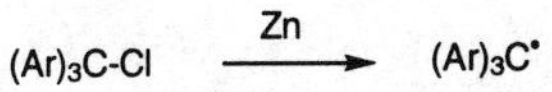

Tetrahedron **1974**, *30*, 2009.

Gould-Jacobs Reaction

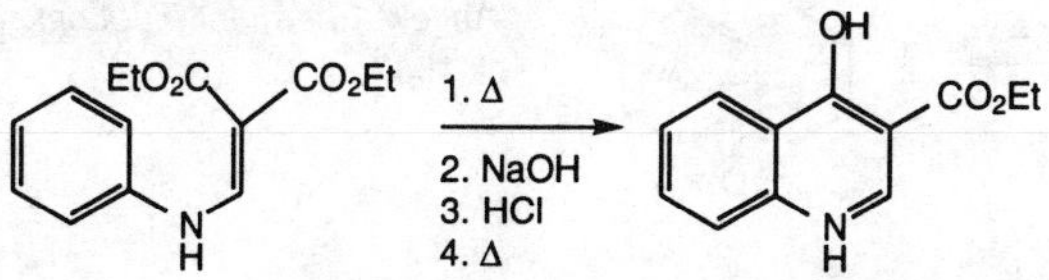

Chem. Rev. **1948**, *43*, 53.

Graebe-Ullmann Synthesis

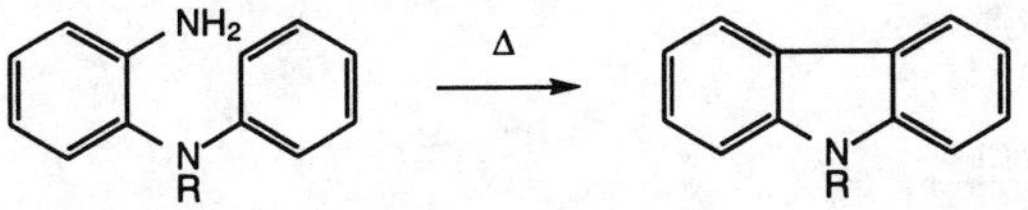

Chem. Rev. **1947**, *40*, 360.

Griess Diazo Reaction

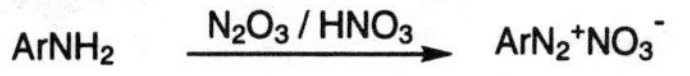

Merck ONR-37.
Russ. Chem. Rev. **1963**, *32*, 59.

Grignard Degradation

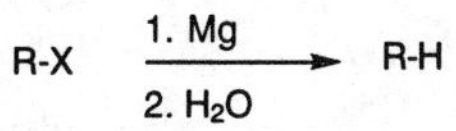

Merck ONR-37.

Grignard Reaction

RMgX + R¹C(=O)R² →(H₃O⁺)→ R¹C(OH)(R)R²

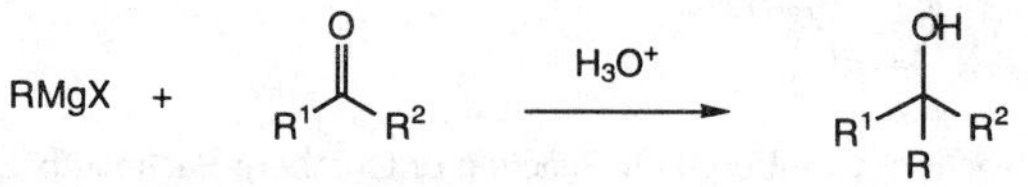

Acc. Chem. Res. **1991**, *24*, 95; 255.
Acc. Chem. Res. **1990**, *23*, 286.
Angew. Chem. Int. Ed. Engl. **1987**, *26*, 990.
Acc. Chem. Res. **1984**, *17*, 109.
Tetrahedron **1984**, *40*, 641.
Synthesis **1981**, 585.
Tetrahedron **1975**, *31*, 2735.
Acc. Chem. Res. **1974**, *7*, 272.
Synthesis **1971**, 347.
Reaction mechanisms: *Pure & Appl. Chem.* **1980**, *52*, 545.
Samarium: *J. Am. Chem. Soc.* **1992**, *114*, 6050.

Note: The formation of the Grignard reagent is also known as the Grignard Reaction.

Grob Fragmentation

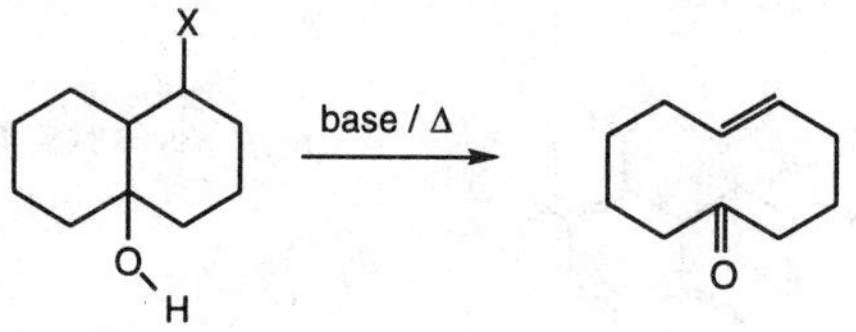

base / Δ

Angew. Chem. Int. Ed. Engl. **1969**, *8*, 535.
Angew. Chem. Int. Ed. Engl. **1967**, *6*, 1.

X= OH, Cl, Br, OTs etc.

Grosheintz-Fischer-Reissert Aldehyde Synthesis (see Reissert Reaction)

Grove's Synthesis (Process)

R-OH →(HCl / ZnCl₂)→ R-Cl

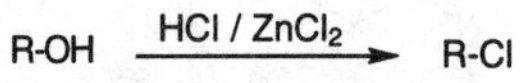

Vogel 382.

Grundmann (Aldehyde) Synthesis

RCOCl →(1. CH₂N₂ / 2. AcOH / 3. H₂ / 4. Pb(OAc)₄)→ RCHO

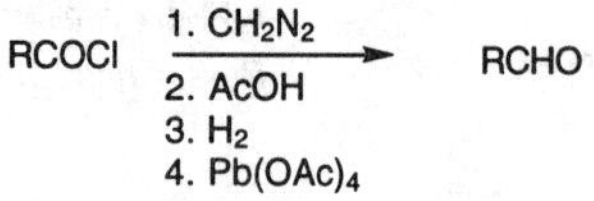

Org. Reactions **1954**, *8*, 225.

Guareschi Reaction

NCCH₂CO₂Et, Δ

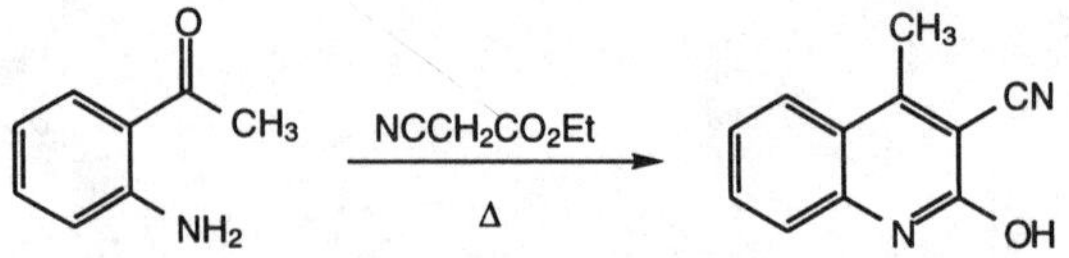

Chem. Rev. **1942**, *30*, 127.

Guareschi-Thorpe Condensation

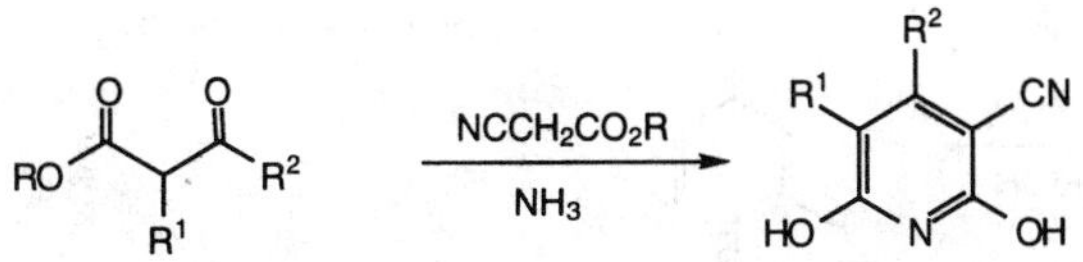

Merck ONR-38.

Guerbet Reaction

$$RCH_2CH_2OH \xrightarrow[\text{2. } Cr_2O_3 \text{ or } Pd / \Delta]{\text{1. Na or K}} RCH_2CH_2CHRCH_2OH$$

Tetrahedron **1967**, *23*, 1723.
J. Org. Chem. **1950**, *15*, 54.
Amines: *J. Org. Chem.* **1960**, *25*, 2126.
Mechanism: *Tetrahedron* **1967**, *23*, 1723.

Gustavson Reaction (see Freund Reaction)

Gutknecht Pyrazine Synthesis

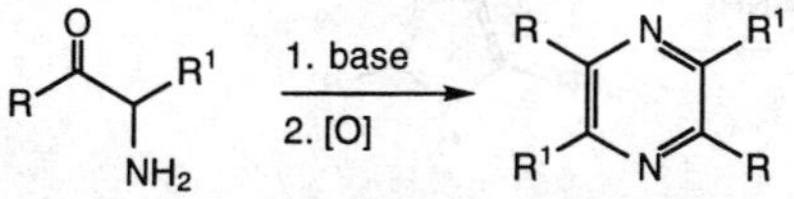

Merck ONR-39.
Chem. Rev. **1947**, *40*, 291.

Haber-Weiss Reaction (see Fenton Reaction)

Haller-Bauer Reaction (Cleavage)

$$R \overset{O}{\underset{}{\|}} R \xrightarrow[\text{2. } H^+]{\text{1. NaNH}_2} R\text{-H} + RCONH_2$$

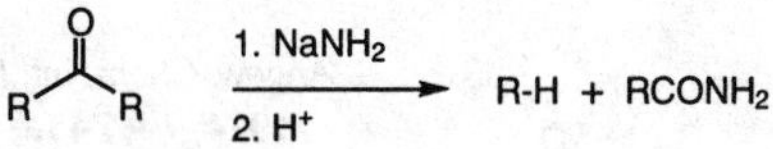

Can. J. Chem. **1992**, *70*, 1274.
Can. J. Chem. **1990**, *68*, 1106.
Org. Prep. Proc. Int. **1990**, *22*, 167.
Org. Reactions **1957**, *9*, 1.

Haloform Reaction

$$R \overset{O}{\underset{}{\|}} CH_3 \xrightarrow{X_2 / HO^-} RCO_2^- + CHX_3$$

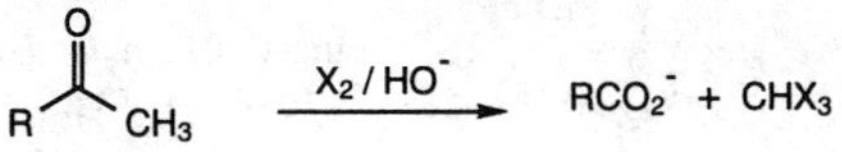

X = halogen

Chem. Ind. (London) **1959**, 1383.

Note: Also known as the Iodoform Reaction and Lieben Iodoform Reaction.

Hammick Reaction

$$\xrightarrow[\Delta]{RCOR^1}$$

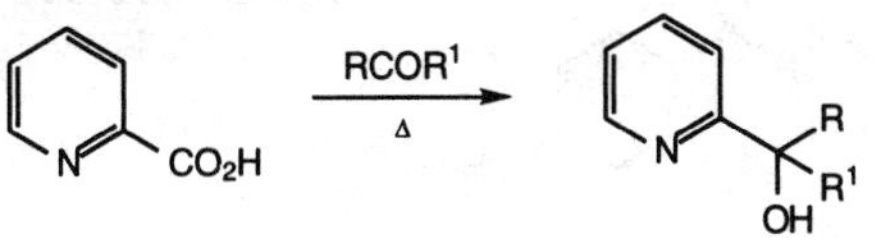

J. Chem. Soc. **1949**, 173; 659.

Hantzsch (Dihydropyridine) Synthesis

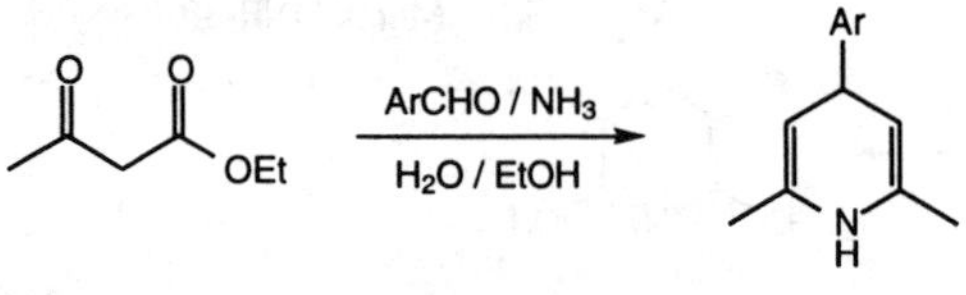

Synthesis **1983**, 761.

Hantzsch (Pyridine) Synthesis

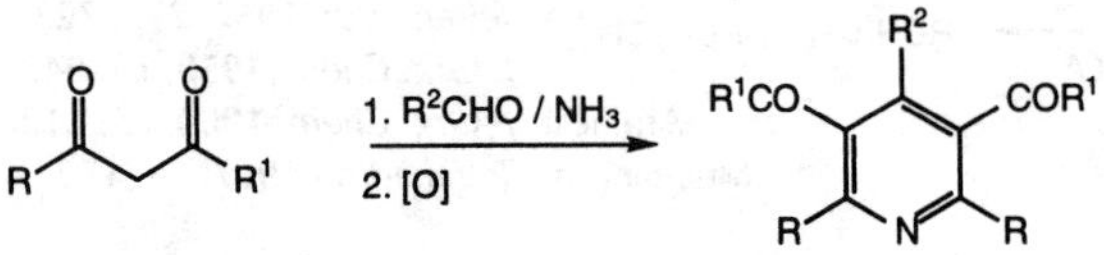

J. Chem. Soc., Perkin Trans. 1 **1975**, 926.
Tetrahedron **1970**, 26, 4655.
Chem. Rev. **1944**, *35*, 94.

Hantzsch (Pyrrole) Synthesis

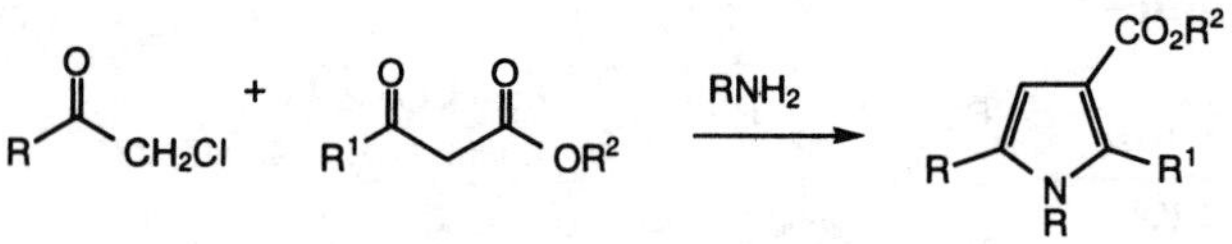

Can. J. Chem. **1970**, *48*, 1689.

Note: The reaction of α-haloketones with thioamide or selenoureas results in thiazoles and selenazoles respectively.

Harries Ozonide Reaction

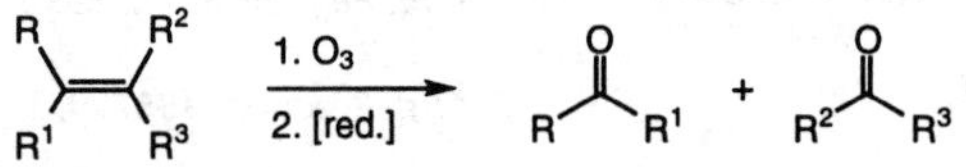

Angew. Chem. Int. Ed. Engl. **1974**, *14*, 745.
Acc. Chem. Res. **1968**, *1*, 313.
Chem. Rev. **1958**, *58*, 925.

Hass Reaction

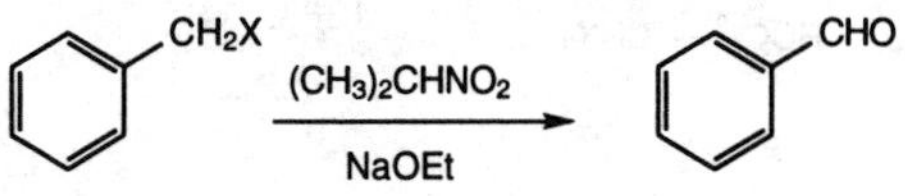

X = halogen

Angew. Chem. Int. Ed. Engl. **1974**, *14*, 734.

Hassner-Ritter Reaction

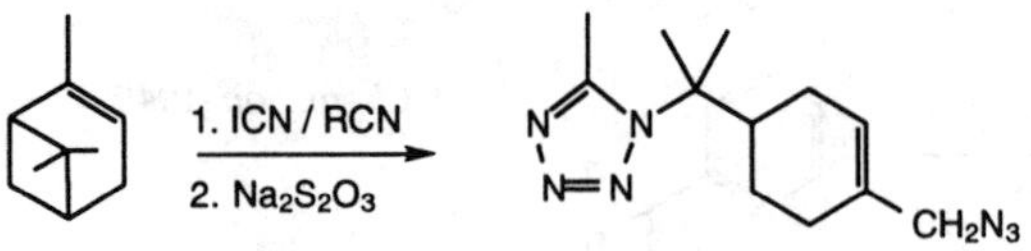

Synthesis **1973**, 356.

Hauser Rearrangement

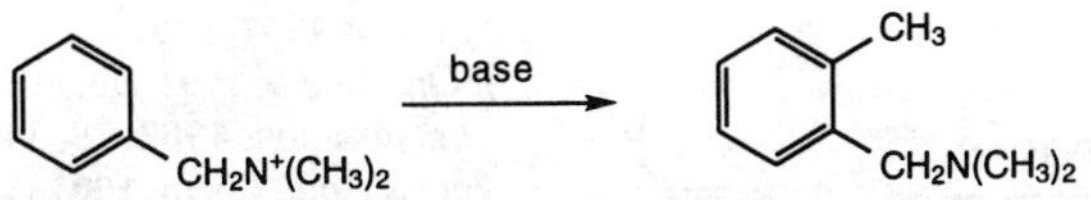

Angew. Chem. **1960**, *72*, 315.
Quart. Rev. **1958**, *12*, 15.

Note: 1. Also known as the Sommelet-Hauser Rearrangement.
2. The Sommelet Rearrangement is acid catalyzed.

Haworth Methylation

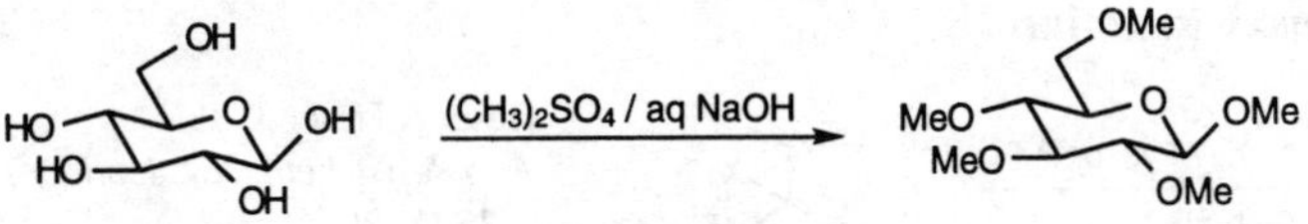

Merck ONR-40.

Haworth Phenanthrene Synthesis (see Haworth Reaction)

Haworth Reaction (Synthesis)

Org. Reactions **1984**, *30*, 124.

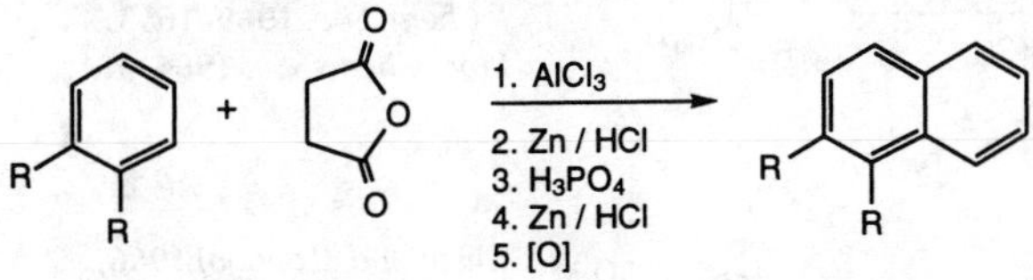

Note: Also known as the Haworth Phenanthrene Synthesis.

Hayashi Rearrangement

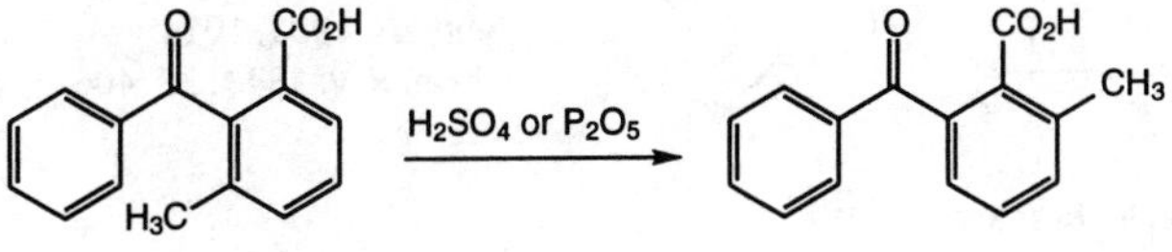

J. Am. Chem. Soc. **1958**, *80*, 3652.
J. Am. Chem. Soc. **1956**, *78*, 3817.

Heck Reaction

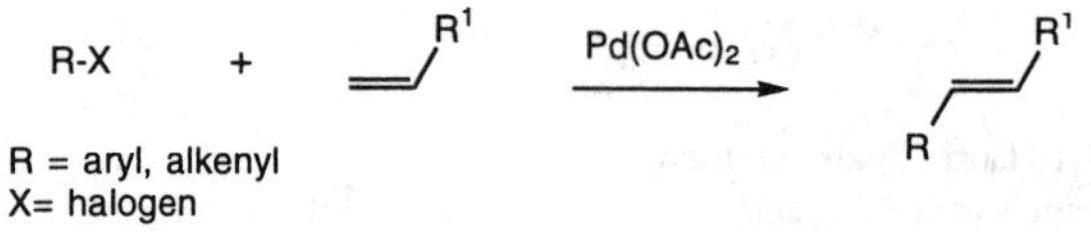

Synthesis **1991**, 539.
J. Org. Chem. **1991**, *56*, 4093.
Org. Reactions **1982**, *28*, 345.
Pure & Appl. Chem. **1981**, *53*, 2323.
Acc. Chem. Res. **1979**, *12*, 146.
Pure & Appl. Chem. **1978**, *50*, 691.

Hell-Volhard-Zelinsky Reaction

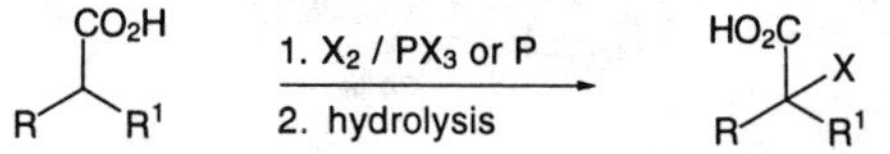

Fieser & Fieser *3*, 35.
Ann Rep. Chem. Soc. (London) **1931**, 96.

Note: Also known as the Hell-Volhard-Zelinskii Reaction or Hell-Volhard-Zelinsky Halogenation

Henbest Reduction

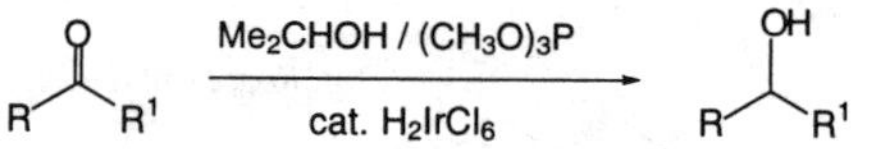

J. Chem. Soc., Chem. Commun. **1970**, 162.
J. Chem. Soc. **1969**, 1653.
Proc. Chem. Soc. **1964**, 361.

Henkel Reaction

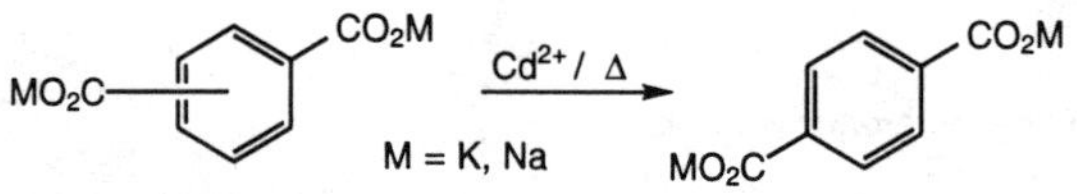

Chem. Ind. (London) **1966**, 1798.
Angew. Chem. **1960**, *72*, 738.

Henry Reaction

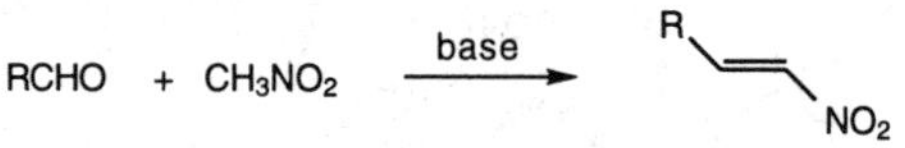

Synthesis **1983**, 1014.
Chem. Rev. **1943**, *32*, 406.

Note: Also known as the Kamlet Reaction.

Herz Reaction

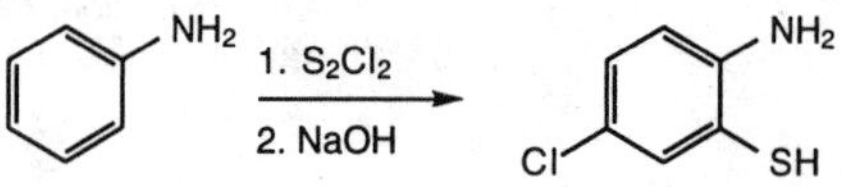

Chem. Rev. **1957**, *57*, 1011.

Hetero-Cope Rearrangement (see Cope Rearrangement)

Hetero-Diels-Alder Reaction (see Diels-Alder Reaction)

Heumann-Pfleger Indigo Synthesis

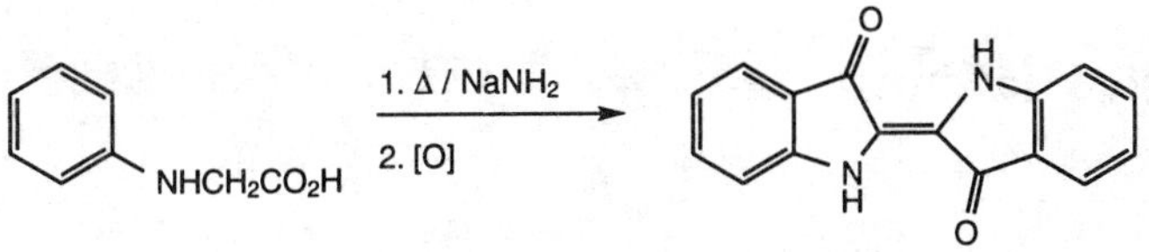

Merck ONR-43.

Hilbert-Johnson Reaction (Coupling)

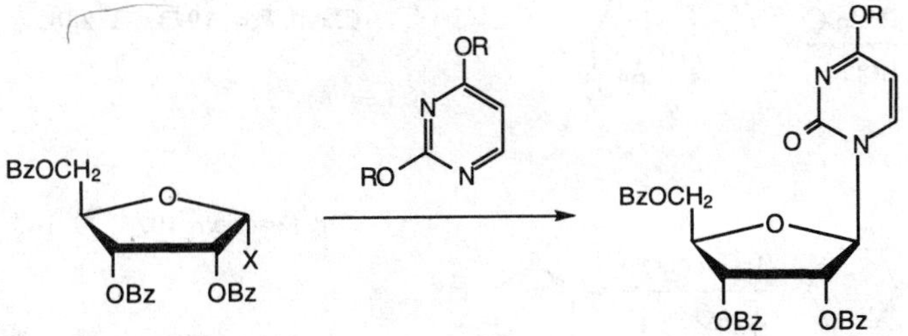

Chem. Ber. **1981**, *114*, 1234.
Adv. Heterocycl. Chem. **1967**, *8*, 115.

Hinsberg Oxindole Synthesis

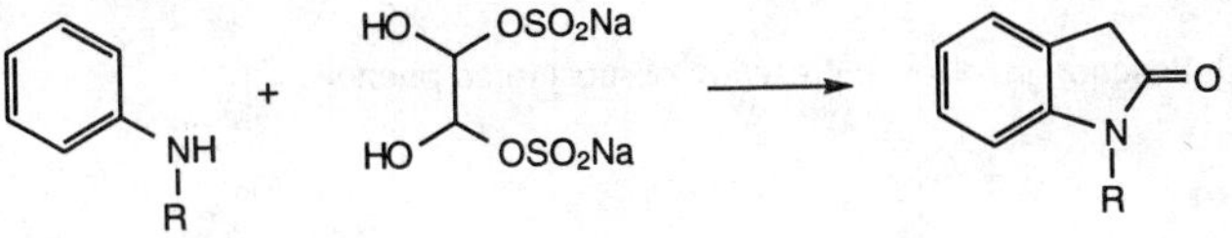

Merck ONR-43.

Hinsberg Reaction

$$RNH_2 \xrightarrow{R^1SO_2Cl} R^1SO_2NHR \xrightarrow[\text{2. conc acid or base}]{\text{1. } R^2X} RR^2NH$$

J. Chem. Ed. **1972**, 49, 287.

Note: The Hinsberg Test is also known as the Hinsberg Reaction.

Hinsberg Sulfone Synthesis

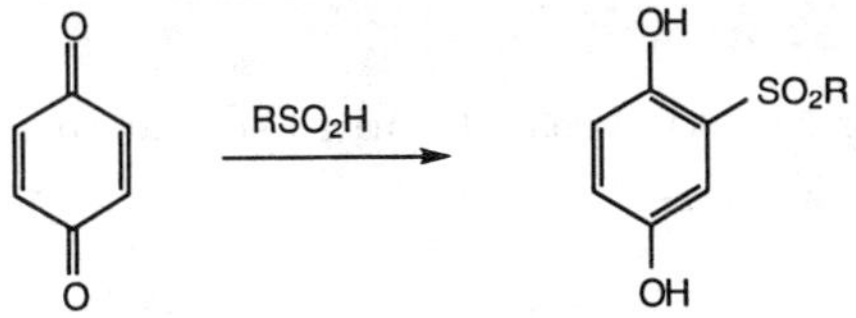

Merck ONR-44.

Hinsberg (Thiophene) Synthesis

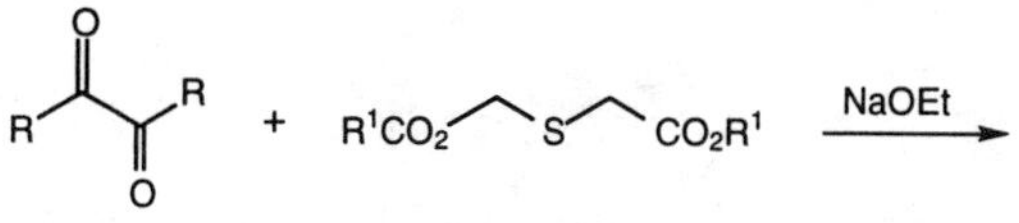

Comp. Org. Chem. **4**, 833.

Hoch-Campbell Aziridine (Ethylenimine) Synthesis

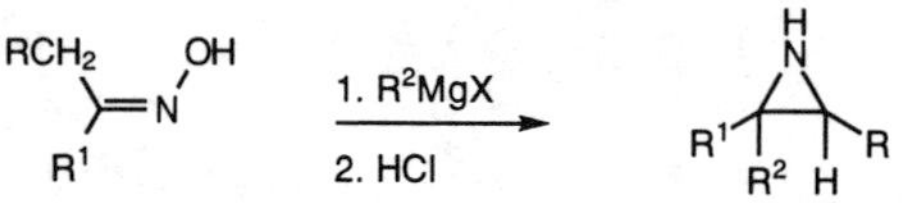

Merck ONR-44.
Chem. Rev. **1973**, *73*, 288.

Hock Cleavage

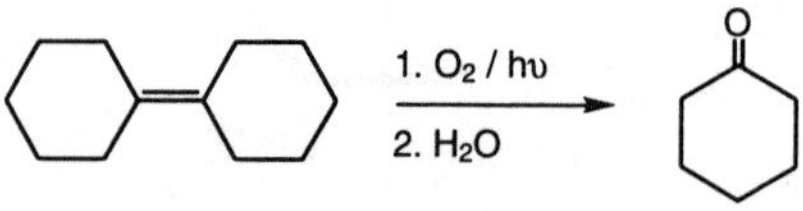

Org. Reactions **1973**, *20*, 160.

Hoesch Reaction (Acylation)

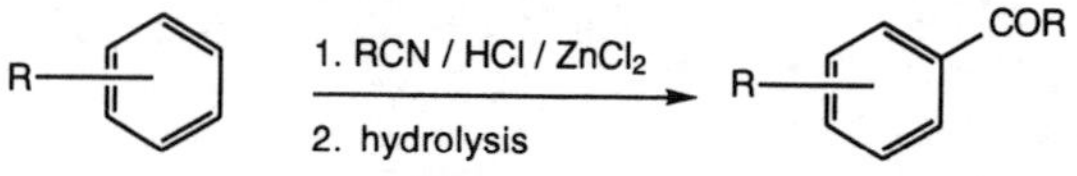

R = OH, OR, NR_2

Russ. Chem. Rev. **1962**, *31*, 615.
Chem. Rev. **1961**, *61*, 179.
Org. Reactions **1949**, *5*, 387.

Note: The Houben-Hoesch Reaction involves the reaction of substituted phenols.

Hofer-Moest Reaction

$$\text{RCOONa} \xrightarrow{\text{electrolysis}} \text{ROH}$$

Chem. Rev. **1967**, *67*, 651.
Quart. Rev. **1952**, *6*, 395.

Hofmann Degradation (see either Hofmann Degradation or Hofmann Rearrangement)

Hofmann Degradation (Elimination)

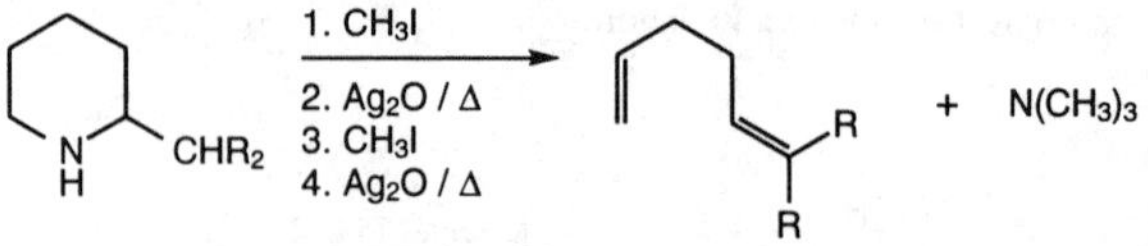

Org. Synth. **1976**, *55*, 3.
Chem. Technol. **1971**, 297.
Org. Reactions **1970**, *18*, 403.
Org. Reactions **1960**, *11*, 317.

Note: Also known as the Hofmann Exhaustive Methylation or Hofmann Decomposition.

Hofmann Isonitrile Synthesis

$$\text{R-NH}_2 \xrightarrow{\text{CHCl}_3\text{ / NaOH}} \text{R-NC}$$

Merck ONR-45.

Note: Also known as the Carbylamine Reaction or Hofmann Carbylamine Reaction.

Hofmann-Löffler Reaction

$$\text{R}^1\text{R}^2\text{R}^3\text{C-H} \quad + \quad \text{R}_2\text{NH}^+\text{Cl} \xrightarrow{h\upsilon \text{ or } \text{Fe}^{2+}} \text{R}^1\text{R}^2\text{R}^3\text{C-Cl}$$

Chem. Rev. **1963** *63*, 55.

Hofmann-Löffler-Freytag Reaction (Rearrangement)

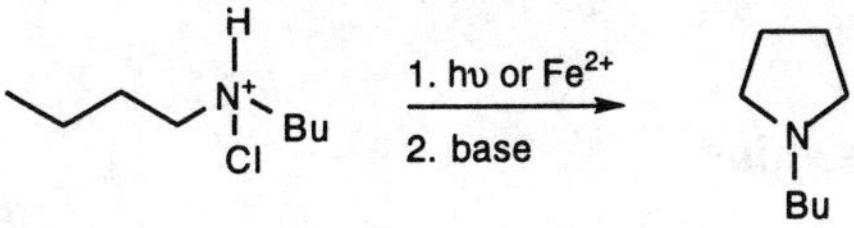

Chem. Rev. **1963** *63*, 55.

Note: Also known as the Hofmann-Löffler Rearrangement.

Hofmann Martius Rearrangement (Reaction)

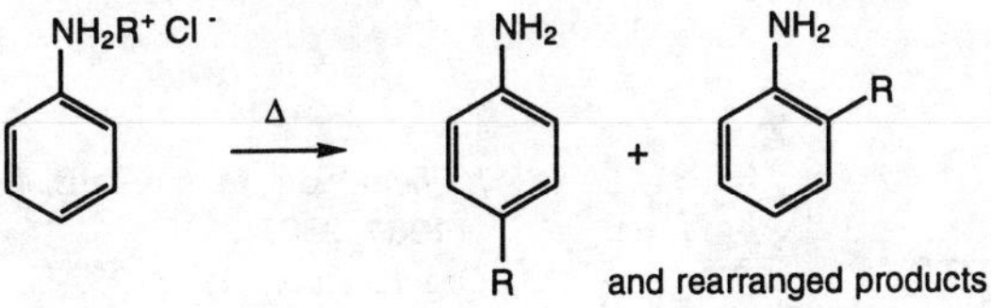

Tetrahedron **1964**, *20*, 2717.
Org. Reactions **1953**, *7*, 134.

Note: Also known as the Hofmann Rearrangement.

Hofmann Rearrangement (Reaction)

$$\text{RCONH}_2 \xrightarrow{\text{Br}_2\text{ / NaOH}} \text{RNH}_2$$

Org. Synth. **1988**, 66, 132.
Tetrahedron **1971**, *27*, 1317.
Org. Reactions **1946**, 3, 267.
Chem. Rev. **1934**, *14*, 219.

Note: Also known as the Hofmann Degradation.

Hofmann-Sand Reactions

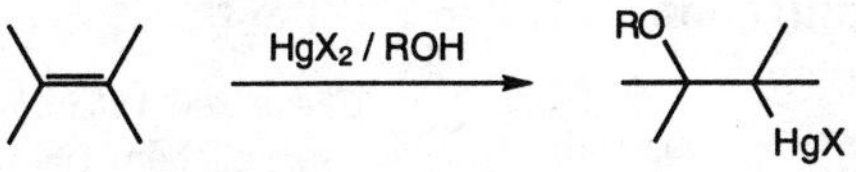

X = halogen, nitrate, acetate or sulfite

Chem. Rev. **1951**, *48*, 7.

Homo Diels-Alder Reaction (see Diels-Alder Reaction)

Hooker Reaction (Synthesis)

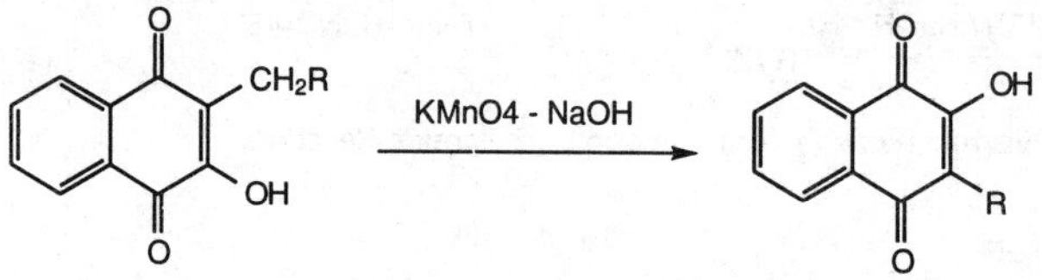

J. Am. Chem. Soc. **1948**, *70*, 3215.
J. Am. Chem. Soc. **1936**, *58*, 1164; 1179.

Note: Also known as the Hooker Oxidation.

Horner (Wadsworth) Emmons Reaction

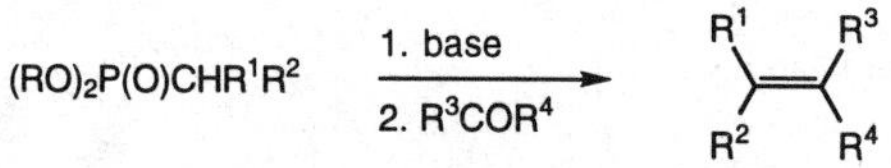

Chem. Res. **1983**, *16*, 411.
Org. Reactions **1977**, *25*, 73.
Chem. Rev. **1974**, *74*, 87.

Note: Also see the Wittig Reaction.

Horner (-Wittig) Reaction

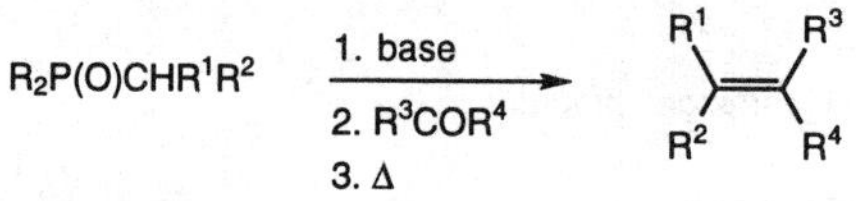

J. Chem. Soc., Perkin Trans. 1 **1985**, 2307.
Org. Reactions **1977**, *25*, 73.

Note: Also see the Wittig Reaction.

Houben-Fischer Synthesis

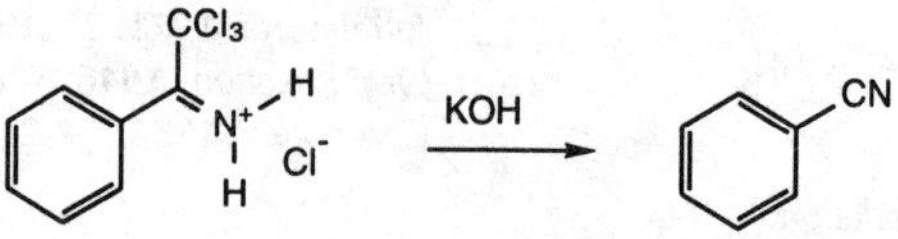

Org. Reactions **1949**, *5*, 390.
Chem. Rev. **1948**, *42*, 221.

Houben-Hoesch Reaction (see Hoesch Reaction)

Huang-Minlon Reduction (Modification)

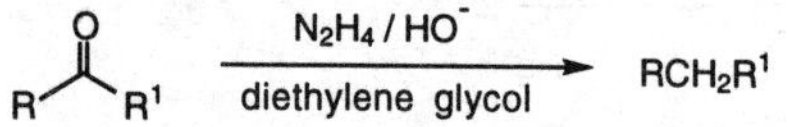

Chem. Rev. **1965** 65, 63.
Org. Reactions **1948**, *4*, 385.

Hubenett Synthesis

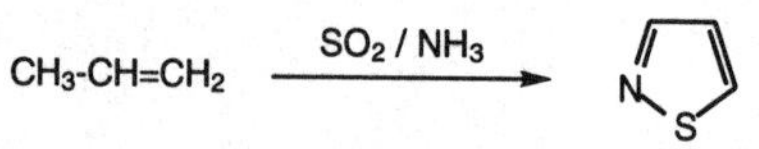

Adv. Heterocycl. Chem. **1965**, 4, 107.
Angew. Chem. Int. Ed. Engl. **1963**, *2*, 714.

Hunsdiecker Reaction (Synthesis)

$$RCO_2Ag + X_2 \xrightarrow{\Delta} R\text{-}X$$

$$RCO_2H + X_2 \xrightarrow[\text{e.g. HgO}]{\text{metal salt}} R\text{-}X$$

X = Cl, Br, I

Org. Reactions **1972**, *19*, 326.
Tetrahedron **1971**, *27*, 5323.
Org. Synth. **1963**, *43*, 9.
Org. Reactions **1957**, *9*, 332.
Chem. Rev. **1956**, *56*, 219.

Note: 1. Also known as the Borodine Reaction or Borodine-Hunsdiecker Reaction.
2. Also see the Cristol-Firth-Hunsdiecker Reaction and Simonini Reaction.

Hurtley Reaction

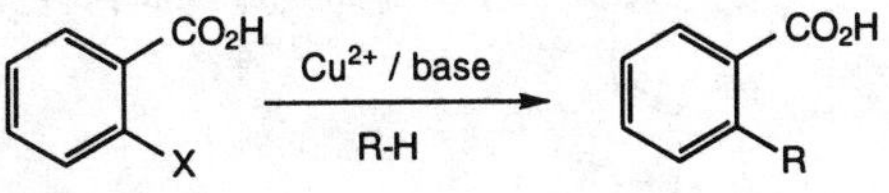

X = halogen

Tetrahedron **1975**, *31*, 2606.

Hydroboration Reaction (see Chapter 6)

Hydroformylation Reaction (see Oxo Process)

Ibuka-Yamamoto Reaction

J. Am. Chem. Soc. **1989**, *111*, 4864.

Indirect Diazotization Method (see Griess Diazo Reaction)

Intramolecular Reactions (see specific Named Reaction)

Iodoform Reaction (see Haloform Reaction)

Ireland-Claisen Rearrangement

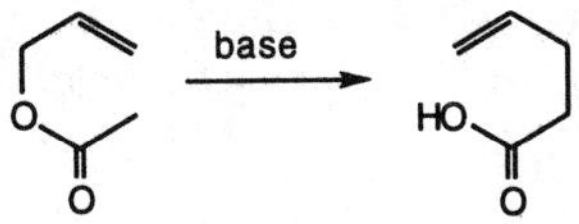

Aldrichimica Acta **1993**, *26*, 17.
J. Org. Chem. **1991**, *56*, 5826.
J. Am. Chem. Soc. **1976**, *98*, 2868.
J. Org. Chem. **1976**, *41*, 986.
Tetrahedron Lett. **1975**, 3975.

Note: Also known as the Ireland Ester Enolate Reaction or Ireland Reaction.

Irvine-Purdie Methylation (see Purdie Methylation)

Ivanov Reaction

RCH_2CO_2H + R^1MgX ⟶ $RCH(MgX)CO_2MgX$

Synthesis **1970**, 615.

Jacobsen (Asymmetric Epoxidation) Reaction

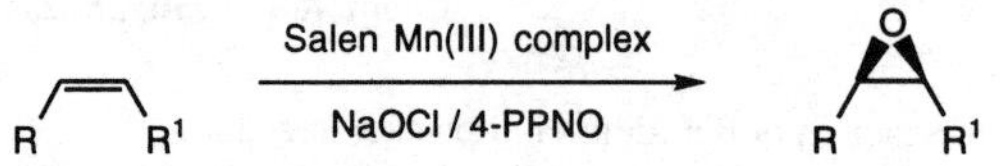

J. Org. Chem. **1992**, *57*, 4321.
J. Am. Chem. Soc. **1991**, *113*, 7063.

Jacobsen Reaction (Rearrangement)

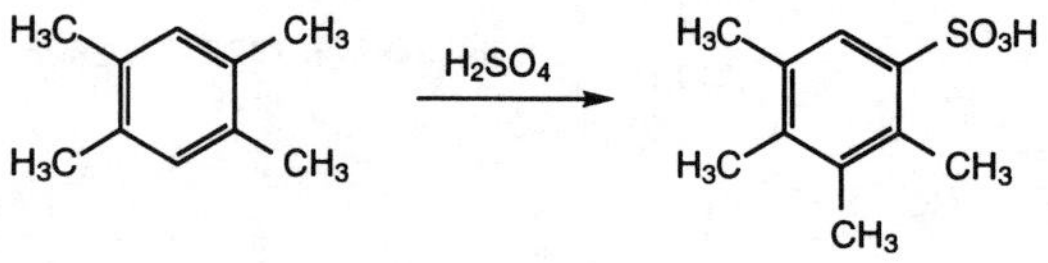

J. Org. Chem. **1962**, *27*, 4408.
Org. Reactions **1942**, *1*, 370.

Janovsky Reaction

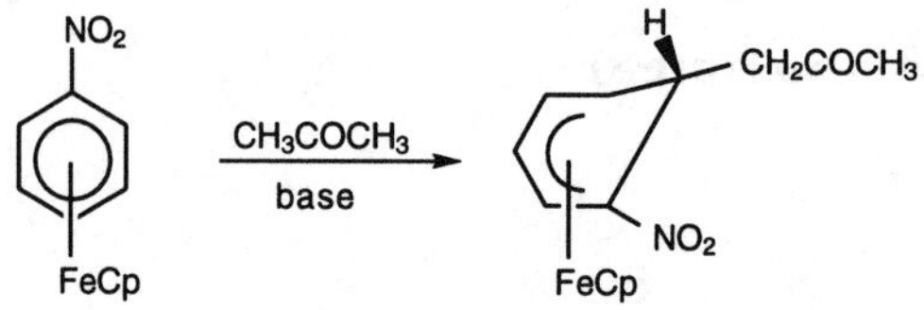

Can. J. Chem. **1986**, *64*, 2030
J. Chem. Soc., Chem. Commun. **1985**, 1296
Russ. Chem. Rev. **1978**, *47*, 1061

Note: Also known as the Yanovsky Reaction or Yanovskii Reaction.

Japp-Klingemann Reaction

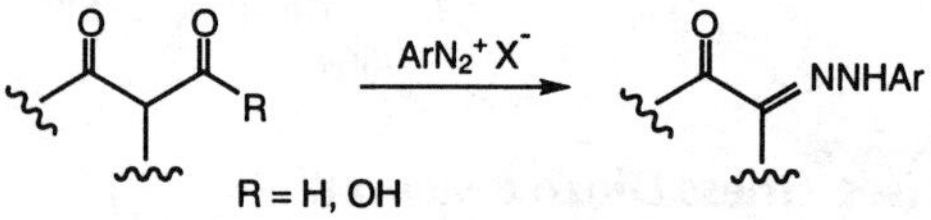

R = H, OH

Tetrahedron **1982**, *38*, 1527.
Chem. Rev. **1969**, *69*, 233.
Org. Reactions **1959**, *10*, 143.

Johnson-Orthoester Rearrangement

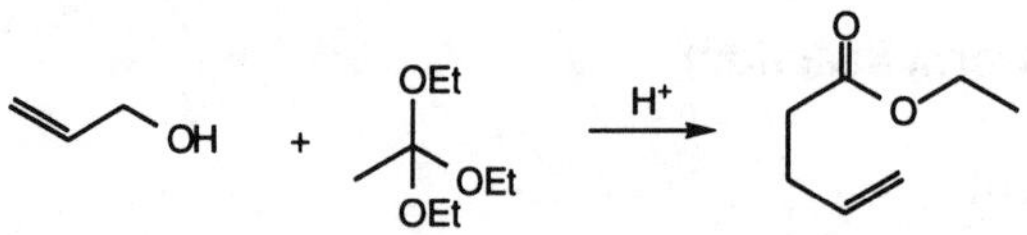

J. Am. Chem. Soc. **1970**, *92*, 741.

Jones Oxidation

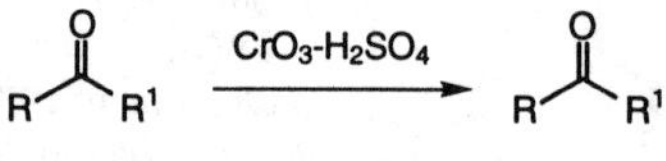

Org. Synth. **1990**, *68*, 175.
J. Chem. Soc. **1953**, 2548.

Jourdan Synthesis

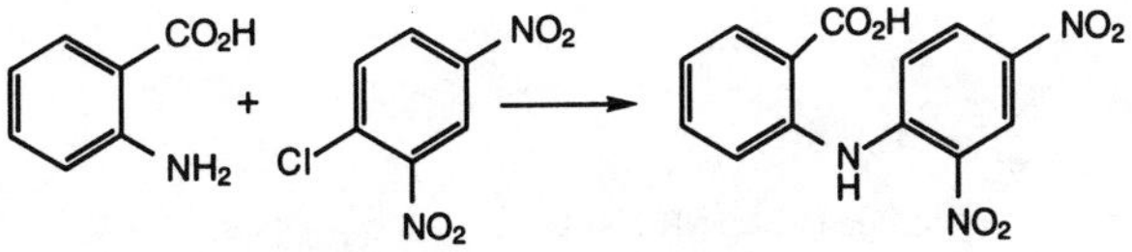

Org. Reactions **1965**, *14*, 19.

Jourdan-Ullmann Synthesis

Synthesis **1985**, 217.

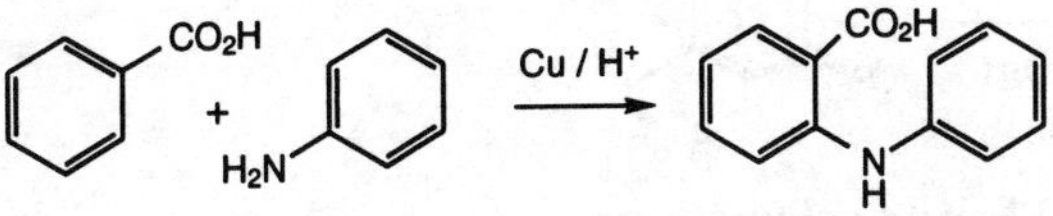

Note: Also known as the Ullmann Reaction.

Julia-Johnson Synthesis

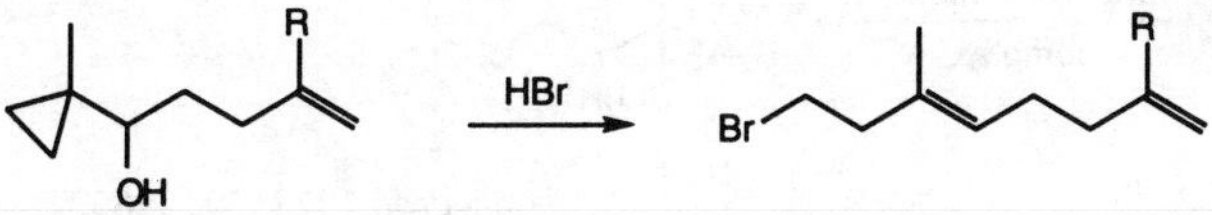

Synthesis **1971**, 178.

Julia Reaction

Tetrahedron Lett. **1989**, *30*, 4833.
Pure & Appl. Chem. **1985**, *57*, 763.
Tetrahedron. **1977**, *33*, 2019.

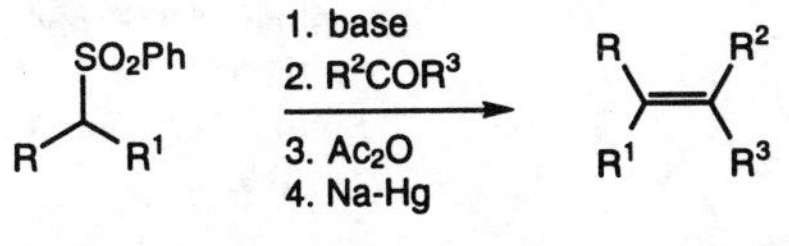

Kalb-Gross Oxidation

J. Org. Chem. **1964**, *29*, 1273.

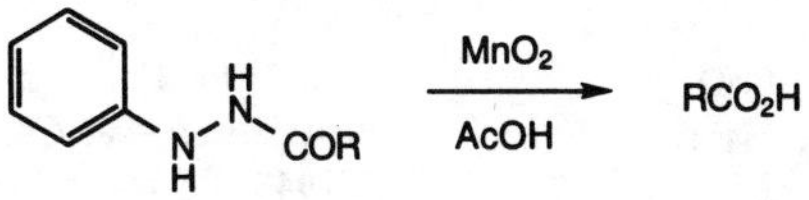

Kaluza Reaction

Comp. Org. Chem. **2**, 514.

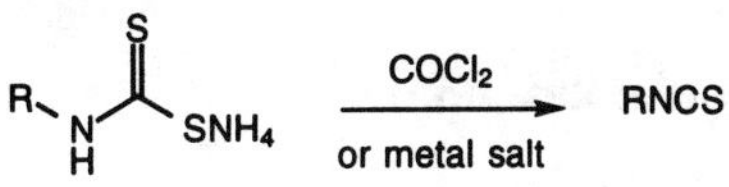

Kamlet Reaction (see Henry Reaction)

Katritzky Pyrylium-Pyridinium Reaction (Method)

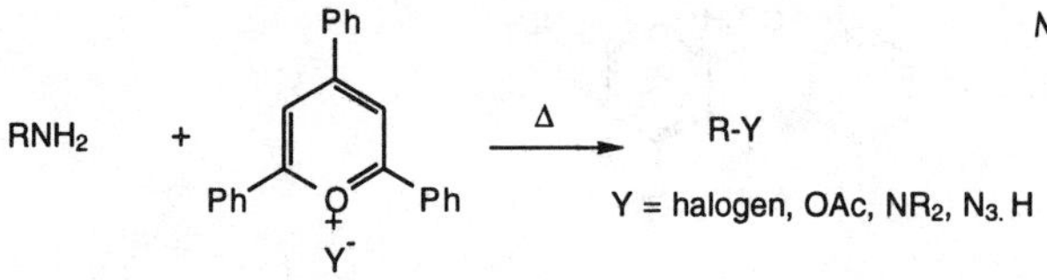

March 313.

Kendall-Mattox Reaction

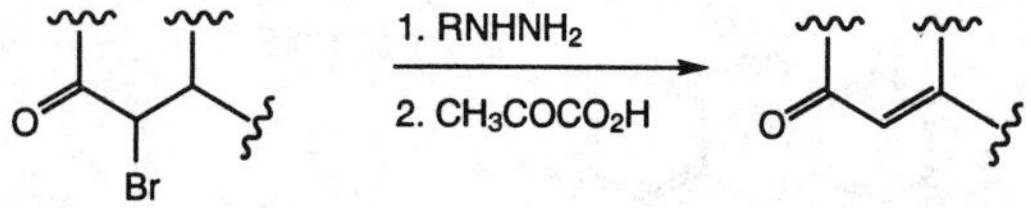

J. Org. Chem. **1963**, *28*, 453.

Khand Reaction (see Pauson-Khand Cycloaddition)

Kharasch Reaction

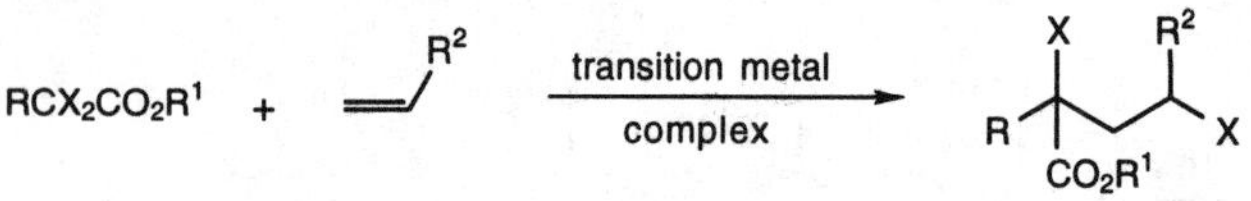

Tetrahedron **1973**, *29*, 2989.
Org. Reactions **1963**, *13*, 91.
Intramolecular: *Tetrahedron* **1988**, *44*, 4671.

Kharasch-Sosnovsky Reaction

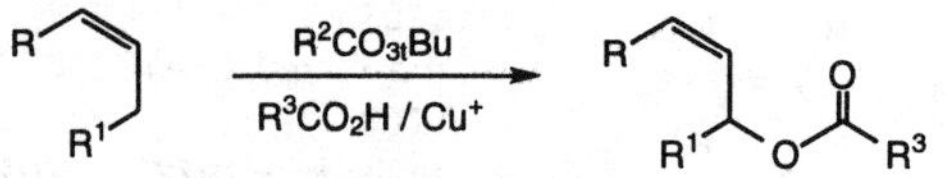

Synthesis **1977**, 894.
Synthesis **1972**, 1
J. Am. Chem. Soc. **1959**, *81*, 5819.

Note: Also known as the Acyloxylation Reaction

Kiliani (-Fischer) Synthesis

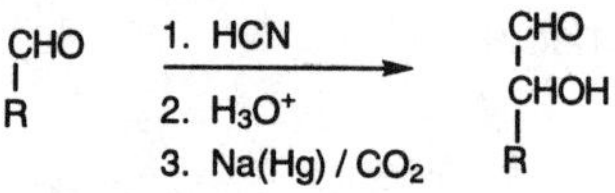

Comp. Org. Chem. **2**, 758.
Adv. Carbohydrat. Chem. **1945**, *1*, 1.

Kimel-Cope Rearrangemant (see Carroll Rearrangement)

Kindler Reaction (see Willgerodt-Kindler Reaction)

Kirsanov Reaction

$$R_3PCl_2 + R^1NH_2 \longrightarrow R_3P{=}NR^1$$

Comp. Org. Chem. **2**, 1302.
Obshch. Khim. **1956**, *26*, 903.

Kishner Cyclopropane Synthesis

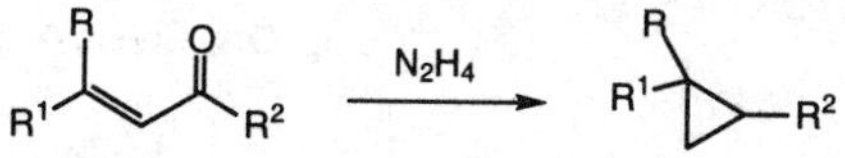

Merck ONR-50.

Kizhner Reduction (see Wolff-Kishner Reduction)

Knoevenagel Condensation (Reaction)

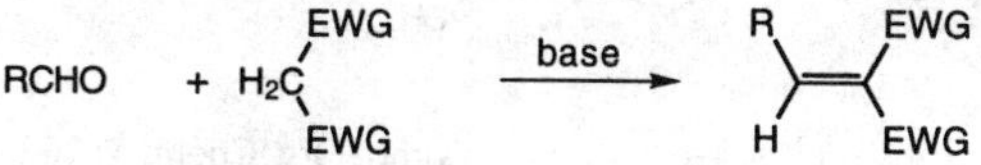

Tetrahedron Lett. **1972**, 663.
Org. Reactions **1967**, *15*, 204.
Chem. Rev. **1944**, *35*, 240.

Knoevenagel Bucherer Reaction

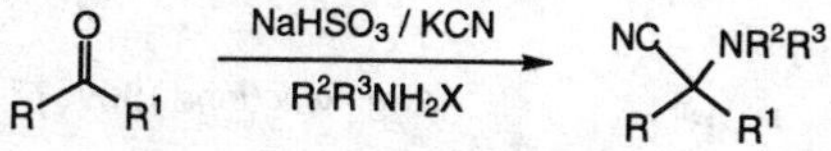

Comp. Org. Chem. **2**, 531.

Knoevenagel Diazotization

$$ArNH_2 + RONO \xrightarrow{HCl} ArN_2^+Cl^- + ROH$$

Merck ONR-51(37).
Quart. Rev. **1952**, 6, 358.

Knoop-Oesterlin Amino Acid Synthesis

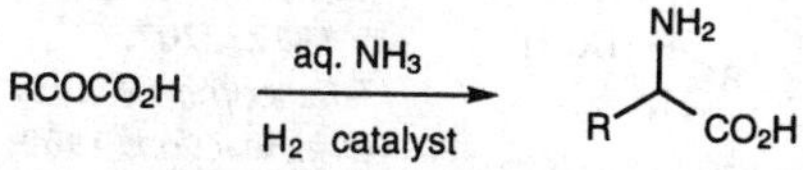

Merck ONR-51.
Chem. Rev. **1963**, *63*, 633.

Knorr-Paal Reaction (see Paal-Knorr Pyrrole Synthesis)

Knorr Pyrazole Synthesis

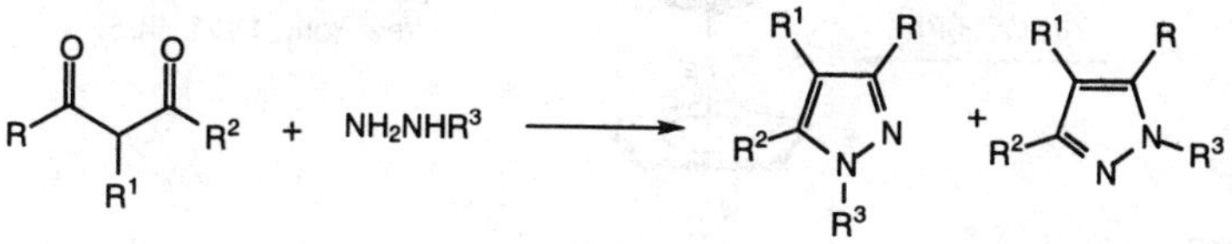

J. Org. Chem. **1971**, *36*, 853.
Org. Synth. **1943**, *Coll. Vol. 2*, 202.

Knorr (Pyrrole) Synthesis

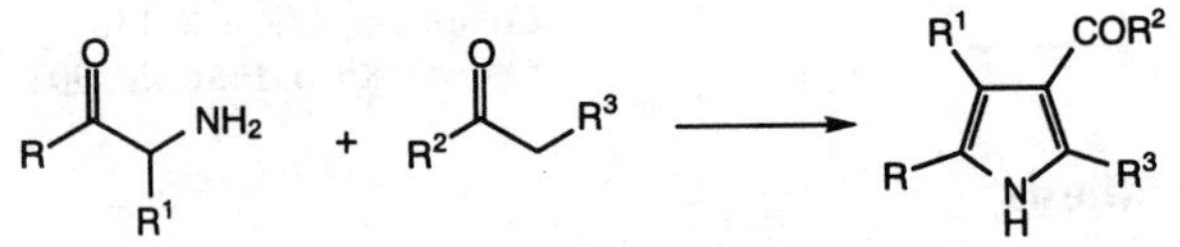

Comp. Org. Chem. **4**, 296.

Knorr (Quinoline) Synthesis

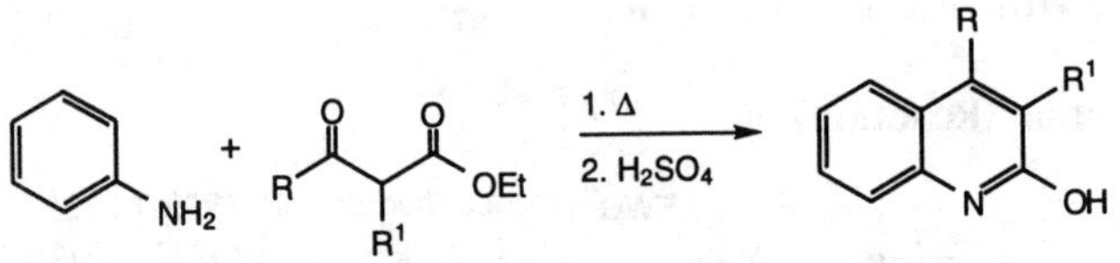

Comp. Org. Chem. **4**, 162.
Chem. Rev. **1944**, 35, 158.

Koch Reaction

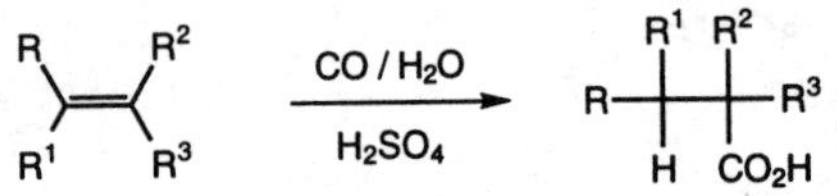

Org. Reactions **1969**, 17, 249.

Koch-Haaf Reaction

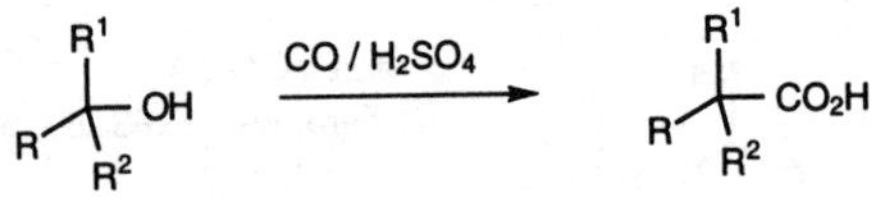

Tetrahedron Lett. **1981**, 22,
2365.
J. Chem. Soc., Perkin Trans. 1
1972, 2707.
Tetrahedron **1971**, 27, 3.
Org. Reactions **1969**, 17, 249.

Kocheshkov Reaction

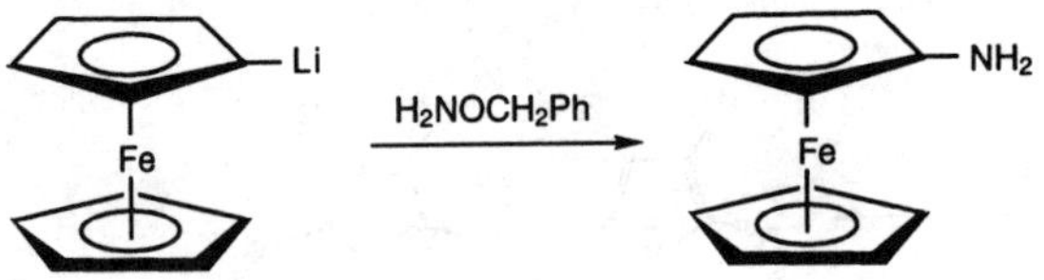

Fieser & Fieser 'Advanced Organic Chemistry', Reinhold,
New York, **1961**, 908.

Kochi Reaction

RCO₂H $\xrightarrow[\text{2. LiCl / }\Delta]{\text{1. Pb(OAc)}_4}$ RCl

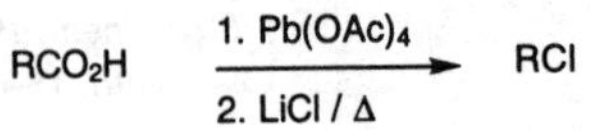

Org. Reactions **1972**, 19, 327.

Koenigs-Knorr Synthesis (Reaction)

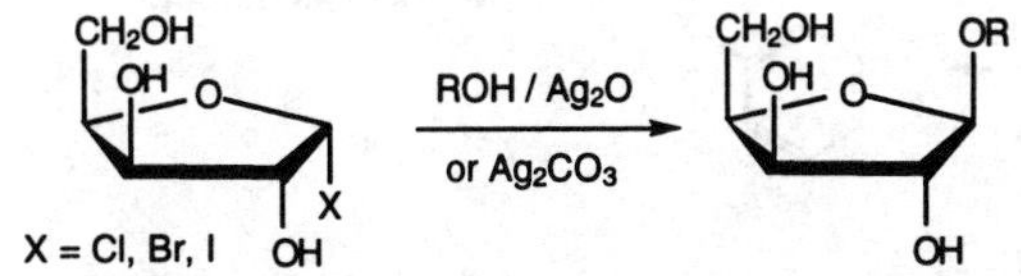

Can. J. Chem. **1990**, *68*, 2045.
Angew. Chem. Int. Ed. Engl. **1986**, *25*, 212.
Adv. Carbohydr. Chem. Biochem. **1977**, *34*, 243.

Note: Also known as the Koenigs-Knorr Method or Koenigs-Knorr Glycosidation.

Kogasin Process (see Fischer-Tropsch Synthesis)

Kolbé Reaction

$$RCO_2^- \xrightarrow{\text{electrolysis}} R\text{-}R$$

J. Am. Chem. Soc. **1978**, *100*, 2239.
Org. Reactions **1972**, *19*, 286.
Synthesis **1971**, 285.
Chem. Rev. **1967**, *67*, 623.
Quart. Rev. **1952**, *6*, 380.
Mixed Kolbé: *Tetrahedron Lett.* **1988**, *29*, 2797; 2801.

Note: Also known as the Kolbé Electrolysis Reaction or Kolbé Electrolysis Synthesis.

Kolbé-Schmitt (Carboxylation) Synthesis

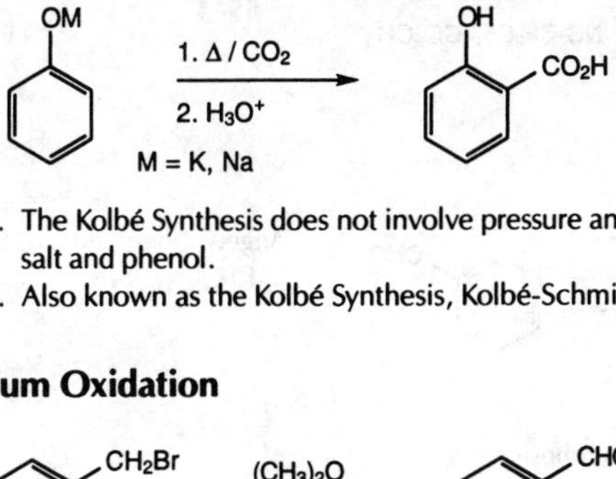

Chem. Rev. **1957**, *57*, 583.
J. Org. Chem. **1954**, *19*, 510.

Note: 1. The Kolbé Synthesis does not involve pressure and results in the synthesis of the disodium salt and phenol.
2. Also known as the Kolbé Synthesis, Kolbé-Schmitt Reaction, or Schmitt Synthesis.

Kornblum Oxidation

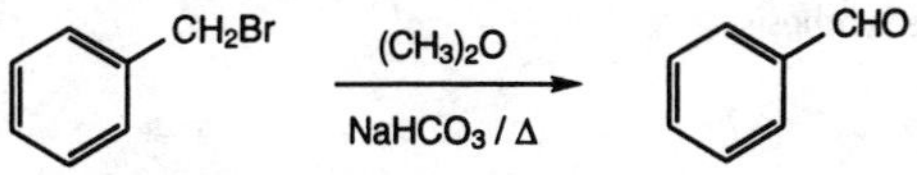

Org. Reactions **1990**, *39*, 297.
Chem. Rev. **1967**, *67*, 247.

Kornblum Reaction

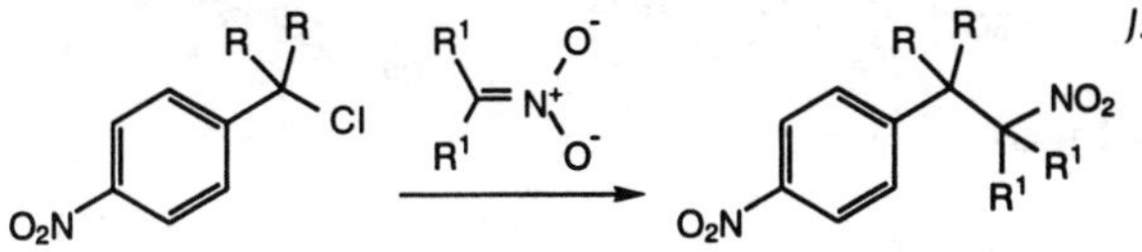

J. Am. Chem. Soc. **1970**, *92*, 5513.

Körner-Contardi Replacement Reaction

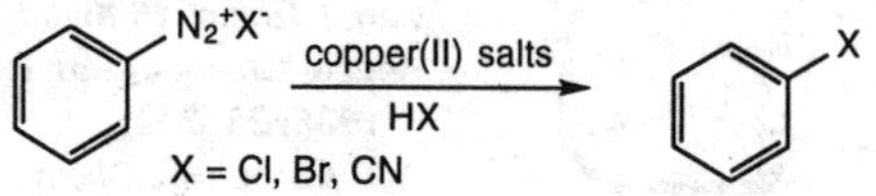

Merck ONR-53(80).

Kostanecki-Robinson Reaction.

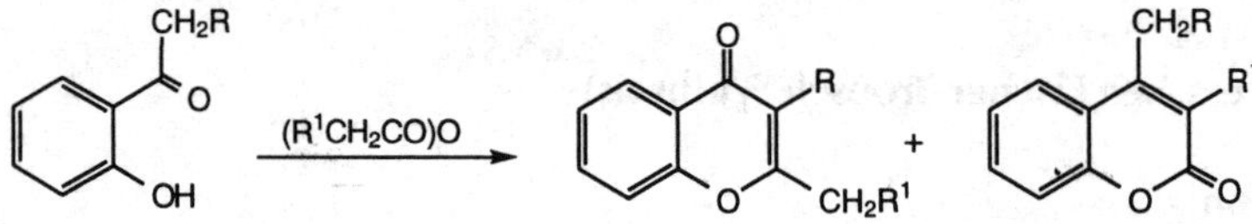

Tetrahedron **1969**, *25*, 715.
Org. Reactions **1953**, *8*, 91.

Note: Also known as Kostanecki Acylation or Bargellini Reaction.

Krafft Degradation

$$(RCH_2CO_2)_2M \xrightarrow[2.\ [O]]{1.\ (AcO)_2M\ /\ \Delta} RCO_2H$$

Merck ONR-53.

Krespan Reaction

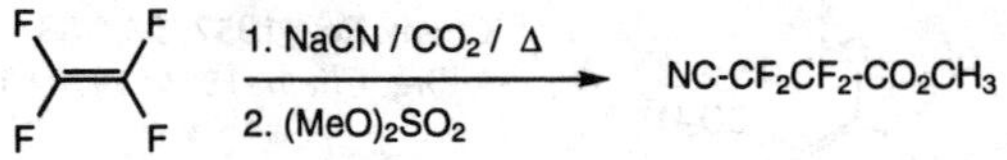

J. Am. Chem. Soc. **1984**, *106*, 5544.

Kröhnke Reaction

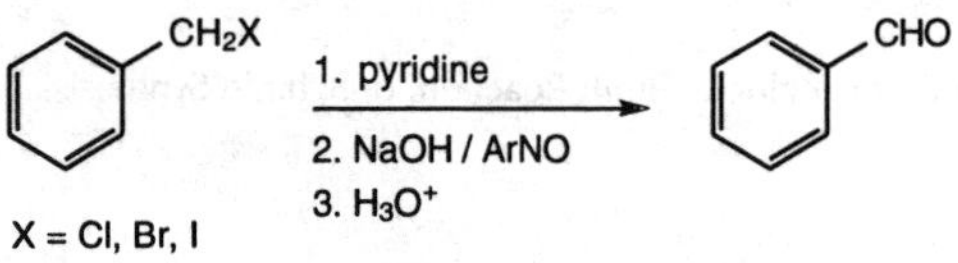

Angew. Chem. Int. Ed. Engl. **1963**, *2*, 380.

Note: Also known as Kröhnke Aldehyde Synthesis.

Kucherov Reaction

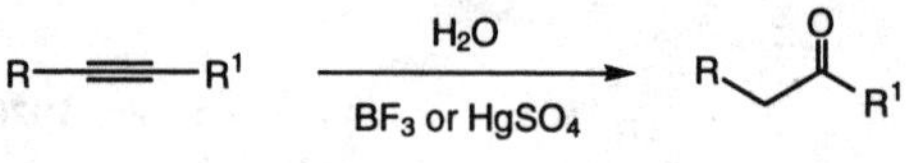

Merck ONR-54.

Kuhn-Roth Oxidation (Method)

$$R(CH_3)_n \xrightarrow[H_2SO_4]{CrO_3} n\ CH_3CO_2H$$

Chem. Ber. **1933**, 66, 1273.

Kuhn-Winterstein Reaction

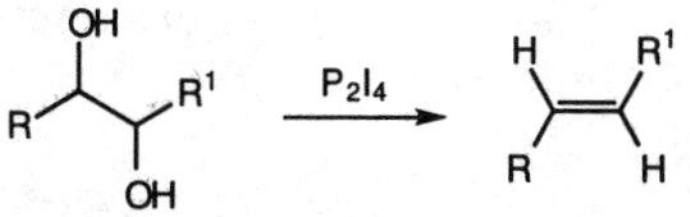

Org. Reactions **1984**, *30*, 504.

Landenburg Rearrangement

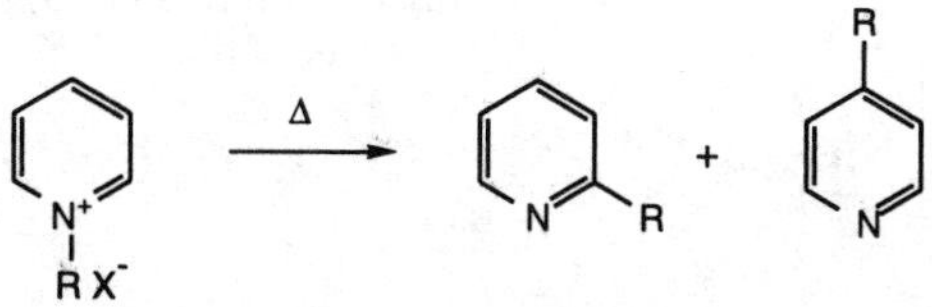

J. Chem. Soc. (C), **1969**, 146.
Org. Reactions **1953**, *7*, 135.

Lander Rearrangement

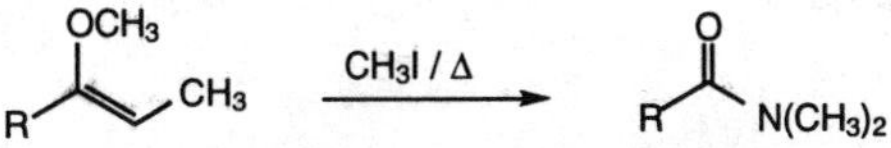

Org. Reactions **1965**, *14*, 24.

Lederer-Manasse Reaction

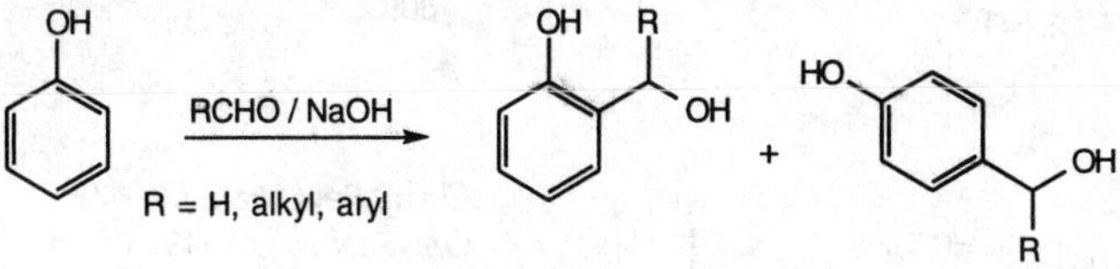

Angew. Chem. Int. Ed. Engl. **1963**, *2*, 373.

Lehmstedt-Tântâsescu Reaction

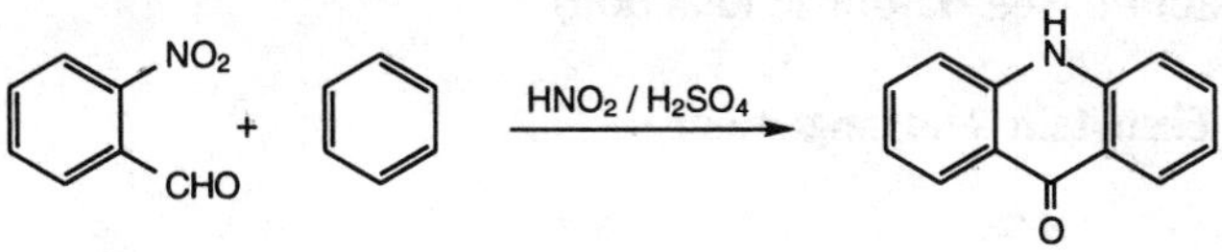

Merck ONR-54.
Bull. Chim. Soc., France **1960**, 698.

Leimgruber-Batcho Indole Synthesis

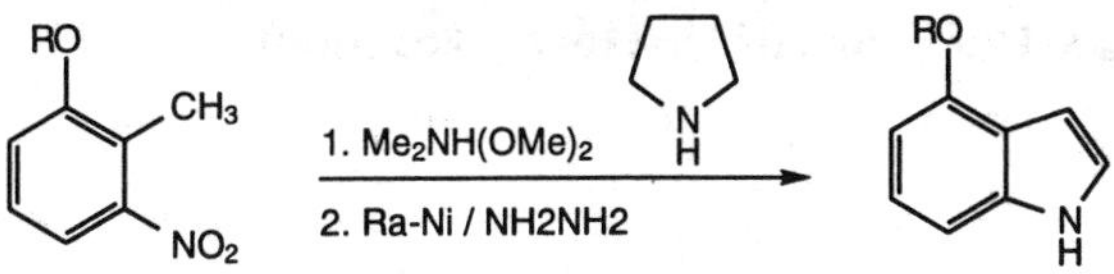

Heterocycles **1984**, *22*, 195.
Org. Synth. **1984**, *63*, 214.
Heterocycles **1984**, *22*, 195.

Lemieux-Johnson Oxidation

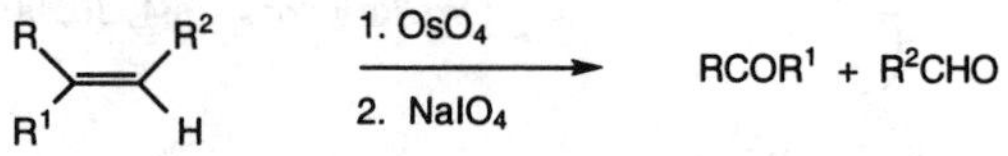

R,R², R¹,H   1. OsO₄   →   RCOR¹ + R²CHO   *J. Org. Chem.* **1956**, *21*, 478.
              2. NaIO₄

Lemieux-von Rudolff Oxidation

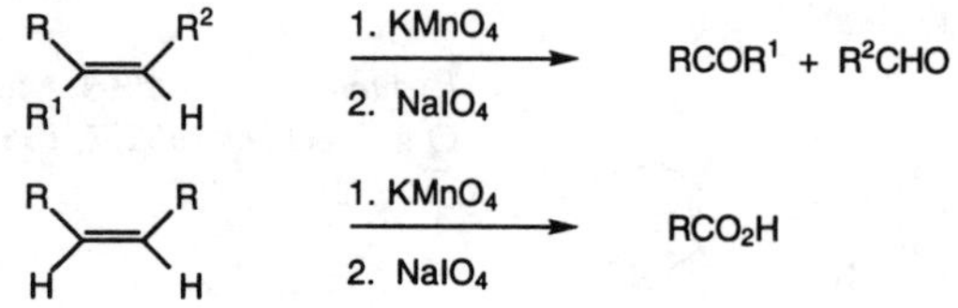

R,R², R¹,H   1. KMnO₄   →   RCOR¹ + R²CHO   *Can. J. Chem.* **1956**, *34*, 1413.
              2. NaIO₄   *Can. J. Chem.* **1955**, *33*, 1701;
                                                    1710; 1714..

R,R, H,H   1. KMnO₄   →   RCO₂H
              2. NaIO₄

Letts Nitrile Synthesis

RCO₂H $\xrightarrow{\text{KSCN}/\Delta}$ RCN   *Chem. Rev.* **1948**, *42*, 264

Leuckart Thiophenol Reaction

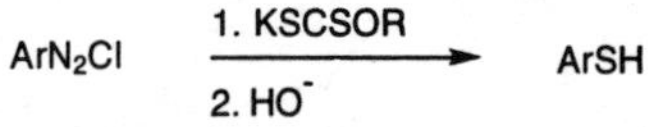

ArN₂Cl   1. KSCSOR   →   ArSH   Merck ONR-55.
          2. HO⁻   *Org. Synth.* **1955**, *Coll. Vol. 3*,
                                      809.

Leuckart Reaction

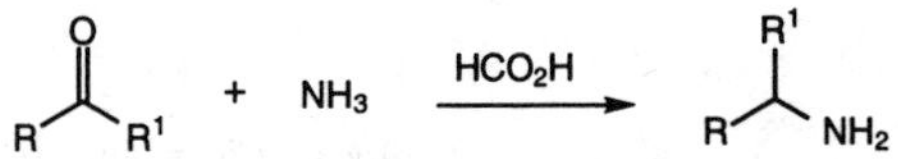

R,O,R' + NH₃ $\xrightarrow{\text{HCO}_2\text{H}}$ R,R¹,NH₂   *Chem. Rev.* **1969**, *69*, 679.
                                                                    Org. Reactions **1949**, *5*, 301.

Note: 1. Also known as the Leuckart-Wallach Reaction.
       2. Also see the Wallach Reaction.

Lieben Iodoform Reaction (see Haloform Reaction)

Lobry de Bruyn-van Ekenstein Rearrangement

CHO, H—OH $\xrightarrow{\text{base}}$ CH₂OH =O   ⇌   CHO, HO—H

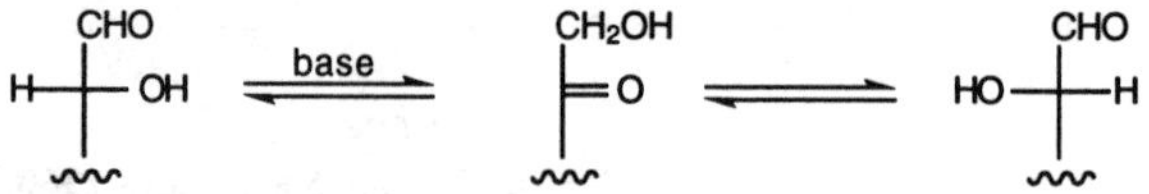

(carbohydrate)

J. Chem. Soc. **1905**, *87*, 570.
Chem. Ber. **1904**, *37*, 4827.

Löffler-Freytag Reaction (see Hofmann Löffler-Freytag Reaction)

Lössen Rearrangement (Reaction)

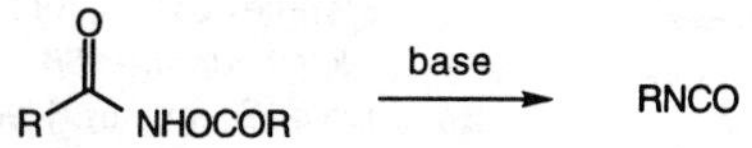

Synthesis **1989**, 61.
Angew. Chem. Int. Ed. Engl. **1974**, *13*, 376.
Synthesis **1971**, 128.
Org. Reactions **1946**, *3*, 267.
Chem. Rev. **1943**, *33*, 209.

Note: Hydrolysis of the the isocyanate to the amine is also known as the Lössen Rearrangement.

Lowe Synthesis

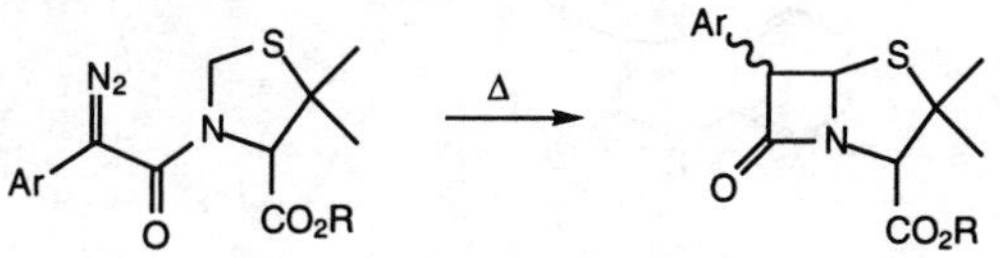

Chem. Rev. **1976**, *76*, 147.

Luche Reaction (Reduction)

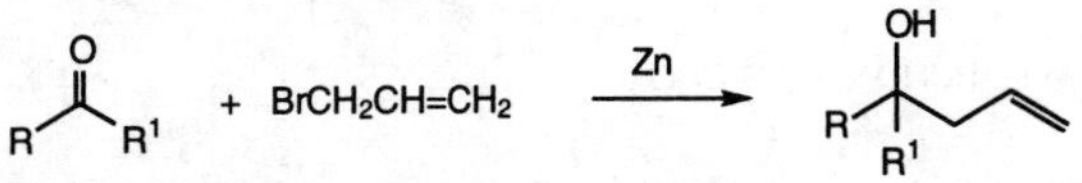

J. Org. Chem. **1991**, *56*, 4333.
J. Org. Chem. **1985**, *50*, 910.

McCluskey Degradation (Cleavage)

$$R_2NR^1 \xrightarrow{PhOCOCl} R_2NCO_2Ph$$

Synthesis **1989**, 1.
J. Chem. Soc. (C) **1967**, 2015.

McCormack Cycloaddition

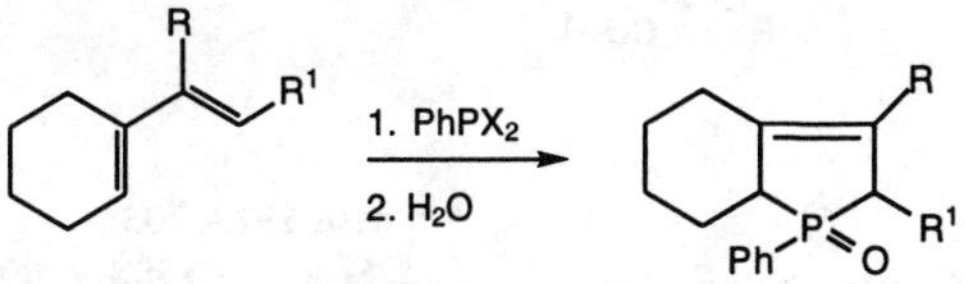

J. Org. Chem. **1976**, *41*, 238.

McFadyen-Stevens Reaction (Reduction)

$$RCOCl \xrightarrow[\text{2. base}]{\text{1. } PhSO_2NHNH_2} RCHO$$

R = aryl or alkyl with no α-hydrogens

J. Org. Chem. **1961**, *26*, 3664.

McLafferty Rearrangement

X,Y,W = C or heteroatom

J. Am. Chem. Soc. **1990**, *112*, 2015.
Chem. Rev. **1974**, *74*, 215.

McMurry (Olefination) Reaction

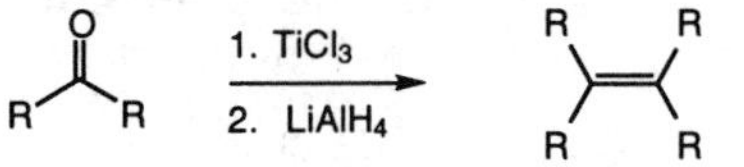

Chem. Rev. **1989**, *89*, 1513.
Synthesis **1989**, 883.
Tetrahedron. **1988**, *44*, 4295.
Acc. Chem. Res. **1983**, *16*, 405.

Note: Also known as the McMurry Olefin Synthesis, McMurry-Fleming Synthesis or McMurry Coupling.

Madelung (Indole) Synthesis

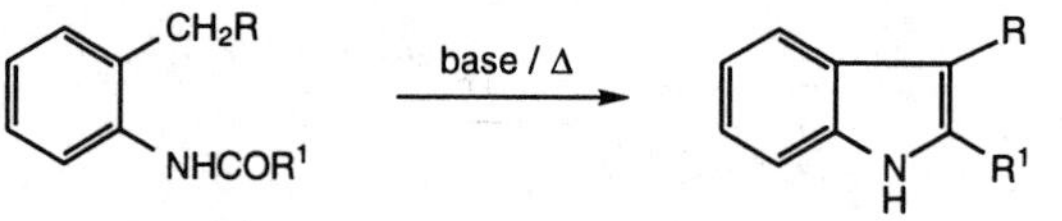

Comp. Org. Chem. **4**, 461.

Malaprade Reaction

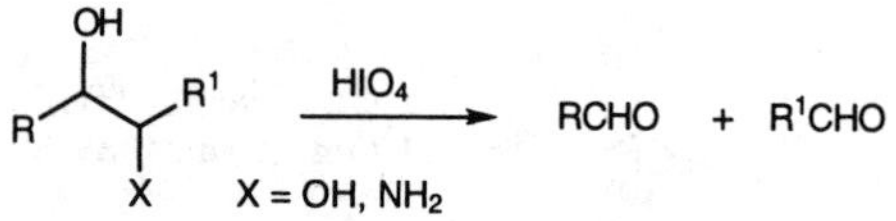

Org. Reactions **1944**, *2*, 341.

Note: Also known as the Periodic Acid Oxidation.

Malonic Ester Synthesis

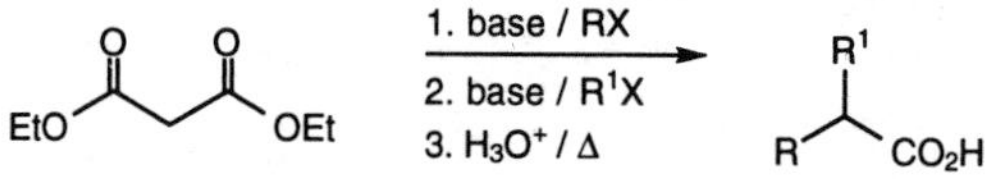

March 412.

Mannich Reaction

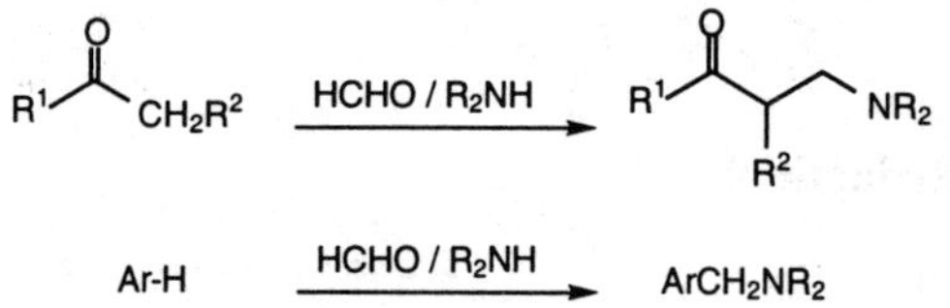

Synthesis **1973**, 703.
Org. Reactions **1953**, *7*, 99.
Org. Reactions **1942**, *1*, 303.

Marbet-Saucy Rearrangement

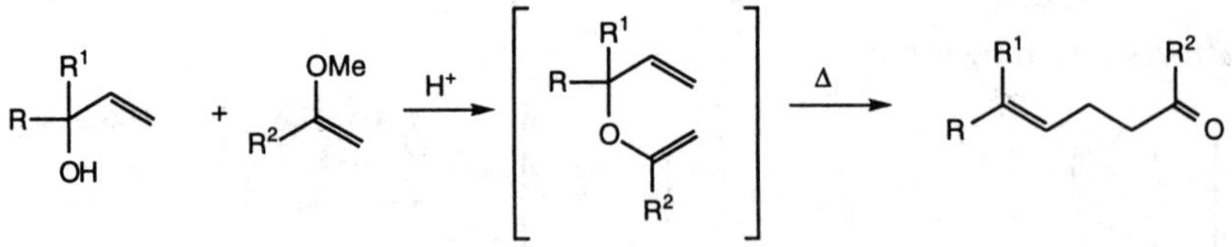

J. Am. Chem. Soc. **1973**, *95*, 553.
Helv. Chim. Acta **1967**, *50*, 2091; 2095..

Marckwald Asymmetric Synthesis

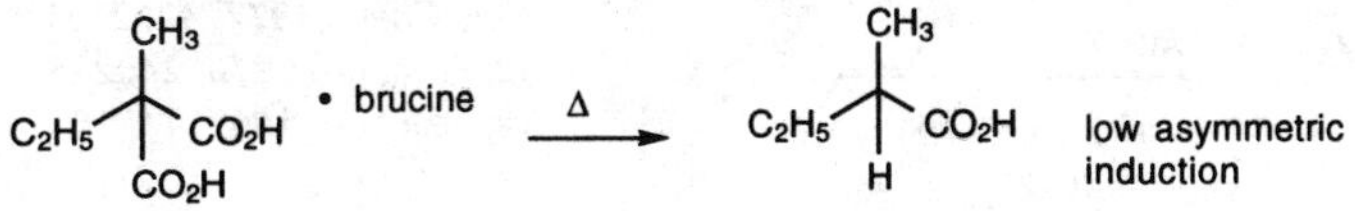

Bull. Chem. Soc., France **1956**, 987.
J. Chem. Soc. **1948**, 1085.

Marckwald Rearrangement

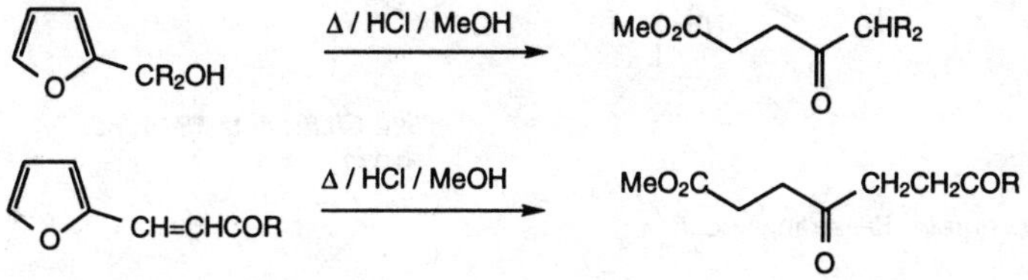

Comp. Org. Chem. **4**, 702.

Marckwald Synthesis

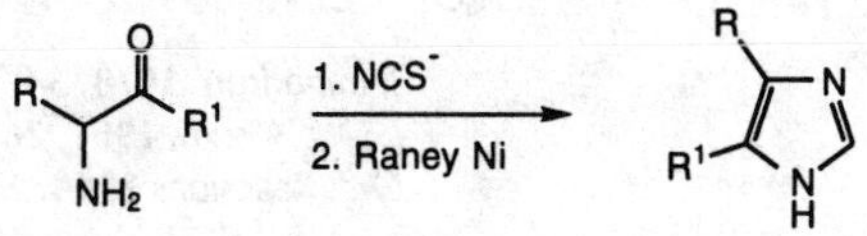

Comp. Org. Chem. **4**, 398.

Martinet Dioxindole Synthesis

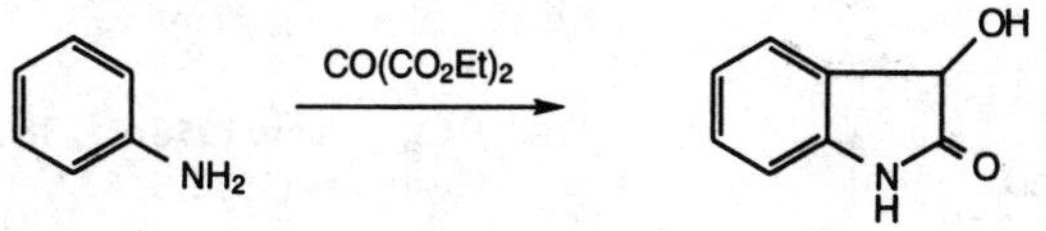

Chem. Rev. **1945**, *37*, 472.

Mascarelli Reaction

Chem. Rev. **1968**, *68*, 209.

Mattox-Kendall Reaction (see Kendall-Mattox Reaction)

Meerwein Arylation Reaction

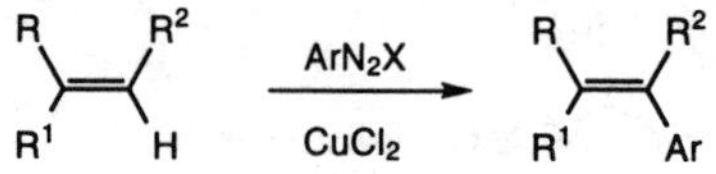

J. Org. Chem. **1977**, *42*, 2431.
Org. Reactions **1976**, *24*, 225.
Org. Reactions **1960**, *11*, 189.

Note: Also known as the Meerwein Reaction.

Meerwein-Eschenmoser Rearrangement

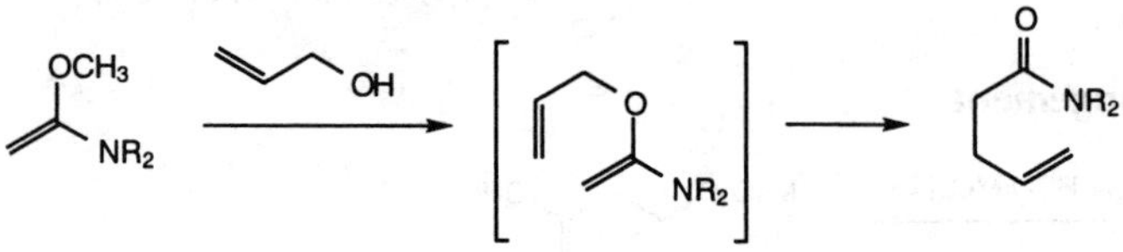

Helv. Chim. Acta **1979**, *62*, 1922.

Note: Also see the Eschenmoser Reaarrangement.

Meerwein-Ponndorf-Verley Reduction (Reaction)

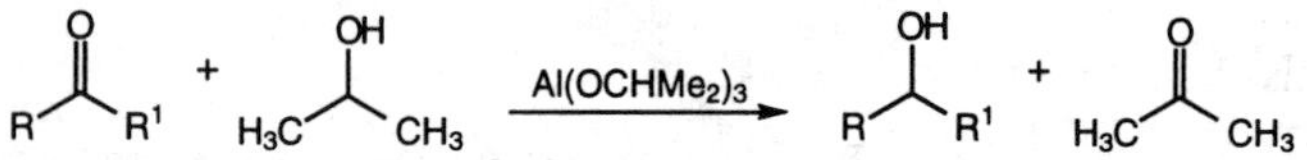

Tetrahedron. **1978**, *34*, 933.
J. Org. Chem. **1961**, *26*, 290.
Org. Reactions **1944**, *2*, 178.

Note: 1. Also known as the Meerwein-Ponndorf Reaction.
2. The reverse reaction is known as the Oppenauer Oxidation.

Meerwein Reaction

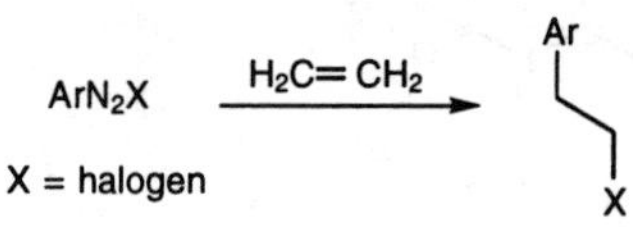

J. Org. Chem. **1956**, *21*, 380.
Quart. Rev. **1952**, *6*, 377.

Note: The Meerwein Reaction also refers to the Meerwein Arylation Reaction.

Meerwein Rearrangement (see Wagner-Meerwein Rearrangement)

Meisenheimer Rearrangement

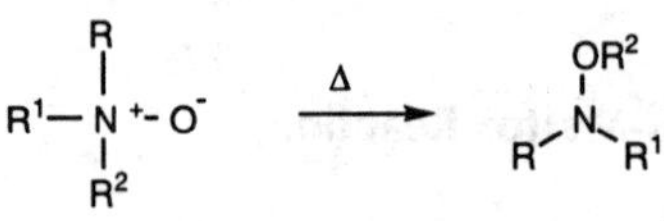

Comp. Org. Chem. **2**, 203.

Menschutkin Reaction

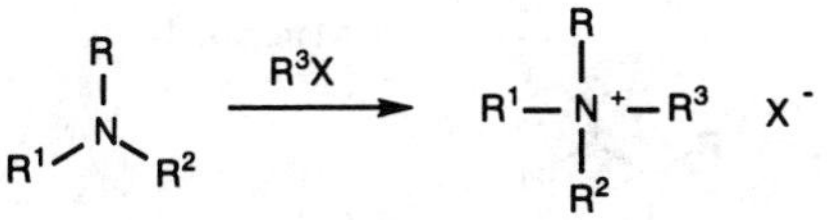

Tetrahedron. **1975**, *31*, 9.

Merrifield Solid-Phase Peptide Synthesis

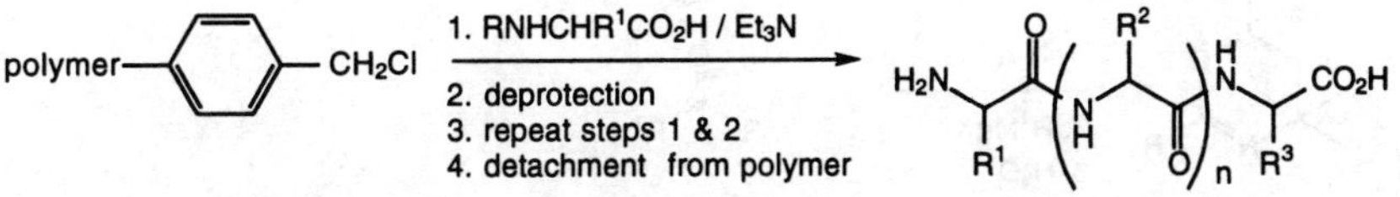

J. Am. Chem. Soc. **1989**, *111*, 8024.
Synthesis **1981**, 413.

Meyer-Hartmann Reaction

$$ArIO + Ar^1IO_2 \xrightarrow{HO^-} ArAr^1I^+ \; IO_3^-$$

Chem. Rev. **1942**, *32*, 249.

Meyer Reaction

$$Na_2SnO_2 + RX \longrightarrow RSnO_2Na$$

Merck ONR-60.
Fieser & Fieser **6**, 272.

Note: 1. Alkali arsenites and plumbites also react.
2. Also known as the Victor Meyer Reaction.

Meyer-Schuster Rearrangement

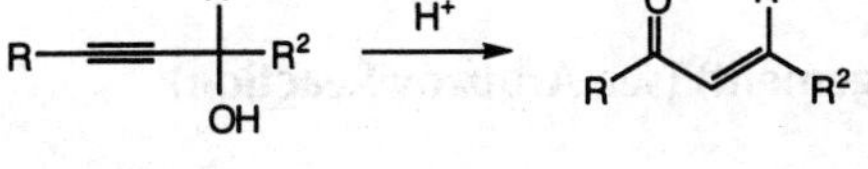

Chem. Rev. **1971**, *71*, 429.
Mechanism: *J. Am. Chem. Soc*. **1989**, *111*, 829.

Meyer Synthesis

$$RX + MNO_2 \longrightarrow RONO + RNO_2$$

Chem. Rev. **1943**, *32*, 374.
J. Am. Chem. Soc. **1939**, *61*, 279.

Note: Also known as the Victor Meyer Synthesis.

Meyers (Aldehyde) Synthesis

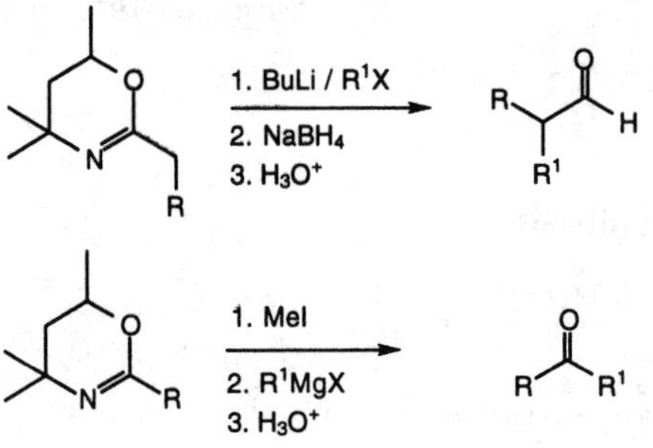

March 424.

Michael Reaction

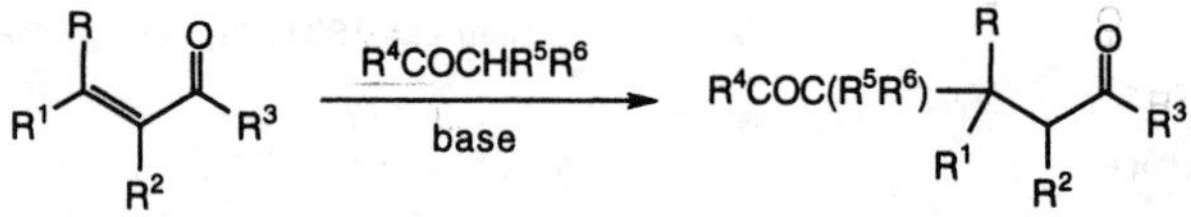

Tetrahedron. **1976**, *32*, 3.
Org. Reactions **1959**, *10*, 179.
Asymmetric Synthesis: Tetrahedron: Asymmetry **1992**,
3, 459.

Note: 1. Also known as the Michael Condensation, Michael Addition, or Michael Method.
2. Now covers most 1,4-additions of nucleophiles to α,β-unsaturated systems.

Michaelis Reaction

$(RO)_2P(O)Na \xrightarrow{R^1X} (RO)_2P(O)R^1$

Pure & Appl. Chem. **1964**, *9*,
307.
Org. Reactions **1951**, *6*, 277.

Michaelis-Arbuzov Reaction (Rearrangement) (see Arbuzov Reaction)

Miescher Degradation

$RR^1CHCH_2CH_2CO_2CH_3 \xrightarrow[\substack{2.\ base \\ 3.\ NBS \\ 4.\ base \\ 5.\ [O]}]{1.\ PhMgBr}$

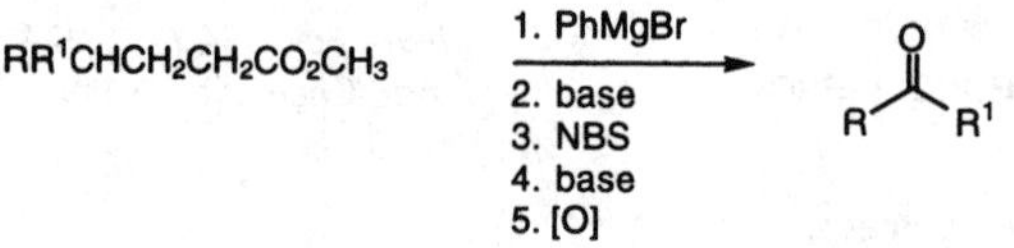

Merck ONR-61.

Mignonac Reaction

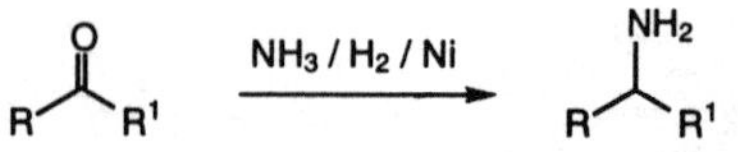

Merck ONR-61.

Milas' Reaction

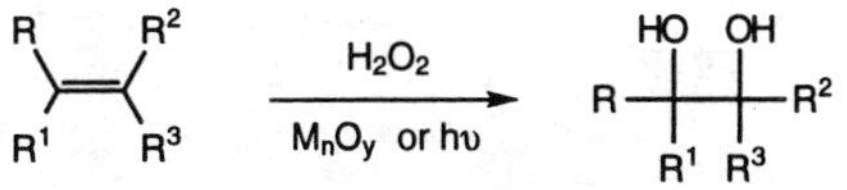

J. Am. Chem. Soc. **1937**, *59*, 543; 2342.

Note: Also known as the Milas Hydroxylation of Olefins.

Mills Reaction

$$ArNH_2 + Ar^1NO \xrightarrow{AcOH} Ar\text{-}N{=}N\text{-}Ar^1$$

March 573.

Mitsunobu Reaction

$$ROH \xrightarrow[\text{2. HX}]{\text{1. }(NCO_2Et)_2\text{ / }Ph_3P} RX$$

Org. Reactions **1992**, *42*, 335.
J. Org. Chem. **1991**, *56*, 670.
J. Org. Chem. **1989**, *54*, 257.
J. Org. Chem. **1987**, *52*, 4235.
Synthesis **1981**, 1.

Miyazaki-Newman-Kwart Rearrangement

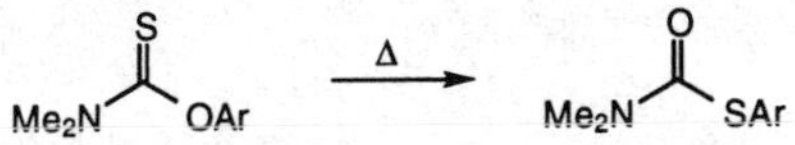

Comp. Org. Chem. **3**, 476.

Note: Also known as the Newman-Kwart Rearrangement.

Moffatt-Mattocks Haloacetylation

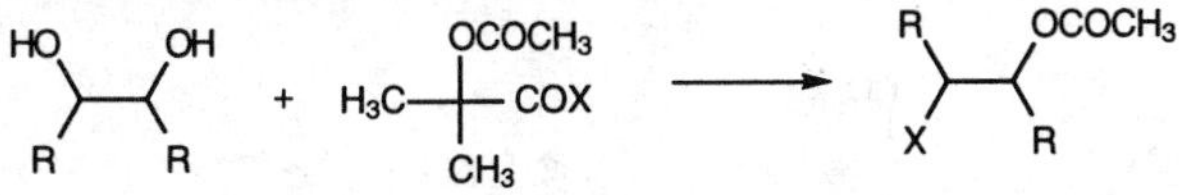

J. Am. Chem. Soc. **1973**, *95*, 4016; 4025.
J. Org. Chem. **1973**, *38*, 3179.

Moffatt (-Pfitzner) Oxidation

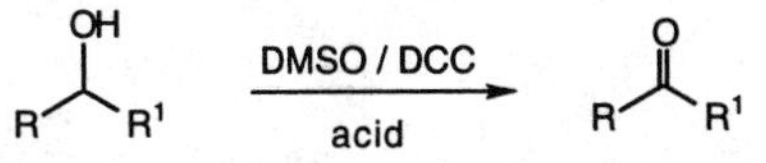

Org. Reactions **1990**, *39*, 297.
Chem. Rev. **1967**, *67*, 247

Note: Also known as the Pfitzner-Moffatt Oxidation.

Morgan-Walls Reaction

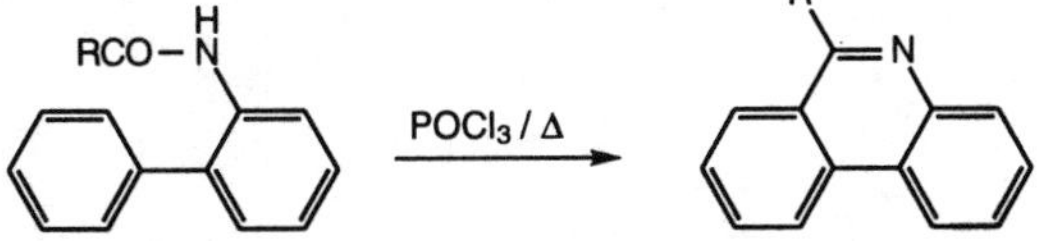

Chem. Rev. **1968**, *68*, 209.
Chem. Rev. **1950**, *46*, 175.

Mozingo Reaction

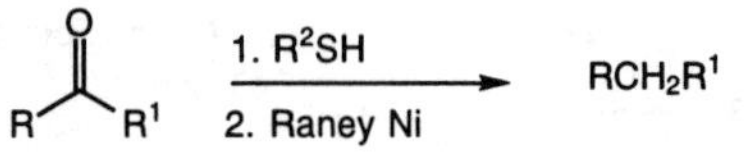

Comp. Org. Chem. **1**, 1080.

Mukaiyama-Michael Reaction

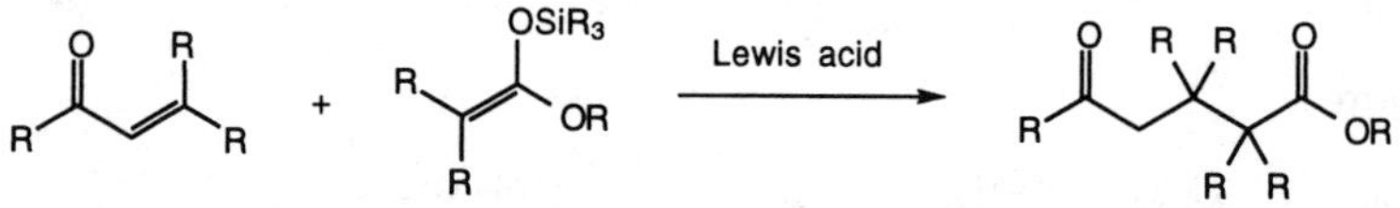

Chem. Lett. **1988**, 91.
Chem. Lett. **1987**, 1691; 1183.
Chem. Lett. **1986**, 1805.
J. Am. Chem. Soc. **1974**, 96, 7503.

Mukaiyama (Type) Aldol Reaction

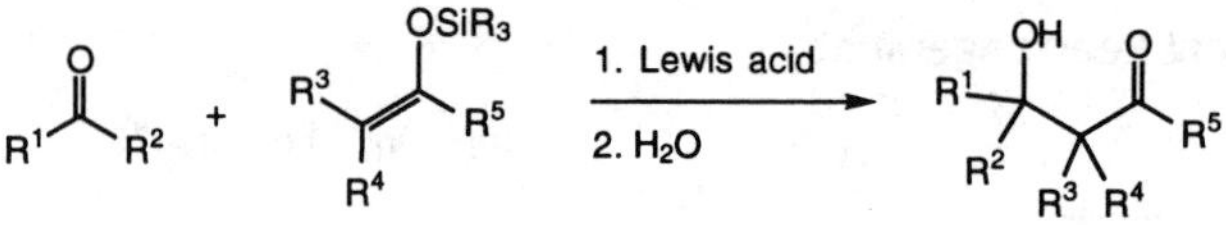

Synthesis **1990**, 569.
Chem. Lett. **1987**, 1121.
Chem. Lett. **1982**, 1903.
Org. Reactions **1982**, *28*, 203.

Mumm Rearrangement

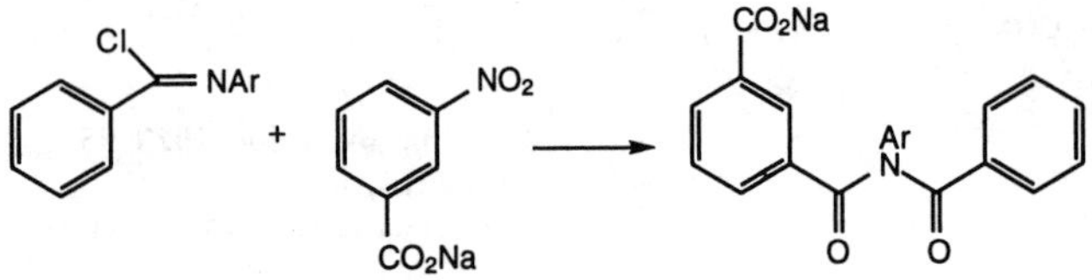

Org. Reactions **1962**, *12*, 31.

Mundy Rearrangement

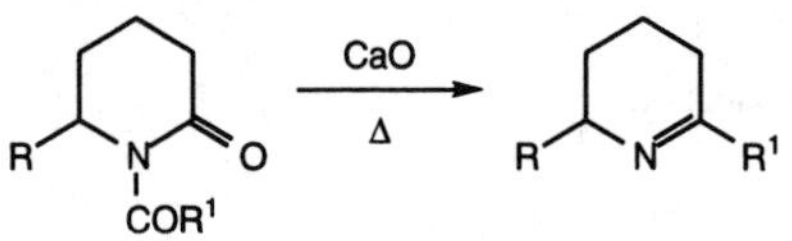

Tetrahedron **1977**, *33*, 1565.

Nametkin Rearrangement

Merck ONR-62.

Nazarov Cyclization

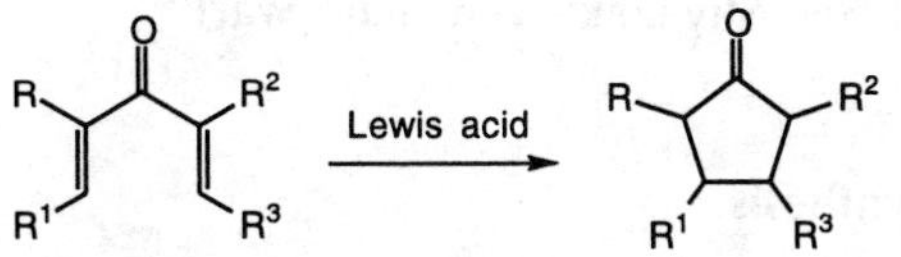

J. Org. Chem. **1991**, *56*, 735.
Synthesis **1983**, 429.
J. Org. Chem. **1979**, *44*, 462.

Silicon Directed: *J. Org. Chem.* **1990**, *55*, 5543.

Neber Rearrangement (Reaction)

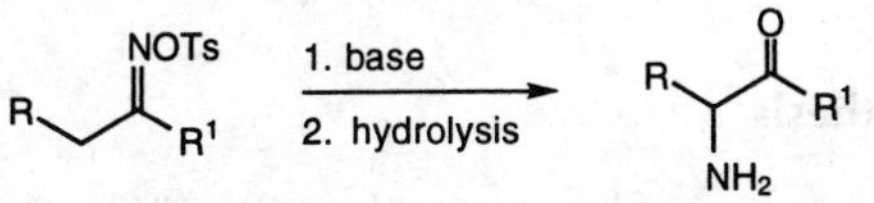

J. Med. Chem. **1972**, *15*, 348.
Chem. Rev. **1964**, *64*, 81.
Org. Reactions **1960**, *11*, 45.

Nef Aldehyde and Ketone Synthesis (see the Nef Reaction)

Nef Reaction

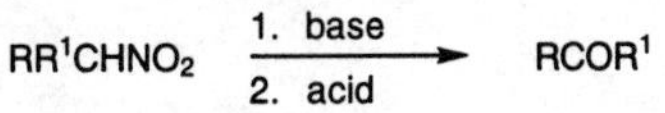

$$RR^1CHNO_2 \xrightarrow[\text{2. acid}]{\text{1. base}} RCOR^1$$

Org. Reactions **1990**, *38*, 655.
Chem. Rev. **1955**, *55*, 137.

Nef Synthesis (Reaction)

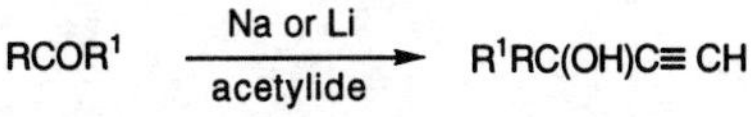

$$RCOR^1 \xrightarrow[\text{acetylide}]{\text{Na or Li}} R^1RC(OH)C{\equiv}CH$$

March 838.

Nencki Reaction

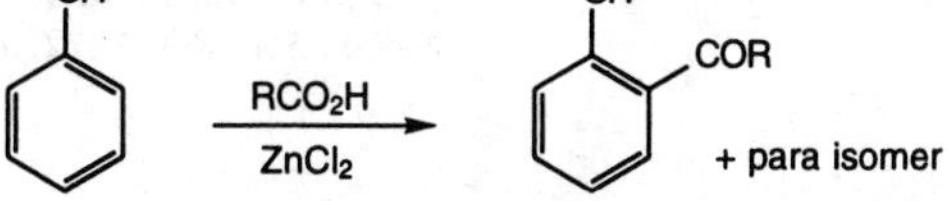

Merck ONR-63.
J. Chem. Soc. **1932**, 918.

Nenitzescu (Indole) Synthesis (Reaction)

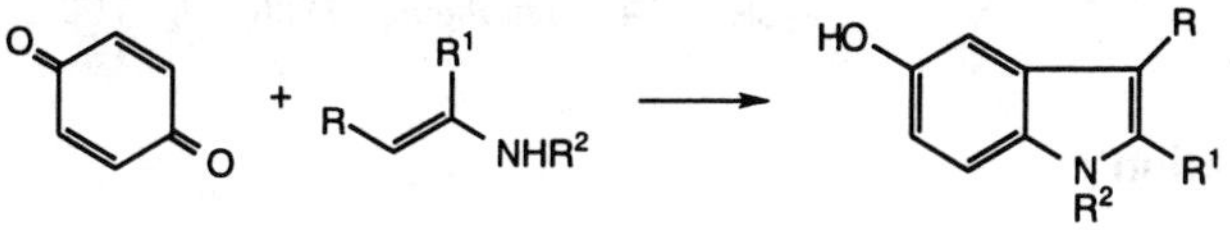

Org. Reactions **1973**, *20*, 337.
Tetrahedron **1973**, *29*, 921.
Tetrahedron **1971**, *27*, 5031.

Nenitzescu Reaction

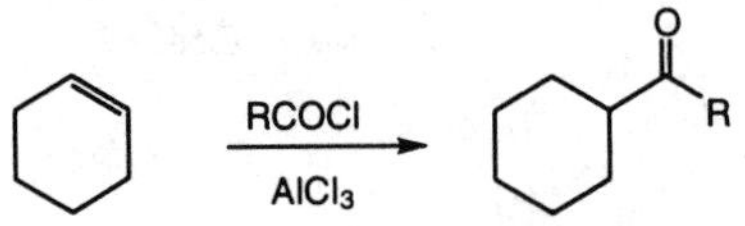

J. Am. Chem. Soc. **1951**, *73*, 3512.

Note: Also known as the Nenitzescu Reductive Acylation.

Newman-Kwart Rearrangement (see Miyazaki-Newman-Kwart Rearrangement)

Niementowski (Quinazoline) Synthesis

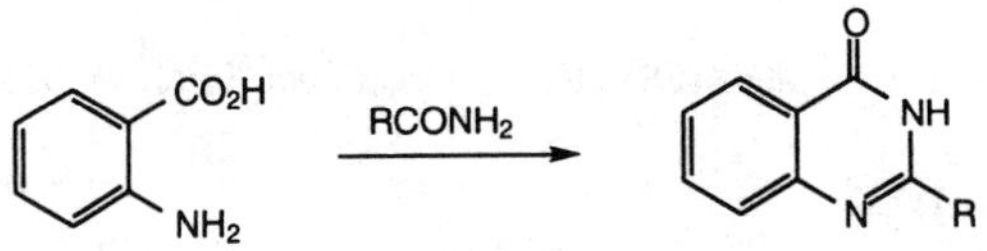

Org. Prep. Proced. Internat. **1973**, *5*, 260.

Niementowski (Quinoline) Synthesis

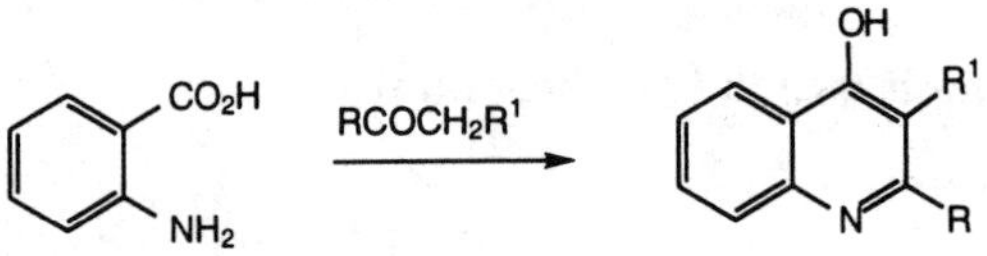

Chem. Rev. **1942**, *30*, 127.

Nierenstein Reaction

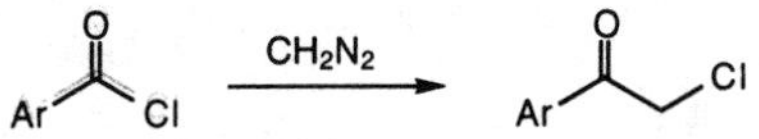

Merck ONR-64.
Org. Reactions **1942**, *1*, 38.

Norrish Type I Cleavage

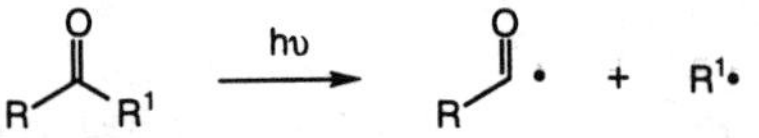

J. Am. Chem. Soc. **1980**, *102*, 314.
J. Org. Chem. **1978**, *43*, 3314.
Chem. Soc. Rev. **1972**, *1*, 465.

Norrish Type II Cleavage

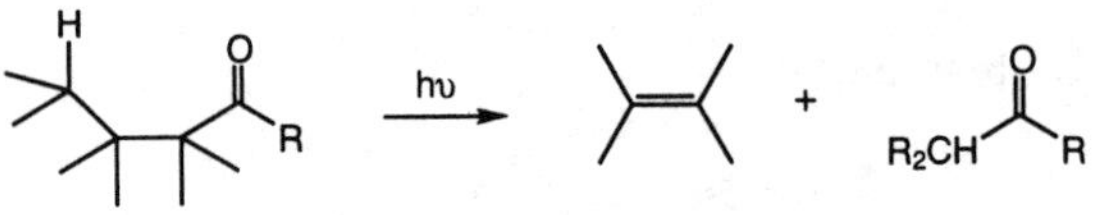

Pure & Appl. Chem. **1988**, *60*, 999.
Tetrahedron **1988**, *44*, 3413.

Norrish Type II Cyclization

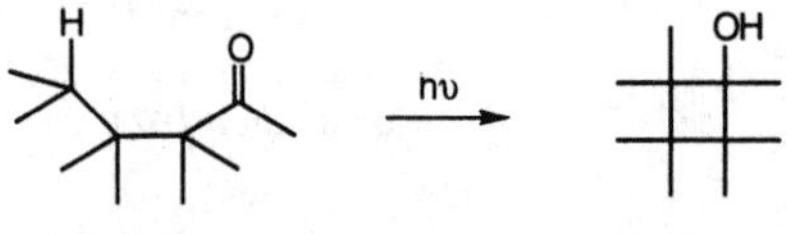

J. Chem. Soc., Chem. Commun. **1990**, 1214.

Nozaki-Kishi Reaction

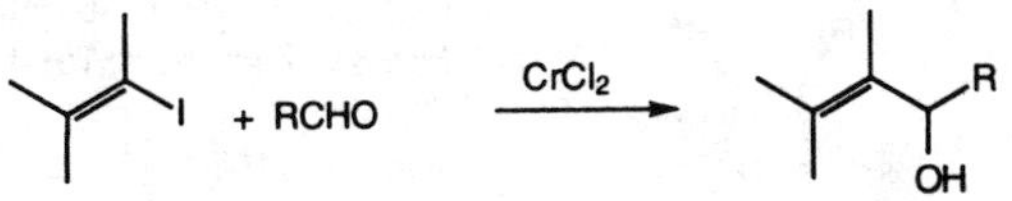

J. Am. Chem. Soc. **1986**, 108, 5644.
Tetrahedron Lett. **1983**, 24, 5281.

Oppenauer Oxidation

Synthesis **1984**, 839.
Org. Reactions **1951**, 6, 207.
Org. Reactions **1944**, 2, 181.

Note: The reverse reaction is known as the Meerwein-Ponndorf-Verley Reduction.

Ortoleva-King Reaction

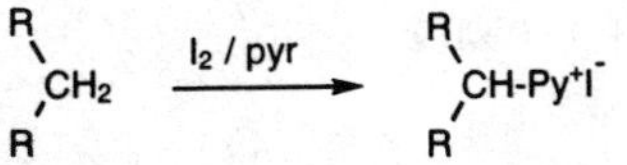

Angew. Chem. Int. Ed. Engl. **1963**, 2, 386.

Orton Rearrangement

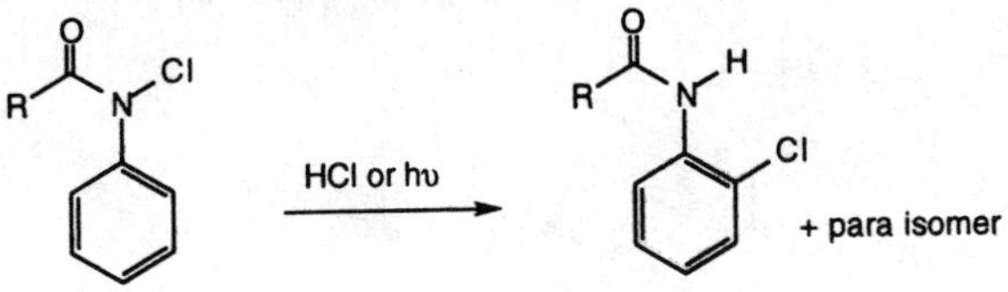

March 503.

+ para isomer

Ostromyslenskii (Ostromisslenskii) Reaction

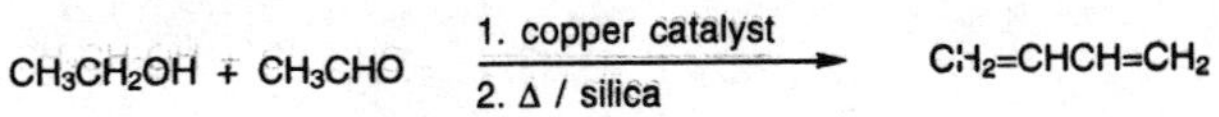

n. Rev. **1945**, 36, 73.

$$CH_3CH_2OH + CH_3CHO \xrightarrow[\text{2. } \Delta \text{ / silica}]{\text{1. copper catalyst}} CH_2=CHCH=CH_2$$

Oxo Process (Reaction)

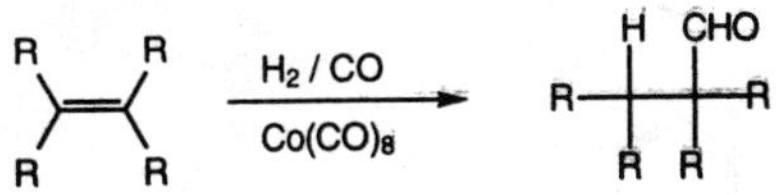

Acc. Chem. Res. **1981**, 14, 259.
Chem. Rev. **1962**, 62, 283.

Note: Also known as the Oxo Synthesis.

Oxy-Cope Rearrangement

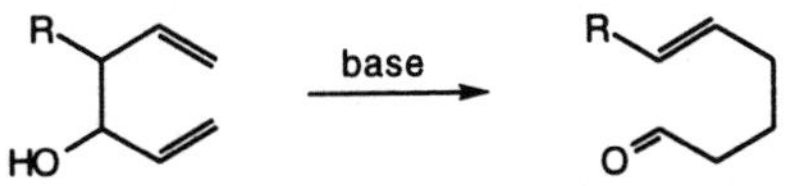

Angew. Chem. Int. Ed. Engl. **1990**, 29, 609.
Org. Reactions **1975**, 22, 61.

Paal-Knorr Synthesis (Reaction)

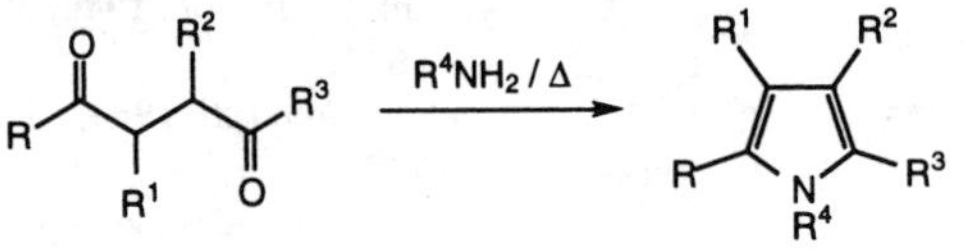

J. Org. Chem. **1991**, *56*, 6924.
J. Heterocyclic Chem. **1968**, *5*, 757.
Chem. Rev. **1963**, *63*, 511.

Paal Synthesis (Reaction)

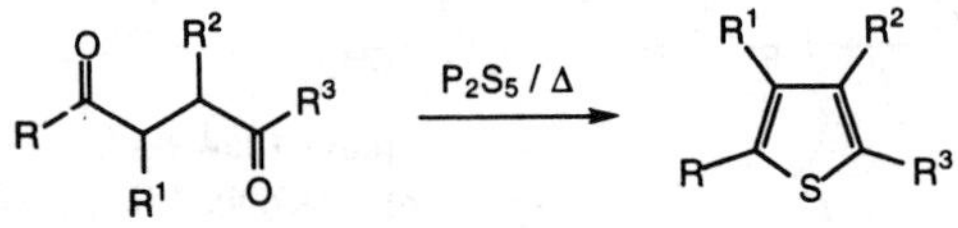

Org. Reactions **1951**, 6, 410.

Note: Also known as the Paal Cyclization, Paal Condensation or Paal-Knorr Synthesis of Thiophenes.

Paolini Reaction

$$R_2NCH_2R^1 \xrightarrow[2.\ H_2O]{1.\ (PhCO_2)_2} R_2NH + R^1CHO$$

Chem. Ber. **1960**, *93*, 363.

Para-Claisen Rearrangement

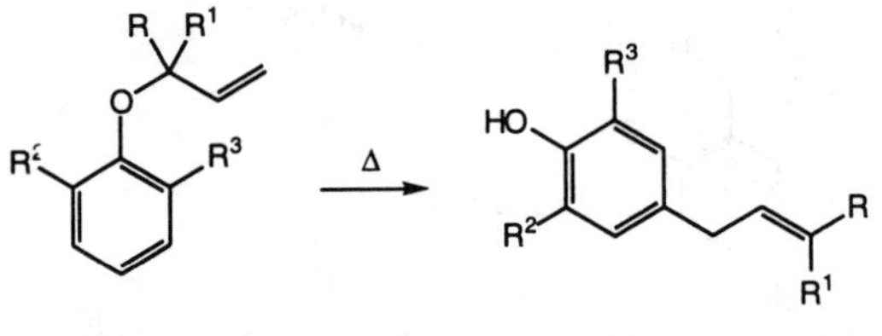

Synthesis **1977**, 589.

Note: Also see the Claisen Rearrangement.

Parham Reaction

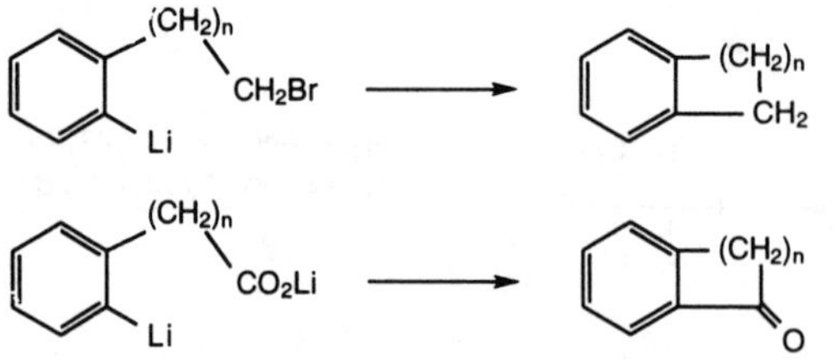

J. Org. Chem. **1978**, *43*, 3800.
J. Org. Chem. **1975**, *40*, 2394.

Note: Also known as the Parham Cycloalkylation and Cycloacylation Reaction.

Parikh-Doering Oxidation

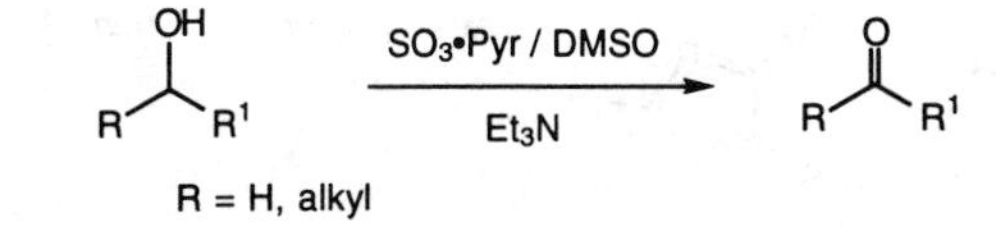

Tetrahedron **1978**, *34*, 1651.
J. Am. Chem. Soc. **1967**, *89*, 5505.

Passerini Reaction

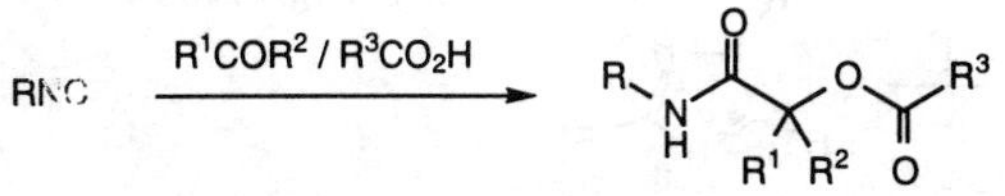

Synthesis **1977**, 332.
Synthesis **1973**, 345.
Angew. Chem. Int. Ed. Engl. **1962**, *1*, 8.

Note: Also see the Ugi Reaction.

Paterno-Büchi Reaction

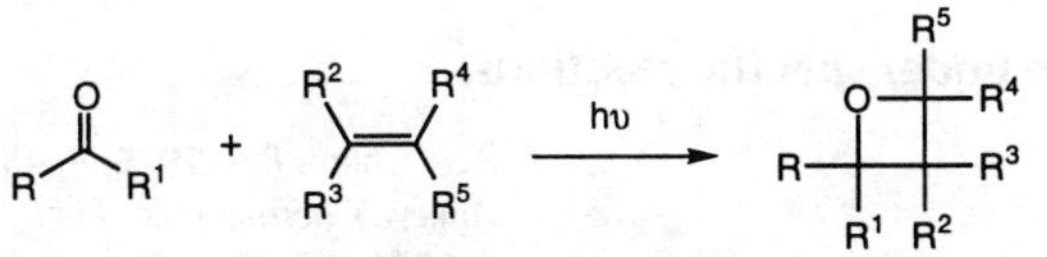

Synthesis **1989**, 152.
J. Chem. Soc., Chem. Commun. **1973**, 374.
J. Am. Chem. Soc. **1954**, *76*, 4327.

Pauson-Khand Cycloaddition

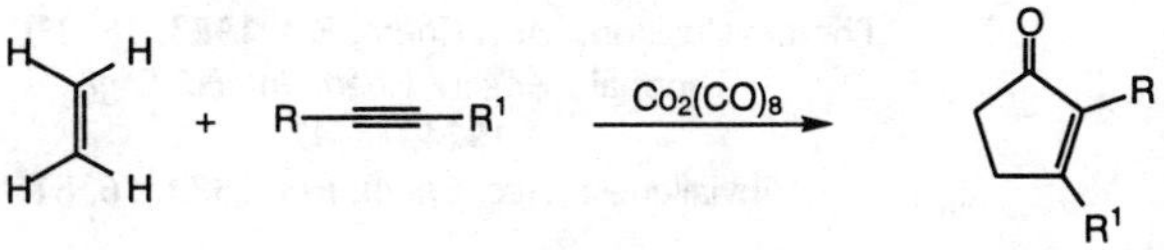

Org. Reactions **1991**, *40*, 1.
Chem. Rev. **1988**, *88*, 1081.
Tetrahedron **1985**, *41*, 5855.
Intramolecular: *J. Am. Chem. Soc.* **1992**, *114*, 5555.

Note: Also known as the Khand Reaction.

Payne Rearrangement

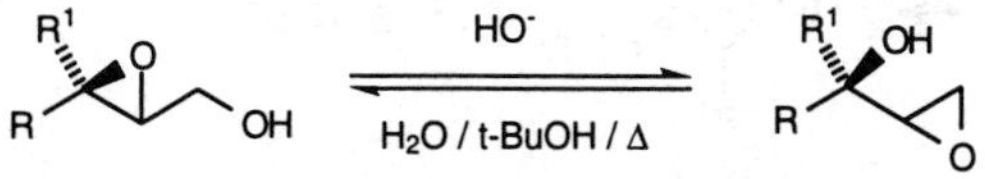

J. Org. Chem. **1983**, *48*, 3761.
J. Org. Chem. **1962**, *27*, 3819.

Pechmann Condensation (Reaction) (see von Pechmann Reaction)

Pechmann Coumarin Synthesis (see von Pechmann Reaction)

Pechmann Pyrazole Synthesis (see von Pechmann Pyrazole Synthesis)

Pechmann Triazole Synthesis

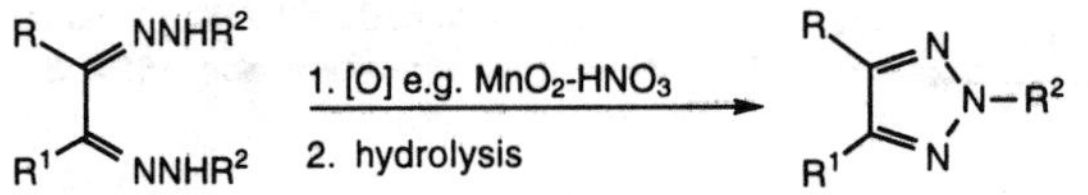

Adv. Heterocyclic Chem. **1974**, *16*, 33.

Pellizzari Reaction

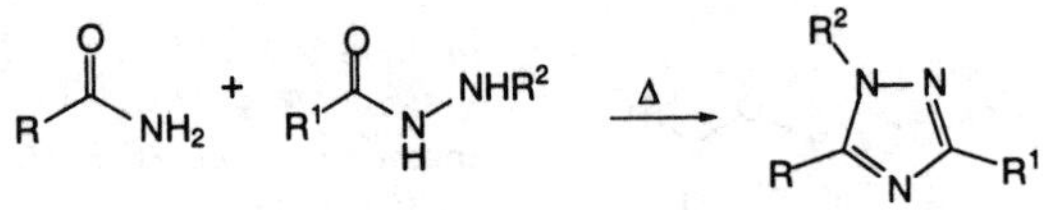

Comp. Org. Chem. **4**, 407.

Pelouze Synthesis

$ROSO_2OK$ or $ROPO_4K$ $\xrightarrow{KCN}$ RCN

Chem. Rev. **1948**, *42*, 192.

Pericyclic Reactions (see also under specific reactions)

Acc. Chem. Res. **1976**, *9*, 453.
Angew. Chem. Int. Ed. Engl. **1974**, *13*, 751.
Survey Progr. Chem. **1973**, *6*, 113.

Cation radicals: *Acc. Chem. Res.* **1987**, *20*, 371.
Photocyclization: *Acc. Chem. Res.* **1983**, *16*, 210.
Thermal: *Angew. Chem. Int. Ed. Engl.* **1974**, *13*, 47.
Vinylallenes: *Acc. Chem. Res.* **1983**, *16*, 81.

Periodic Acid Oxidation (see Malaprade Reaction)

Perkin Alicyclic Synthesis

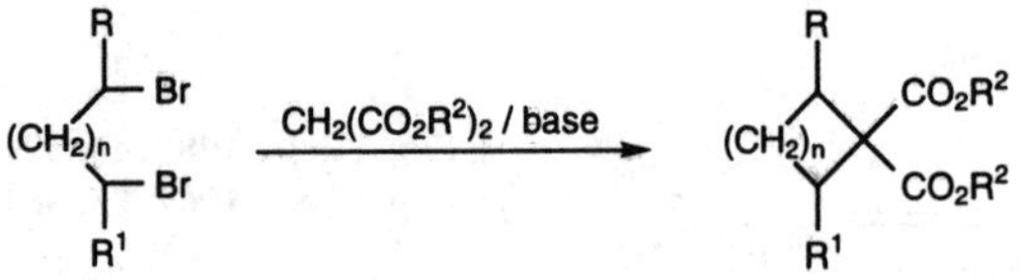

Merck ONR-67.

Perkin Reaction

ArCHO + Ac_2O $\xrightarrow{\text{base} / \Delta}$ $ArCH{=}CHCO_2H$

Synthesis **1972**, 263.
J. Chem. Soc. (B) **1969**, 647.
Org. Reactions **1942**, *1*, 210.

Perkin Rearrangement

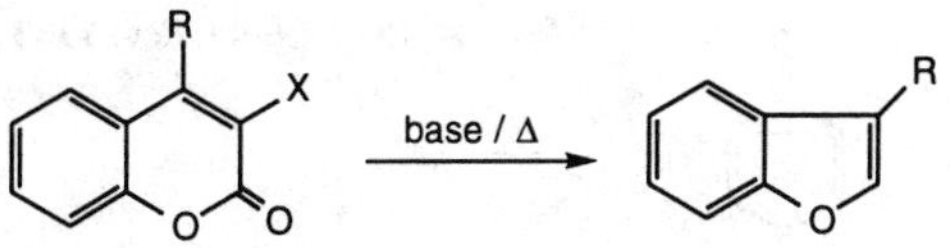

Merck ONR-67.

Perkow (Perkov) Reaction

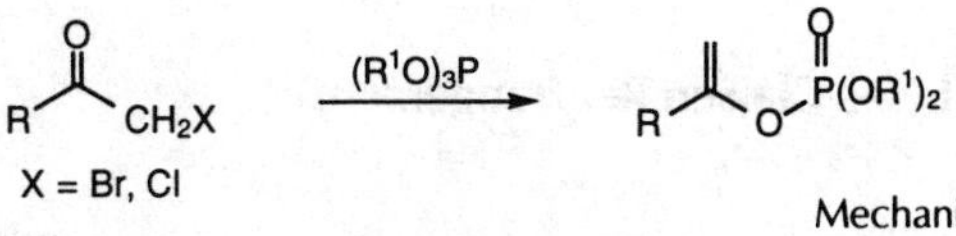

Tetrahedron Lett. **1979**, 2913.
Pure & Appl. Chem. **1964**, *9*, 307.
Chem. Rev. **1961**, *61*, 607.
Mechanism: *J. Org. Chem.* **1963**, *28*, 345.

Peterson (Olefination) Reaction

SYNLETT **1991**, 764.
Org. Reactions **1990**, *38*, 1.
Synthesis **1984**, 384.
Acc. Chem. Res. **1977**, *10*, 442.

Note: Also known as the Peterson Elimination or Silyl-Wittig Reaction.

Petrenko-Kritschenko Piperidone Synthesis

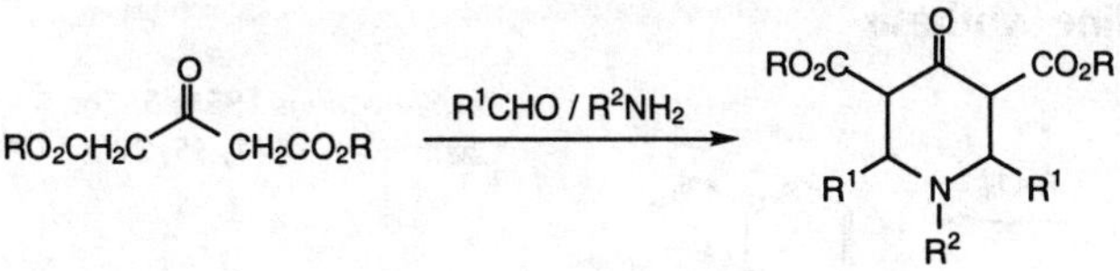

Merck ONR-68.

Pfau-Plattner Azulene Synthesis

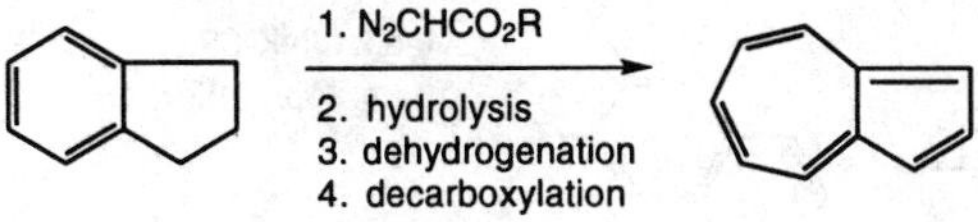

Merck ONR-68.

Pfitzinger Synthesis (Reaction)

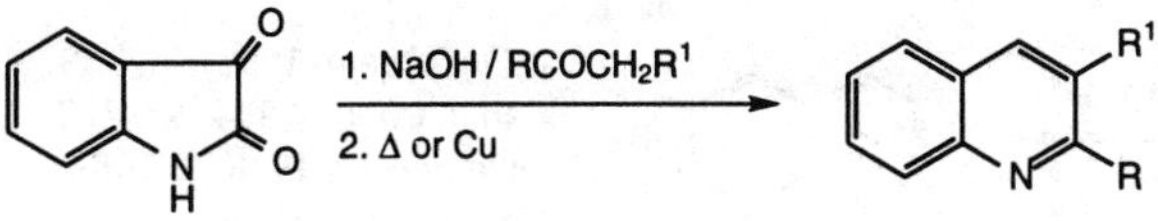

Chem. Rev. **1944**, *35*, 152.

Pfitzner-Moffatt Oxidation (see Moffatt Oxidation)

Phenol-Dienone Rearrangement

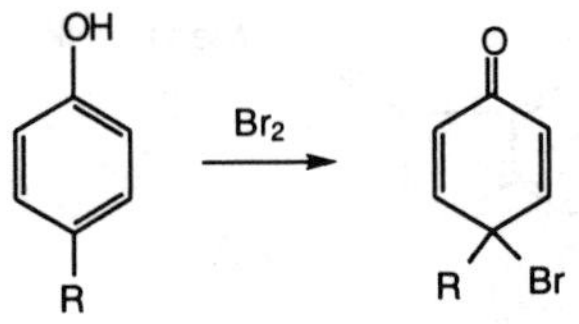

Russ. Chem. Rev. **1963**, *32*, 75.

Photo-Beckmann Rearrangement (see Beckmann Rearrangement)

Photo-Claisen Rearrangement (see Claisen Rearrangement)

Photo-Fries Rearrangement

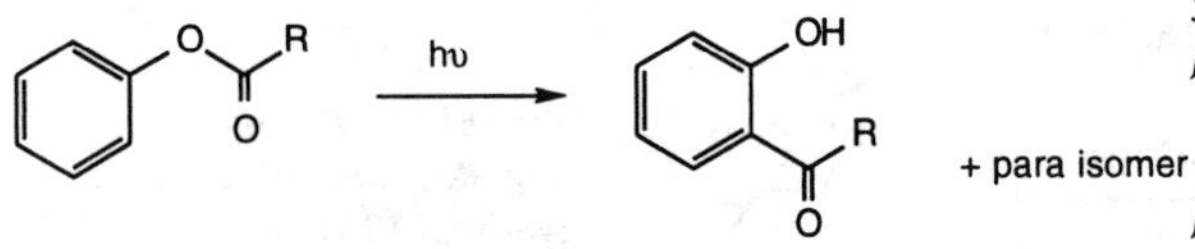

Synthesis **1987**, 882.
J. Am. Chem. Soc. **1974**, *96*, 449.
J. Chem. Soc. **1963**, 1781

Note: Also see the Fries Rearrangement.

Photo-Smiles Rearrangement (see Smiles Rearrangement)

Pictet-Gams Isoquinoline Synthesis

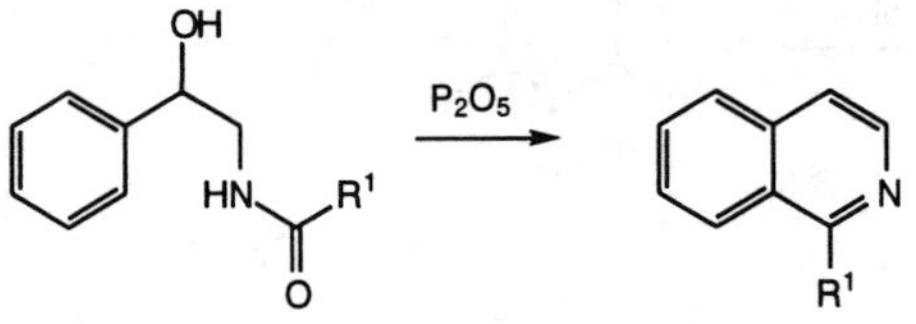

Org. Reactions **1951**, 6, 76.
Chem. Rev. **1944**, *35*, 220.

Pictet-Hubert Reaction

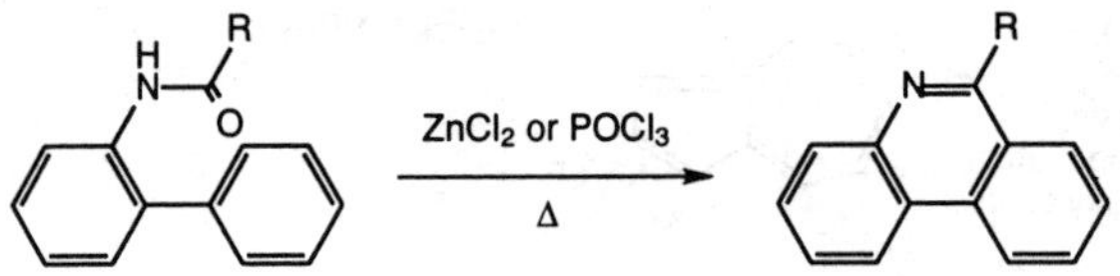

Merck ONR-69.
Chem. Rev. **1950**, *46*, 175.

Pictet-Spengler (Isoquinoline) Synthesis

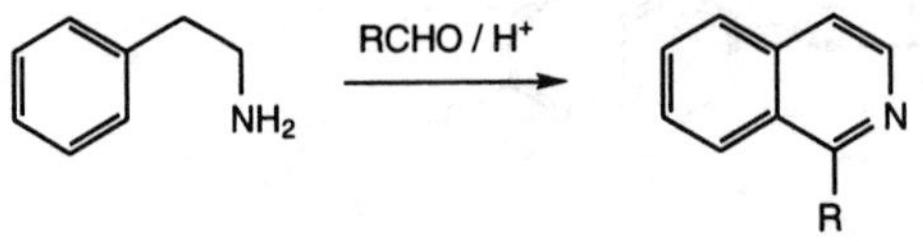

J. Org. Chem. **1991**, *56*, 359.
Can. J. Chem. **1984**, *62*, 2721.
Heterocycles **1978**, 9, 1089.
Org. Reactions **1951**, 6, 151.

Piloty Alloxazine Synthesis

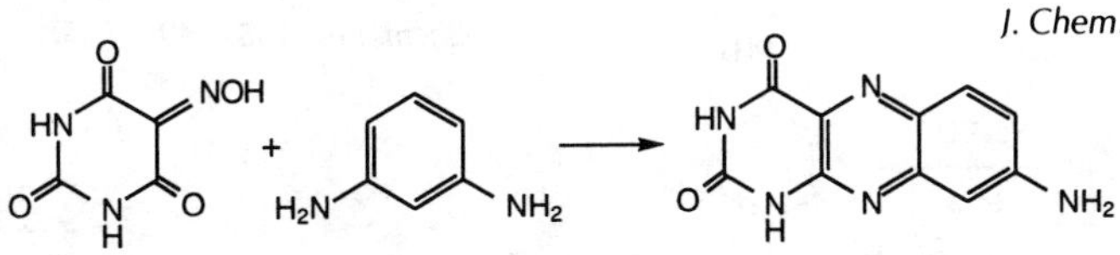

J. Chem. Soc. **1948**, 1926.

Piloty (-Robinson) Pyrrole Synthesis

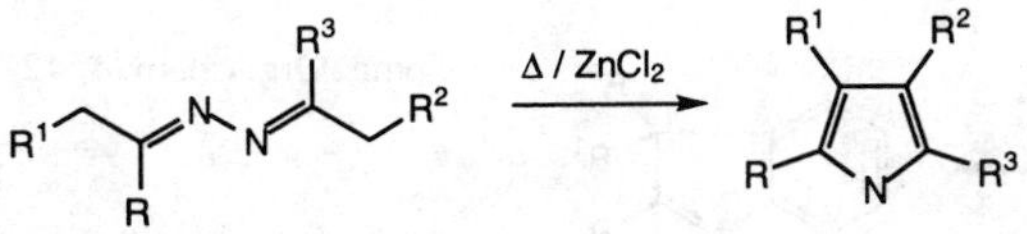

Comp. Org. Chem. **4**, 300.
Chem. Rev. **1969**, 69, 246.

Pinacol Rearrangement

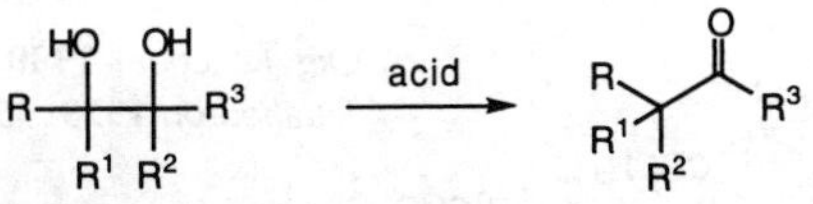

Quart. Rev. **1960**, *14*, 357.

Pinner (Ortho Ester) Reaction

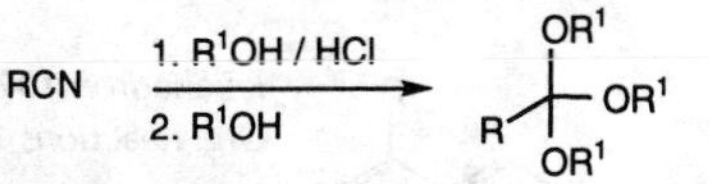

Chem. Rev. **1961**, *61*, 179.

Pinner Rearrangement

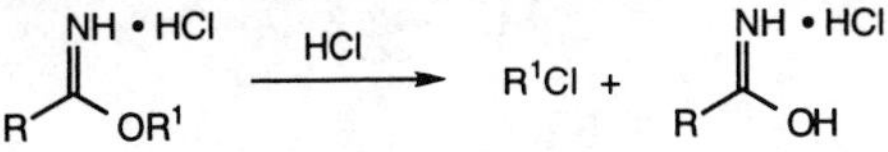

Russ. Chem. Rev. **1962**, *31*, 615.
Chem. Rev. **1961**, *61*, 179.

Pinner Synthesis

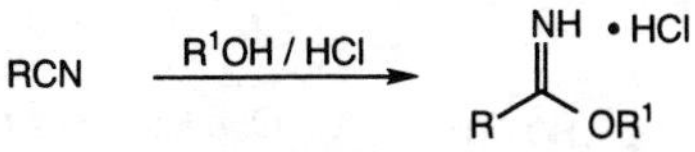

Russ. Chem. Rev. **1962**, *31*, 615.
Chem. Rev. **1961**, *61*, 179.

Thiols: *Synthesis* **1972**, 622.

Pinner (Triazine) Synthesis

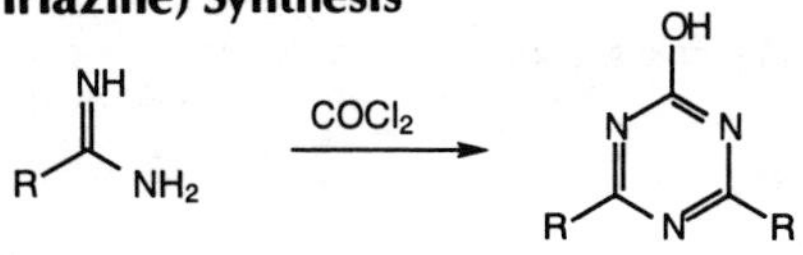

Merck ONR-71.
J. Am. Chem. Soc. **1956**, *78*, 2447.

Piria Reaction

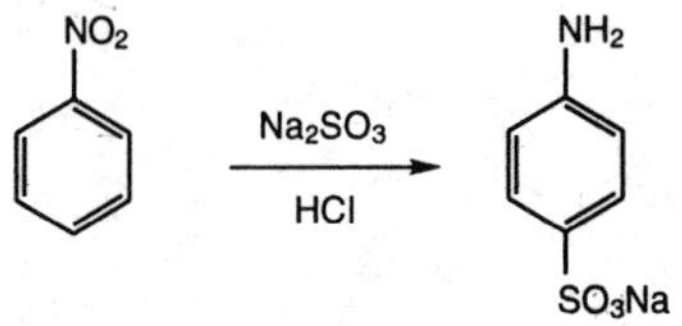

Chem. Rev. **1951**, *49*, 398.

Plancher Rearrangement

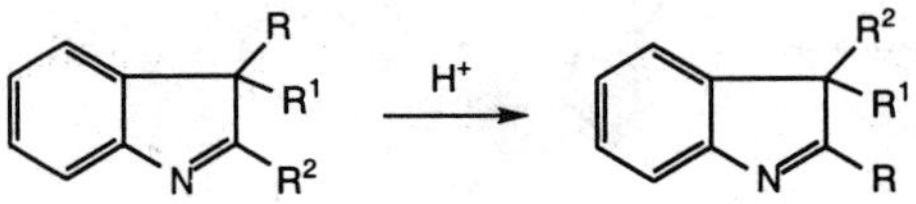

Comp. Org. Chem. **4**, 427.

Polonovski (Polonovsky) Reaction

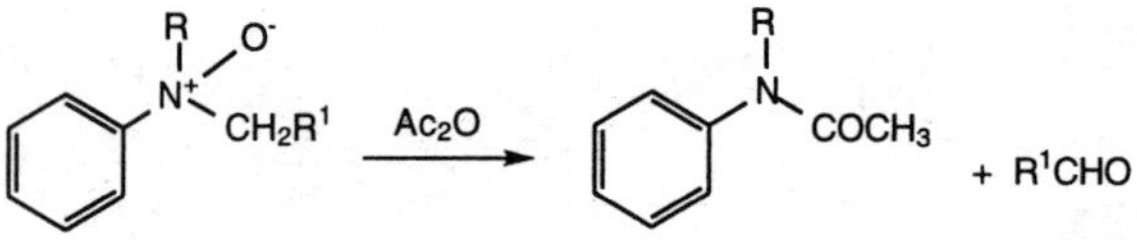

Org. Reactions **1990**, *39*, 85.
Tetrahedron **1979**, *35*, 2175.

Pomeranz-Fritsch Reaction

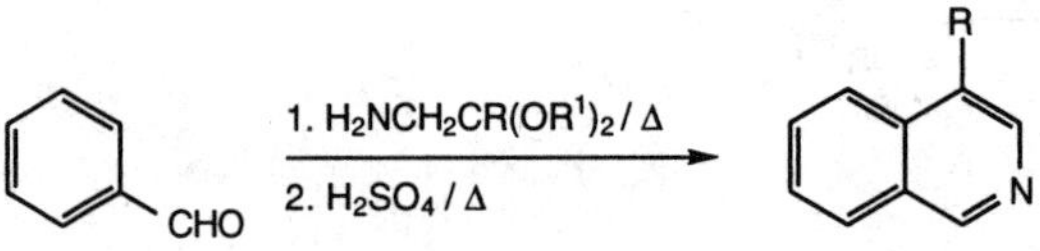

Tetrahedron **1971**, *27*, 1253.
Org. Reactions **1951**, 6, 191.

Ponzio Reaction

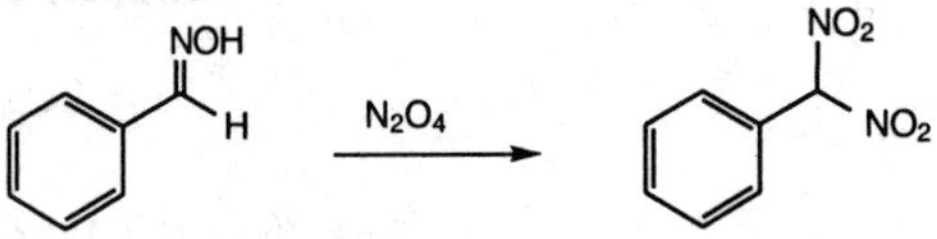

Chem. Rev. **1945**, *36*, 183.

Prévost Reaction (Glycolization)

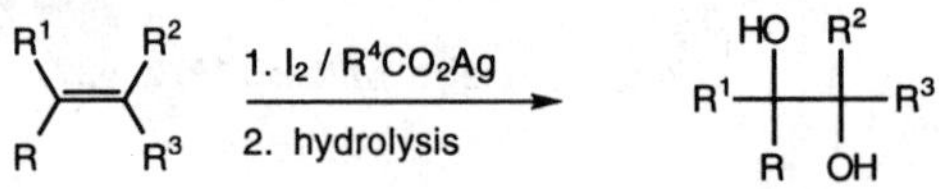

Org. Reactions **1957**, 9, 350.

Note: Also see the Woodward *cis*-Hydroxylation Reaction.

Prileschajew (Prilezhaev) Reaction

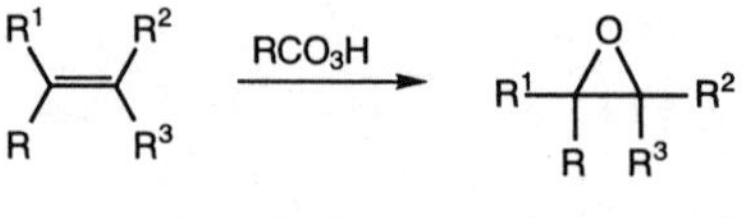

Org. Reactions **1953**, *7*, 378.
Chem. Rev. **1949**, *45*, 16.

Note: Also known as the Prilezhaev Epoxidation or Prileschaiev Reaction.

Prins Reaction

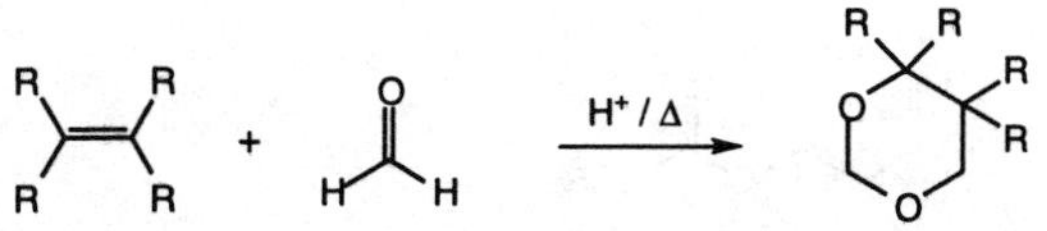

Synthesis **1977**, 661.
J. Org. Chem. **1970**, *35*, 2419.
Russ. Chem. Rev. **1968**, *37*, 17.
Chem. Rev. **1952**, *51*, 505.

Mechanism: *Tetrahedron Lett.* **1972**, 2797.
J. Am. Chem. Soc. **1963**, *85*, 47.

Pschorr Reaction (Synthesis)

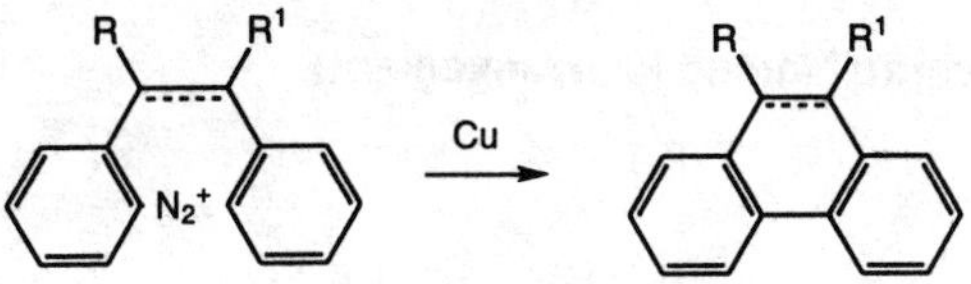

Tetrahedron **1980**, *36*, 3327.
Chem. Rev. **1976**, *76*, 532.
J. Heterocycl. Chem. **1971**, *8*, 341.
Org. Reactions **1959**, *10*, 21.
Org. Reactions **1957**, 9, 410.

Note: Also known as the Pschorr Ring Closure.

Pummerer Rearrangement (Reaction)

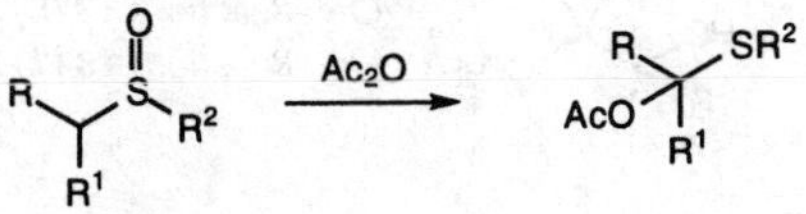

Org. Reactions **1991**, *40*, 157.
Adv. Org. Chem. **1969**, 6, 356.

Purdie Methylation

$$\mathrm{ROH} \xrightarrow{\mathrm{MeI\,/\,Ag_2O}} \mathrm{ROCH_3}$$

Comp. Org. Chem. **1**, 820.

Quelet Reaction

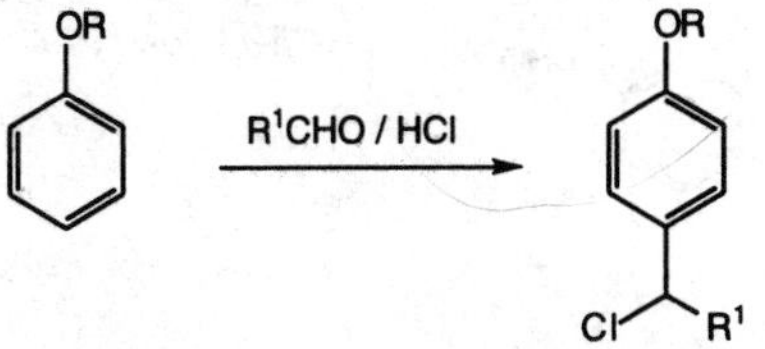

Merck ONR-74.

Raecke Process (Reaction) (see Henkel Reaction)

Ramberg-Bäcklund Reaction

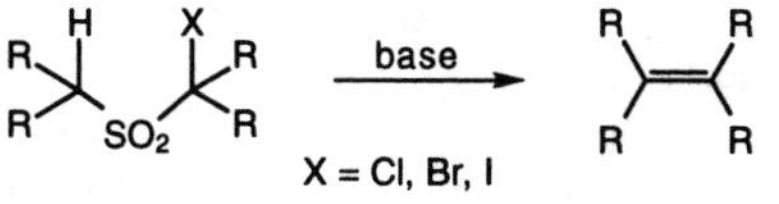

J. Am. Chem. Soc. **1989**, *111*, 779.
Org. Reactions, **1977**, *25*, 1.
Tetrahedron **1977**, *33*, 2019.
Acc. Chem. Res. **1968**, *1*, 209.

Mechanism: *Acc. Chem. Res.* **1970**, *3*, 285.

Rap-Stoermer Condensation

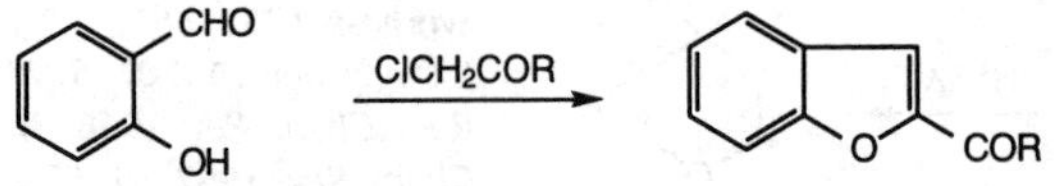

Rasoda Reaction (Synthesis)

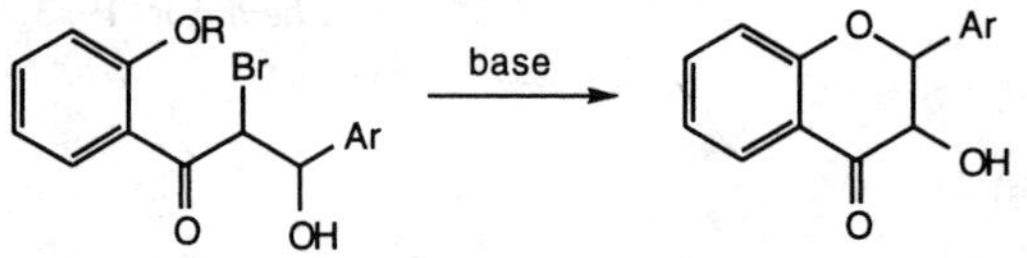

Tetrahedron **1980**, *36*, 1987.

Rearrangements (see under specific Named Rearrangement)

Reed Reaction

$$RH \xrightarrow{SO_2\ /\ Cl_2\ /\ h\nu} RSO_2Cl$$

March 637.

Reformatsky Reaction

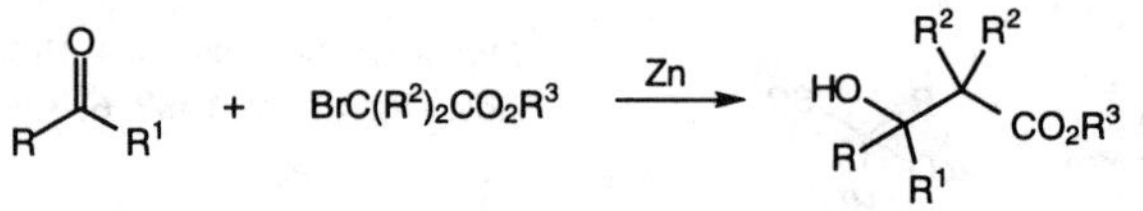

Synthesis **1989**, 571.
Org. Reactions **1975**, *22*, 423.
Org. Reactions **1942**, *1*, 1.

Reilly-Hickinbottom Rearrangement

$$ArNHR \xrightarrow{MCl_2} RNH_2 + ArNH_2$$

March 504.

Reimer-Tiemann Reaction

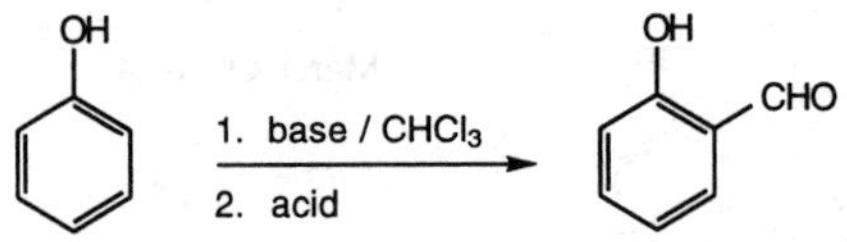

Org. Reactions **1982**, *28*, 1.
Chem. Rev. **1960**, *60*, 169.

Reissert Indole Synthesis

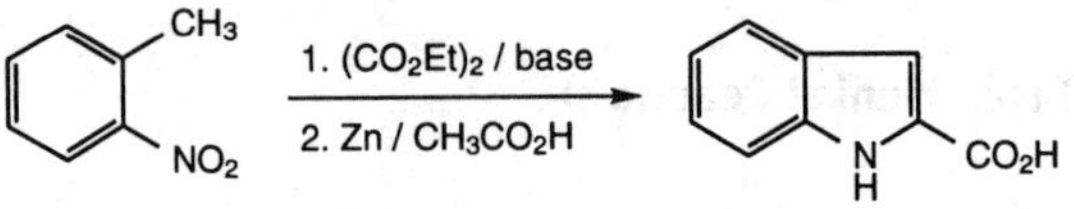

Org. Synth. **1963**, *43*, 40.

Reissert Reaction (Addition)

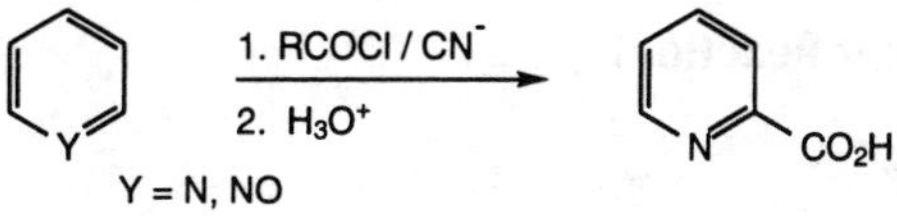

Heterocycl. **1973**, *1*, 165.
Chem. Rev. **1955**, *55*, 511.

Note: The *N*-acylated 2-cyano intermediates are known as Reissert Compounds.

Reppe Reaction

$$RC\equiv CH \xrightarrow[Ni(CO)_4]{CO / H_2O} RHC{=}CHCO_2H$$

Angew. Chem. Int. Ed. Engl. **1966**, *5*, 649.

Retro-Aldol Reaction (see Aldol Reaction)

Retro-Baeyer-Villiger Rearrangement (see Baeyer-Villiger Rearrangement)

Retro-Diels-Alder Reaction (see Diels-Alder Reaction)

Retro-Ene Reaction (see Ene Reaction)

Retro-Mannich Reaction (see Mannich Reaction)

Retropinacol Rearrangement

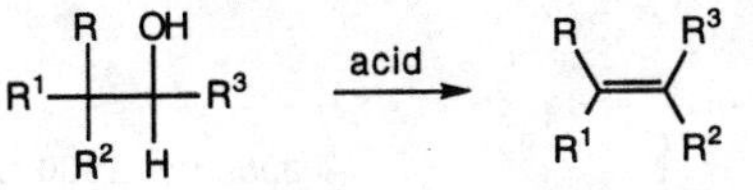

Merck ONR-75.

Retro-Smiles Rearrangement (see Smiles Rearrangement)

Reverdin Rearrangement (Reaction)

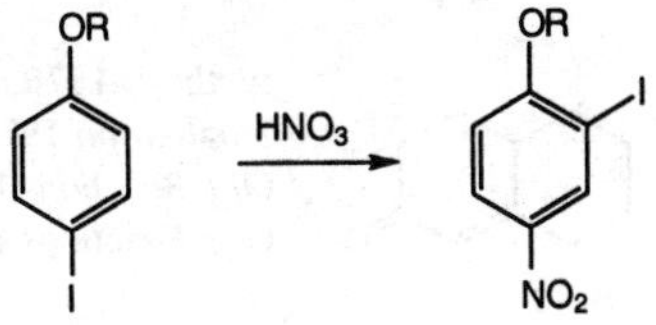

Merck ONR-76.
J. Chem. Soc. **1916**, 109; 1078.

Richey-Story Rearrangement

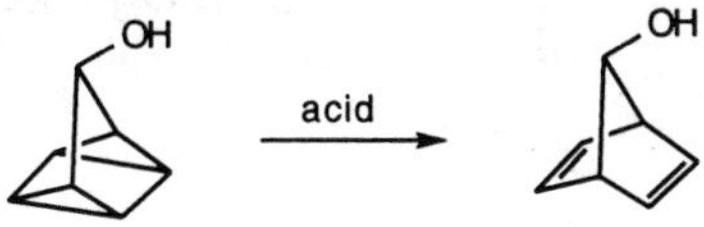

J. Am. Chem. Soc. **1964**, *86*, 527.
J. Am. Chem. Soc. **1963**, *85*, 3057.

Richter (Cinnoline) Synthesis (see von Richter (Cinnoline) Synthesis)

Richter Reaction (see von Richter Reaction)

Riehm (Quinoline) Synthesis

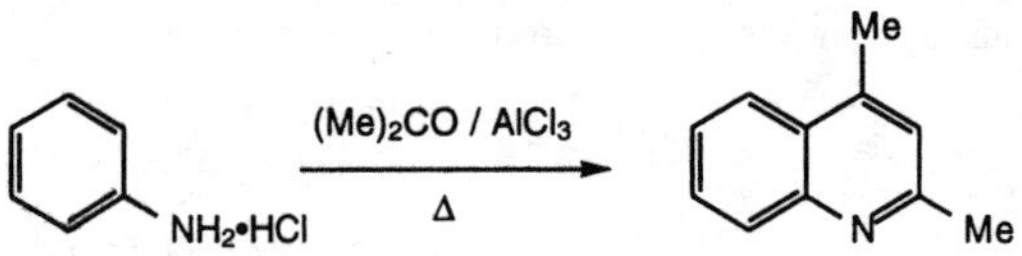

Merck ONR-77.
J. Am. Chem. Soc. **1951**, *73*, 975.

Riemschneider Reaction

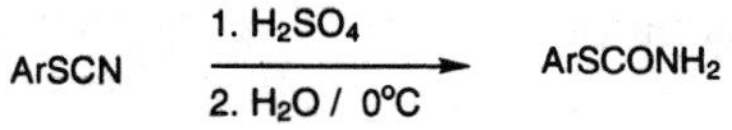

J. Am. Chem. Soc. **1953**, *75*, 6067.
J. Am. Chem. Soc. **1951**, *73*, 5905.

Note: Also known as the Riemschneider Thiocarbamate Synthesis.

Riley Oxidation (Reaction)

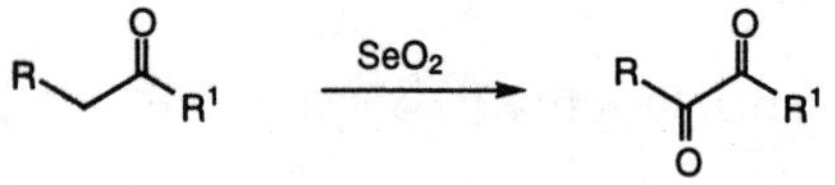

Org. Reactions **1976**, *24*, 261.
Org. Reactions **1949**, *5*, 331.

Rimini Aldehyde Reaction (see Angeli Rimini Aldehyde Reaction)

Ritter Reaction

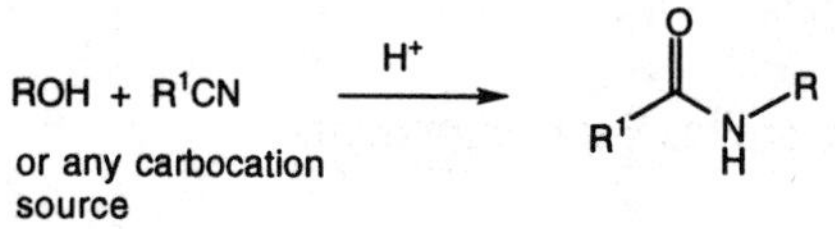

Tetrahedron **1980**, *36*, 1279.
Synthesis **1979**, 549.
Tetrahedron **1978**, *34*, 2815.
Org. Reactions **1969**, *17*, 213.

Robinson Annelation Reaction

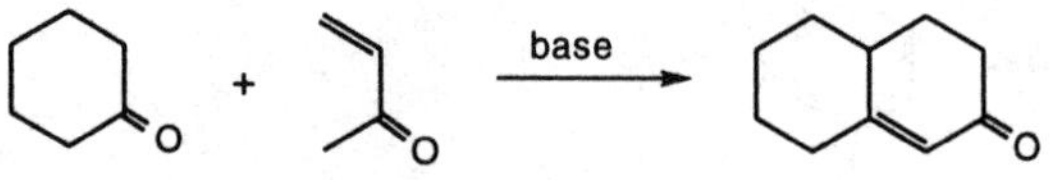

Synthesis **1976**, 777.
Tetrahedron **1976**, *32*, 3.
Org. Reactions **1968**, *16*, 1.
Org. Reactions **1959**, *10*, 179.

Robinson-Gabriel Synthesis

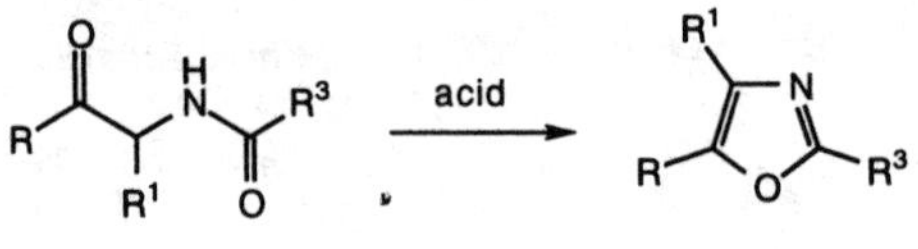

Comp. Org. Chem. **4**, 974.

Robinson-Schöpf Reaction

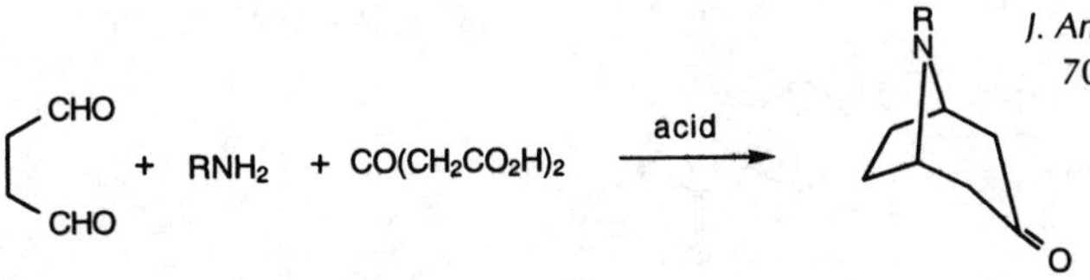

J. Am. Chem. Soc. **1979**, *101*, 7032.

Roelen Reaction

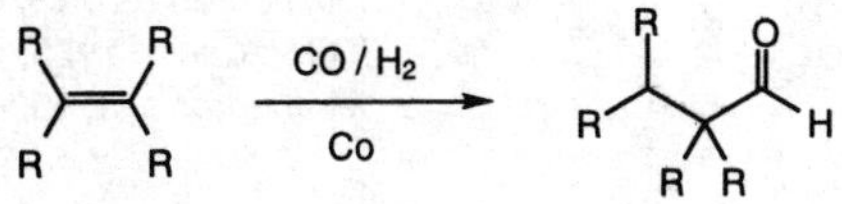

Angew. Chem. Int. Ed. Engl. **1966**, 5, 648.

Rosenmund Reaction

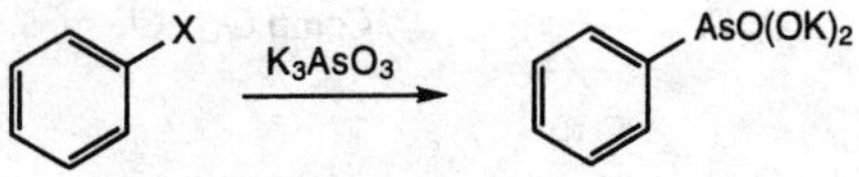

Org. Reactions **1944**, 2, 431

Rosenmund Reduction

$$\text{RCOCl} \xrightarrow{H_2\,/\,\text{Pd-BaSO}_4} \text{RCHO}$$

J. Org. Chem. **1991**, 56, 5159.
Chem. Rev. **1953**, 52, 245.
Org. Reactions **1948**, 4, 362.

Rosenmund-von Braun Synthesis (Reaction)

$$\text{ArX} \xrightarrow{\text{CuCN}\,/\,\Delta} \text{ArCN}$$

Chem. Rev. **1948**, 42, 207.
J. Org. Chem. **1941**, 6, 795.

Rothemund Reaction

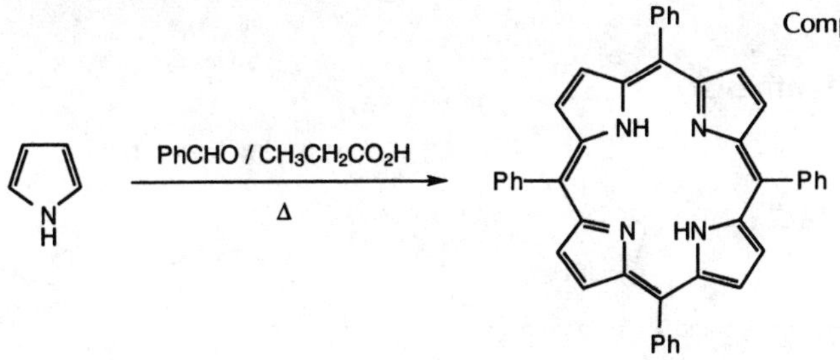

Comp. Org. Chem. **4**, 333.

Rowe Rearrangement

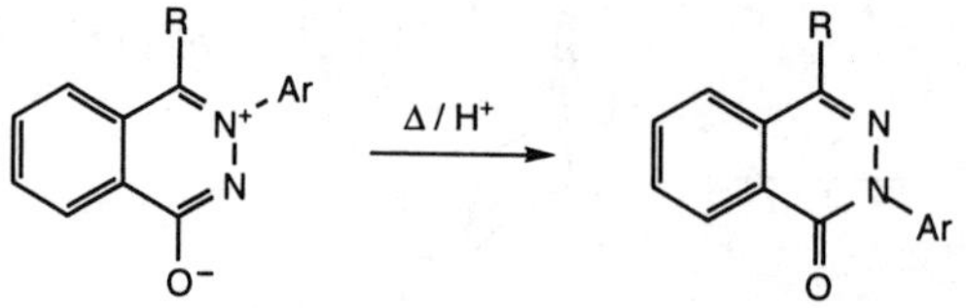

Chem. Rev. **1948**, *43*, 487.

Rubottom Reaction

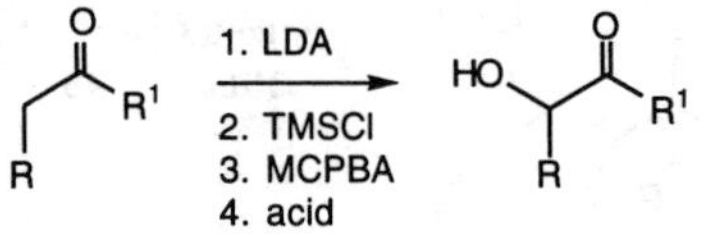

Tetrahedron Lett. **1978**, 4603.
J. Org. Chem. **1975**, *40*, 3427.
Tetrahedron Lett. **1974**, 4319.

Ruff (-Fenton) Degradation

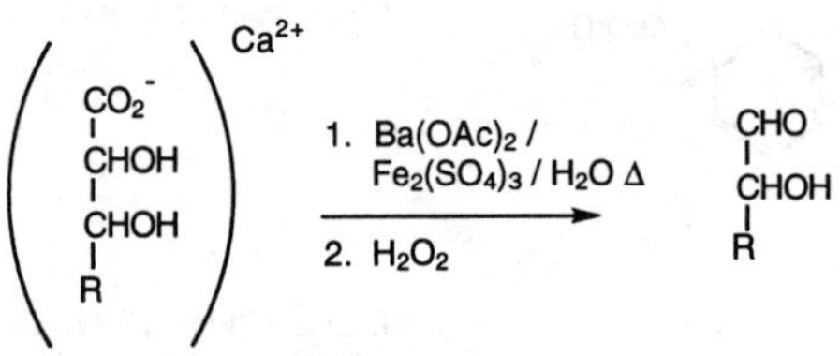

One carbon degradation of sugars.

Comp Org. Chem. **5**, 695.

Rühlmann Reaction (see Bouveault-Blanc Condensation)

Rupe Rearrangement (Reaction)

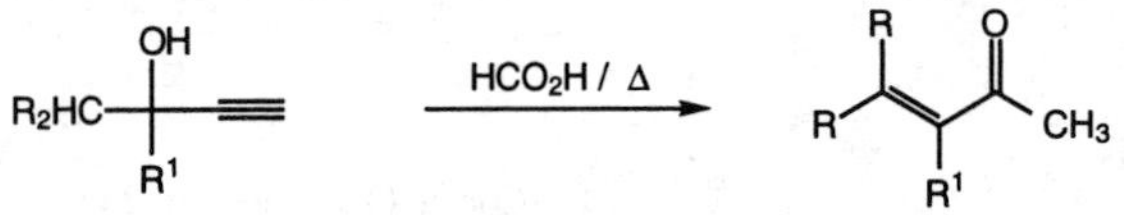

Synthesis **1981**, 473.
Chem. Rev. **1971**, *71*, 429.
Chem. Rev. **1969**, *69*, 675.

Ruzicka Cyclization (Synthesis)

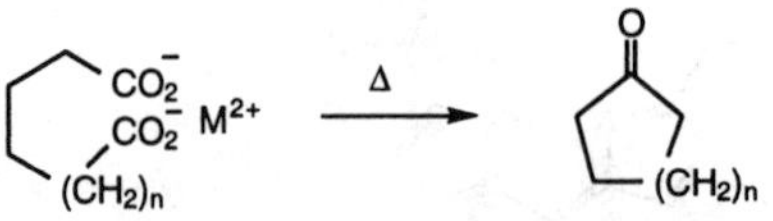

M = Ca, Ba for small rings and Th for large rings

Note: Also known as the Ruzicka Large Ring Synthesis.

Comp. Org. Chem. **1**, 99.

Sabatier-Senderens Reduction

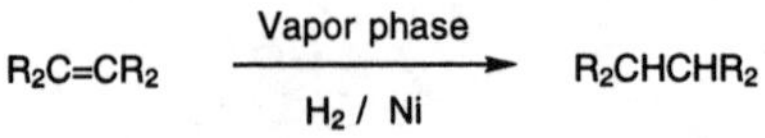

Merck ONR-80.

Sakurai Reaction

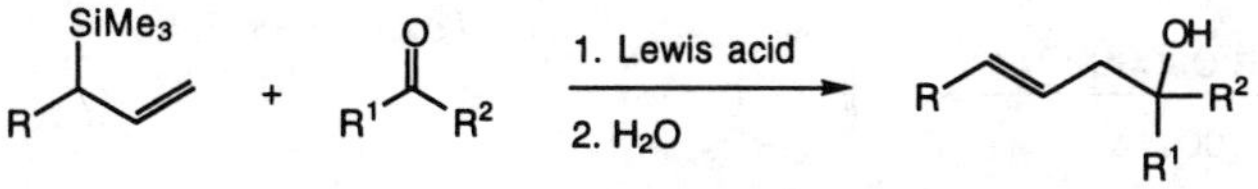

Pure & Appl. Chem. **1982**, 1.
Tetrahedron. Lett. **1980**, *21* 1357.

Sandmeyer Diphenylurea Isatin Synthesis

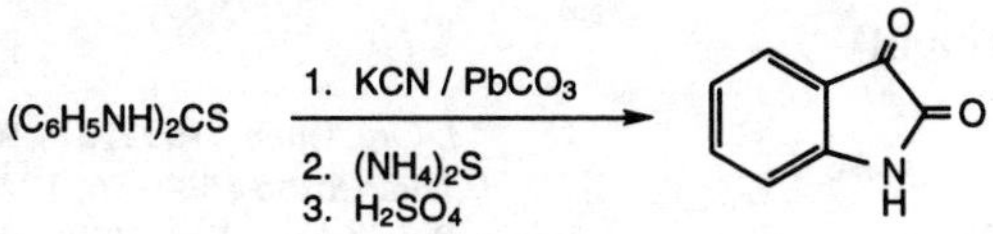

Merck ONR-80.

Sandmeyer Isonitrosoacetanilide Isatin Synthesis

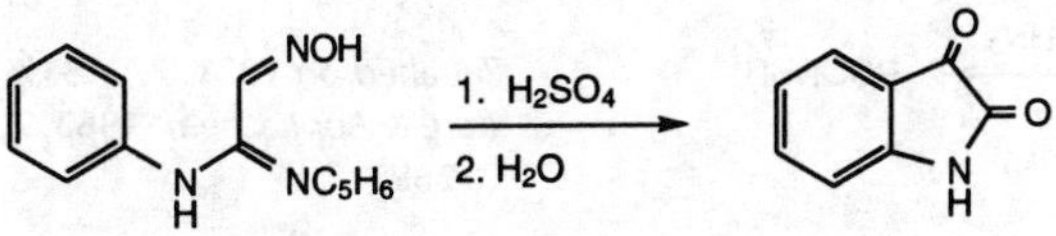

Merck ONR-80.
J. Org. Chem. **1956**, *21*, 171.

Sandmeyer Reaction

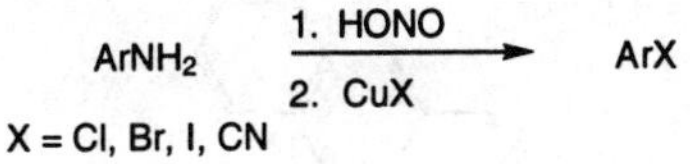

J. Chem. Soc., Chem. Commun. **1984**, 1523.
Tetrahedron Lett. **1977**, 3519.
J. Chem. Soc. **1966**, 1249.
Org. Reactions **1960**, *11*, 210.
Angew. Chem. **1953**, *65*, 155.
Quart. Rev. **1952**, 6, 358.

Sarett Oxidation

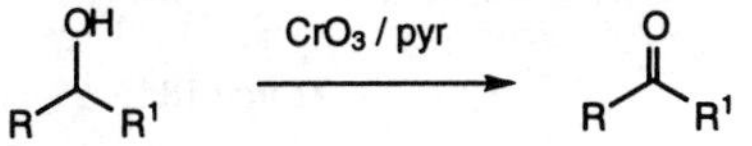

J. Org. Chem. **1963**, *28*, 323.
J. Am. Chem. Soc. **1953**, *75*, 422.

Saucy-Marbet Rearrangement (see Marbet-Saucy Rearrangement)

Schiemann Reaction

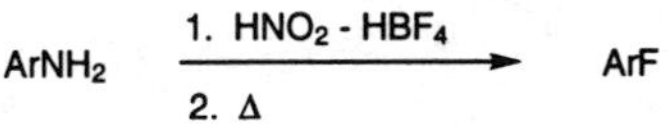

Chem. Ber. **1990**, *123*, 1673.
Org. Reactions **1949**, *5*, 193.

Schlittler-Müller Reaction

1. $(R^1O)_2CHCHO / \Delta$
2. H_2SO_4 / Δ

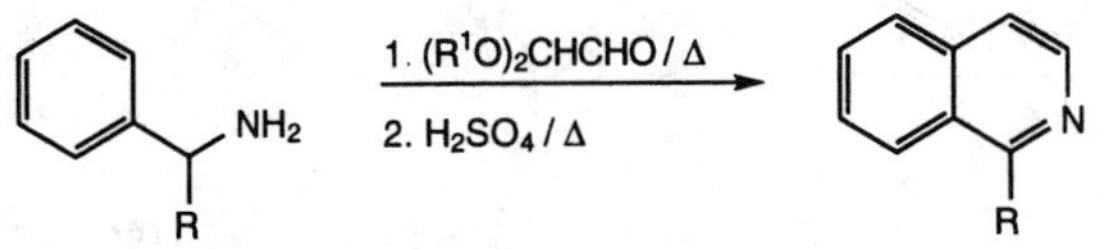

Org. Reactions **1951**, 6, 191.

Schmidlin Ketene Synthesis

$$CH_3COCH_3 \xrightarrow{\Delta} H_2C=C=O$$

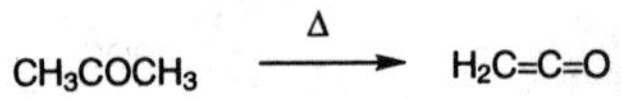

Merck ONR-81.
Org. Reactions **1946**, 3, 109.

Schmidt Reaction (Rearrangement)

$$RCO_2H \xrightarrow{H_2SO_4 / HN_3} RNH_2$$

$$RCHO \xrightarrow{H_2SO_4 / HN_3} RCN$$

$$RCOR^1 \xrightarrow{H_2SO_4 / HN_3} RCONHR^1$$

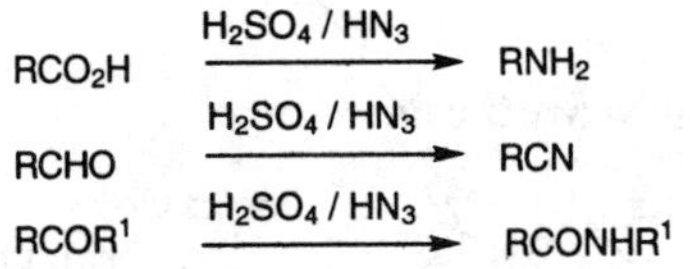

J. Org. Chem. **1992**, 57, 1635.
Tetrahedron **1980**, 36, 1279.
Russ. Chem. Rev. **1978**, 47, 1084.
Russ. Chem. Rev. **1971**, 40, 835.
Tetrahedron **1971**, 27, 1317.
Pure & Appl. Chem. **1963**, 7, 269.

Schmitt Synthesis (see Kolbé-Schmitt Synthesis)

Scholl Reaction

Lewis acid

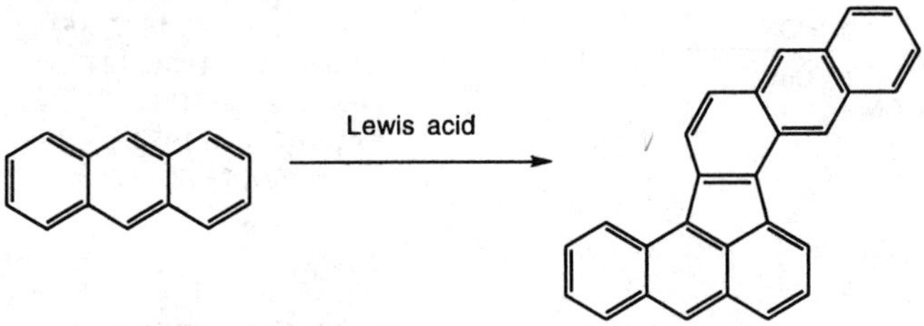

March 484.

Schönberg Reaction

$(CH_3O)_3P / \Delta$

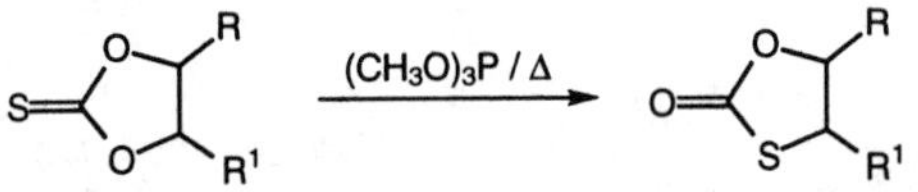

Org. Reactions **1984**, 30, 477.

Schöpf Condensation

pH 7

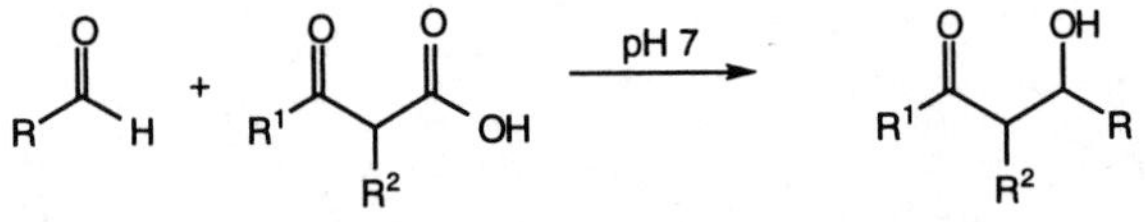

Org. Reactions **1968**, 16, 32.

Schorigin (Shorygin) Reaction

$$Hg(R)_2 \xrightarrow[\text{2. } CO_2]{\text{1. Na}} RCO_2Na$$

Merck ONR-82.

Note: The reaction of alkyl sodium compounds with carbon dioxide is known as the Wanklyn Reaction.

Schotten-Baumann Reaction (Procedure)

Chem. Rev. **1953**, *52*, 272.

Schwenk-Papa Reduction

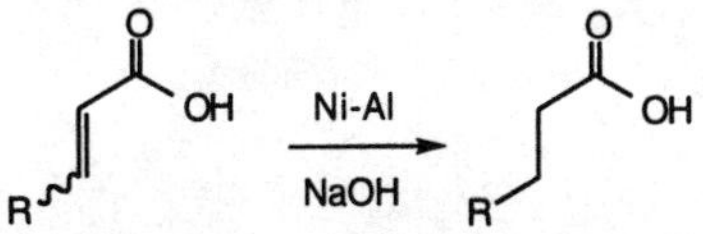

J. Org. Chem. **1944**, *9*, 1.

Desulfurization: *Org. Reactions* **1962**, *12*, 356.

Selenium Dioxide Oxidation (see Riley Oxidations)

Semidine Rearrangement (see Benzidine Rearrangement)

Semmler-Wolff Reaction

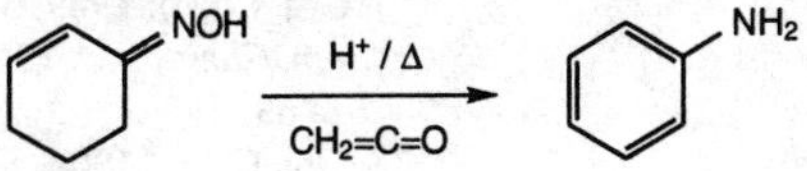

Synthesis **1980**, 483; 887.

Serini Reaction

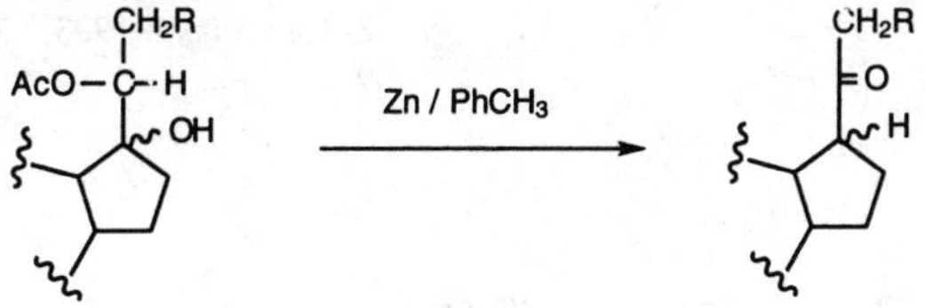

J. Chem. Soc. **1949**, 1671.

Shapiro Reaction

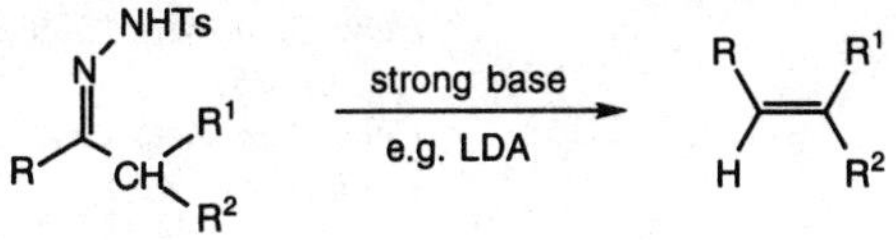

Org. Reactions **1990**, *39*, 3.
Acc. Chem. Res. **1983**, *16*, 55.
Org. Reactions **1976**, *23*, 405.

Sharpless (Asymmetric) Dihydroxylation

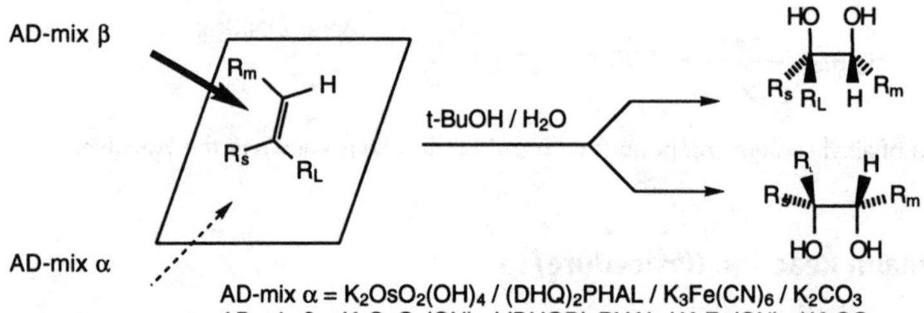

AD-mix α = $K_2OsO_2(OH)_4$ / $(DHQ)_2PHAL$ / $K_3Fe(CN)_6$ / K_2CO_3
AD-mix β = $K_2OsO_2(OH)_4$ / $(DHQD)_2PHAL$ / $K_3Fe(CN)_6$ / K_2CO_3

J. Org. Chem. **1992**, *57*, 2768.
J. Am. Chem. Soc. **1988**, *110*, 1968.

Sharpless (Asymmetric) Epoxidation

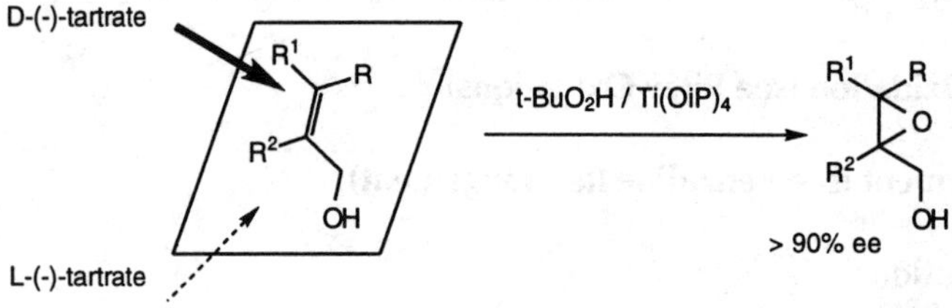

J. Org. Chem. **1989**, *54*, 1295.
J. Am. Chem. Soc. **1987**, *109*, 5765.
J. Org. Chem. **1986**, *51*, 1922, 3710.
Synthesis **1986**, 89.
Pure & Appl. Chem. **1983**, *55*, 589.

Siegrist Reaction

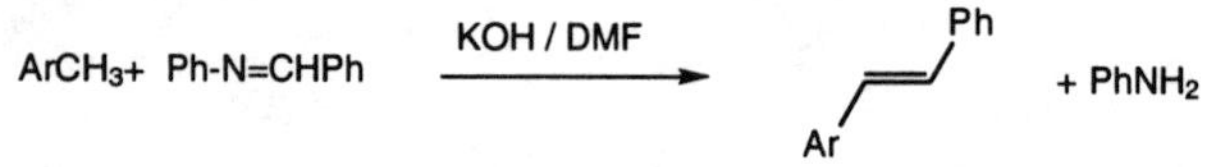

Helv. Chim. Acta **1969**, *52*, 1282; 2521.
Helv. Chim. Acta **1967**, *50*, 906.

Siloxy-Cope

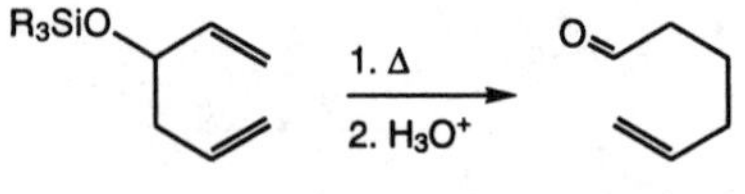

Org. Reactions **1975**, *22*, 64.
J. Am. Chem. Soc. **1972**, *94*, 7074.

Silyl-Wittig Reaction (see Peterson Reaction)

Simmons-Smith Reaction

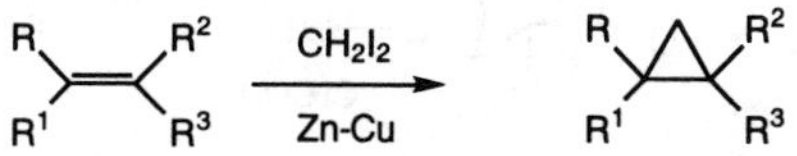

Can. J. Chem. **1985**, *63*, 2969.
Org. Reactions **1973**, *20*, 1.

Note: Substituted dihalomethanes may also be used.

Simonini Reaction

$$RCO_2Ag \xrightarrow{I_2 / \Delta} RCO_2R$$

Tetrahedron **1971**, *27*, 5323.
Org. Reactions **1957**, *9*, 322.
Chem. Rev. **1956**, *56*, 259.

Note: Also see the Cristol-Firth-Hunsdiecker Reaction and Hunsdiecker Reaction.

Simonis Chromone Cyclization

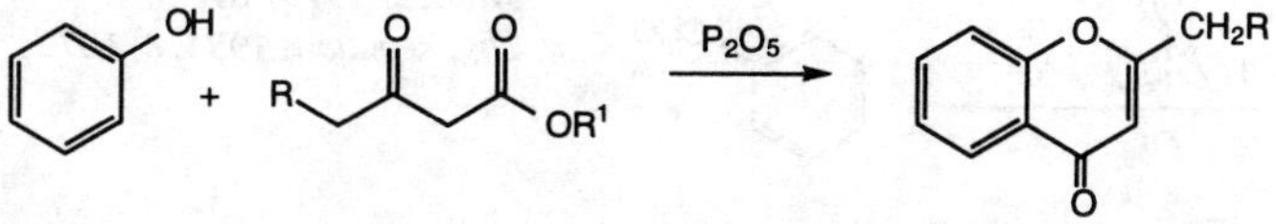

Org. Reactions **1953**, *7*, 15.

Skraup (Quinoline) Synthesis (Reaction)

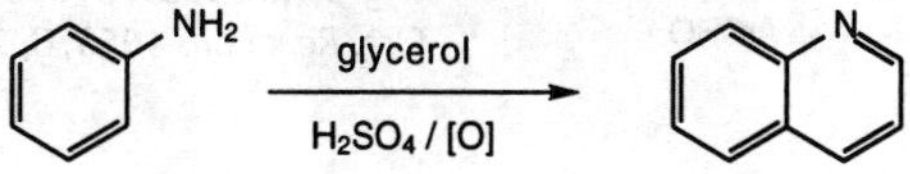

Synthesis **1976**, 691.
Org. Reactions **1953**, *7*, 59.
Chem. Rev. **1944**, *35*, 152.

Smiles (-Truce) Rearrangement

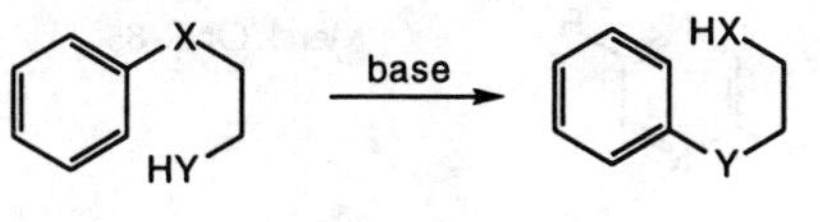

X = O, SO_2 Y = OH, NH_2

J. Am. Chem. Soc. **1989**, *111*, 4224.
Synthesis **1977**, 33.
Org. Reactions **1970**, *18*, 99.
Chem. Rev. **1951**, *49*, 362.

Photo-Smiles: *Acc. Chem. Res.* **1983**, *16*, 285.
Retro-Smiles: *Org. Reactions* **1970**, *18*, 108.

Note: Also known as the Truce Rearrangement or Truce-Smiles Rearrangement.

Smirnov-Zamkov Reaction

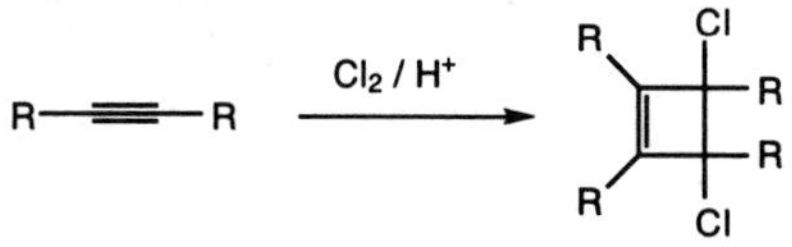

Org. Reactions **1962**, *12*, 17.

Smith Degradation

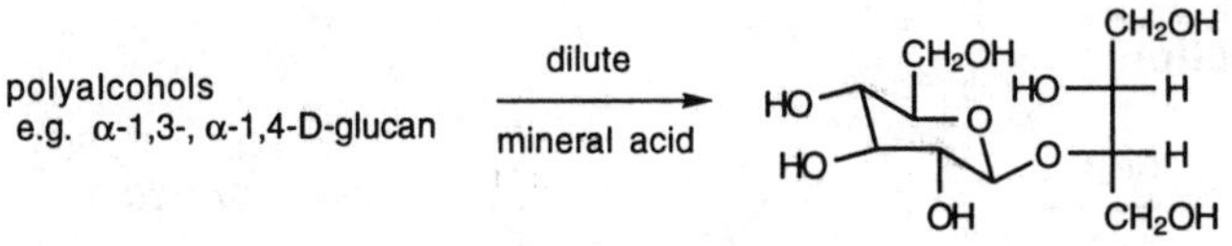

Comp. Org. Chem. **5**, 765.

Sommelet-Hauser Rearrangement

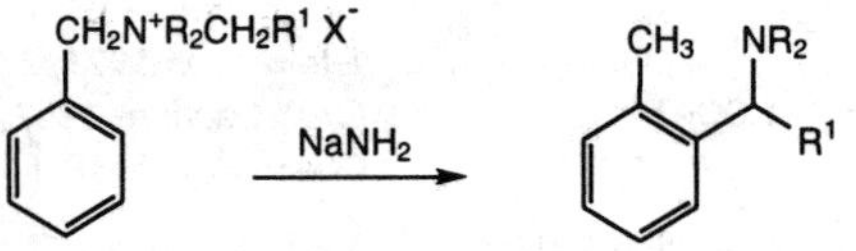

Synthesis **1991**, 117.
Org. Reactions **1970**, *18*, 403.

Sommelet Reaction

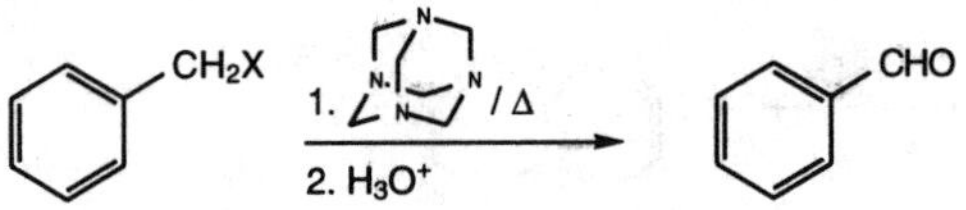

Synthesis **1979**, 164.
Org. Reactions **1954**, *8*, 197.

Sommelet Rearrangement (see Sommelet-Hauser Rearrangement)

Sonn-Müller Reaction (Method)

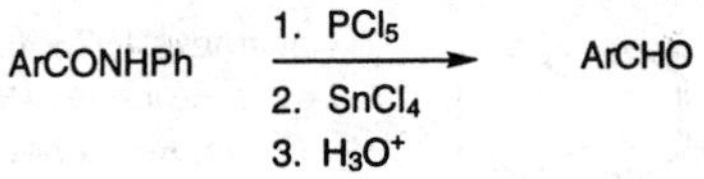

Org. Synth. **1955**, *3*, 818.
Org. Reactions **1953**, *8*, 240.

Staedel-Rügheimer Pyrazine Synthesis

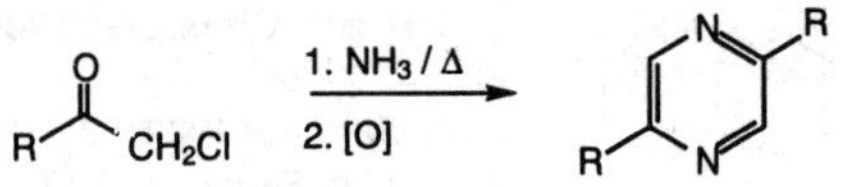

Merck ONR-85.

Staudinger Reaction

$$R_3P + R\text{-}N_3 \longrightarrow R_3P{=}N\text{-}R$$

Tetrahedron **1992**, *48*, 1353.
Synthesis **1985**, 202.
Tetrahedron **1981**, *37*, 437.

Note: The Staudinger Reaction refers to two reactions.

Staudinger Reaction

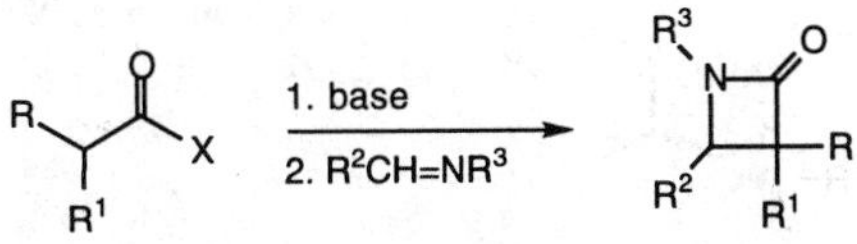

J. Am. Chem. Soc. **1992**, *114*, 6249.

Note: The Staudinger Reaction refers to two reactions.

Stephens-Castro Coupling (see Castro-Stephens Coupling)

Stephen Reaction

$$RCN \xrightarrow[\text{3. } H_3O^+]{\text{1. } HCl\ /\ SnCl_4} RCHO$$

Russ. Chem. Rev. **1962**, *31*, 615.
Org. Synth. **1955**, *8*, 246.

Note: Also known as the Stephen Aldehyde Synthesis or Stephen Reduction.

Stetter Reaction

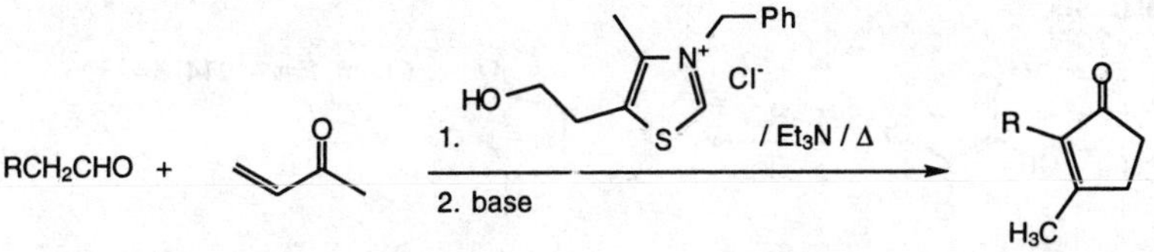

Org. Synth. **1987**, *65*, 26.

Stevens Rearrangement

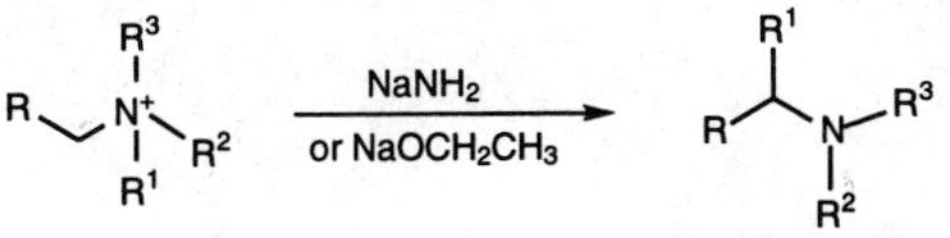

Synthesis **1973**, 98.
Chem. in Britain **1971**, *7*, 287.
Org. Reactions. **1970**, *18*, 403.
Angew. Chem. Int. Ed. Engl. **1962**, *1*, 155.
Org. Reactions. **1960**, *11*, 356.
Sulfonium salts: *J. Am. Chem. Soc.* **1989**, *111*, 7149.

Stieglitz Rearrangement

$$Ar_3CNHX \xrightarrow{\text{base}} Ar_2C{=}NAr$$

$$Ar_3CNHOH \xrightarrow{PCl_5} Ar_2C{=}NAr$$

March 989.

Stiles-Sisti Reaction

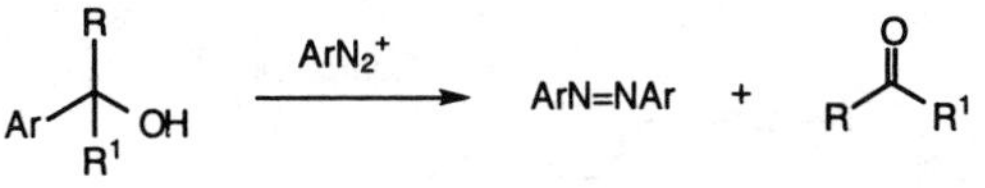

March 509.

Still-Wittig Rearrangement

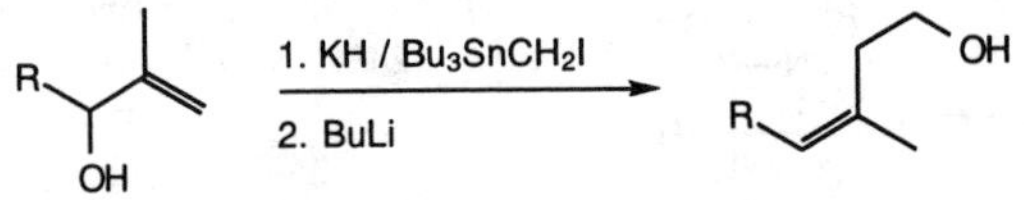

J. Am. Chem. Soc. **1978**, *100*, 1927.

Stobbe Condensation

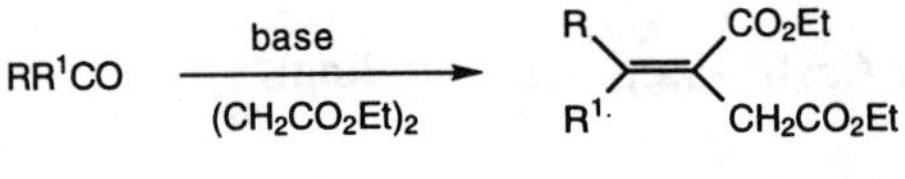

Heterocycl. **1977**, 6, 715.
Org. Reactions **1951**, 6, 1.

Stollé-Becker Synthesis

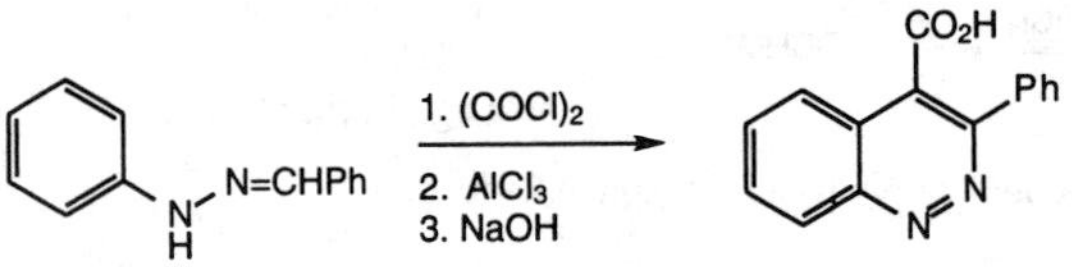

Comp. Org. Chem. **2**, 702.

Stollé Synthesis

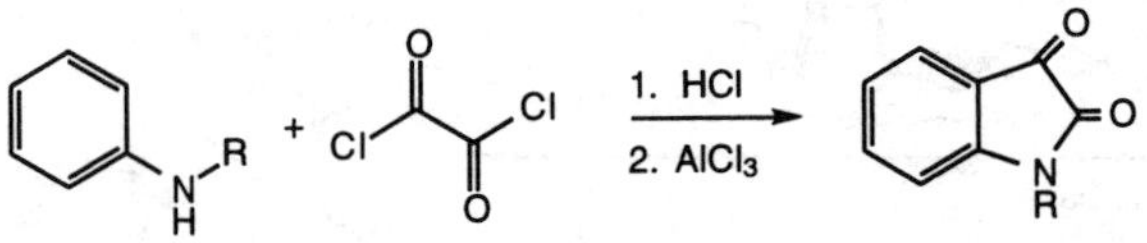

Chem. Rev. **1944**, *34*, 396.

Stork Enamine Reaction (Synthesis)

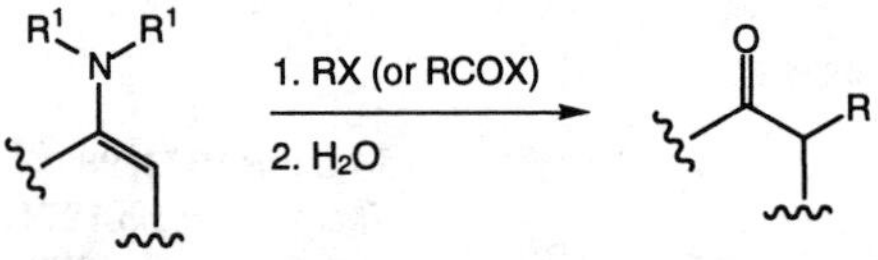

Tetrahedron **1980**, *38*, 1975.

Story Synthesis

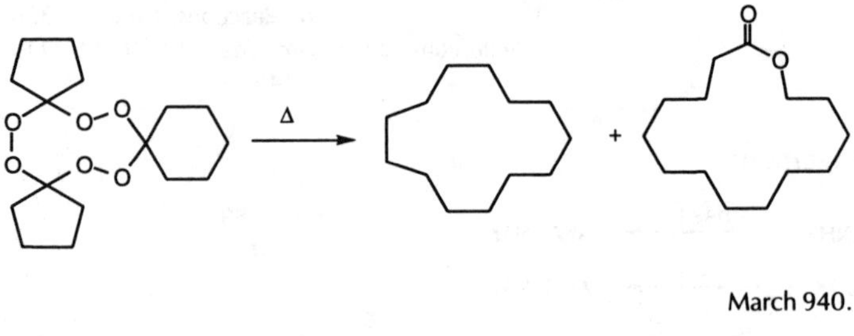

March 940.

Straus Coupling

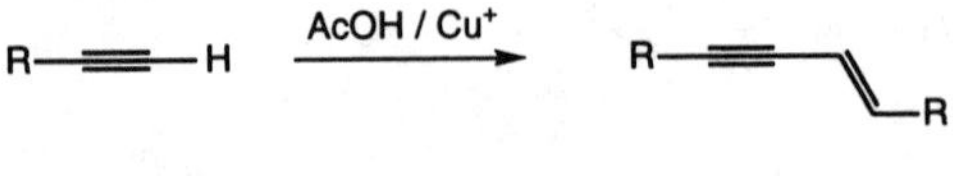

Comp. Org. Chem. **1**, 207.

Straus Coupling

$$Ph\text{---}\equiv\text{---}H \xrightarrow{\text{air}/\Delta} Ph\text{---}\equiv\text{---}\equiv\text{---}Ph$$

Fieser & Fieser 'Advanced Organic Chemistry', Reinhold, New York, **1961**, 223.

Strecker (Amino Acid) Synthesis

$$RCHO \xrightarrow[\text{2. hydrolysis}]{\text{1. CN}^-/\text{NH}_4\text{Cl}} \underset{R}{\overset{NH_2}{\diagup}}CO_2H$$

Note: Also known as the Strecker Reaction.

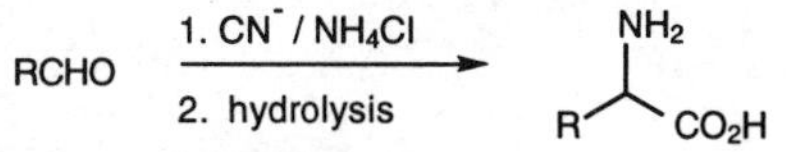

J. Med. Chem. **1973**, *16*, 901.
Chem. Ber. **1971**, *104*, 3594.
Chem. Rev. **1948**, *42*, 236.

Strecker Degradation

$$\underset{R}{\overset{NH_2}{\diagup}}CO_2H + \underset{O}{\overset{O}{\diagdown}}\underset{R^2}{\overset{R^1}{\diagup}} \xrightarrow{\Delta} RCHO$$

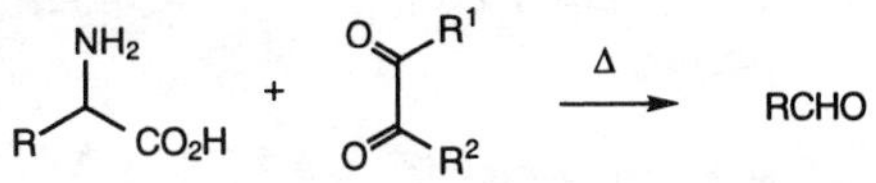

Chem. Rev. **1952**, *50*, 261.

Strecker Sulfite Alkylation

$$RX + M_2SO_3 \xrightarrow{CH_3I} RSO_3M$$

Merck ONR-88.

Swarts Reaction

$$RCX_3 + SbF_3 \xrightarrow{SbX_5} RCF_3$$
X = halogen

Org. Reactions **1944**, *2*, 53.

Swern Oxidation

$$\underset{R}{\overset{OH}{\diagup}}R^1 \xrightarrow[R_3N]{(\text{COCl})_2/\text{DMSO}} \underset{R}{\overset{O}{\diagup}}R^1$$

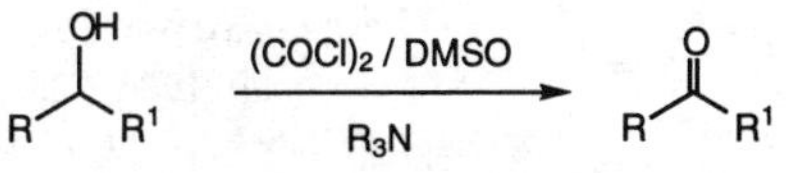

Org. Reactions **1990**, *39*, 297.
Synthesis **1981**, 165.
J. Org. Chem. **1978**, *43*, 2480.
Tetrahedron **1978**, *34*, 1651.

Tafel Rearrangement

$$CH_3COCRR^1CO_2R^2 \xrightarrow[\text{EtOH}/\text{H}_2\text{SO}_4]{\text{electrolysis}} CH_3CH_2CH_2CHRR^1$$

Merck ONR-88.

Tântâsescu Reaction (see Lehmstedt-Tântâsescu Reaction)

ter Meer Reaction

$$\underset{R}{\overset{NO_2}{\diagup}}Cl \xrightarrow{NO_2^-} \underset{R}{\overset{NO_2}{\diagup}}NO_2$$

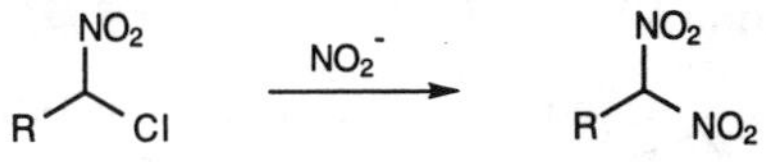

J. Am. Chem. Soc.. **1956**, *78*, 4980.

Teuber Reaction

Chem. Rev. **1971**, *71*, 229.

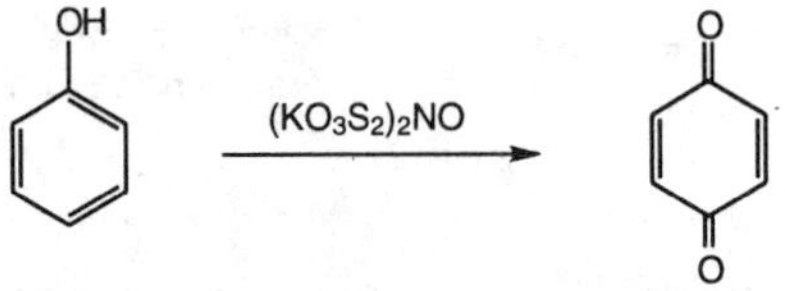

Thiele-Winter Acetoxylation (Reaction)

Org. Reactions **1972**, *19*, 199.

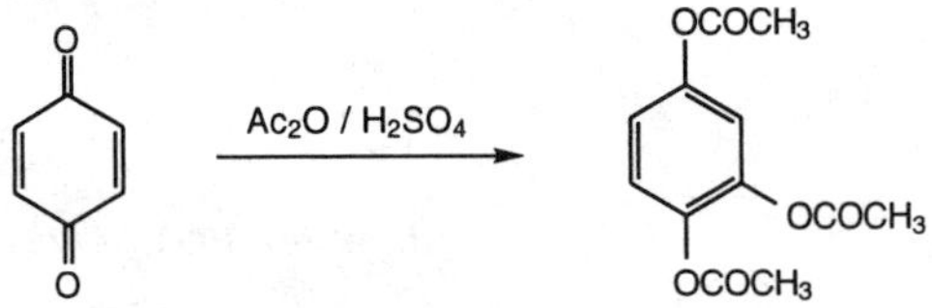

Note: Also known as the Thiele Reaction or Thiele Acetoxylation.

Thio-Claisen Rearrangement

J. Chem. Soc., Chem. Commun. **1981**, 1153.
J. Am. Chem. Soc. **1976**, *98*, 7084.
Org. Reactions **1975**, *22*, 46.

Thio-Wittig Reaction

J. Chem. Soc., Chem. Commun. **1985**, 1061.
J. Chem. Soc., Chem. Commun. **1984**, 525.
Angew. Chem. Int. Ed. Engl. **1977**, *16*, 418.

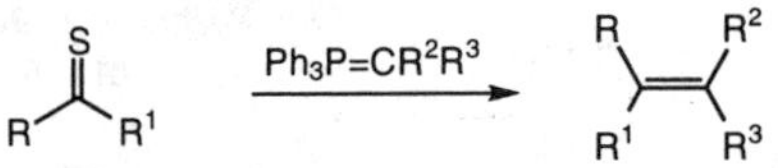

Note: Also see the Wittig Reaction.

Thorpe Reaction (Condensation)

Org. Reactions **1967**, *15*, 1.

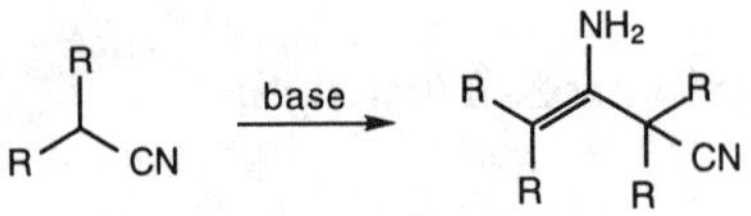

Thorpe-Ziegler Reaction (Condensation)

Adv. Org. Chem. **1970**, *7*, 1.
Org. Reactions **1967**, *15*, 28.

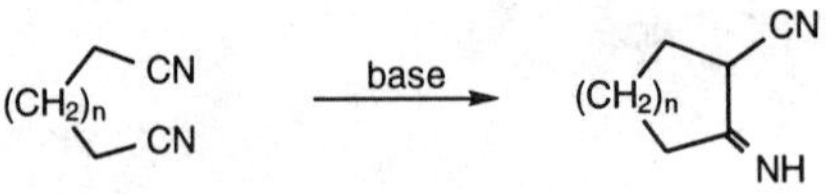

Note: Also known as the Thorpe-Ziegler Method.

Tiemann Rearrangement (Reaction)

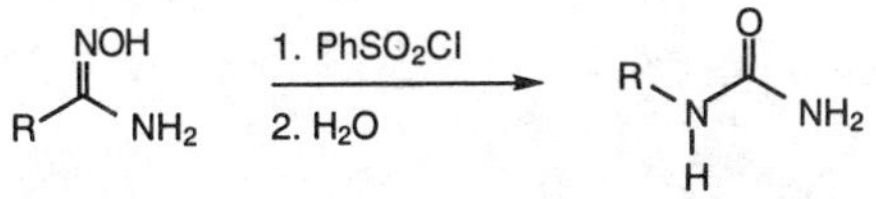

MERCK ONR-89.
Org. Reactions **1946**, *3*, 366.

Tiffeneau-Demyanov Reaction (Ring Expansion)

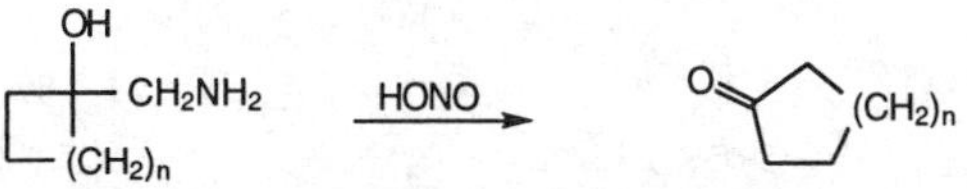

Tetrahedron **1973**, *29*, 1941.
Org. Reactions **1960**, *11*, 157.

Note: Also known as the Tiffeneau Reaction, Tiffeneau Rearrangement, or Tiffeneau Ring Enlargement.

Tipson-Cohen Reaction

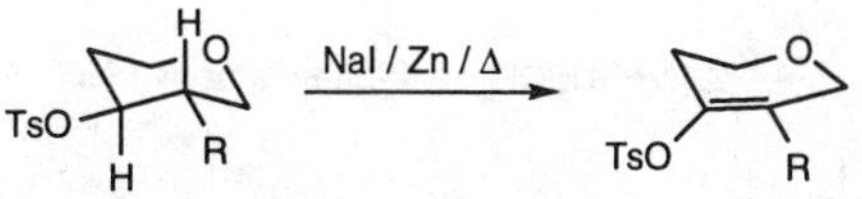

Org. Reactions **1984**, *30*, 499.

Tishchenko Reaction

$$\mathrm{RCHO} \xrightarrow{\mathrm{Al(OEt)_3}} \mathrm{RCO_2CH_2R}$$

J. Org. Chem. **1991**, *56*, 5159.

Todd-Atherton Reaction (see Atherton-Todd Reaction)

Tollens' Reaction

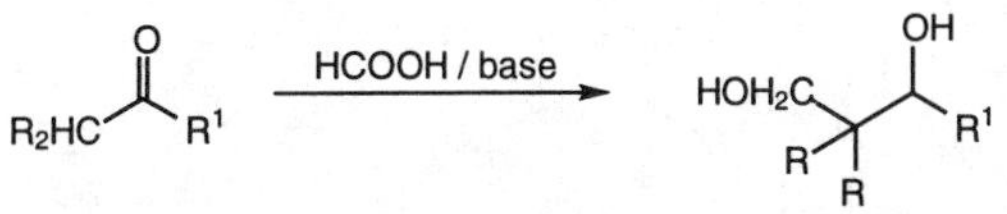

Comp. Org. Chem. **1**, 673.

Torgov Reaction

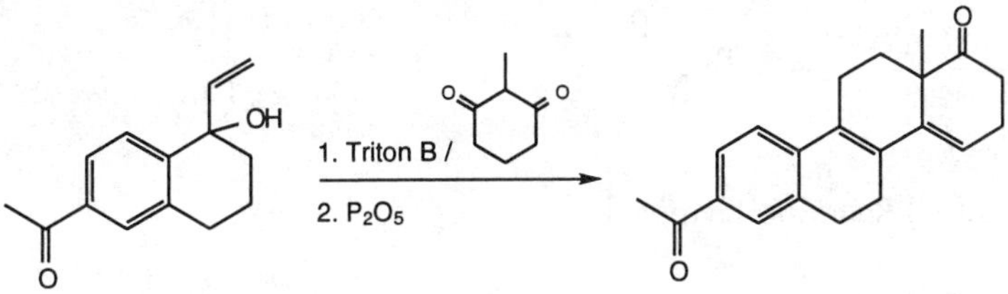

Bull. Soc. Chim. France **1969**, 4561.

Transamination Reaction

$$\mathrm{RCHNH_2CO_2H} + \mathrm{R^1COCO_2H} \longrightarrow \mathrm{R^1CHNH_2CO_2H} + \mathrm{RCHO}$$

J. Org. Chem. **1941**, *6*, 867.

Traube Purine Synthesis

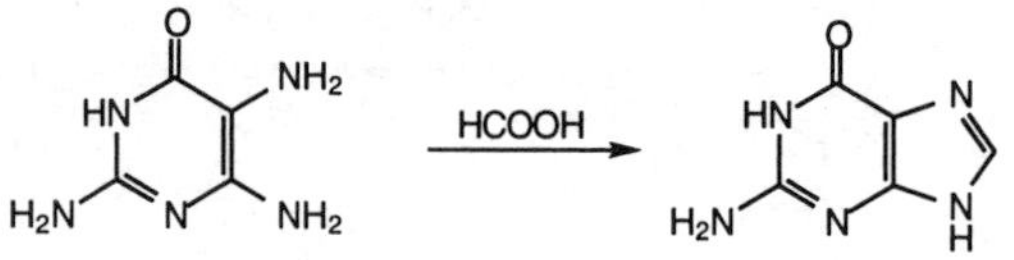

Comp. Org. Chem. **4**, 499.

Treibs Reaction

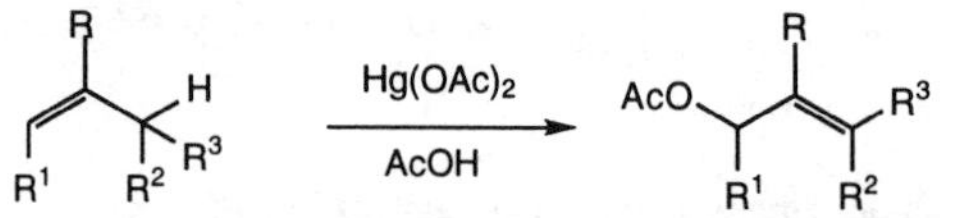

Comp. Org. Chem. **1**, 589.

Truce (Smiles) Rearrangement (see Smiles Rearrangement)

Tscherniac-Einhorn Reaction

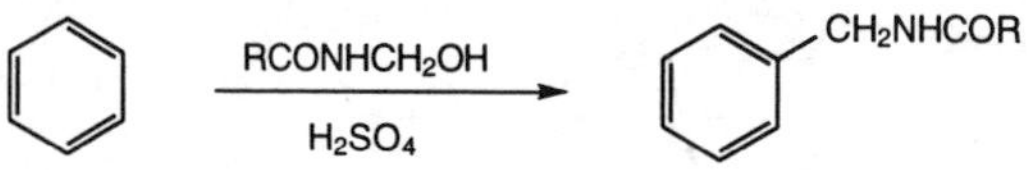

Org. Reactions **1965**, *14*, 63.

Tschitschibabin (Amination) Reaction (see Chichibabin Reaction)

Tschugaeff Reaction (see Chugaev Reaction)

Turpin Reaction

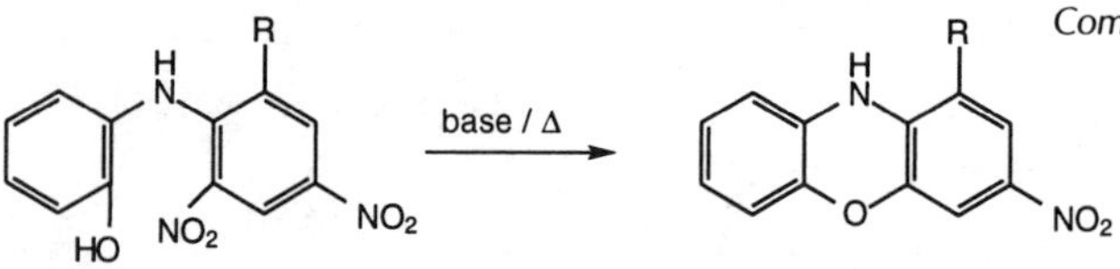

Comp. Org. Chem. **4**, 1068.

Ugi Reaction

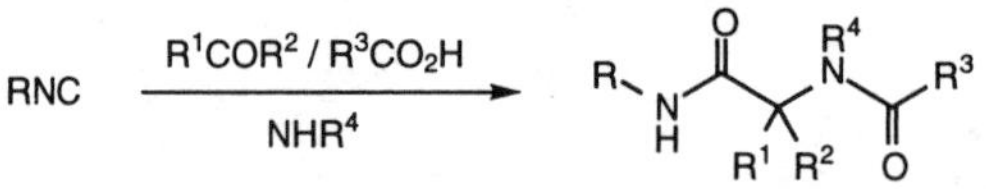

Angew. Chem. Int. Ed. Engl. **1962**, *1*, 8.
Chem. Ber. **1961**, *94*, 1116.

Ullmann (Coupling) Reaction (Synthesis)

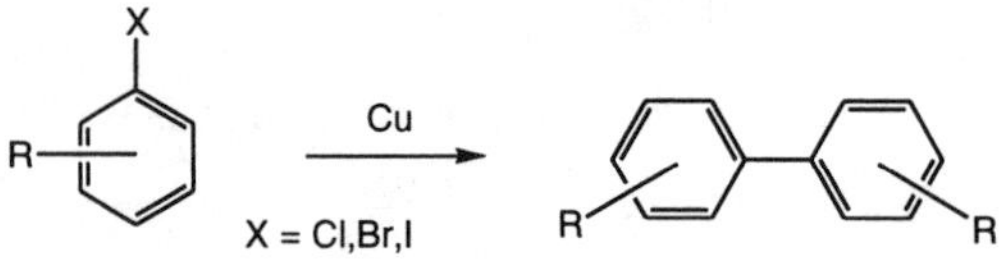

J. Am. Chem. Soc. **1991**, *113*, 1427.
Synthesis **1974**, 9.
Russ. Chem. Rev. **1972**, *41*, 1046.
Org. Reactions. **1965**, *14*, 19.

Chem. Rev. **1964**, *64*, 613.
Org. Reactions. **1944**, *2*, 243.

Note: 1. Unsymmetrical biaryls can be prepared when one of the reactants is activated or by reacting an aryl salt with a cuprate.
2. The Ullmann Reaction also refers to the Jourdan-Ullmann Reaction.

Ullmann Ether Synthesis

$$\mathrm{ArOH} + \mathrm{ArX} \xrightarrow{\text{base / Cu}^{2+}} \mathrm{ArOAr}$$

Russ. Chem. Rev. **1974**, *43*, 679.

Urech Hydantoin Reaction

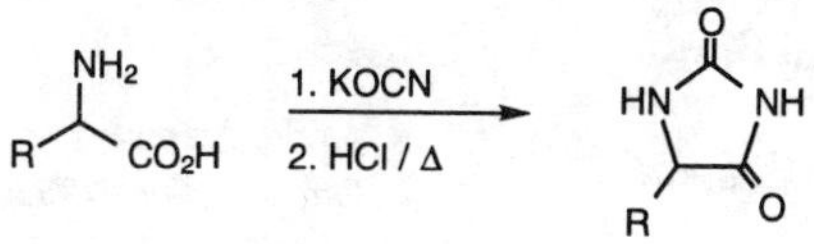

Merck ONR-91.
Chem. Rev. **1950**, *46*, 406.

van Alphen Rearrangement

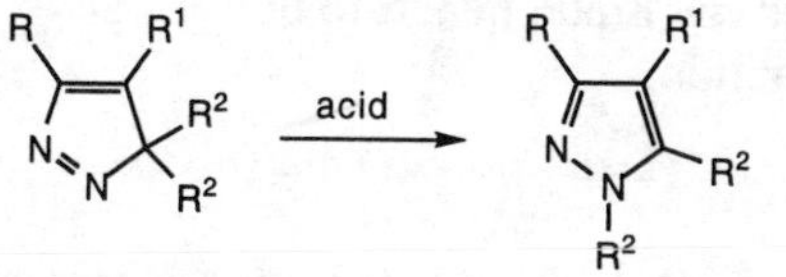

Tetrahedron **1975**, *31*, 1677.

Varrentrapp Reaction

$$\mathrm{CH_3(CH_2)_nCH{=}CH(CH_2)_nCO_2H} \xrightarrow{\text{KOH / }\Delta} \mathrm{CH_3(CH_2)_{2n}CO_2K}$$

Can. J. Chem. **1961**, *39*, 1730.
Tetrahedron **1960**, *8*, 221.

Victor Meyer Synthesis (see Meyer Synthesis)

Vilsmeier (Haack) Reaction

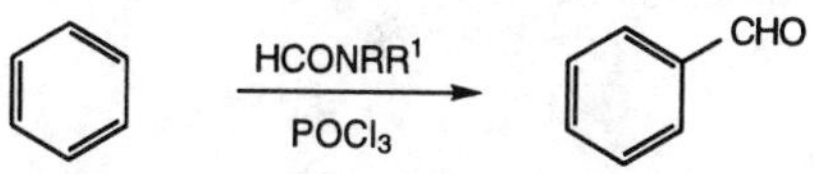

Tetrahedron **1992**, *48*, 3659.
Chem. Ind. (London) **1973**, 870.
Bull. Soc. Chim. France **1962**, 1989.

Note: 1. Nucleophilic aromatic compounds only.
2. Also known as the Vilsmeier Formylation.

Vinylcyclopropane-Cyclopentane Rearrangement

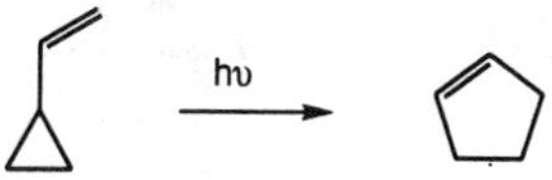

Org. Reactions **1985**, *33*, 247.

Voight (Amination) Reaction

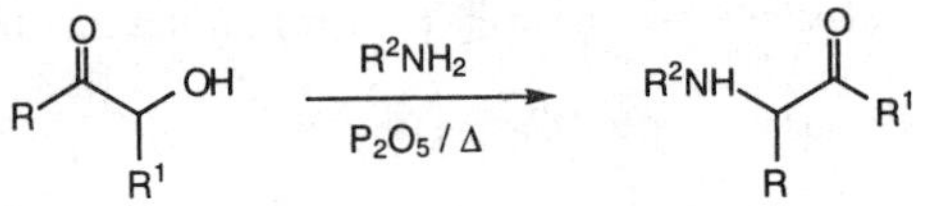

Merck ONR-92.
J. Org. Chem. **1956**, *21*, 49.

Volhard-Erdmann Synthesis (Cyclization)

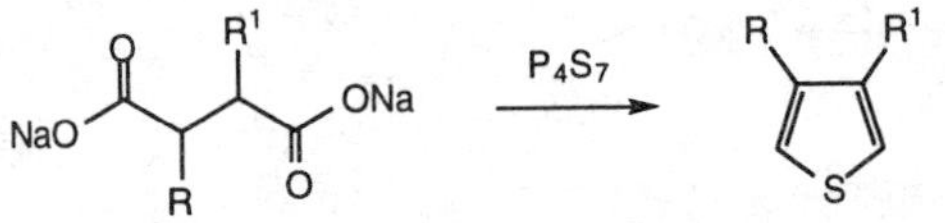

Org. Reactions **1951**, 6, 410.

von Braun Cyanogen Bromide Reaction

$$R_3N \xrightarrow{BrCN} RBr + R_2NCN$$

Synthesis **1989**, 1.
Org. Reactions **1953**, 7, 198.

Note: Also known as the von Braun Degradation or von Braun Reaction.

von Braun Degradation (see either von Braun Reaction or von Braun Cyanogen Bromide Reaction)

von Braun Reaction

$$RNHCOR^1 \xrightarrow{PX_5 \ or \ SOCl_2} RX + R^1CN$$

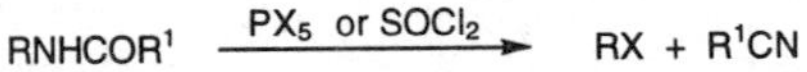

Tetrahedron **1980**, 36, 1279.
Tetrahedron **1973**, 29, 3309.
Org. Reactions **1953**, 7, 198.

Note: Also known as the von Braun Degradation and von Braun Amide Degradation.

von Braun-Rudolph Synthesis

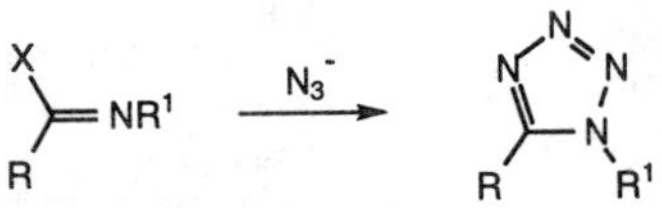

Comp. Org. Chem. **2**, 480.

Note: 1. Also known as the von Pechmann Pyrazole Synthesis.

von Pechmann Reaction

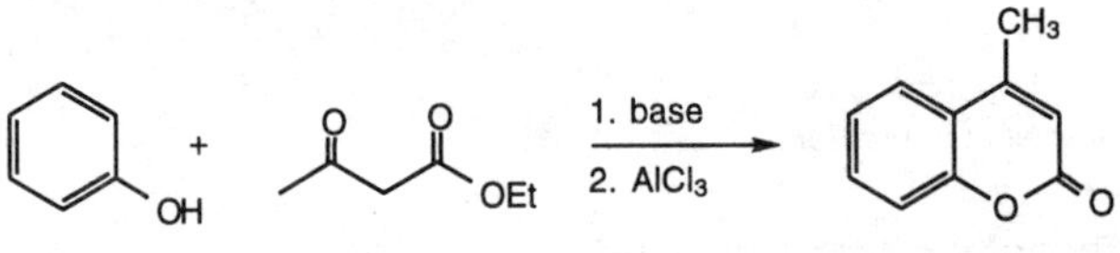

Synthesis **1981**, 524.
Synthesis **1980**, 715.
Org. Reactions **1953**, 7, 1.
Chem. Rev. **1945**, 36, 10.

Note: Also known as the Pechmann Reaction or Pechmann Cyclization.

von Pechmann Pyrazole Synthesis

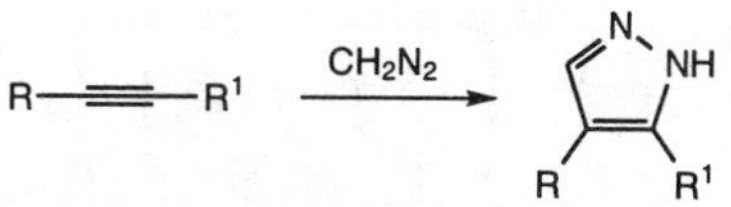

Merck ONR-66.

von Richter (Cinnoline) Synthesis

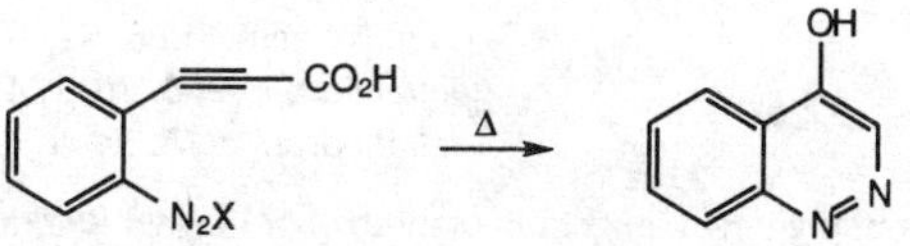

Chem. Rev. **1958**, *37*, 270.

von Richter Reaction (Rearrangement)

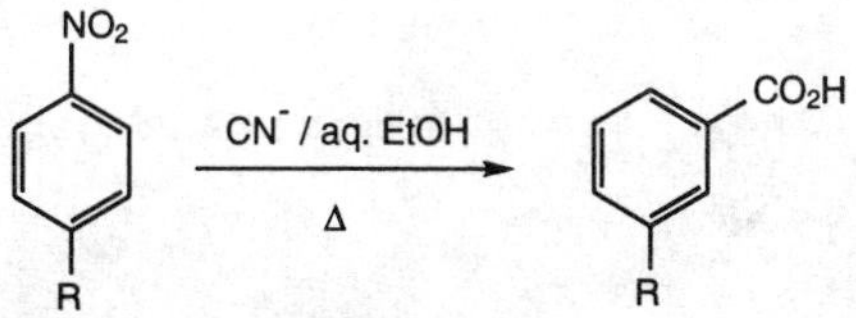

March 603.

von Rudloff Oxidation

$$RHC{=}CHR \xrightarrow{NaIO_4 \text{ / cat. } KMnO_4} RCHO + RCO_2H$$

Comp. Org. Chem. **5**, 620.

Wacker Oxidation (Reaction)

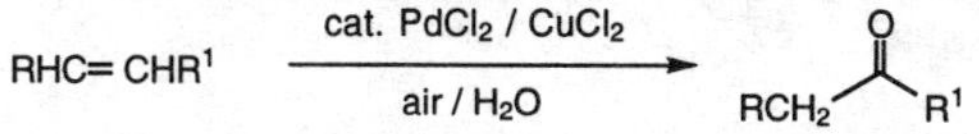

J. Org. Chem. **1990**, *55*, 2924.
Synthesis **1984**, 369.
Synthesis **1970**, 225.

Note: Also known as the Wacker Process.

Wadsworth-Emmons Reaction (see Wittig Reaction)

Wadsworth-Horner-Emmons Olefination Reaction (see Wittig Reaction)

Wagner-Jauregg Reaction

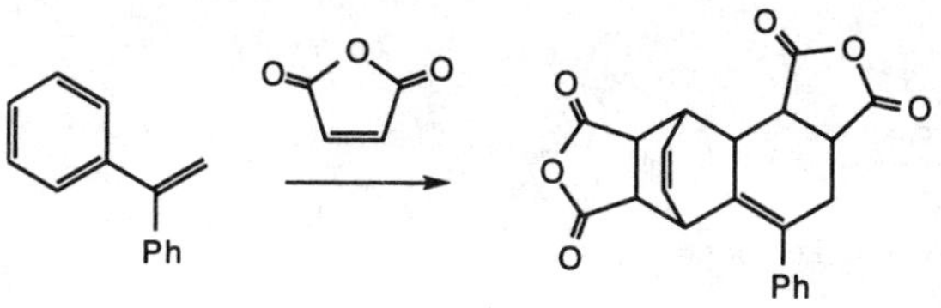

Synthesis **1980**, 769.

Wagner-Meerwein Rearrangement

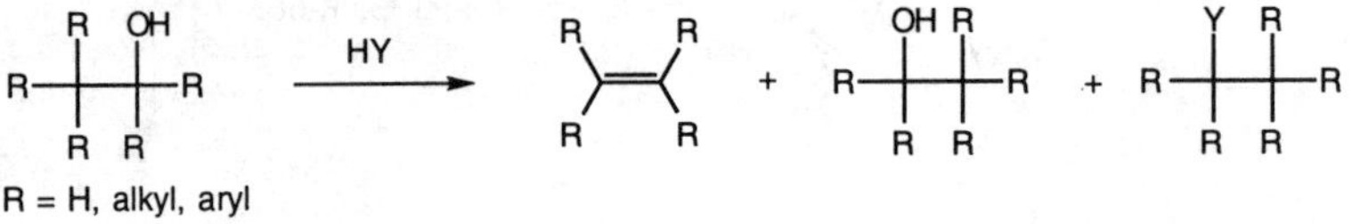

J. Am. Chem. Soc. **1990**, *112*, 8941.
Can. J. Chem. **1986**, *64*, 1081.
Tetrahedron **1980**, *36*, 2745.
Tetrahedron **1978**, *34*, 475.

Note: 1. Also known as the Meerwein Rearrangement (migration of methyl) or Wagner Rearrangement (change in ring).
2. Rearrangement occurs when there are no α-hydrogens adjacent to the alcohol.

Walden Inversion

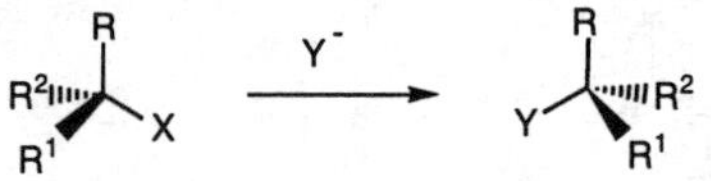

Chimia **1974**, *28*, 1.

Walk Rearrangement

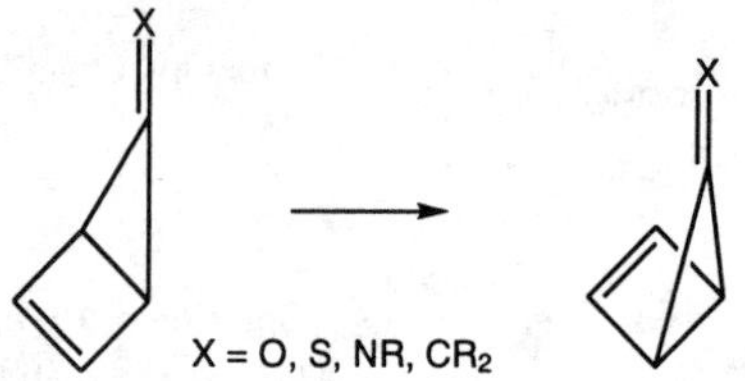

J. Am. Chem. Soc. **1989**, *111*, 4643.
Tetrahedron **1982**, *38*, 567.

Wallach Degradation

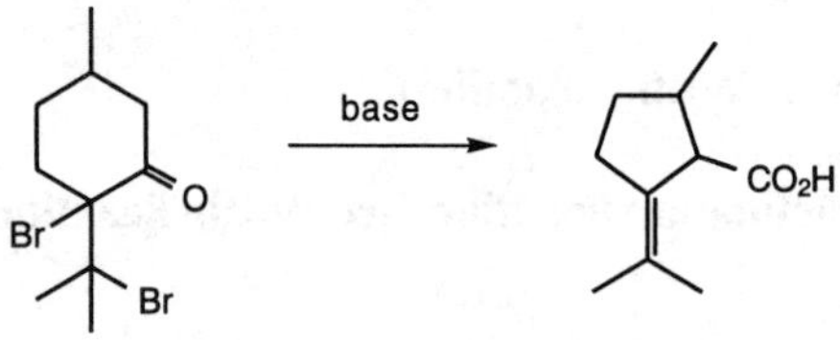

Org. Reactions **1960**, *11*, 280.

Wallach Reaction

$R^1R^2NH \xrightarrow{R_2CO\,/\,HCO_2H} R^1R^2NCHR_2$

Chem. Rev. **1969**, *69*, 679.

Note: Also known as the Leukart-Wallach Reaction.

Wallach Rearrangement (Transformation)

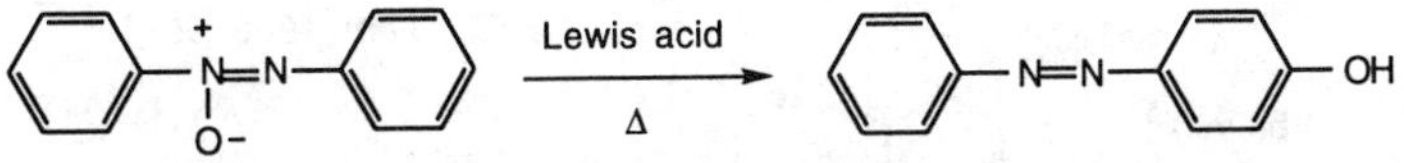

Can. J. Chem. **1986**, *64*, 1108.
Acc. Chem. Res. **1975**, *8*, 132.

Wallach Synthesis

Comp. Org. Chem. **4**, 397.

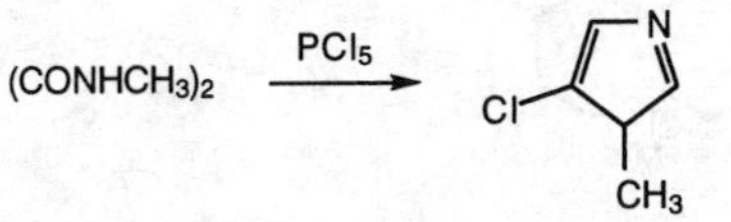

Watanabe-Conlon Transvinylation

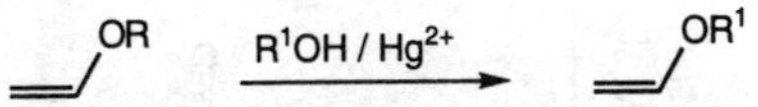

J. Am. Chem. Soc. **1957**, *79*, 2828.

Wanklyn Reaction (see Schorigin Reaction)

Water-Gas Shift Reaction

$$CO + H_2O \xrightarrow{\text{cat.}} H_2 + CO_2$$

J. Am. Chem. Soc. **1993**, *115*, 118.
Adv. Organomet. Chem. **1988**, *28*, 139.
Acc. Chem. Res. **1981**, *14*, 31.

Weerman Degradation

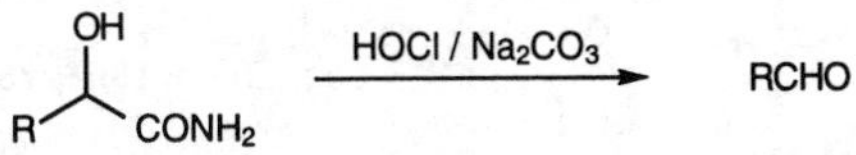

Org. Reactions **1946**, *3*, 275.

Weiss Reaction

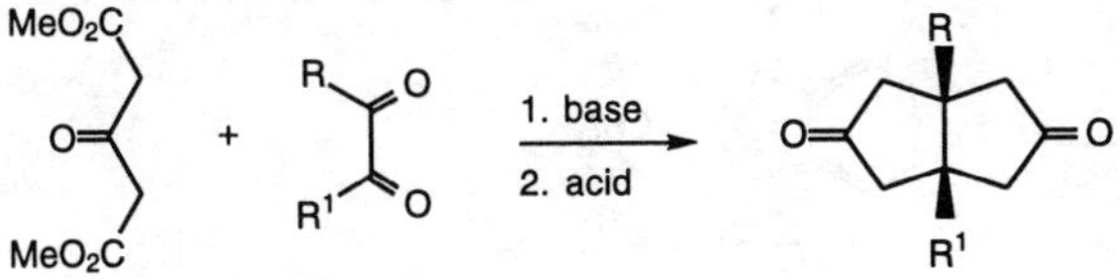

Tetrahedron **1991**, *47*, 3665.

Wessely-Moser Rearrangement

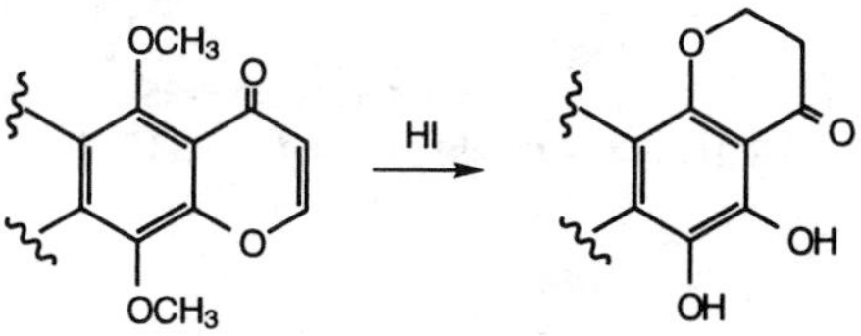

Comp. Org. Chem. **4**, 681; 755.

Wessely Oxidation

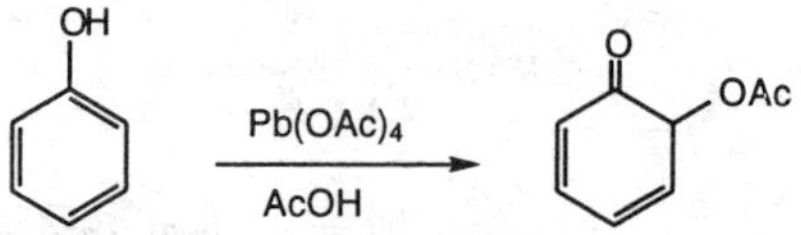

Can. J. Chem. **1988**, 66, 1.

Westphalen (-Lettré) Rearrangement

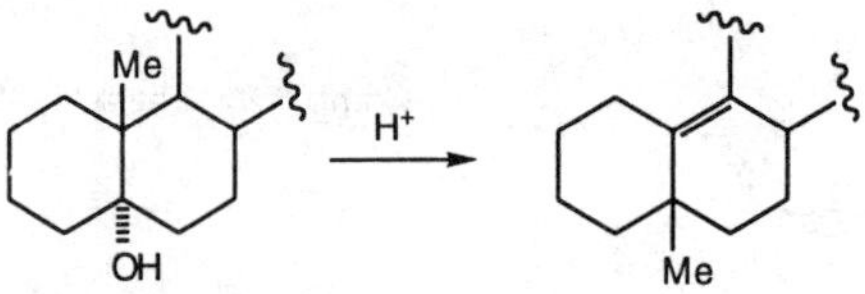

Can. J. Chem. **1987**, 65, 595.
Tetrahedron **1965**, 21, 1567.

Weygand Degradation

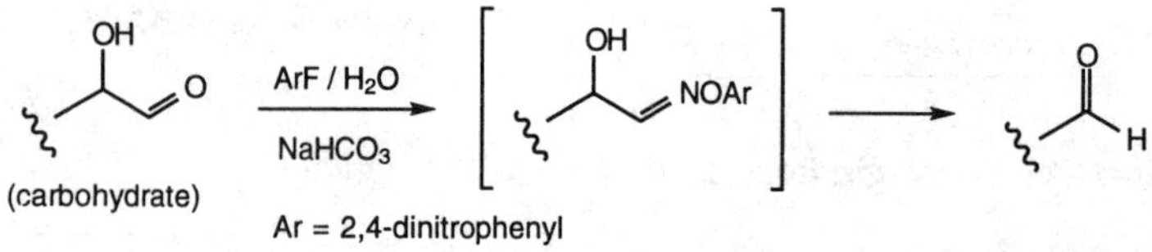

Fieser & Fieser 'Advanced Organic Chemistry', Reinhold, New York, **1961**, 945.

Wharton Reaction (Fragmentation)

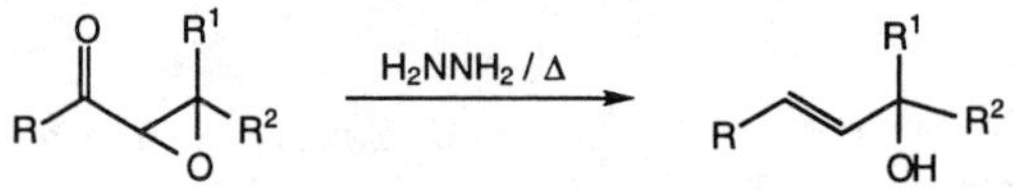

J. Chem. Soc., Chem. Commun. **1987**, 1720.
Tetrahedron **1978**, 34, 1261.
Helv. Chim. Acta **1970**, 53, 531.
J. Org. Chem. **1961**, 26, 3615.

Wheeler Reaction

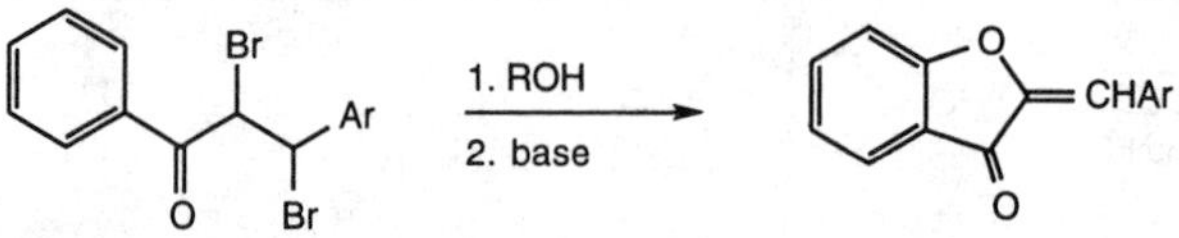

Tetrahedron **1979**, 35, 875.

Note: Also known as the Wheeler Aurone Synthesis.

Whiting Reaction

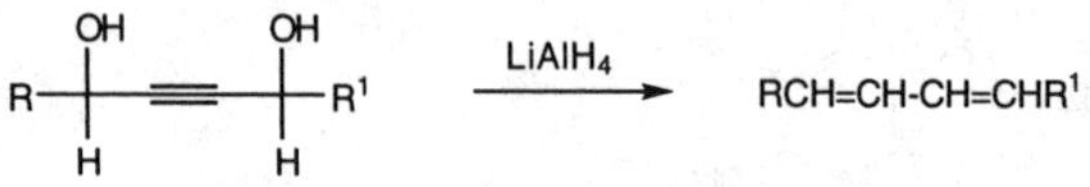

Merck ONR-95.
J. Chem. Soc. **1954**, 4006.

Wichterle Reaction

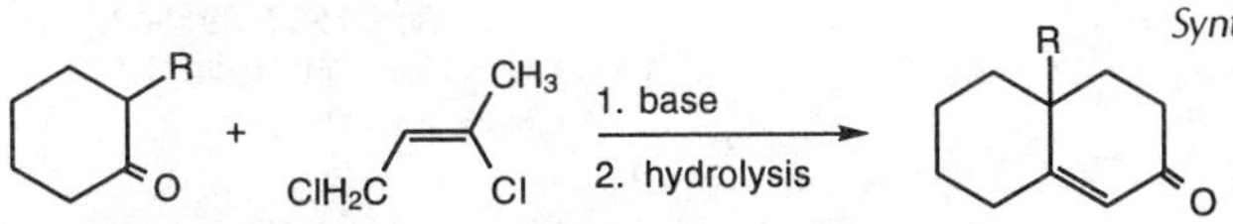

Synthesis **1976**, 777.

Widequist Reaction

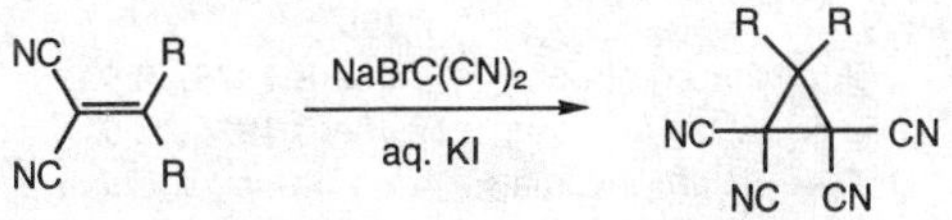

J. Org. Chem. **1966**, *31*, 2784.
J. Org. Chem. **1963**, *28*, 220.

Widman-Stoermer (Cinnoline) Synthesis

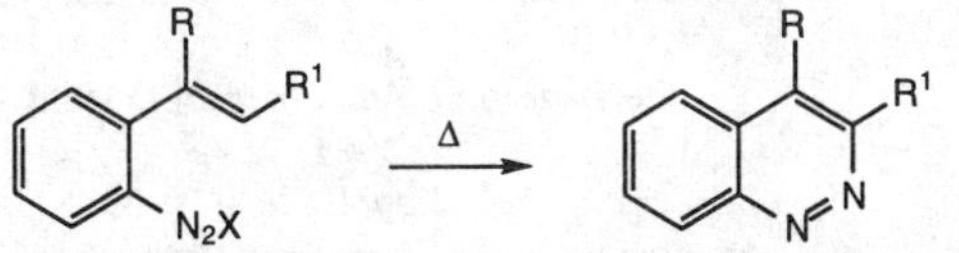

Tetrahedron **1971**, *27*, 2913.
Org. Reactions **1959**, *10*, 21.

Wieland Degradation (see Barbier-Wieland Degradation)

Willgerodt-Kindler Reaction

$$ArCOCHR(CH_2)_nCH_3 \xrightarrow{\text{Sulfur - RR}^1\text{NH}} ArCHR(CH_2)_nCSNH_2$$

Synthesis **1975**, 358.
Angew. Chem. Int. Ed. Engl. **1966**, *5*, 457.
Angew. Chem. Int. Ed. Engl. **1964**, *3*, 19.
Chem. Rev. **1961**, *61*, 52.
Org. Reactions **1946**, *3*, 83.

Willgerodt Reaction

$$ArCOCHR(CH_2)_nCH_3 \xrightarrow{\text{aq. }(NH_4)_2S_x} ArCHR(CH_2)_nCONH_2$$

Synthesis **1975**, 358.
Org. Reactions **1946**, *3*, 83.

Note: The ammonium salt of the acid rather than the amide may also be produced.

Williamson (Ether) Synthesis

$$RX + R^1OM \longrightarrow ROR^1$$

Comp. Org. Chem. **1**, 620.

Note: Also known as the Williamson Reaction.

Wittig Reaction

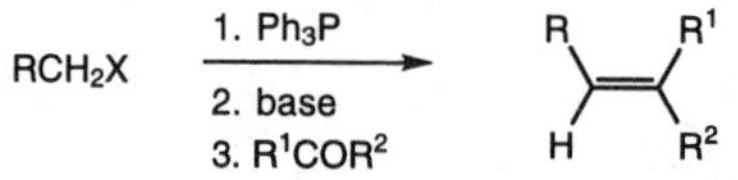

Chem. Rev. **1989**, *89*, 863.
Chem. Soc. Rev. **1988**, *17*, 1.
Chem. Rev. **1974**, *74*, 87.
Top. Stereochem. **1970**, *5*, 1.
Org. Reactions **1965**, *14*, 270.
Pure & Appl. Chem. **1964**, *9*, 271.
Pure & Appl. Chem. **1964**, *9*, 255.

Aromatic compounds: *Synthesis* **1975**, 765.
Bis-Wittig Reactions: *Synthesis* **1975**, 765.
Carbohydrates: *Adv. Carbohydr. Chem.* **1972**, *27*, 227.
Intramolecular Wittig Reaction: *Tetrahedron* **1980**, *36*, 1717.
Industrial chemistry: *Angew. Chem. Int. Ed. Engl.* **1977**, *16*, 423.
Mechanism: *J. Am. Chem. Soc.* **1990**, *112*, 3905.
Can. J. Chem. **1985**, *63*, 426.
Stereochemistry: *Chem. Rev.* **1989**, *89*, 863.
Wadsworth-Emmons Reaction: *Acc. Chem. Res.* **1983**, *16*, 411.
Wittig-Horner: *Chem. Ber.* **1958**, *91*, 61.
Wittig-Horner-Wadsworth-Emmons Reaction: *J. Am. Chem. Soc.* **1961**, *83*, 1733.

Note: Also see the Aza-Wittig, Horner, Horner-Wadsworth-Emmons Thio-Wittig and Wittig-Schlosser Reactions.

Wittig Rearrangement

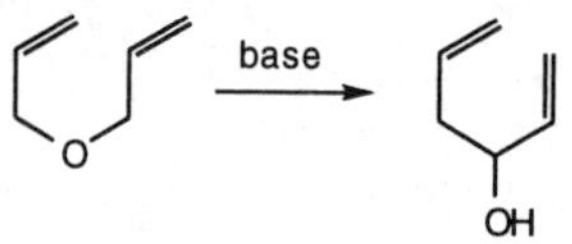

Synthesis **1991**, 594.
J. Org. Chem. **1991**, *56*, 1185.
Chem. Rev. **1986**, *86*, 885.

Wittig-Schlosser Reaction

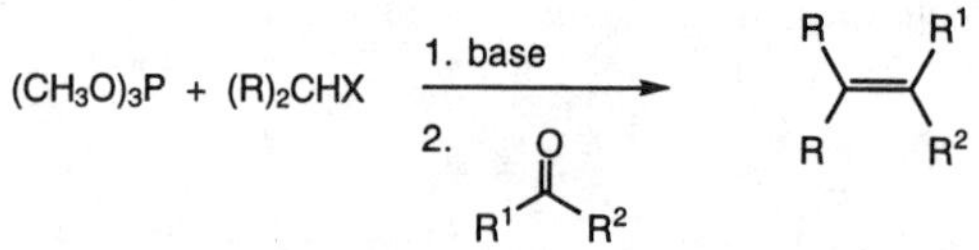

J. Am. Chem. Soc. **1980**, *102*, 352.

Note: Also see the Wittig Reaction.

Wohl Degradation

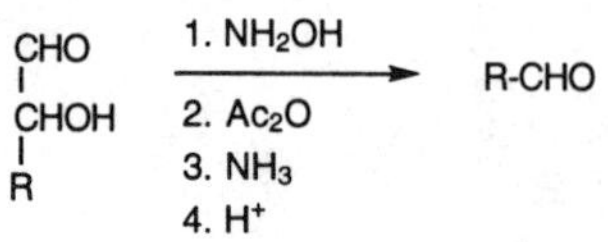

Comp. Org. Chem. **5**, 696.

Wohl-Ziegler Reaction (Bromination)

$$RCH_2CH{=}CHCH_2R^1 \xrightarrow{NBS} RCHBrCH{=}CHCH_2R^1 + RCH{=}CHCHBrCH_2R^1$$

R = alkyl, aryl

Chem. Rev. **1948**, *43*, 271.

Wöhler Synthesis

$$NH_4CNO \longrightarrow CO(NH_2)_2 \xrightarrow{C_2H_5OH} H_2NCO_2C_2H_5$$

Chem. Soc. Rev. **1978**, *7*, 1.
Chem. Rev. **1965**, *65*, 567.

Wolff-Kishner Reduction

$$RCOR^1 \xrightarrow[HO^-]{H_2NNH_2} RCH_2R^1$$

Chem. Rev. **1965**, *65*, 63.
J. Chem. Soc. **1963**, 1855.
Org. Reactions **1948**, *4*, 378.

Wolff Rearrangement

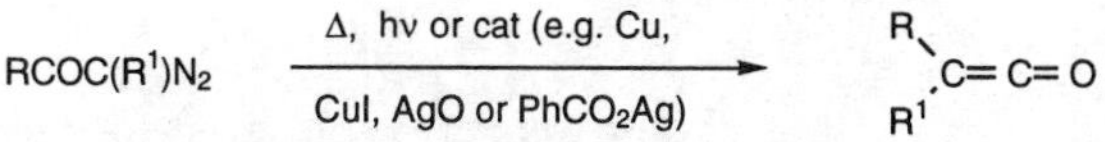

Chem. Rev. **1983**, *83*, 519.
Angew. Chem. Int. Ed. Engl. **1975**, *14*, 32.
Tetrahedron **1971**, *27*, 1317.
Russ. Chem. Rev. **1967**, *36*, 260.

Wolffenstein-Böters Reaction (Procedure)

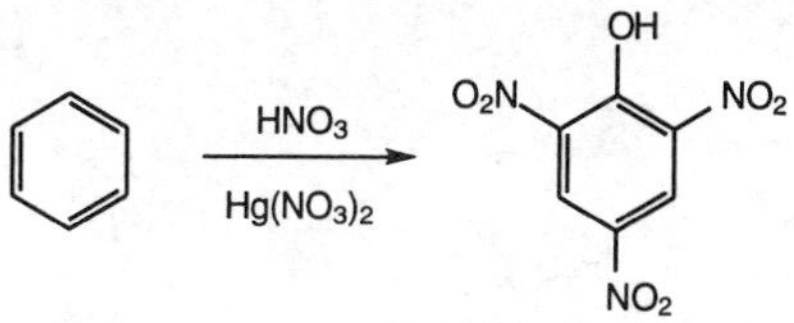

Merck ONR-97.
J. Am. Chem. Soc. **1947**, *69*, 773

Note: Also known as the Oxy-Nitration Process.

Woodward *cis*-Hydroxylation Reaction (Method)

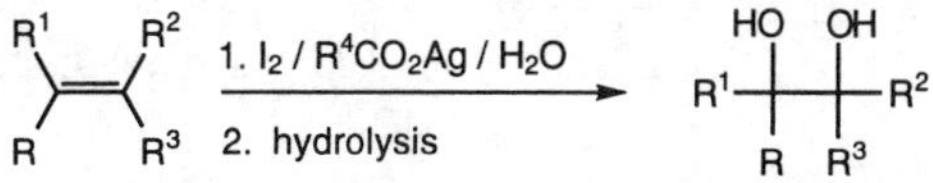

Org. Reactions **1957**, *9*, 350.

Note: Also known as the Woodward Modification of the Prévost Reaction.

Wurtz-Fittig Reaction (Coupling)

$$ArX + RX \xrightarrow{Na} Ar\text{-}R$$

March 399.

Silanes: *Tetrahedron* **1983**, *39*, 6425.

Note: In the Wutz-Fittig Reaction one reactant is alkyl and the other aryl (also see the Wurtz Reaction and Fittig Reaction).

Wurtz Reaction (Coupling)

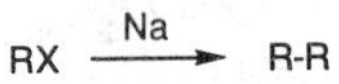

Synthesis **1973**, 86.
Org. Reactions **1970**, *18*, 60.

Note: In the Wutz Reaction both reactants are alkyl (also see the Wurtz-Fittig Reaction and Fittig Reaction).

Ynamine-Claisen Rearrangement

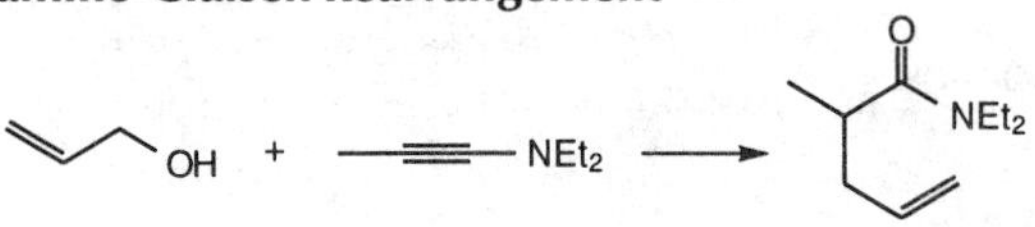

J. Org. Chem. **1979**, *44*, 882.
Tetrahedron **1966**, *22*, 6425.

Note: Also known as the Ficini-Barara Rearrangement.

Zaitsev Reaction

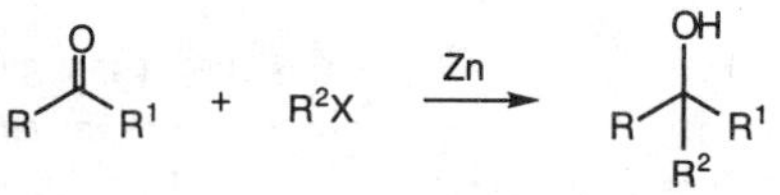

Comp. Org. Chem. **3**, 991.

Ziegler Alkylation

March 599.

Ziegler Cyclization (Method)

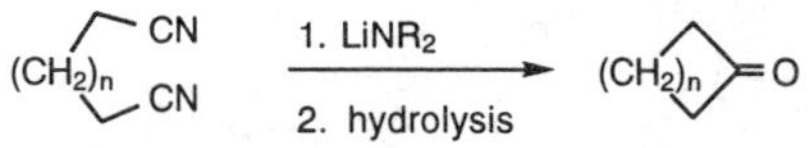

Merck ONR-99.

Ziegler-Natta Polymerization (Process)

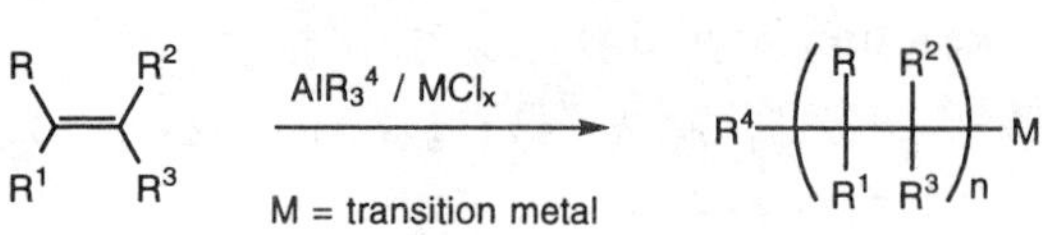

J. Am. Chem. Soc. **1990**, *112*, 4953.
Angew. Chem. **1960**, *72*, 829.
Angew. Chem. **1959**, *71*, 623.

Ziegler-Schwartz Synthesis

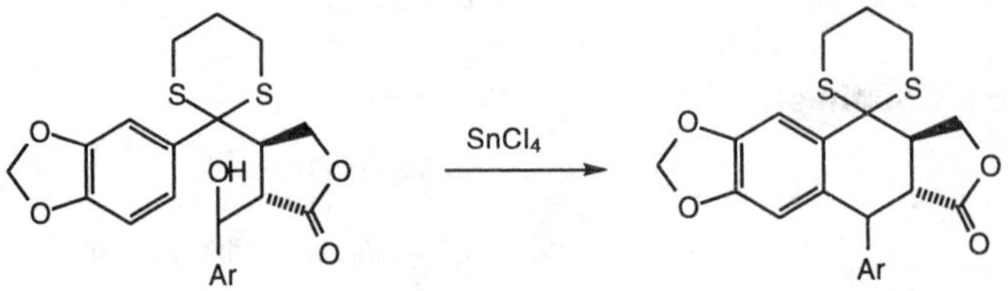

J. Org. Chem. **1978**, *43*, 985.

Ziegler-Suhl Synthesis

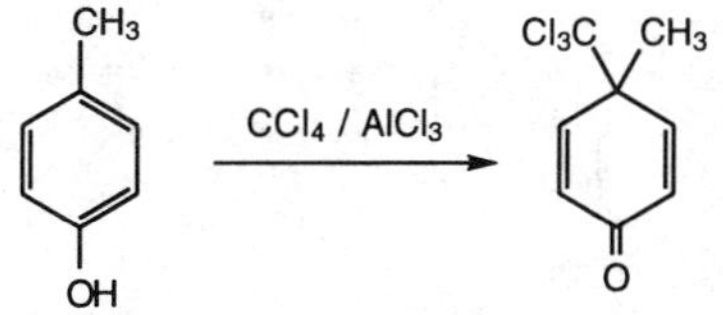

J. Org. Chem. **1954**, *19*, 978.

Ziegler Synthesis

$$AlEt_3 \xrightarrow[2.\ O_2\,/\,H_2O]{1.\ CH_2{=}CH_2} ROH$$

Angew. Chem. Int. Ed. Engl. **1966**, *5*, 650.

Ziegler-Thorpe Synthesis (see Thorpe-Ziegler Reaction)

Zimmermann Reaction

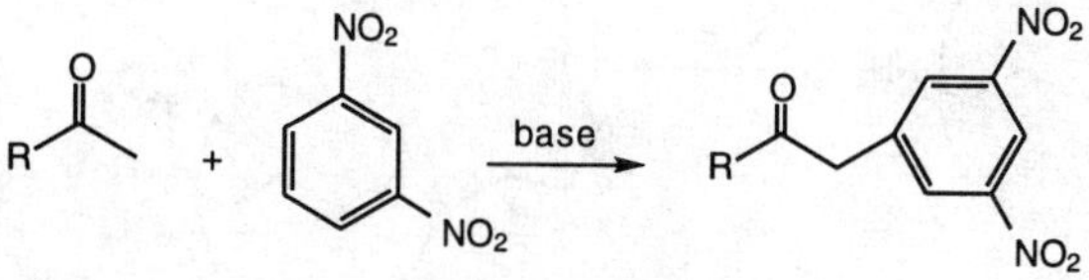

Tetrahedron **1962**, *18*, 1131.

Zimmermann Rearrangement

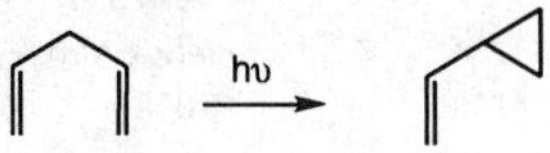

J. Am. Chem. Soc. **1989**, *111*, 1007.

Note: Also known as the Di-π-Methane Rearrangement.

Zincke Disulfide Cleavage

$$ArSR \xrightarrow{X_2} ArSX$$

$R = SR^1, H \quad X = Br, Cl$

Merck ONR-100.
Chem. Rev. **1946**, *39*, 283.

Zincke Nitration (Method)

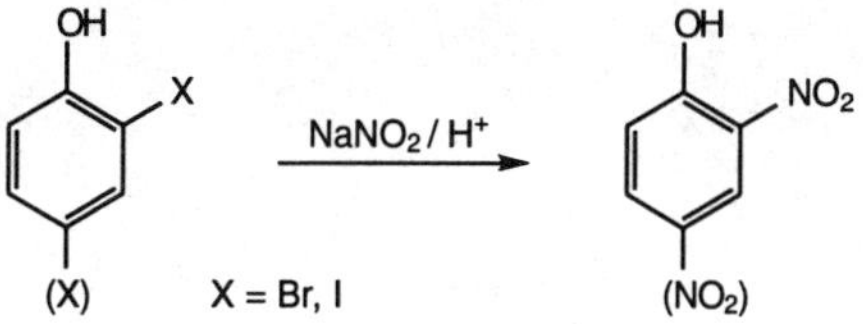

J. Am. Chem. Soc. **1944**, *66*, 1872.
J. Am. Chem. Soc. **1933**, *55*, 2125.

Zincke-Suhl Reaction

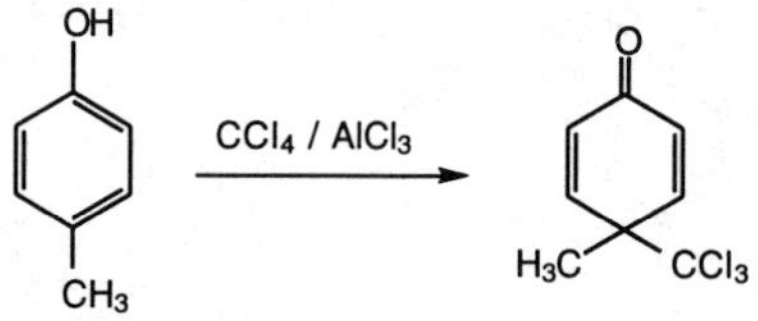

Merck ONR-100.
J. Am. Chem. Soc. **1959**, *81*, 6450.

Zinin Reduction

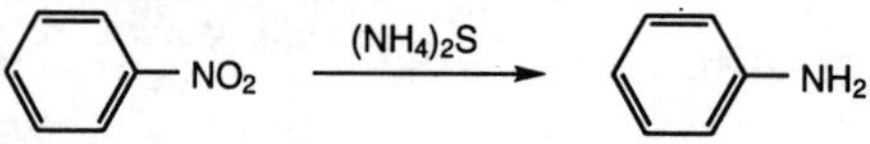

Org. Reactions **1973**, *20*, 455.

Zip Reaction

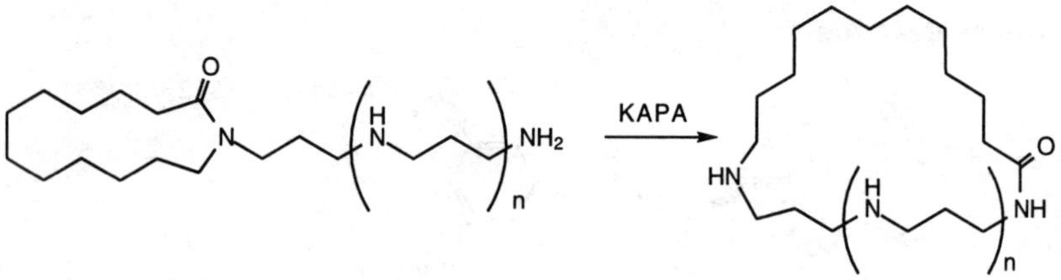

Angew. Chem. Int. Ed. Engl. **1978**, *17*, 200.
Helv. Chim. Acta **1979**, *62*, 811.

Zipper Reaction

$$R(CH_2)_mC\equiv C(CH_2)_nCH_3 \xrightarrow[H_2N(CH_2)_3NH_2]{LiNH(CH_2)_3NH_2\ /\ KOtBu} RCH_2(CH_2)_{m+n}C\equiv CH$$

R = H, OH

Org. Synth. **1988**, 66, 129.
J. Am. Chem. Soc. **1975**, *97*, 891.

Chapter 5

Named Reagents

This chapter contains an alphabetical list of named reagents. Named reactions have not been included, and should be accessed by the list in Chapter 4. For the purposes of this chapter, a named reagent is derived from a person, a reaction, or a recognized trade name (e.g., Alpine Borane®). Note registered or trade marked acronyms appear in Chapter 1. Chiral shift reagents appear under their trademark, "Resolve."

References have not been provided as we believe that, in the vast majority of cases, the reader will only wish to discern the chemical name. If further information is desired, it may often be accessed from reference texts such as March or Fieser and Fieser.

Note, that in a few cases, one named reagent can refer to more than one chemical. The named reagents have been listed in alphabetical order and it is for the reader to decide which one is correct in that specific context. In some cases, the named reagent may have been used incorrectly in the literature. The interpretation of the named reagent is noted, but care should be taken to ensure that interpretation of that reagent is correct.

Ace Chloride	α-chloroethylchloroformate
Adamite	zinc arsenate
Adams' Catalyst	pre-hydrogenated platinum dioxide
Adams' Reagent	zinc cyanide
Adamsite	phenarsazine chloride
Adkins Catalyst	copper-chromium oxide
Adogen® 464	methyltrialkyl(C_8-C_{10})ammonium chloride
Aldritiol™-2	2,2′-dipyridyl disulfide, 2,2′-dithiodipyridine
Aldritiol™-4	4,4′-dipyridyl disulfide, 4,4′-dithiodipyridine
Aliquat® 336	tricaprylylmethylammonium chloride
Aliquat® 336 Nitrate	tricaprylylmethylammonium nitrate
Alpine-Boramine™	*N*,*N*′-bis(monoisopinocampheylborane)-*N*,*N*,*N*′,*N*′-tetramethylethylenediamine
Alpine-Borane®	*B*-isopinocampheyl-9-borabicyclo[3.3.1] nonane
Alpine-Hydride®	lithium *B*-isopinocampheyl-9-borabicyclo [3.3.1]nonyl hydride
Amadori Reagents	aminoketoses
Amberlite®	ion exchange resin
Amberlyst	ion-exchange macroreticular resin
Ammonia Color Reagent	dipotassium tetraiodomercurate(II) in dilute sodium hydroxide (also known as Nessler's Reagent)
Anderson-Shapiro Reagent	(4*R*,5*R*)-2-chloro-4,5-dimethyl-1,3,2-dioxaphospholane 2-oxide
Armstrong Acid	1,5-naphthalenedisulfonic acid
Arnstein Tripeptide	δ-D-homoglutamyl-L-cysteyl-D-valine
Attenburrow Oxide	'active' manganese dioxide
Badische Acid	7-amino-1-naphthalenesulfonic acid
Ballester-Molinet-Castañer Reagent	sulfur monochloride-aluminum chloride-sulfuryl chloride; BMC or Reagent BMC
Bang's Reagent	aqueous solution of copper sulfate, potassium carbonate, potassium hydrogen carbonate and potassium chloride
Barfoed's Reagent	aqueous solution of cupric acetate and acetic acid
Barnett Catalyst	chlorine-sulfur dioxide
Barton Esters	2-mercaptopyridine *N*-oxide *O*-esters
Bates Reagent	μ-oxo-bis[tris(dimethylamino)-phosphonium] bistetrafluoroborate
Baudisch's Reagent	aqueous solution of ammonium nitrosophenylhydroxylamine; Cuperron
Baum's Acid	1-naphthol-2-sulfonic acid
Baumann-Fromm Disulfides	1,2-dithiol-3-one

Baumann-Fromm Trithiones	1,2-dithiole-3-thione
Beckmann Mixture	potassium dichromate in dilute sulfuric acid
Belleau's Reagent	2-ethoxyethoxycarbonyl-1,2-dihydroquinoline; EEDQ
Benedict's Solution (qualitative)	aqueous solution of copper sulfate, sodium citrate and sodium carbonate
Benedict's Solution (quantitative)	aqueous solution of potassium thiocyanate, potassium ferrocyanide, sodium (potassium) citrate, copper sulfate and sodium carbonate
Benzidine Solution	*p*-diaminodiphenyl (benzidine) in glacial acetic acid
Bergmann Reagent	benzyloxyformyl chloride; 2,4-dinitro-5-fluoroaniline
Bertrand's Reagents	aqueous copper sulfate; aqueous Rochelle salt and sodium hydroxide; aqueous ferric sulfate and sulfuric acid; aqueous potassium permanganate
Besthorn's Hydrazone	3-methyl-2-benzothiazolinone hydrazone hydrochloride
Bial's Reagent	orcinol ferric chloride spray reagent
Bisnor-S	dimer of norbornadiene
Biuret Reagent	allophanamide
BMC Reagent	see Ballester-Molinet-Castañer Reagent
Bolton-Hunter Reagent	3-(4-hydroxyphenyl)propionic acid *N*-hydroxysuccinimide ester (also known as the Rudinger Reagent)
Borax	sodium tetraborate
Bower-Compound	6,7-epoxy-3,7-dimethyl-1-[3,4-(methylenedioxy)phenoxy]-2-nonene
Brady's Reagent	acidic solution of 2,4-dinitrophenylhydrazine in aqueous ethanol
Brassard's Diene	1,3-dimethoxy-1-trimethylsiloxybuta-1,3-diene
Bredereck's Reagent	*tert*-butoxy-bis(dimethylamino)methane
Brodie Solution	sodium chloride-sodium choleate
Broenner's Acid	see Brönner's Acid
Bromosolve	α-methyl-*p*-bromobenzylamine hydrochloride
Brönner's Acid	6-amino-2-naphthalenesulfonic acid
Brown and Brown Catalyst	platinum catalyst formed by the treatment of chloroplatinic acid with sodium borohydride
Brown's Reagent	*S*-(-)-2-[(phenylamino)carbonyloxy]propionic acid
Bruke's Reagent	aqueous solution of potassium iodide saturated with mercuric iodide
Buchner Acids	cycloheptatriene carboxylic acids

Bunte Salts	*S*-alkylthiosulfate salts
Burgess Reagent	(methoxycarbonylsulfamoyl)triethylammonium hydroxide
Burrow's Solution	aluminum acetate solution
Burton's Reagent	copper-dibromodifluoromethane-*N*,*N*-dimethylacetamide
Buse's Reagent	2-methyl-2-trimethylsiloxypente-3-one
"Cage" Synthon	1,4,4a,8a-tetrahydro-*endo*-1,4-methanonaphthalene-5,8-dione
Carbitol®	2-(2-ethoxyethoxy)ethanol
Carnot's Reagent	sodium bismuth thiosulfate in alcohol
Caro's Acid	potassium monopersulfate triple salt (peroxymonosulfuric acid)
Carr-Price Reagent	antimony trichloride
Carrel-Dakin Solution	sodium hypochlorite solution
Cassel's Green	barium manganate
Cassella's Acid	2-naphthol-7-sulfonic acid
Cassella's Acid F	7-amino-2-naphthalenesulfonic acid
CBS Catalyst	(*R*)-oxazaborolidine
CBS Reagent	see Corey's "CBS" Reagent
Celite	silicon dioxide
Channing's Solution	potassium triiodomercurate(II) solution
Chevreul's Salt	cuprous sulfite
Chicago Acid	1-amino-8-naphthol-2,4-disulfonic acid
Chirald®	(2*S*,3*R*)-(+)-4-dimethylamino-1,2-diphenyl-3-methyl-2-butanol
Chiraphos	2,3-bis(diphenylphosphino)butane
Chloride Reagent	aqueous solution of silver nitrate, potassium nitrate and ammonia
Christmas Factor	Factor IX; antihemophilic factor B
Claisen's Alkali	35% potassium hydroxide in aqueous methanol (20%)
Cleland Reagent Chiral	(-)-1,4-dithio-L-erythritol
Cleland Reagent Racemic	1,4-dithio-D,L-erythritol; DTE
Clerici's Solution	thallous furmarate and malonate
Cleve's Acid (1,6)	5-amino-2-naphthalenesulfonic acid
Cleve's Acid (1,7)	8-amino-2-naphthalenesulfonic acid; 1-naphthylamine-7-sulfonic acid
Colcemid®	(*S*)-(-)-demecolcine
Collin's Reagent	chromium trioxide bispyridine complex
Collman Reagent	di-sodium tetracarbonylferrate dioxane complex
Cookeite	lithium aluminum silicate
Corey Lactone	hexahydro-4-hydroxymethyl-5-(4-phenylbenzoyloxy)cyclopenta[*b*]furan-2-one

Corey's "CBS" Reagent	borane-chiral *B*-methyloxazaborolidine
Corey's Reagent	pyridinium chlorochromate
Corey-Kim Reagent	*N*-chlorosuccinimide-dimethyl sulfoxide
Corey-Seebach Reagent	ethyl benzoylformate
Cori Ester	glucose-1-phosphate
Cornforth Reagent	chromium trioxide-pyridine-water
Cream of Tartar	potassium bitartrate
Cuperron	aqueous solution of ammonium nitrosophenylhydroxylamine
Dahl's Acid	6-amino-1-naphthalenesulfonic acid
Dahl's Acid II	1-naphthylamine-4,6-disulfonic acid
Dahl's Acid III	1-naphthylamine-4,7-disulfonic acid
Dakin's Solution (modified)	sodium hypochlorite solution
Danishefsky's Diene	*trans*-1-methoxy-3-trimethylsiloxy-1,3-butadiene
Darco®	activated carbon
Darvon Alcohol	4-dimethylamino-3-methyl-1,2-diphenyl-2-butanol
Davis Reagent	*trans*-2-(phenylsulfonyl)-3-phenyloxaziridine
Davy Reagent Methyl	2,4-bis(methylthio)-1,3-dithia-2,4-diphosphetane-2,4-disulfide
Delta Acid	7-amino-2-naphthalenesulfonic acid
Dess-Martin Reagent	1,1,1-triacetoxy-1,1-dihydro-1,2-benziodoxol-3(1*H*)-one
Devarda's Alloy	see Devarda's Metal
Devarda's Metal	alloy of copper, aluminum, and zinc
Dewar Benzenes	bicyclo[2.2.0]hexadienes
Dewar Pyridine	2-azabicyclo[2.2.0]hexadienes
Dewar Pyridone	*N*-substituted 2-azabicyclo[2.2.0]hex-2-one-4-enes
Diamide	1,1′-azobis(*N*,*N*-dimethylformamide)
Diatomaceous Earth	silicon dioxide
DIAZALD®	*N*-methyl-*N*-nitroso-*p*-toluenesulfonamide
Diels Acid	3,4-seco-Δ^5-cholesten-3,4-dioic acid
Diels-Alder Catalyst(s)	Lewis acid catalysts
Diels Hydrocarbon	3′-methyl-1,2-cyclopentenophenanthrene
Diphos	1,2-bis(diphenylphosphino)ethane
Dippel's oil	bone oil
Disulfiram	tetraethylthiuram disulfide
Dithizone	diphenylthiocarbazone
D,L-Djenkolic Acid	D,L-*S*,*S*′-methylenebiscysteine
Dobbin's Reagent	mercuric chloride-potassium iodide solution
Dobell's Solution	sodium borate solution
Doctor's Solution	aqueous solution of lead plumbite

Donovan's Solution	solution of arsenic and mercuric iodides
Duthaler-Hafner Reagent	[(4*R*,5*R*) or (4*S*,5*S*)-2,2-dimethyl-1,3-dioxolan-4,5-bis(diphenylmethoxyl)]cyclopentadienyl-chlorotitanium
Eapine	*B-iso*-2-ethylapopinocampheyl-9-borabicyclo [3.3.1]nonyl
Eaton's Reagent (Eaton's Acid)	phosphorous pentoxide in methanesulfonic acid
Ebert-Merz α-Acid	2,7-naphthalenedisulfonic acid
Ebert-Merz β-Acid	2,6-naphthalenedisulfonic acid
Edetic Acid	ethylenediaminetetraacetic acid
Ehrlich's Reagent	4-dimethylaminobenzaldehyde
Ehrlich's Solution	4-dimethylaminobenzaldehyde in concentrated hydrochloric acid
Ellman's Reagent	5,5′-dithio-bis(2-nitrobenzoic acid)
Epsilon Acid	6-naphthylamine-8-sulfonic acid
Esbach's Reagent	aqueous solution of picric acid and citric acid
Eschenmoser's Salt	*N*,*N*-dimethylmethyleneammonium iodide
Eschka's Mixture	magnesium oxide-sodium carbonate
Etard's Reagent	chromium oxychloride
Etard's Salt	cuprous sulfite (hemihydrate)
Ethocel	ethyl cellulose
Ethylal	formaldehyde diethyl acetal
Evan's Reagent	(*S*)-*N*-propionyl-4-*tert*-butyl-2-oxazolidinone
Evans' Oxazolidinone	(4*S*)-(-)-4-isopropyl-2-oxazolidinone
Ewer-Pick Acid	1,6-naphthalenedisulfonic acid
FAR	fluoroalkylamine reagent
Fehling's Reagent I	cupric sulfate solution
Fehling's Reagent II	alkaline potassium sodium tartrate solution
Fehling's Solution	aqueous solution of copper sulfate, sodium tartrate and sodium hydroxide
Fenton's Reagent	ferrous or ferric salts-hydrogen peroxide
Ferene®	3-(2-pyridyl)-5,6-bis(5-sulfo-2-furyl)-1,2,4-triazine, disodium salt
Ferene® Triazine	5,6-di-2-furyl-3-(2-pyridyl)-1-2,4-triazine
Ferron	7-iodo-8-hydroxyquinoline-5-sulfonic acid
FerroZinc® Iron Reagent	3-(2-pyridyl)-5,6-diphenyl-1,2,4-triazine-*p*,*p*′-disulfonic acid, mono sodium salt
Fétizon's Reagent	silver(I) carbonate-celite®
Feulgen's Stain Solution	fuchsin sulfurous acid; leucosulfonic acid
Fiberfrax®	aluminum silicate
Fieser's Reagent	chromium trioxide-acetic acid
Fieser's Solution	sodium anthraquinone-β-sulfonate-sodium hydrosulfite, in an aqueous solution of potassium hydroxide

Fischer Chromium Carbene Complex	pentacarbonylchromium methylene complex
Fischer Speier Catalyst	hydrochloric acid
Fischer's Salt	cobalt potassium nitrate
Fischer's Yellow	cobaltic potassium nitrite
Fischer-Reagent	iodine, sulfur dioxide, and pyridine in methanol
Fisher Base	1,3,3-trimethyl-2-methyleneindoline
Fiske & Subbarow Reducer	1-amino-2,2-naphthol-4-sulfonic acid-sodium bisulfite-sodium sulfite
Fleming's Silylcuprate Reagent	bis(dimethylphenylsilyl)cyanocuprate, dilithium salt
Florisil	activated magnesium silicate
Folin's Amino Acid Reagent	1,2-naphthaquinone-4-sulfonic acid
Folin-Ciocalteu's Phenol Reagent	2N phenol solution
Folin Solution	aqueous solution of ammonium sulfate, uranium acetate and acetic acid
Folin-Wu Reagent	phosphate-molybdate solution
Fowler's Solution	potassium arsenite solution
Fraude's Reagent	perchloric acid
Fraxin	8-(β-D-glucopyranosyloxy)-7-hydroxy-6-methoxycoumarin
Fremy's Salt	potassium nitrosodisulfonate
Fremy's Salt	potassium bifluoride
Freund's Compound	methyl isothiocyanate sulfide
Fritz-Schenk Reagent	acetic anhydride and perchloric acid in ethyl acetate
Froehde's Reagent	sodium molybdate dissolved in concentrated sulfuric acid
Fuchsin (Aldehyde) Reagent	a solution of fuchsin in water that is saturated with sulfur dioxide
Gabriel Reagent	di-*tert*-butyl iminodicarboxylate
Gamma Acid	2-amino-8-naphthol-6-sulfonic acid
Geissman-Weiss Lactone	2-oxa-6-azabicyclo[3.3.0]octan-3-one
Gibb's Reagent	2,6-dichloroquinone-4-chloroimide
Giblaar Poison	sodium fluoroacetate
Gilchrist's Reagent	methyl (3-bromo-2-oximido)pyruvate
Gilman's Catalyst	alloy of magnesium and copper treated with iodine
Girard's Reagents	see Girard's Reagent D, P or T
Girard's Reagent D	2-dimethylaminoethylhydrazone hydrochloride
Girard's Reagent P	1-(carboxymethyl)pyridinium chloride hydrazide
Girard's Reagent T	(carboxymethyl)trimethylammonium chloride hydrazide

Glauber's Salt	sodium sulfate decahydrate
Gold's Reagent	dimethylamino(methyleneaminomethylene)-dimethylammonium chloride
Gomberg Reagent	magnesium-magnesium iodide
Graham's Salt	sodium hexametaphosphate
Gram's Iodine Solution	iodine in alcohol
Griess Reagent	aqueous solution of sulfanilic acid, α-naphthylamine, and acetic acid
Grignard Reagent	alkyl or aryl magnesium halide
Günzberg Reagent	alcohol solution of vanillin and phloroglucinol
Hageman Factor	blood coagulation factor
Hagemann's Ester	4-carbethoxy-3-methyl-2-cyclohexen-1-one
Hager's Reagent	saturated solution of picric acid in water
Hamilton's Reagent	iron(II)-hydrogen peroxide-catechol
Hanker-Yates Reagent	1,4-phenylenediamine hydrochloride-pyrocatechol
Hantzsch Amides	1,4-dihydro-2,6-dimethyl-3,5-bis[(*R*-amino)carbonyl]pyridine
Hantzsch Esters	1,4-dihydro-2,6-dimethyl-3,5-bis[(*R*-carboxylates]pyridine
Hanus Solution	iodine and bromine in glacial acetic acid
Harden-Young Ester	fructose 1,6-diphosphate
Hartshorn Salt	ammonium carbonate
Hatchett's Brown	cupric ferrocyanide
Hauser's Salt	zirconium sulfate
Heimgartner Reagent	2,4-bis(*p*-tolylthio)-1,3-dithia-2,4-diphosphetane-2,4-disulfide
Helicin	salicylaldehyde-β-D-glucoside
Helmgartner's Reagent	2,4-bis(4-methylphenylthio)-1,3-dithia-2,4-diphosphetane-2,4-disulfide
Henbest's (Iridium) Catalyst	iridium tetrachloride-trimethyl phosphite
Herzberg's Paper	Congo red paper
Heyn's Catalyst	10% platinum on charcoal
Hill Reagent	2,6-dichlorophenolindophenol
Hoagland Solution	iron ethylenediaminetetraacetic acid solution
Hofmann's Violet	triethylrosaniline hydrochloride
Horner-Wadsworth-Emmons Reagents	alkyl phosphonates
Huber's Reagent	aqueous solution of ammonium molybdate and potassium ferrocyanide
Hubl's Reagent	mercuric chloride and iodine in alcohol
Hudson-Weber Reagents	*N*-iodoacetyl-*N'*-(5-sulfo-1-naphthyl)ethylenediamine and *N*-iodoacetyl-*N'*-(8-sulfo-1-naphthyl)ethylenediamine

Hunig's Base	diisopropylethylamine; DIPEA
Itsuno's Reagent	borane-(*S*)-2-amino-3-methyl-1,1-diphenyl-butan-1-ol
Ivanov (Ivanoff) Reagent	salt formed by reaction of arylacetic acids (or salts) with isopropylmagnesium halides
Jacquemart's Reagent	aqueous solution of mercuric nitrate and nitric acid
Jones Reagent	chromium trioxide-sulfuric acid
Julion's Carbon Chloride®	hexachlorobenzene
K-Selectride®	potassium tri-*sec*-butylborohydride
Kalle's Acid	1-naphthylamine-2,7-disulfonic acid
Kammerer's Porphyrin	protoporphyrin-IX
Karl-Fischer-Reagent	iodine, sulfur dioxide and pyridine in methanol
Kastle-Meyer Reagent	phenolphthalein
Kemp's TriAcid	1,3,5-trimethyl-1,3,5-cyclohexanetricarboxylic acid
Kendall's Compound A	11-dehydrocorticosterone
Kendall's Compound B	corticosterone
Kendall's Compound C	allopregnane-3α,11β,17α,21-tetrol-20-one
Kendall's Compound D	5α-pregnane-3β,17α,20β,21-pentol
Kendall's Compound E	cortisone
Kendall's Compound F	hydrocortisone
Kendall's Compound G	5α-pregnane-3β,17α,21-triol-11,20-dione
Kendall's Compound H	5α-pregnane-3β,21-diol-11,20-dione
Kendall's Deoxy Compound B	deoxycorticosterone
Kharasch Reagent	2,4-dinitrobenzenesulfenyl chloride
Kieselguhr	diatomaceous earth; silicon dioxide
Kiliani Mixture	sodium dichromate in dilute sulfuric acid
Kjeldahl Catalyst Tablets (copper)	tablet of copper and anhydrous sodium sulfate
Kjeldahl Catalyst Tablets (selenium)	tablet of selenium and anhydrous sodium sulfate
Kjeldahl Catalyst Tablets (sodium sulfate)	tablet of anhydrous sodium sulfate
Kjeldahl Catalyst Tablets (titanium)	tablet of hydrated cupric sulfate-titanium dioxide and anhydrous sodium sulfate
Kjeldahl Mixture	sodium hydroxide-sodium thiosulfate
Klein's Reagent	saturated solution of borotungstate
Knorr's Pyrrole	2,4-dimethyl-3,5-dicarbethoxypyrrole
Koch Acid	8-amino-1,3,6-naphthalenetrisulfonic acid
Kojic Acid	5-hydroxy(tosyloxy)iodobenzene
König Salt	*N*-cyanopyridinium bromide
Koppescharr's Solution	bromine solution (conc.)
Koser's Reagent	[hydroxy(tosyloxy)iodo]benzene

Koshland's Reagent I	2-hydroxy-5-nitrobenzyl bromide
Koshland's Reagent II	2-methoxy-5-nitrobenzyl bromide
Koshland's Reagent III	2-bromo-2′-hydroxy-5′-nitroacetanilide
Kovac's Complex	*N,N*-dicyclohexylcarbodiimide pentachlorophenol complex
Kryptocyanine	1,1′-diethyl-4,4′-carbocyanine iodide
Kryptoflex® 21	1,4,10-trioxa-7,13-diazacyclopentadecane
Kryptoflex® 211	4,7,13,18-tetraoxa-1,10-diazabicyclo[8.5.5] eicosine
Kryptoflex® 222	4,7,13,16,21,24-hexaoxa-1,10-diazabicyclo [8.8.8]hexacosine
Kryptoflex® 222B	5,6-benzo-4,7,13,16,21,24-hexaoxa-1,10-diazabicyclo[8.8.8]hexacosine
KS-Selectride®	potassium trisiamylborohydride
Kunz Reagent	2-(triphenylphosphonio)ethyl chloroformate chloride
Kurrol's Salt	sodium phosphate
Kynurenic Acid	4-hydroxyquinoline-2-carboxylic acid
L-Cleland's Reagent	L-dithiothreitol
L-Selectride®	lithium tri-*sec*-butylborohydride
Lac Sulfur	sulfur, precipitated
Lalancette's Reagent	sodium borohydride-sulfur
Laurent's Acid	1-naphthylamine-5-sulfonic acid
Lawesson Reagent, (LR)	[2,4-bis(4-methoxyphenyl)-2,4-dithioxo-1,3,2,4-dithiadiphosphetane;4-methoxyphenylthiophosphoric cyclic di(thioanhydride)]
Lawsone	2-hydroxy-1,4-naphthoquinone
Lazier Catalyst	copper chromite
Leishman's Stain	eosin methylene blue
Lemieux-von Rudloff Reagent	potassium carbonate-sodium metaperiodate-potassium permanganate solution
Leuch's Anhydrides	anhydride intermediate formed by the reaction of amino acids with phosgene
Lewis Acid	electron pair acceptor
Lewis Base	electron pair donor
Lindlar Catalyst	~5% palladium on calcium carbonate poisoned with lead
Lindlar Catalyst Poison	2,2′-(ethylenedithio)diethanol
Lintner's Starch	starch ~2% by wt. solution in water
Lipshutz Reagents	cyanocuprates
Lochmann's Base	*n*-butyllithium/potassium *tert*-butoxide—as a heterogeneous suspension in hexane
Lohmann's Enzyme	creatine kinase

Lomant's Reagent	3,3′-dithiodipropionic acid bis(*N*-hydroxysuccinimide ester)
Lophine	2,4,5-triphenylimidazole
LS-Selectride®	lithium trisiamylborohydride
Lucas Reagent	zinc chloride in concentrated hydrochloric acid
Luff-Schoorl Reagent	aqueous solution of cupric sulfate, citric acid and sodium carbonate
Lugol's Solution	5% iodine solution
McMurry's Reagent	titanium(III) chloride-lithium aluminum hydride
M-PYROL®	1-methyl-2-pyrrolidinone
McQuillin Catalyst	rhodium chloride-pyridine-dimethyl formamide-borohydride complex
Maddrell's Salt	sodium hexametaphosphate
Magic Acid	fluorosulfuric acid-antimony pentafluoride
Magic Methyl	methyl fluorosulfonate
Magneson's Reagent (I)	4-(4-nitrophenylazo)resorcinol
Magneson's Reagent II	4-(4-nitrophenylazo)-1-naphthol
Man-Tan	dihydroxyacetone
Mander's Reagent	methyl cyanoformate
Mannich Bases	β-dialkylamino ketones
Mannich Reagent	dimethyl(methylene)ammonium iodide
Mannich Reagents	dialkyl(methylene)ammonium salts
Marfey's Reagent	*N*-α-(2,4-dinitro-5-fluorophenyl)-L-alaninamide
Margaric Acid	*N*-heptadecanoic acid
Marignac's Salt	potassium stannosulfate
Marme's Reagent	potassium tetraiodocadmate
Marquis Reagent	aqueous formaldehyde and concentrated sulfuric acid
Marshall's Reagent	*N*-(1-naphthyl)ethylenediamine dihydrochloride
Martin Ligand	2-(bromo)phenyl-1-trifluoromethyl-2,2,2-trifluoroethoxide
Martin Sulfurane Dehydrating Agent	bis[α,α-bis(trifluoromethyl)benzenemethanolato]diphenylsulfur
Mattock's Bromide	2-acetoxy-2-methylpropanoyl bromide
Mauve Factor Reagent	2,4-dimethyl-3-ethylpyrrole
Mayer's Reagent	aqueous solution of mercuric chloride and potassium iodide
Meerwein Reagent	*O*-methyldibenzofuranium fluoroborate
Meerwein Salt	trialkyloxonium tetrafluoroborate
Meerwein's Reagent	triethyloxonium tetrafluoroborate
Meisenheimer Complexes	intermediate formed from the nucleophilic addition to an aromatic ring.
Meldrum's Acid	2,2-dimethyl-1,3-dioxane-4,6-dione
Mennige	lead(II,III) oxide

Merbromin	mercurochrome
Merrifield's Peptide Resin	chloromethylated styrene-divinylbenzene copolymer
Methyl Carbitol®	2-(2-methoxyethoxy)ethanol
Methyl Meldrum's Acid	2,2,5-trimethyl-1,3-dioxane-4,6-dione
Meyer's Reagent	*N*-methyl-*N*-(2-pyridyl)formamide
Meyers' Chiral Auxiliary I	(+)-3-isopropyl-7*a*-methyltetrahydropyrrolo [2,1-*b*]oxazol-5(6*H*)-one
Meyers' Chiral Auxiliary II	(+)-3-isopropyl-7*a*-phenyltetrahydropyrrolo [2,1-*b*]oxazol-5(6*H*)-one
Michler's Hydrol	4,4'-bis(dimethylamino)benzhydrol
Michler's Ketone	4,4'-bis(dimethylamino)benzophenone
Midland's Reagent	9-borabicyclo[3.3.1]nonane
Miescher-Wieland Ketone	(±)-8a-methyl-3,4,8,8a-tetrahydro-1,6(2*H*,7*H*)-naphthalenedione
Miescher-Wieland Ketone Chiral	*S*(+)-8a-methyl-3,4,8,8a-tetrahydro-1,6-(2*H*,7*H*)-naphthalenedione
Millon's Reagent	aqueous solution of mercuric nitrate, nitrous acid-nitric acid
Millon's Reagent	dihydroxomercuric ammonium hydroxide; mercury-nitric acid-water
Mitsunobu Reagent	triphenylphosphine-diethyl azadicarboxylate
Modified McMurry's Reagent	titanium(III) chloride-lithium aluminum hydride
Mohr's Salt	ammonium ferrous sulfate
Monsel's Solution	ferric subsulfate
Mori Reagent	4,4-dimethylhex-5-ene-3-one
Morland Salt	guanidine tetrathiocyanatodiamine-chromate(III)
Morton Oxide	'active' manganese dioxide
Mosher Amide	any amine reacted with Mosher's Acid Chloride
Mosher Esters	α-methoxy-α-(trifluoromethyl)phenylacetates
Mosher's Acid	α-methoxy-α-trifluoromethylphenylacetic acid;
Mosher's Acid Chloride	α-methoxy-α-trifluoromethylphenylacetic acid chloride
Mukaiyama's Reagent	titanium(IV) chloride-zinc
Murahashi Reagent	(*N*-methylphenylamino)triphenylphosphonium iodide
Mustard Gas	bis(2-chloroethyl)sulfide
Mustard Oil	allyl isothiocyanate
Muthmann's Liquid	1,1,2,2-tetrabromoethane
N-Selectride®	sodium tri-*sec*-butylborohydride
Nakayama Reagent	1,3-benzodithiolylium tetrafluoroborate
Nazarov Reagent	ethyl 3-oxo-4-pentenoate; methyl-5-methoxy-3-oxovalerate
Nazarov Reagents	γ,δ-unsaturated-β-keto esters

NB-Enantride™	[2-[2-(benzyloxy)ethyl]-6,6-dimethylbicyclo[3.1.1]-3-nonyl]-9-boratabicyclo[3.1.1]nonane
Neimann's Compound	L-1-keto-3-carbomethoxy-1,2,3,4-tetrahydroisoquinoline; KCTI
Nessler's Reagent	dipotassium tetraiodomercurate(II) in dilute sodium hydroxide
Neuberg Ester	fructose-6-phosphoric acid
Neville Winther Acid	1-naphthol-4-sulfonic acid sodium salt
New Weinreb-Nahm-Wittig Reagent	*N*-methoxy-*N*-methyl-2-(triphenylphosphoranylidene) acetamide
Ninhydrin Reagent	1,2,3-triketohydrinene monohydrate
Nishimura Catalyst	rhodium oxide-platinum oxide
Nitresolve	α-methyl-*p*-nitrobenzylamine hydrochloride
Noe's Reagent	bis[(2*R*,3a*S*,4*R*,7a*S*) or (2*S*,3a*R*,4*S*,7a*R*) octahydro-7,8,8-trimethyl-4,7-menthanobenzofuran-2-yl] ether
Norit A™	activated carbon
Normant Reagents	organocopper-magnesium halide salts
Noyori's BINAL-H Reagent	lithium (1,1′-binaphthalene-2,2′-diolato)-(ethanolato)hydridoaluminate
Noyori's Reagent	2,2′-dihydroxy-1,1′-binaphthyl-lithium aluminum hydride
NPS-Amino Acids	*N*-*o*-nitrophenylsulfenyl-amino acids
Nujol	mineral oil
Nylander Reagent	aqueous solution of potassium sodium tartrate-potassium hydroxide and bismuth subnitrate
Nystead Reagent	cyclo-dibromodi-μ-methylene[μ-(tetrahydrofuran)]trizinc in tetrahydrofuran
Obermayer's Reagent	ferric chloride in concentrated hydrochloric acid
O'Donnell's Schiff Bases	α-imino esters
Oesper's Reagent	ferrous ethylenediammonium sulfate
Olah's Reagent	pyridinium poly(hydrogen fluoride)
ortho-Nitresolve	α-methyl-*o*-nitrobenzylamine hydrochloride
Oxone®	potassium monopersulfate triple salt
PADAC® β-Lactamase Substrate	7-(thienyl-2-acetamido)-3-[2-(4-*N*,*N*-dimethylaminophenylazo)pyridinium methyl]-3-cephem-4 carboxylic acid
Parikh-Doering Reagent	dimethyl sulfoxide-sulfur trioxide-pyridine-triethylamine
Patton and Reeder's Reagent	[3-hydroxy-4-(2-hydroxy-4-sulfo-1-naphthylazo)naphthalene-2-carboxylic acid
Pavy's Solution	aqueous solution of copper sulfate-potassium sodium tartrate, sodium hydroxide and ammonia

Payne's Reagent	peroxybenzimidic acid
Pearl Ash	potassium carbonate
Pearlman's Catalyst	palladium hydroxide on activated charcoal
Peri Acid	1-naphthylamine-8-sulfonic acid
Pfitzner-Moffatt Reagent	dicyclohexylcarbodiimide-orthophosphoric acid in dimethyl sulfoxide
Pfitzner-Moffatt Reagent	manganese dioxide
Phenyl Mustard Oil	phenyl isothiocyanate
Phenyl-Meldrum's Acid (5-Phenyl-Meldrum's Acid)	2,2-dimethyl-5-phenyl-1,3-dioxane-4,6-dione
Pirkle's Alcohol	(*R*)-(-)-2,2,2-trifluoro-1-(9-anthryl)ethanol
Pirrung's Reagent	2,6-dimethylphenoxypropanoate
Plummer's Inhibitor	D,L-2-mercaptomethyl-3-guanidinoethylthio-propanoic acid
Prévost's Reagent	silver iodobenzoate
Prins Dimer	dimer of hexachlorocyclopentadiene
Proton Sponge®	1,8-bis(dimethylamino)naphthalene, *N,N,N',N'*-tetramethyl-1,8-naphthalene-diamine
Purdie's Reagent	methyl iodide-silver oxide
Purpald®	4-amino-3-hydrazino-5-mercapto-1,2,4-triazole; 4-amino-5-hydrazino-4*H*-1,2,4-triazole-3-thiol
Purple Benzene	tetrabutylammonium permanganate in benzene
Racemic Acid	*dl*-tartaric acid
Raney Alloy	nickel-aluminum alloy
Raney Catalyst	Raney Alloy leached with alkali
Raney Cobalt (Catalyst)	cobalt-aluminum alloy leached with alkali
Raney Nickel	Raney Alloy leached with alkali
Ratcliffe Reagent	chromium trioxide-pyridine on celite
Raybin's Reagent	5-diazouracil
Reagent BMC	sulfur monochloride-aluminum chloride-sulfuryl chloride
Reformatsky Reagent	zinc-ethyl trichloroacetate
Reichardt's Betaine	see Reichardt's Dye
Reichardt's Dye	2,6-diphenyl-4-(2,4,6-triphenylpyridinio) phenolate
Reichstein's Substance A	5α-pregnane-3β,11β,17α,20β,21-pentol
Reichstein's Substance C	5α-pregnane-3α,11β,17α,21-tetrol-20-one
Reichstein's Substance D	5α-pregnane-3β,17α,21-triol-11,20-dione
Reichstein's Substance Dehydro-C	allotetrahydrocortisone
Reichstein's Substance E	4-pregnene-11β,17α,20β,21-tetrol-3-one
Reichstein's Substance EPI 'E'	4-pregnene-11β,17α,20α,21-tetrol-3-one

Reichstein's Substance EPI 'U'	4-pregnene-17α,20α,21-triol-3,11-dione
Reichstein's Substance Fa	cortisone
Reichstein's Substance G	adrenosterone
Reichstein's Substance H	corticosterone
Reichstein's Substance J	5α-pregnane-3β,17α,20β-triol
Reichstein's Substance K	allopregnane-3β,17α,20β,21-tetrol
Reichstein's Substance L	allopregnane-3β,17α,-diol-20-one
Reichstein's Substance M	hydrocortisone
Reichstein's Substance N	5α-pregnane-3β,21-diol-11,20-dione
Reichstein's Substance O	allopregnane-3β,17α,20α,-triol
Reichstein's Substance P	5α-pregnane-3β,17α,21-triol-20-one
Reichstein's Substance Q	deoxycorticosterone
Reichstein's Substance R	5α-pregnane-3β,11β,21-triol-20-one
Reichstein's Substance S	cortexolone (11-deoxycortisol; 11-deoxy-17-hydroxycorticosterone)
Reichstein's Substance T	4-pregnene-20,21-diol-3,11-dione
Reichstein's Substance U	4-pregnene-17α,20β,21-triol-3,11-dione
Reichstein's Substance V	5α-pregnane-3β,11β,17α,21-tetrol-20-one
Reichstein's Substance X	*d*-aldosterone
Reinecke Salt	ammonium tetrathiocyanatodiamine-chromate(III)
Reissert Compound	2-benzoyl-1,2-dihydroisoquinoline-1-carbonitrile
Reissert Compounds	2-acyl-1-cyano-1,2-dihydroisoquinolines; 2-acyl-1-cyano-1,2,3,4-tetrahydroisoquinolines
Reppe Catalyst	nickel tetracarbonyl
Resolve-Al AgFOD®	(6,6,7,7,8,8,8-heptafluoro-2,2-dimethyl-3,5-octanedionato)silver
Resolve-Al DyFOD™	tris(6,6,7,7,8,8,8-heptafluoro-2,2-dimethyl-3,5-octanedionato)dysprosium(III)
Resolve-Al Dy™	tris(2,2,6,6-tetramethyl-3,5-heptanedionato) dysprosium(III)
Resolve-Al ErFOD™	tris(6,6,7,7,8,8,8-heptafluoro-2,2-dimethyl-3,5-octanedionato)erbium(III)
Resolve-Al EuFOD™	tris(6,6,7,7,8,8,8-heptafluoro-2,2-dimethyl-3,5-octanedionato)europium(III)
Resolve-Al GdFOD™	tris(6,6,7,7,8,8,8-heptafluoro-2,2-dimethyl-3,5-octanedionato)gadolinium(III)
Resolve-Al Gd™	tris(2,2,6,6-tetramethyl-3,5-heptanedionato) gadolinium(III)
Resolve-Al HoFOD™	tris(6,6,7,7,8,8,8-heptafluoro-2,2-dimethyl-3,5-octanedionato)holmium(III)
Resolve-Al Ho™	tris(2,2,6,6-tetramethyl-3,5-heptanedionato)holmium(III)

Resolve-Al La™	tris(2,2,6,6-tetramethyl-3,5-heptanedionato) lanthanum(III)
Resolve-Al PrFOD™	tris(6,6,7,7,8,8,8-heptafluoro-2,2-dimethyl-3,5-octanedionato)praseodynium(III)
Resolve-Al Pr™	tris(2,2,6,6-tetramethyl-3,5-heptanedionato) praseodynium(III)
Resolve-Al YbFOD™	tris(6,6,7,7,8,8,8-heptafluoro-2,2-dimethyl-3,5-octanedionato)ytterbium(III)
Resolve-Al Yb™	tris(2,2,6,6-tetramethyl-3,5-heptanedionato) ytterbium(III)
Resolve-Al™	tris(2,2,6,6-tetramethyl-3,5-heptanedionato) europium(III)
Reychler's Acid	camphorsulfonic acid
Rice's Bromine Solution	aqueous solution of bromine and sodium bromide
Riegel's Paper	Congo red paper
Ringer's Solution	sodium chloride, potassium chloride and calcium chloride in water
Robison Ester	glucose-6-phosphoric acid
Rochelle Salt	potassium sodium tartrate
Rogojski's Salt	cuprous sulfate
Rohr's Salt	ammonium iron(II) sulfate
Rohrbach's Solution	barium mercuric iodide
Rondeau's Reagent	tris(6,6,7,7,8,8,8-heptafluoro-2,2-dimethyl-3,5-octanedionato)praseodynium(III); Resolve-Al PrFOD™
Rongalite	hydroxymethanesulfinic acid, monosodium salt
Rose's Alloy	alloy of bismuth, lead and tin
Rosenstiehl's Green	barium manganate
Rudinger Reagent	3-(4-hydroxyphenyl)propionic acid *N*-hydroxysuccinimide ester; Bolton-Hunter Reagent
Rudolf Guercke Acid	1-naphthol-3,6-disulfonic acid
Rydon-Landaeur Reagent	triphenyl phosphite-methyl iodide
Sage's Dry Bleach	1,3-dichloro-5,5-dimethylhydantoin
Saigo's Reagent	1,1′-(azadicarbonyl)dipiperidine
Sal Ammoniac	block ammonium chloride
Salkowski Reagent	ferric chloride in sulfuric acid
Sandwich Compounds	metallocenes
Sanger's Reagent	2,4-dinitrofluorobenzene; 1-fluoro-2,4-dinitrobenzene
Sarett Reagent	chromium trioxide-pyridine
Sawicki's Reagent	3-methyl-2-benzothiazolinonehydrazone hydrochloride
Schaeffer's Acid	2-naphthol-6-sulfonic acid

Schaeffer's Salt	sodium 2-naphthol-6-sulfonate
Schardinger α-Dextrin	α-cyclodextrin
Schardinger β-Dextrin	β-cyclodextrin
Schardinger Dextrin	cyclodextrin
Schardinger γ-Dextrin	γ-cyclodextrin
Scheibler's Reagent	aqueous solution of phosphotungstic acid
Schiff Bases	ketimines
Schiff's Reagent	fuchsin sulfurous acid; leucosulfonic acid
Schmidt's Deaminase	5′-adenylic acid deaminase
Schneider-Hanze Reagent	*N*-methylmorpholine oxide-hydrogen peroxide
Schoelkopf's Acid	1-naphthol-4,8-disulfonic acids; 1-naphthylamine-8-sulfonic acids
Schoenberg's Reagent	*N*-[bis(*p*-methoxyphenyl)methylene]-benzylamine
Schultze's Reagent (Solution)	aqueous solution of potassium chlorate and nitric acid
Schwartz Reagent	zirconocene hydrochloride; bis(cyclopentadienyl)zirconium chloride hydride
Schweitzer's Reagent	aqueous solution of cupric hydroxide and ammonia
Schweizer's Reagent	vinyltriphenylphosphonium bromide; vinyltributylphosphonium bromide
Seignette Salt	sodium potassium tartrate
Selectride	tri-*sec*-butylborohydride, lithium or potassium salt
Shardinger Dextrin	cyclic oligosaccharides
Shardinger's Enzyme	xanthine oxidase
Sharpless Reagent	tertraisopropyl orthotitanate-diethyl tartrate, and *tert*-butyl hydroperoxide
Sheppard's Polymer	copolymer of *N*-acryl sarcosine methyl ester, *N*,*N*′-bisacrylylethylene diamine and *N*,*N*-dimethylacrylamide
Shiori-Yamada Reagent	diphenylphosphoryl azide-benzene-water
Sievers' Reagent	tris(6,6,7,7,8,8,8-heptafluoro-2,2-dimethyl-3,5-octanedionato)europium(III); Resolve-Al EuFOD™
Silver FOD	(6,6,7,7,8,8,8-heptafluoro-2,2-dimethyl-3,5-octanedionato)silver; Resolve-Al AgFOD™
Silvex	2-(2,4,5-trichlorophenoxy)propionic acid
Simmons-Smith Reagent	bis(iodomethyl)zinc-zinc iodide
Smith's Sulfone	2-*p*-sulfanilylanilinoethanol
Snyder's Reagent	4,7-dihydroxy-1,10-phenanthroline
Sörensen's phosphate	sodium phosphate, dibasic dihydrate

Sörensen's Salt	di-sodium hydrogen orthophosphate
Sorrel Salt	potassium binoxalate
Steglich's Reagent	4,6-diphenylthieno[3,4-*d*]-1,3-dioxol-2-one 5,5-dioxide
Stile's Reagent	magnesium methyl carbonate; MMC
Stoke's Reagent	aqueous solution of ferrous sulfate, tartaric acid (ammonia is added just prior to use)
Stuart Factor	factor X
Stuart-Prower Factor	factor X
Sulfan®	sulfur trioxide, stabilized
Super Hydride	lithium triethylborohydride
Super-Deuteride®	lithium triethylborodeuteride
Super-Hydride®	lithium triethylborohydride
Superhydride	sodium triethylborohydride
Swern Reagent	dimethyl sulfoxide-oxalyl chloride
Tebbe Reagent	μ-chloro-μ-methylene[bis(cyclopentadienyl)-titanium]dimethylaluminum
Tetralin®	1,2,3,4-tetrahydronaphthalene
Thiele Reagent	chromium trioxide-acetic anhydride
Thiomichler's Ketone	4,4′-bis(dimethylamino)thiobenzophenone
Thiram	tetramethylthiuram disulfide
Thomas Flour	calcium phosphate, tetrabasic
Thomas Phosphate	calcium phosphate, tetrabasic
Thoulet's Solution	potassium triiodomercurate(II) solution
Tillman Reagent	sodium salt of *N*-(3,5-dichloro-4-hydroxy-phenyl)-*p*-benzoquinone imine
Tiron®	4,5-dihydroxy-*m*-benzenedisulfonic acid
Tobias Acid	2-naphthylamine-1-sulfonic acid
Tollen's Reagent	diamminesilver(I) hydroxide/sodium hydroxide solution
Trapp's Mixture	diethyl ether-tetrahydrofuran-pentane
Traut's Reagent	2-iminothiolane hydrochloride
Triton B	benzyltrimethylammonium hydroxide; BTAM
Tröger's Base	2,8-dimethyl-6*H*,12*H*-5,11-methanodibenzo [*b*,*f*][1,5]diazocine
Trolox®	6-hydroxy-2,5,7,8-tetramethylchroman-2-carboxylic acid
Tsuda's Reagent	*N*-(2-diethylaminoethyl)-1-naphthylamine oxalate
Tween® 20	polyoxyethylene (20) sorbitan monolaurate
Tween® 40	polyoxyethylene (20) sorbitan monopalmitate
Tween® 60	polyoxyethylene (20) sorbitan monostearate
Tween® 80	polyoxyethylene (20) sorbitan monooleate

Tween® 85	polyoxyethylene sorbitanmonooleate
Twitchell's Reagent	naphthalene-oleic acid-sulfuric acid
U-Catalysts	see Urushibara Catalysts
Udenfriend's Reagent	iron(II)-EDTA-ascorbic acid-O_2
Uffelmann's Reagent	ferric chloride in phenol
Urushibara Catalysts	nickel, cobalt or iron catalyst leached with alkali, formed by a displacement reaction from the metal halide salt by zinc
Uvinul® D49	2,2′-dihydroxy-4,4′-dimethoxybenzophenone
Van Boom's Reagent	2-chloro-4*H*-1,3,2-benzodioxaphosporin-4-one
van Urk's Reagent	4-(dimethyl)aminobenzaldehyde in dilute hydrochloric acid and ethanol
Vaska's Compound	carbonylbis(triphenylphosphine)iridium(I) chloride
Vedejs Reagent	oxodiperoxymolybdenum(pyridine) (hexamethylphosphoric triamide)
Versen-Ol®	*N*-(2-hydroxyethyl)ethylenediaminetriacetic acid
Viehe's Salt	dimethyl(dichloromethylene)-ammonium chloride
Vilsmeier Reagent	(chloromethylene)dimethylammonium chloride
Vilsmeier-Haack Reagent	(chloromethylene)dimethylammonium chloride
Vitride	sodium bis(2-methoxyethoxy)aluminum hydride
von Braun's Epimer Reagent	di(*p*-α-methylhydrazinophenyl)methane
Wagner's Reagent (Solution)	aqueous solution of potassium iodide and iodine
Wagner's Solution	aqueous solution of citric acid and salicylic acid
Walborsky's Reagent	1,1,2,2-tetramethylbutyl isocyanide
Waldenstrom Ester	uroporphyrin III ester
Wang Polymer	4-(benzyloxy)benzyl alcohol polymer bound
Wang/Richard's Reagent	dimethyl 3,3′-dithiopropionimidate dihydrochloride
Wanzlick's Reagent	*N*,*N*′-diphenylethylenediamine
Weiler Dianion	dianion of methyl acetoacetate
Weinreb Amide	*N*-methoxy-*N*-methyl amides
Weinreb-HEW Reagent	diethyl (*N*-methoxy-*N*-methylcarbamoylmethyl)phosphonate
Welter's Bitter Yellow	picric acid
Wheland Intermediates	carbocation intermediate in electrophilic aromatic substitution
Wichterle Reagent	1,3-dichloro-2-butene

Wieland-Miescher Ketone	9-methyl-$\Delta^{5(10)}$-octalin-1,6-dione
Wijs' Chloride ([Special]Solution)	iodine monochloride in glacial acetic acid
Wilke Catalyst	'naked nickel'
Wilkinson's Catalyst	tris(triphenylphosphine)rhodium(I) chloride
Wild's Reagent	lithium-ammonia-ethanol
Windaus-Grundmann Ketone	[1*R*-[1α(*R**),3aβ,7aα]]-octahydro-1-(5-hydroxy-1,5-dimethylhexyl)7a-methyl-4*H*-inden-4-one
Winterstein's Acid	(*S*)-3-(*N*,*N*-dimethylamino)-3-phenylpropanoic acid
Wintersteiner's Compound A	allopregnane-3β,11β,17α,20β,21-pentol
Wintersteiner's Compound B	allopregnane-3β,17α,21-triol-11,20-dione
Wintersteiner's Compound D	allopregnane-3α,11β,17α,21-tetrol-20-one
Wintersteiner's Compound F	cortisone
Wintersteiner's Compound G	allopregnane-3β,17α,-diol-20-one
Witt-Utermann Solution	potassium hydroxide in aqueous ethanol
Woelm-Alumina	basic alumina
Woodward's Reagent K	2-ethyl-5-phenylisoxazolium-3′-sulfonate; *N*-ethyl-5-phenylisoxazolium-3′-sulfonate
Woodward's Reagent L	2-*tert*-butyl-5-methylisoxazolium perchlorate
Wurster's Blue	tetramethyl-*p*-phenylenediamine
Wurster's Reagent	*N*,*N*,*N*′,*N*′-tetramethyl-1,4-phenylenediamine
Yamada Reagent	diphenyl phosphorazidate
Yamaguchi's Reagent	2,4,6-trichlorobenzoyl chloride
Yarovenko Reagent	*N*,*N*-diethyl-2-chloro-1,1,2-trifluoroethylamine
Zeise's Salt	potassium trichloro(ethylene)platinate(II)
Zerewitinoff Reagent	methylmagnesium iodide in *n*-butyl ether
Ziegler-Natta Catalyst	titanium tetrachloride-triethylaluminum
Ziram	dimethyldithiocarbamic acid, zinc salt
Zybill's Complexes	iron tetracarbonyl silyl-hexamethylphosphoramide complexes

Chapter 6

Review References

The purpose of this chapter is to provide a rapid entry to the literature through review citations. To help in this process, only keywords are used, with subtopics as appropriate. Only common journals have been cited, although there are a few exceptions to this rule when a particularly important review has appeared in a less-well known journal. Authors have not been included solely on space limitations. The majority of the references are after 1970, but again key reviews prior to this date have been included. The listing is not intended to help with the establishment of synthetic methods, for example, it is often not possible to tell if alkenes are prepared from ketones or the other way round by the use of keywords alone. If synthetic methods are required, the reader is referred to many of the excellent books on the subject. Although the emphasis is on organic chemistry, some topics of general interest have been included.

Although many topics have been cross-referenced, brevity has demanded that this not be done in every instance. If general reviews are required, first, look under that topic. For many subjects, subtopics have been derived from a general, basic, chemical background. For example, physical and analytical methodologies related to nmr, including the use of heteroatoms such as nitrogen-15, are listed under nmr. If a review on the nmr of a specific compound is required, look under that compound for the subtopic nmr. As another example, organometallic chemistry covers general topics and reviews that deal with a number of elements. If a specific element is required, such as nickel, look under nickel chemistry and organonickel chemistry.

To avoid undue fragmentation functional groups have been renamed to the IUPAC system. Thus, alkenes have been used in place of olefins. For reactive intermediates, metals, and natural products, etc., the naming system of the original paper has been retained.

Some subjects are peripheral to main stream organic chemistry. Analytical methods have only been listed under the title of that methodology. The use of a technique for a specific class of compounds is given under that class. Natural products have been collected under that general heading, with the exception of class of compounds that have been thoroughly investigated, such as steroids, terpenes, etc. Medicinal chemistry has also been used as a general category for topics that relate to medicinal uses of compounds rather than their organic chemistry.

The use of the suffix, ff, after a page number denotes that this is a symposia-in-print or papers from a conference on that subject. In these cases, the topics will be broad for the keyword definition, but papers within the series may only deal with very specific aspects of the subject.

Named reactions, including rearrangements, are in Chapter 4. Thus, if a general review on the hetero-Diels Alder reaction is required, the reader is referred to this chapter. However, if a review on the use of aldehydes in a hetero-Diels Alder reaction is desired, look in this chapter under aldehydes and the appropriate subtopic.

Absolute stereochemistry	
Asymmetric synthesis	*Acc. Chem. Res.* **1985**, *18*, 280.
Abzymes	*See catalytic antibodies.*
Acetal formation	
Carbohydrates	*Chem. Rev.* **1979**, *79*, 491.
Acetal hydrolysis	
Proton transfer	
Reaction mechanisms	*Pure & Appl. Chem.* **1982**, *54*, 1853.
Acetalization	
Alkenes	*Acc. Chem. Res.* **1990**, *23*, 49.
	Synthesis **1981**, 501.
Asymmetric synthesis	*Pure & Appl. Chem.* **1988**, *60*, 49.
	Tetrahedron: Asymmetry **1990**, *1*, 477.
Coupling reactions	*Synthesis* **1987**, 1043.
Hydrolysis	*Acc. Chem. Res.* **1984**, *17*, 305.
	Chem. Rev. **1974**, *74*, 581.
Reaction mechanisms	*Pure & Appl. Chem*, **1982**, *54*, 1837.
Acetamide complexes	*Chem. Soc. Rev.* **1988**, *17*, 181.
α-Acetamido sulfides	*Tetrahedron* **1984**, *40*, 1633.
Acetic formic anhydride	*Tetrahedron* **1990**, *46*, 1081.
Acetoacetic ester condensation	*Org. Reactions* **1942**, *1*, 266.
η^2-Acyl group	*Chem. Rev.* **1988**, *88*, 1059.
Acetylcholinesterase	
Reaction mechanisms	*Chem. Rev.* **1987**, *87*, 955.
Acids	
Sulfoxides	*Acc. Chem. Res.* **1973**, 6, 132.
Aconite alkaloids	*Chem. Soc. Rev.* **1977**, 6, 413.
Historical perspective	*Tetrahedron* **1985**, *41*, 485.
Acotinase	*Acc. Chem. Res.* **1980**, *13*, 345.
Actinide chemistry	*Angew. Chem. Int. Ed. Engl.* **1991**, *30*, 1069.
	Chem. Soc. Rev. **1988**, *17*, 69.
Actinomycins	*Chem. Rev.* **1974**, *74*, 625.
	Angew. Chem. Int. Ed. Engl. **1975**, *14*, 375.
Activated metals	*Top. Curr. Chem.* **1975**, *59*, 1.
Acyclic stereochemistry	*Tetrahedron* **1984**, *40*, 2197ff.
Acyclic systems	
Asymmetric synthesis	*Tetrahedron* **1980**, *36*, 3.
Acyl anion equivalents	*See cyanohydrins, dithianes, and umpolung*
Acyl chlorides	
Amines	*Synthesis* **1989**, 1.
Acyl complexes	
Organometallic chemistry	*Chem. Soc. Rev.* **1988**, *17*, 147.

Acyl cyanides	*Angew. Chem. Int. Ed. Engl.* **1982**, *21*, 36.
Acyl isocyanates	
Addition reactions	
Unsaturated compounds	*Synthesis* **1982**, 433.
Cycloaddition reactions	*Synthesis* **1974**, 461.
Acyl migrations	*Russ. Chem. Rev.* **1968**, *37*, 587.
Acyl peroxides	*Acc. Chem. Res.* **1983**, *16*, 304.
Acyl transfer reactions	*Chem. Rev.* **1975**, *75*, 627.
Elimination-addition mechanisms	*Chem. Rev.* **1975**, *75*, 627.
Reaction mechanisms	*Acc. Chem. Res.* **1989**, *22*, 387.
Acylamines	
Chlorinations	*Angew. Chem.* **1962**, *74*, 848.
Acylation	*Quart. Rev.* **1963**, *17*, 160.
Activated esters	*SYNLETT* **1991**, 755.
Alcohols	
Asymmetric synthesis	*Angew. Chem. Int. Ed. Engl.* **1973**, *12*, 25.
Alkynes	*SYNLETT* **1991**, 755.
4-Dialkylaminopyridines	*Angew. Chem. Int. Ed. Engl.* **1978**, *17*, 569.
DMAP catalysis	*Chem. Soc. Rev.* **1983**, *12*, 129.
Heterocycles	*Angew. Chem. Int. Ed. Engl.* **1973**, *12*, 119.
Intramolecular	*Angew. Chem. Int. Ed. Engl.* **1973**, *12*, 119.
	Org. Reactions **1944**, *2*, 114.
Isocyanates	*Chem. Soc. Rev.* **1975**, *4*, 231.
Ketenes	*Chem. Soc. Rev.* **1975**, *4*, 231.
Ketones	*Org. Reactions* **1954**, *8*, 59.
Cyclic ketones	*Org. Reactions* **1944**, *2*, 114.
Nucleophilic	*Angew. Chem. Int. Ed. Engl.* **1969**, *8*, 639.
	Tetrahedron **1976**, *32*, 1943.
Peptides	*Chem. and Ind. (London)* **1974**, 723.
Reaction mechanisms	*Chem. Soc. Rev.* **1975**, *4*, 231.
Acylcyclopentenes	
Isomerization	*Tetrahedron* **1976**, *32*, 641.
Acylneuraminic acids	*Angew. Chem. Int. Ed. Engl.* **1973**, *12*, 127.
Acyloins	*Org. Reactions* **1948**, *4*, 256.
Acyloxylation	*Synthesis* **1972**, 1.
	Synthesis **1973**, 567.
Acylsilanes	*Chem. Soc. Rev.* **1990**, *19*, 147.
	Synthesis **1989**, 647.
Adamantane	
Rearrangements	*Chem. Soc. Rev.* **1974**, *3*, 479.
Addition reactions	*Chem. Rev.* **1975**, *75*, 439.
	See also dipolar addition reactions and electrophilic additions.

Acyl isocyanates	
Unsaturated compounds	*Synthesis* **1982**, 433.
Aldehydes	
Alkenes	*Angew. Chem. Int. Ed. Engl.* **1976**, *15*, 639.
Alkenes	*Acc. Chem. Res.* **1974**, *7*, 155.
	Tetrahedron **1980**, *36*, 701.
Aldehydes	*Angew. Chem. Int. Ed. Engl.* **1976**, *15*, 639.
Alkynes	*Adv. Org. Chem.* **1969**, 6, 213.
Reductions	*Chem. Rev.* **1942**, *31*, 77.
Electrophilic	*Acc. Chem. Res.* **1981**, *14*, 227.
Nucleophilic	
Reaction mechanisms	*Tetrahedron* **1989**, *45*, 4017.
Organopalladium reagents	*Chem. Rev.* **1989**, *89*, 1433.
Singlet oxygen	*Pure Appl. Chem.* **1971**, *27*, 635.
Sulfenyl chlorides	*Acc. Chem. Res.* **1979**, *12*, 282.
Alkynes	*Angew. Chem. Int. Ed. Engl.* **1967**, 6, 423.
Alkenes	*Adv. Org. Chem.* **1969**, 6, 213.
	Chem. Rev. **1942**, *31*, 77.
Azides	*Chem. Rev.* **1969**, *69*, 353.
Asymmetric synthesis	*Acc. Chem. Res.* **1971**, *4*, 9.
Base catalysed	
Hydrocarbons	*Acc. Chem. Res.* **1974**, *7*, 155.
Benzyne	*Acc. Chem. Res.* **1974**, *7*, 6.
	Acc. Chem. Res. **1974**, *7*, 301.
Bromo compounds	*Synthesis* **1977**, 145.
Butadiene	
Palladium catalysis	*Acc. Chem. Res.* **1973**, 6, 8.
Carboxylic acid derivatives	*Synthesis* **1970**, 99.
	See also conjugate additions, and organocopper reagents.
Conjugated systems	*Chem. Rev.* **1959**, *59*, 329.
Cyclohexene	
Photochemistry	*Acc. Chem. Res.* **1969**, *2*, 33.
1,3-Dipolar	*Synthesis* **1982**, 701.
Ketocarbenes	*Angew. Chem.* **1961**, *73*, 368.
Electrochemical	
Aniodic	*Russ. Chem. Rev.* **1975**, *44*, 999.
Free radical	
Unsaturated compounds	*Chem. Rev.* **1962**, *62*, 599.
	Org. Reactions **1963**, *13*, 91.
Grignard reagents	
Ketones	
Reaction mechanisms	*Acc. Chem. Res.* **1974**, *7*, 272.
Nitroalkenes	*Acc. Chem. Res.* **1984**, *17*, 109.

Addition reactions, cont'd.

Hydrocarbons	*Synthesis* **1974**, 309.
Base catalysed	*Acc. Chem. Res.* **1974**, *7*, 155.
Isocyanates	*Acc. Chem. Res.* **1969**, *2*, 186.
Isocyanides	*Synthesis* **1969**, 65.
Ketenes	*Tetrahedron* **1986**, *42*, 2587.
Ketones	
Grignard reagents	
Reaction mechanisms	*Acc. Chem. Res.* **1974**, *7*, 272.
Organometallic compounds	*Chem. Rev.* **1975**, *75*, 521.
Metal complexes	
Unsaturated compounds	*Adv. Organometal. Chem.* **1967**, *5*, 237.
Nitroalkenes	
Grignard reagents	*Acc. Chem. Res.* **1984**, *17*, 109.
Organocopper reagents	
Conjugate additions	*Org. Reactions* **1972**, *19*, 1.
Organocuprates	
Alkyne acetals	*Pure & Appl. Chem.* **1983**, *55*, 1759.
Dienals	*Pure & Appl. Chem.* **1983**, *55*, 1759.
Dienones	*Pure & Appl. Chem.* **1983**, *55*, 1759.
Electron transfer reactions	*Acc. Chem. Res.* **1976**, *9*, 59.
Organometallic compounds	
Ketones	*Chem. Rev.* **1975**, *75*, 521.
Organopalladium reagents	
Alkenes	*Chem. Rev.* **1989**, *89*, 1433.
Phosphorus esters	*Synthesis* **1979**, 81.
Photochemistry	
Dienes	*Pure & Appl. Chem.* **1988**, *60*, 1009.
	Synthesis **1980**, 165.
	Synthesis **1980**, 769.
Naphthalene nitriles	*Pure & Appl. Chem.* **1988**, *60*, 1009.
Polar additions	
Alkenes	*Angew. Chem. Int. Ed. Engl.* **1964**, *3*, 245.
Isocyanates	*Acc. Chem. Res.* **1969**, *2*, 186.
Radical chemistry	
Reaction mechanisms	*Angew. Chem. Int. Ed. Engl.* **1982**, *21*, 401.
Selectivity	*Acc. Chem. Res.* **1983**, *16*, 328.
Singlet oxygen	
Alkenes	*Pure Appl. Chem.* **1971**, *27*, 635.
Thermochemistry	
Arenes	*Synthesis* **1980**, 165.
	Synthesis **1980**, 769.
Thiocyanogen	*Org. Reactions* **1946**, *3*, 240.
Transition metal compounds	*Acc. Chem. Res.* **1969**, *2*, 10.

Unsaturated compounds	*Synthesis* **1970**, 99.
	Synthesis **1977**, 145.
Acyl isocyanates	*Synthesis* **1982**, 433.
Free radical	*Angew. Chem. Int. Ed. Engl.* **1970**, *9*, 273.
	Russ. Chem. Rev. **1972**, *41*, 516.
	Synthesis **1977**, 145.
Aerosols	*Angew. Chem. Int. Ed. Engl.* **1991**, *30*, 466.
Agrochemicals	*Angew. Chem. Int. Ed. Engl.* **1981**, *20*, 151.
	Angew. Chem. Int. Ed. Engl. **1991**, *30*, 1193.
	Pure & Appl. Chem. **1986**, *58*, 365.
Alamethicin	*Acc. Chem. Res.* **1981**, *14*, 356.
Alcohol dehydrogenase	*Acc. Chem. Res.* **1986**, *19*, 321.
Alcohols	*Synthesis* **1981**, 501.
Acylations	
Asymmetric synthesis	*Angew. Chem. Int. Ed. Engl.* **1973**, *12*, 25.
Alkenes	*Org. Reactions* **1963**, *13*, 1.
Bridgehead cage alcohols	*Chem. Rev.* **1989**, *89*, 1035.
Carbonylation	*Russ. Chem. Rev.* **1973**, *42*, 202.
Conformations	*Chem. Soc. Rev.* **1976**, *5*, 411.
Deoxygenation	*Tetrahedron* **1983**, *39*, 2609.
Esters	*Org. Reactions* **1954**, *8*, 1.
Nucleophilic substitution	*Org. Reactions* **1983**, *29*, 1.
O-D-alcohols	*Synthesis* **1972**, 254.
Organoboranes	*Acc. Chem. Res.* **1969**, *2*, 65.
Organotin compounds	*Tetrahedron* **1985**, *41*, 643.
Oxidative cyclizations	*Synthesis* **1970**, 209.
Resolutions	*Org. Reactions* **1944**, *2*, 376.
Strained alcohols	*Chem. Rev.* **1989**, *89*, 1035.
Ziegler synthesis	*Angew. Chem. Int. Ed. Engl.* **1966**, *5*, 650.
Aldehydes	
Alkenes	
Hetero-Diels Alder reaction	*Angew. Chem. Int. Ed. Engl.* **1976**, *15*, 639.
Alkylations	
Nitrogen derivatives	*Synthesis* **1983**, 517.
Allylmetal compounds	
Asymmetric synthesis	*Acc. Chem. Res.* **1987**, *20*, 243.
	Angew. Chem. Int. Ed. Engl. **1982**, *21*, 555.
Carbanions	*Tetrahedron* **1988**, *44*, 4653.
Carboxylic acids	*Org. Reactions* **1954**, *8*, 218.
Ketones	*Angew. Chem. Int. Ed. Engl.* **1980**, *19*, 171.
Coupling reactions	*Synthesis* **1979**, 633.
Gattermann synthesis	*Org. Reactions* **1957**, *9*, 37.
Hetero-Diels Alder reaction	*Angew. Chem. Int. Ed. Engl.* **1976**, *15*, 639.

Aldehydes, cont'd.	
Ketones	
Carboxylic acids	*Angew. Chem. Int. Ed. Engl.* **1980**, *19*, 171.
Nucleophilic additions	
Alkenes	*Org. Reactions* **1991**, *40*, 407.
Organoboranes	*Acc. Chem. Res.* **1969**, *2*, 65.
Oxidations	*Chem. Rev.* **1954**, *54*, 325.
α-Thioalkylations	*Synthesis* **1987**, 589.
Transition metal complexes	*Acc. Chem. Res.* 984, *17*, 326.
Unsaturated aldehydes	*Acc. Chem. Res.* **1988**, *21*, 47.
Alicyclic-aromatic	
Isomerization	*Chem. Rev.* **1943**, *33*, 89.
Aliphatic compounds	
Azo compounds	
Photochemistry	*Pure & Appl. Chem,* **1980**, *52*, 2621.
Diazonium compounds	*Angew. Chem., Int. Ed. Engl.* **1976**, *15*, 251.
Elimination reactions	*Angew. Chem. Int. Ed. Engl.* **1967**, 6, 534.
Nitrosations	*Org. Reactions* **1953**, *7*, 327.
Nucleophilic substitution	
Reaction mechanisms	*Chem. Soc. Rev.* **1990**, *19*, 83.
Rearrangements	*Chem. Rev.* **1956**, *56*, 753.
Strain	*Angew. Chem., Int. Ed. Engl.* **1985**, *24*, 529.
Substitution reactions	
S_N2	*Acc. Chem. Res.* **1973**, 6, 46.
	Acc. Chem. Res. **1976**, 9, 281.
Alkali metals	
Organometallic compounds	*Acc. Chem. Res.* **1989**, *22*, 152.
Unsaturated compounds	
Additions	*J. Organometal. Chem.* **1973**, *47*, 225.
Alkali metal complexes	
Macrocycles	*Acc. Chem. Res.* **1978**, *11*, 49.
	Pure & Appl. Chem. **1986**, *58*, 1485.
Alkali metal fluorides	*Synthesis* **1983**, 169.
Alkaloids	*Acc. Chem. Res.* **1968**, *1*, 78.
	Acc. Chem. Res. **1976**, *9*, 319.
	Acc. Chem. Res. **1977**, *10*, 193.
	Acc. Chem. Res. **1977**, *10*, 227.
	Acc. Chem. Res. **1984**, *17*, 289.
	Pure & Appl. Chem. **1981**, *53*, 1141.
Aconite alkaloids	*Chem. Soc. Rev.* **1977**, 6, 413.
Historical perspective	*Tetrahedron* **1985**, *41*, 485.
Aporphinoid alkaloids	
Biogenetic pathways	*Tetrahedron* **1984**, *40*, 4795.

Biosynthesis	*Acc. Chem. Res.* **1969**, *2*, 59.
	Acc. Chem. Res. **1970**, *3*, 151.
	Acc. Chem. Res. **1971**, *4*, 100.
	Acc. Chem. Res. **1972**, *5*, 148.
Bis-indole alkaloids	*Tetrahedron* **1982**, *38*, 223.
Cephalotaxaus alkaloids	*Acc. Chem. Res.* **1975**, *8*, 158.
Daphniphyllum alkaloids	
Biomimetic synthesis	*Pure & Appl. Chem.* **1990**, *62*, 1911.
Biosynthesis	*Pure & Appl. Chem.* **1989**, *61*, 289.
Diterpene alkaloids	*Pure & Appl. Chem.* **1986**, *58*, 719.
Ellipticine	*Tetrahedron* **1986**, *42*, 2389.
Ergot alkaloids	*Tetrahedron* **1980**, *36*, 3123.
Biosynthesis	*Tetrahedron* **1976**, *32*, 873.
Indole alkaloids	*Acc. Chem. Res.* **1970**, *3*, 151.
	Acc. Chem. Res. **1984**, *17*, 35.
	Tetrahedron **1983**, *39*, 3627ff.
Isoquinoline alkaloids	*Pure & Appl. Chem.* **1986**, *58*, 685.
Photochemistry	*Acc. Chem. Res.* **1972**, *5*, 212.
Macrolactam alkaloids	*Pure & Appl. Chem.* **1981**, *53*, 1141.
Morphinandienone alkaloids	*Chem. Rev.* **1971**, *71*, 47.
Nitrones	*Acc. Chem. Res.* **1979**, *12*, 396.
NMR	*Acc. Chem. Res.* **1974**, *7*, 46.
Phenolic couplings	*Synthesis* **1972**, 657.
Photochemistry	*Chem. Rev.* **1980**, *80*, 269.
Pyridocarcazole alkaloids	*SYNLETT* **1991**, 289.
Pyrrolizidine alkaloids	*SYNLETT* **1990**, 433.
Biosynthesis	*Chem. Soc. Rev.* **1989**, *18*, 375.
	Pure & Appl. Chem. **1985**, *57*, 453.
Quinolizidine alkaloids	
Biosynthesis	*Pure & Appl. Chem.* **1985**, *57*, 453.
Synthesis	*Pure & Appl. Chem.* **1981**, *53*, 1271.
Thiaspirane alkaloids	*Acc. Chem. Res.* **1980**, *13*, 39.
Yohimbinoid alkaloids	*Pure & Appl. Chem.* **1986**, *58*, 685.
Alkanes	
Activation	*Chem. Rev.* **1990**, *90*, 403.
	SYNLETT **1991**, 597.
	Angew. Chem. Int. Ed. Engl. **1978**, *17*, 909.
	Acc. Chem. Res. **1975**, *8*, 113.
	Acc. Chem. Res. **1985**, *18*, 302.
	Pure & Appl. Chem. **1985**, *57*, 1897.
	SYNLETT **1991**, 597.
Alkylations	*Chem. Rev.* **1945**, *37*, 323.
Cycloalkanes	*Russ. Chem. Rev.* **1966**, *35*, 448.

Alkanes, cont'd.	
Chlorination	
Photochemistry	*Acc. Chem. Res.* **1990**, *23*, 219.
Cycloalkanes	
Conformation	*Acc. Chem. Res.* **1989**, *22*, 8.
Enzymes	*Chem. Rev.* **1990**, *90*, 1343.
Functionalization	*Chem. Soc. Rev.* **1982**, *11*, 283.
	Pure & Appl. Chem. **1990**, *62*, 1539.
	Pure & Appl. Chem. **1991**, *63*, 1567.
	Pure & Appl. Chem. **1984**, *56*, 13.
Organometallic chemistry	*Chem. Rev.* **1985**, *85*, 245.
	Pure & Appl. Chem, **1980**, *52*, 649.
Oxidative additions	*Pure & Appl. Chem.* **1984**, *56*, 13.
Oxygenation	*Angew. Chem. Int. Ed. Engl.* **1978**, *17*, 909.
Superacid media	*Chem. Rev.* **1982**, *82*, 591.
Alkene complexes	*Angew. Chem. Int. Ed. Engl.* **1982**, *21*, 889.
Alkenediazonium salts	*Angew. Chem. Int. Ed. Engl.* **1979**, *18*, 259.
Alkenes	*Acc. Chem. Res.* **1988**, *21*, 47.
	Angew. Chem. Int. Ed. Engl. **1972**, *11*, 964.
	Angew. Chem. Int. Ed. Engl. **1976**, *15*, 459.
	Chem. Soc. Rev. **1975**, *4*, 155.
	Org. Reactions **1962**, *12*, 57.
	Org. Reactions **1976**, *23*, 405.
	Org. Reactions **1977**, *25*, 73.
	Org. Reactions **1991**, *40*, 407.
	Pure & Appl. Chem. **1988**, *60*, 39.
	Pure & Appl. Chem. **1988**, *60*, 79.
	Synthesis **1970**, 393.
Acetalization	*Acc. Chem. Res.* **1990**, *23*, 49.
Addition reactions	*Acc. Chem. Res.* **1974**, *7*, 155.
	Tetrahedron **1980**, *36*, 701.
Aldehydes	*Angew. Chem. Int. Ed. Engl.* **1976**, *15*, 639.
Alkynes	*Adv. Org. Chem.* **1969**, *6*, 213.
Reductions	*Chem. Rev.* **1942**, *31*, 77.
Polar additions	*Angew. Chem. Int. Ed. Engl.* **1964**, *3*, 245.
Singlet oxygen	*Pure Appl. Chem.* **1971**, *27*, 635.
Alcohols	*Org. Reactions* **1963**, *13*, 1.
Aldehydes	
Hetero-Diels Alder reaction	*Angew. Chem. Int. Ed. Engl.* **1976**, *15*, 639.
Alkyl halides	*Pure & Appl. Chem.* **1985**, *57*, 1827.
Palladium catalysis	*Acc. Chem. Res.* **1979**, *12*, 146.
Aminations	*Pure & Appl. Chem.* **1985**, *57*, 1827.
Autoxidation	*Tetrahedron* **1983**, *39*, 703.
Bromination	*Acc. Chem. Res.* **1990**, *23*, 87.

Bromine azide	*Acc. Chem. Res.* **1971**, *4*, 9.
Carbanions	
Elimination reactions	*Acc. Chem. Res.* **1970**, *3*, 281.
Carbenes	
Reaction mechanisms	*Angew. Chem. Int. Ed. Engl.* **1990**, *29*, 1371.
Carbometallations	*Synthesis* **1981**, 841.
Carbonyl compounds	*Chem. Rev.* **1989**, *89*, 1513.
Circular dichroism	*Tetrahedron* **1976**, *32*, 2475.
Cleavage reactions	*Angew. Chem. Int. Ed. Engl.* **1978**, *17*, 150.
Coupling reactions	*Org. Reactions* **1984**, *32*, 375.
	Pure & Appl. Chem. **1980**, *52*, 2417.
Oxidative	*Acc. Chem. Res.* **1985**, *18*, 120.
Cyclizations	*Acc. Chem. Res.* **1968**, *1*, 1.
Cycloaddition reactions	*Chem. Rev.* **1988**, *88*, 793.
Organosulfur chemistry	*Tetrahedron* **1988**, *44*, 6755.
Cyclopropanes	*Acc. Chem. Res.* **1984**, *17*, 56.
Photochemistry	*Pure & Appl. Chem.* **1980**, *52*, 2669.
	Pure & Appl. Chem. **1990**, *62*, 1597.
Cycloalkenes	*Acc. Chem. Res.* **1980**, *13*, 213.
	Synthesis **1972**, 235.
Intramolecular Wittig	*Tetrahedron* **1980**, *36*, 1717.
Cyclopolymerization	
1,4-Dienes	*Pure Appl. Chem.* **1970**, *23*, 255.
Cyclopropanes	*Chem. Rev.* **1987**, *87*, 411.
Cycloadditions	*Acc. Chem. Res.* **1984**, *17*, 56.
Diazoacetic esters	*Org. Reactions* **1970**, *18*, 217.
Dimerization	*Synthesis* **1974**, 539.
	Chem. Rev. **1986**, *86*, 353.
	Pure & Appl. Chem. **1980**, *52*, 729.
Diols	*Org. Reactions* **1984**, *30*, 457.
1,2-Diols	*Chem. Rev.* **1980**, *80*, 187.
Dipolar cycloadditions	*Org. Reactions* **1988**, *36*, 1.
Enones	
Cycloadditions	
Photochemistry	*Synthesis* **1970**, 287.
Epoxidations	
Asymmetric synthesis	*Angew. Chem. Int. Ed. Engl.* **1991**, *30*, 403.
Peroxyacids	*Org. Reactions* **1953**, *7*, 378.
Friedel-Crafts Acylation	*Chem. Soc. Rev.* **1972**, *1*, 73.
Heterocyclic intermediates	*Tetrahedron* **1990**, *46*, 3321.
Hetero-Diels Alder reaction	*Angew. Chem. Int. Ed. Engl.* **1976**, *15*, 639.
Hofmann elimination	*Org. Reactions* **1960**, *11*, 317.
Host-guest chemistry	*Tetrahedron* **1987**, *43*, 3839.

Alkenes, cont'd.	
Hydroboration	*Org. Reactions* **1963**, *13*, 4.
Hydroformylation	*Acc. Chem. Res.* **1981**, *14*, 259.
Hydrogenation	*Chem. Rev.* **1954**, *54*, 575.
	Synthesis **1973**, 457.
Reaction mechanisms	*Acc. Chem. Res.* **1969**, *2*, 289.
Hydroxylation	
Peracids	*Org. Reactions* **1953**, *7*, 378.
Insertion reactions	
Transition metal catalysis	*Top. Curr. Chem.* **1976**, *67*, 107.
Iodine azide	*Acc. Chem. Res.* **1971**, *4*, 9.
Isomerization	*Acc. Chem. Res.* **1986**, *19*, 78.
	Chem. Rev. **1955**, *55*, 625.
	Synthesis **1969**, 97.
	Synthesis **1970**, 405.
	Tetrahedron **1974**, *30*, 1861.
Inversion	*Tetrahedron* **1980**, *36*, 577.
Photochemical	*Pure & Appl. Chem.* **1988**, *60*, 989.
Ketones	
Titanium coupling	*Acc. Chem. Res.* **1983**, *16*, 405.
Oligomerization	*Ann. New York Acad. Sci.* **1969**, *147*, 614.
	Chem. Rev. **1991**, *91*, 613.
Organic halides	
Heck reaction	*Acc. Chem. Res.* **1979**, *12*, 146.
Organoiron compounds	*Pure & Appl. Chem.* **1982**, *54*, 145.
Organometallic chemistry	*Chem. Rev.* **1988**, *88*, 1047.
	Acc. Chem. Res. **1987**, *20*, 65.
	Chem. Rev. **1973**, *73*, 163.
	Pure & Appl. Chem. **1988**, *60*, 65.
Organopalladium reagents	
Addition reactions	*Chem. Rev.* **1989**, *89*, 1433.
Organophosphorus compounds	*Org. Reactions* **1988**, *36*, 175.
Organoselenium compounds	*Acc. Chem. Res.* **1984**, *17*, 28.
Organosilicon chemistry	*Acc. Chem. Res.* **1977**, *10*, 442.
Oxidations	*Acc. Chem. Res.* **1985**, *18*, 358.
	Angew. Chem. Int. Ed. Engl. **1962**, *1*, 80.
	Pure & Appl. Chem. **1981**, *53*, 2389.
Mercuric salts	*Synthesis* **1971**, 527.
Palladium catalysis	*Synthesis* **1984**, 369.
Oxypalladation	*Acc. Chem. Res.* **1990**, *23*, 49.
Ozonolysis	*Acc. Chem. Res.* **1983**, *16*, 42.
Reaction mechanisms	*Acc. Chem. Res.* **1968**, *1*, 313.
Palladium chemistry	
Substitution reactions	*Synthesis* **1973**, 524.

Peroxyacids	*Tetrahedron* **1976**, *32*, 2855.
Photoadditions	*Chem. Rev.* **1988**, *88*, 1453.
Photochemistry	*Angew. Chem. Int. Ed. Engl.* **1986**, *25*, 661.
	Chem. Soc. Rev. **1974**, *3*, 329.
	Pure & Appl. Chem, **1980**, *52*, 2609.
	Pure & Appl. Chem. **1982**, *54*, 1633.
	Pure & Appl. Chem. **1988**, *60*, 1017.
NMR	*Pure & Appl. Chem.* **1988**, *60*, 933.
Photooxygenation	*Org. Reactions* **1973**, *20*, 133.
Polymerization	*Angew. Chem. Int. Ed. Engl.* **1983**, *22*, 440.
	Chem. Rev. **1958**, *58*, 541.
	Pure & Appl. Chem. **1986**, *58*, 1179.
Organolanthanides	*Acc. Chem. Res.* **1985**, *18*, 51.
Pyramidalized alkenes	*Chem. Rev.* **1989**, *89*, 1095.
Radical chemistry	*Angew. Chem. Int. Ed. Engl.* **1983**, *22*, 753.
	Org. Reactions **1963**, *13*, 91.
	Org. Reactions **1963**, *13*, 150.
Rearrangements	
Photochemistry	*Angew. Chem. Int. Ed. Engl.* **1978**, *17*, 150.
Reductions	
Alkynes	
Addition reactions	*Chem. Rev.* **1942**, *31*, 77.
Rhodium chemistry	*Acc. Chem. Res.* **1968**, *1*, 186.
Singlet oxygen	
Ene reaction	*Acc. Chem. Res.* **1980**, *13*, 419.
Reaction mechanisms	*Chem. Rev.* **1979**, *79*, 359.
Sulfenyl chlorides	
Addition reactions	*Acc. Chem. Res.* **1979**, *12*, 282.
Trisubstituted	*Synthesis* **1971**, 175.
Alkenones	
Cycloalkenones	*Acc. Chem. Res.* **1987**, *20*, 72.
Alkenyl esters	*Acc. Chem. Res.* **1988**, *21*, 229.
Alkenylations	*Synthesis* **1975**, 685.
Alkenyloxazolidines	
Asymmetric synthesis	*Pure & Appl. Chem.* **1988**, *60*, 1689.
3-Alkoxyacrolein	*Synthesis* **1987**, 1.
Alkoxyaluminohydrides	
Reductions	*Synthesis* **1972**, 217.
Alkoxycarbonyl compounds	*Angew. Chem. Int. Ed. Engl.* **1984**, *23*, 556.
Alkoxylipids	*Angew. Chem. Int. Ed. Engl.* **1979**, *18*, 493.
1-Alkoxy-1-siloxycyclopropanes	
Homoenolate equivalents	*Pure & Appl. Chem.* **1988**, *60*, 115.
Alkoxysulfonium ylides	
Oxidations	*Org. Reactions* **1990**, *39*, 297.

Alkyl anion synthons	
Nitroalkenes	*Synthesis* **1988**, 833.
Alkyl bromides	
Bromination	*Acc. Chem. Res.* **1984**, *17*, 160.
Alkyl cations	
Rearrangements	*Pure & Appl. Chem.* **1963**, *7*, 203.
Alkyl groups	
Rearrangements	*Chem. Rev.* **1965**, *65*, 697.
Alkyl halides	
Alkenes	*Pure & Appl. Chem.* **1985**, *57*, 1827.
Palladium catalysis	*Acc. Chem. Res.* **1979**, *12*, 146.
Asymmetric synthesis	*Synthesis* **1969**, 112.
Coupling reactions	*Pure & Appl. Chem.* **1980**, *52*, 669.
Halogen-lithium exchange	*Acc. Chem. Res.* **1982**, *15*, 300.
Heterolytic reactions	*Chem. Rev.* **1969**, *69*, 33.
Hydrolysis	*Russ. Chem. Rev.* **1974**, *43*, 305.
Naphthalene radical anions	*Acc. Chem. Res.* **1971**, *4*, 400.
Nucleophiles	*Acc. Chem. Res.* **1988**, *21*, 414.
Pyrolysis	*Chem. Rev.* **1969**, *69*, 33.
Alkyl radicals	
Rearrangements	*Pure & Appl. Chem.* **1963**, *7*, 203.
Alkyl transfers	
Organometallic reagents	*Acc. Chem. Res.* **1974**, *7*, 351.
Alkylations	*Acc. Chem. Res.* **1974**, *7*, 85.
Aldehydes	
Nitrogen derivatives	*Synthesis* **1983**, 517.
Alkanes	*Chem. Rev.* **1945**, *37*, 323.
Cycloalkanes	*Russ. Chem. Rev.* **1966**, *35*, 448.
Alkylbenzenes	*Chem. Rev.* **1939**, *25*, 329.
Allyl alkylations	
Palladium catalysis	*Acc. Chem. Res.* **1980**, *13*, 385.
Ambident anions	*Synthesis* **1970**, 1.
Amines	*Org. Reactions* **1953**, *7*, 99.
Ammonium salts	*Org. Reactions* **1953**, *7*, 99.
Aromatic amines	*Angew. Chem.* **1957**, *69*, 124.
Aromatic compounds	*Chem. Rev.* **1987**, *87*, 1277.
Asymmetric synthesis	*Tetrahedron: Asymmetry* **1991**, *2*, 1.
Carbanions	*Chem. and Ind. (London)* **1973**, 937.
Carbenes	*Chem. and Ind. (London)* **1973**, 881.
Carbonyl compounds	*Angew. Chem. Int. Ed. Engl.* **1982**, *21*, 96.
	Chem. Rev. **1987**, *87*, 1277.
Asymmetric synthesis	*Pure & Appl. Chem.* **1988**, *60*, 1597.
Conjugate additions	*Org. Reactions* **1990**, *38*, 225.
β-Dicarbonyl compounds	*Org. Reactions* **1969**, *17*, 155.

DMAP catalysis — *Chem. Soc. Rev.* **1983**, *12*, 129.
Enolates
Asymmetric synthesis — *Pure & Appl. Chem.* **1981**, *53*, 1109.
Peptides — *Angew. Chem. Int. Ed. Engl.* **1988**, *27*, 1624.
Esters — *Org. Reactions* **1957**, *9*, 107.
Synthesis **1979**, 561.
Free radical — *Chem. and Ind. (London)* **1973**, 874.
Heterocycles — *Acc. Chem. Res.* **1977**, *10*, 186.
Ketones
Nitrogen derivatives — *Synthesis* **1983**, 517.
Metal complexes
Unsaturated compounds — *Russ. Chem. Rev.* **1975**, *44*, 552.
Nitriles — *Org. Reactions* **1957**, *9*, 107.
Nitro compounds — *Russ. Chem. Rev.* **1966**, *35*, 719.
Organocopper compounds — *Chem. and Ind. (London)* **1973**, 931.
Organometallic chemistry — *Pure & Appl. Chem.* **1982**, *54*, 189.
Organonickel compounds — *Chem. and Ind. (London)* **1973**, 931.
Peptides — *Angew. Chem. Int. Ed. Engl.* **1988**, *27*, 1624.
Spiro compounds
Intramolecular — *Synthesis* **1974**, 383.

Alkylbenzenes
Alkylations — *Chem. Rev.* **1939**, *25*, 329.

2-Alkyl-1,3-cyclopentanediones — *Synthesis* **1989**, 477.

2-Alkyl-3-hydroxy-2-cyclopenten-1-ones — *Synthesis* **1989**, 477.

Alkylidenecarbenes — *Acc. Chem. Res.* **1982**, *15*, 348.

Alkylidenephosphoranes — *Synthesis* **1974**, 775.

Alkyllithiums
Lithioalkenes — *Org. Reactions* **1990**, *39*, 1.

Alkylsulfamoyl chlorides — *Angew. Chem. Int. Ed. Engl.* **1981**, *20*, 151.

Alkyltransition metal complexes
Photochemistry — *Angew. Chem. Int. Ed. Engl.* **1984**, *23*, 766.

Alkynes — *Acc. Chem. Res.* **1976**, *9*, 358.
Acc. Chem. Res. **1987**, *20*, 65.
Org. Reactions **1949**, *5*, 1.
Chem. Rev. **1984**, *84*, 545.
See also propargylic compounds
Acylation reactions — *SYNLETT* **1991**, 755.
Addition reactions — *Angew. Chem. Int. Ed. Engl.* **1967**, 6, 423.
Alkenes — *Chem. Rev.* **1942**, *31*, 77.
Alkyne acetals
Organocuprates — *Pure & Appl. Chem.* **1983**, *55*, 1759.

Alkynes, cont'd.	
Carbometallations	*Synthesis* **1981**, 841.
Coupling reactions	*Org. Reactions* **1984**, *32*, 375.
Cyclizations	*Angew. Chem. Int. Ed. Engl.* **1969**, *8*, 727.
Transition metal catalysis	*Acc. Chem. Res.* **1977**, *10*, 1.
Cycloaddition reactions	*Tetrahedron* **1981**, *37*, 3765.
Transition metal catalysis	*Chem. Rev.* **1988**, *88*, 1081.
Cycloalkynes	
Organometallic chemistry	*Chem. Rev.* **1988**, *88*, 1047.
Cyclotrimerization	*Pure Appl. Chem.* **1973**, *33*, 489.
Diazoacetic esters	*Org. Reactions* **1970**, *18*, 217.
Electrophilic additions	*Acc. Chem. Res.* **1981**, *14*, 227.
Equivalents	
Cycloadditions	*Tetrahedron* **1984**, *40*, 2585.
Hydroboration	*Org. Reactions* **1963**, *13*, 20.
Hydrogenations	*Synthesis* **1973**, 457.
Isomerization	*Quart. Rev.* **1970**, *24*, 585.
Nitriles	*Angew. Chem. Int. Ed. Engl.* **1978**, *17*, 505.
Nucleophilic substitution	*Acc. Chem. Res.* **1976**, *9*, 358.
Oligomerization	
Palladium catalysis	*Acc. Chem. Res.* **1976**, *9*, 93.
Organoiron chemistry	
Nucleophilic addition	*Acc. Chem. Res.* **1988**, *21*, 229.
Organometallic chemistry	*Chem. Rev.* **1988**, *88*, 1047.
	Pure & Appl. Chem. **1990**, *62*, 1021.
	Chem. Rev. **1973**, *73*, 163.
Oxidations	*Chem. Rev.* **1970**, *70*, 267.
Reductions	*Russ. Chem. Rev.* **1962**, *31*, 569.
Alkenes	*Chem. Rev.* **1942**, *31*, 77.
Alkynyl carboxylate esters	*Acc. Chem. Res.* **1991**, *24*, 304.
Alkynyl sulfonate esters	*Acc. Chem. Res.* **1991**, *24*, 304.
Alkynylsilanes	
Cyclization reaction	*Chem. Rev.* **1986**, *86*, 857.
Allene oxides	*Tetrahedron* **1980**, *36*, 2269.
Allenes	*Tetrahedron* **1984**, *40*, 2805.
Asymmetric synthesis	*Synthesis* **1973**, 25.
Electrophilic additions	
Reaction mechanisms	*Chem. Rev.* **1983**, *83*, 263.
Heteroallenes	*Chem. Rev.* **1978**, *78*, 569.
Organopalladium chemistry	*Pure & Appl. Chem.* **1990**, *62*, 1867.
Allenic ketones	*Tetrahedron* **1980**, *36*, 331.
Allergic contact dermatitis	*Pure & Appl. Chem.* **1990**, *62*, 1251.
Allyl alkylations	
Palladium catalysis	*Acc. Chem. Res.* **1980**, *13*, 385.

Allyl anions	*J. Organometal. Chem.* **1974**, 69, 1.
π-Allyl metal compounds	*Chem. Rev.* **1973**, *73*, 487.
	Chem. Soc. Rev. **1985**, *14*, 93.
	Org. Reactions **1972**, *19*, 115.
	Pure & Appl. Chem. **1989**, *61*, 1673.
Photochemistry	*Acc. Chem. Res.* **1986**, *19*, 268.
Allyl cations	
Cycloadditions	*Angew. Chem. Int. Ed. Engl.* **1973**, *12*, 819.
	Angew. Chem. Int. Ed. Engl. **1984**, *23*, 1.
Allyl compounds	
Nucleophilic substitution	*Tetrahedron* **1980**, *36*, 1901.
Allyl oxo carboxylates	
Elimination reactions	*Acc. Chem. Res.* **1987**, *20*, 140.
Allylamines	*Synthesis* **1983**, 685.
Antifungal compounds	*Angew. Chem. Int. Ed. Engl.* **1987**, *26*, 320.
Hydrogen migrations	
Asymmetric synthesis	*Synthesis* **1991**, 665.
Allylations	
Carbonyl compounds	
Palladium catalysts	*Acc. Chem. Res.* **1987**, *20*, 140.
Allylboranes	*Pure & Appl. Chem.* **1987**, *59*, 895.
	Pure & Appl. Chem. **1988**, *60*, 123.
Allylcobalt	*Angew. Chem. Int. Ed. Engl.* **1973**, *12*, 964.
Allylic alkylations	
Organometallic chemistry	*Pure & Appl. Chem.* **1982**, *54*, 189.
Palladium catalysis	*Acc. Chem. Res.* **1980**, *13*, 385.
Allylic compounds	
Asymmetric methods	*Pure & Appl. Chem.* **1985**, *57*, 1845.
Isomerization	*Pure & Appl. Chem.* **1985**, *57*, 1845.
NMR	
Spin-spin coupling	*Chem. Rev.* **1976**, *76*, 593.
Polymerizations	*Chem. Rev.* **1958**, *58*, 807.
	Tetrahedron **1983**, *39*, 2733.
Substitution reactions	*Chem. Rev.* **1956**, *56*, 753.
Allylic rearrangement	
Enzymes	*Chem. Rev.* **1990**, *90*, 1203.
Allylic strain	*Chem. Rev.* **1989**, *89*, 1841.
Allylmetal compounds	
Aldehydes	
Asymmetric synthesis	*Acc. Chem. Res.* **1987**, *20*, 243.
	Angew. Chem. Int. Ed. Engl. **1982**, *21*, 555.
	Acc. Chem. Res. **1987**, *20*, 243.
	Chem. Rev. **1989**, *89*, 257.

Allylpyrophosphates	
Cyclizations	*Acc. Chem. Res.* **1985**, *18*, 220.
Metabolism	*Tetrahedron* **1980**, *36*, 1109.
Allylsilanes	*Acc. Chem. Res.* **1988**, *21*, 200.
	Org. Reactions **1965**, *14*, 52.
	Pure & Appl. Chem. **1982**, *54*, 1.
	SYNLETT **1989**, 1.
	Synthesis **1990**, 969.
	Synthesis **1990**, 1101.
Electrophilic substitution	*Org. Reactions* **1989**, *37*, 57.
Intramolecular addition reactions	*Synthesis* **1988**, 263.
Synthons	*Acc. Chem. Res.* **1988**, *21*, 200.
Allylstannanes	
Cycloaddition reactions	*SYNLETT* **1991**, 1.
Allylsulfoxides	*Acc. Chem. Res.* **1974**, *7*, 147.
Alumina surfaces	
Synthetic methods	*Angew. Chem. Int. Ed. Engl.* **1978**, *17*, 487.
Ambident anions	
Alkylations	*Synthesis* **1970**, 1.
Amides	
Alkenylations	*Synthesis* **1975**, 685.
Aromatics	
Orthometalations	*Chem. Rev.* **1990**, *90*, 879.
Hydrolysis	*Quart. Rev.* **1970**, *24*, 553.
	Tetrahedron **1975**, *31*, 2463.
Metal complexes	*Chem. Rev.* **1982**, *82*, 385.
Photochemistry	*Pure & Appl. Chem.* **1988**, *60*, 941.
Radical chemistry	*Tetrahedron* **1978**, *34*, 3241.
Sulfurization	*Angew. Chem. Int. Ed. Engl.* **1966**, *5*, 451.
Amidoalkylations	*Org. Reactions* **1965**, *14*, 52.
	Synthesis **1970**, 49.
	Synthesis **1984**, 85.
	Synthesis **1984**, 181.
Amidomethylations	*Angew. Chem.* **1957**, *69*, 463.
Amination	
Alkenes	*Pure & Appl. Chem.* **1985**, *57*, 1917.
	Tetrahedron **1983**, *39*, 703.
Carbanions	*Chem. Rev.* **1989**, *89*, 1947.
Heterocycles	*Org. Reactions* **1942**, *1*, 91.
Mesitylenesulfonylhydroxylamine	*Synthesis* **1977**, 1.
Oxaziridines	*Synthesis* **1991**, 327.
Amine oxides	
Pyrolysis	*Org. Reactions* **1960**, *11*, 317.

Amines	
Acyl chlorides	*Synthesis* **1989**, 1.
Alkylations	*Org. Reactions* **1953**, *7*, 99.
α-Amidoalkylations	*Angew. Chem. Int. Ed. Engl.* **1975**, *14*, 15.
	Org. Reactions **1965**, *14*, 52.
Aromatic amines	
Reductions	*Org. Reactions* **1944**, *2*, 262.
Aromatic compounds	
Organolithium compounds	*Acc. Chem. Res.* **1982**, *15*, 306.
Basicity	*Chem. Rev.* **1989**, *89*, 1215.
Chlorinations	*Angew. Chem.* **1962**, *74*, 848.
Circular dichroism	*Chem. Rev.* **1983**, *83*, 359.
Conformations	*Chem. Soc. Rev.* **1976**, *5*, 411.
Dealkylations	*Synthesis* **1989**, 1.
Diazotizations	
Heterocyclic	*Chem. Rev.* **1975**, *75*, 241.
Heterocycles	*Angew. Chem. Int. Ed. Engl.* **1984**, *23*, 420.
α-Metalloamines	*Chem. Rev.* **1984**, *84*, 471.
Nitrous acid	*Acc. Chem. Res.* **1971**, *4*, 315.
Oxidations	
Aromatic compounds	
Electrochemistry	*Acc. Chem. Res.* **1969**, *2*, 175.
Photoreductions	*Chem. Rev.* **1973**, *73*, 141.
Piperidines	*Synthesis* **1984**, 895.
Primary amines	
Diazotization	*Chem. Rev.* **1975**, *75*, 241.
Pyrilium cations	*Tetrahedron* **1980**, *36*, 679.
Protecting groups	*Acc. Chem. Res.* **1973**, *6*, 191.
Fluorenylmethoxycarbonyl	*Acc. Chem. Res.* **1987**, *20*, 401.
Reductions	*Org. Reactions* **1944**, *2*, 262.
Photochemistry	*Chem. Rev.* **1973**, *73*, 141.
Reductive alkylations	*Org. Reactions* **1948**, *4*, 174.
Solvation	*Acc. Chem. Res.* **1971**, *4*, 107.
Strain effects	*Chem. Rev.* **1989**, *89*, 1215.
Thioacetylations	*Angew. Chem. Int. Ed. Engl.* **1966**, *5*, 458.
Aminimides	*Chem. Rev.* **1973**, *73*, 255.
Aminium radicals	*Chem. Rev.* **1978**, *78*, 243.
Amino acids	*Acc. Chem. Res.* **1988**, *21*, 294.
	Angew. Chem. Int. Ed. Engl. **1982**, *21*, 584.
	Angew. Chem. Int. Ed. Engl. **1983**, *22*, 816.
	Angew. Chem. Int. Ed. Engl. **1991**, *30*, 238.
	Tetrahedron **1991**, *47*, 6079.
Chirons	*Angew. Chem. Int. Ed. Engl.* **1991**, *30*, 1531.

Amino acids, cont'd.	
Complexes	
Magnesium chemistry	*Angew. Chem. Int. Ed. Engl.* **1990**, *29*, 1090.
Cyclopropane amino acids	*Tetrahedron* **1990**, *46*, 2231.
2,5-Diketopiperazines	*Pure & Appl. Chem.* **1983**, *55*, 1799.
α,α-Disubstituted amino acids	
3-Amino-2H-azirines	*Angew. Chem. Int. Ed. Engl.* **1991**, *30*, 238.
N-Hydroxy-α-amino acids	*Chem. Rev.* **1986**, *86*, 697.
Polymerization	
Enzymes	*Acc. Chem. Res.* **1973**, *6*, 361.
Synthesis	*Angew. Chem. Int. Ed. Engl.* **1978**, *17*, 176.
	Tetrahedron **1988**, *44*, 5253ff.
Vitamin B_6	*Acc. Chem. Res.* **1989**, *22*, 115.
Amino imino boranes	*Angew. Chem. Int. Ed. Engl.* **1988**, *27*, 1603.
Aminoacyl RNA transferase	*Acc. Chem. Res.* **1973**, *6*, 299.
Aminoacyl transfer RNA	
Peptides	*Acc. Chem. Res.* **1977**, *10*, 239.
Aminoacyl tRNA synthetase	*Angew. Chem. Int. Ed. Engl.* **1981**, *20*, 217.
Proofreading	*Acc. Chem. Res.* **1987**, *20*, 79.
α-Aminoaldehydes	*Tetrahedron* **1980**, *36*, 2359.
Asymmetric synthesis	*Chem. Rev.* **1989**, *89*, 149.
α-Aminoalkylations	
Umpolung	*Angew. Chem. Int. Ed. Engl.* **1975**, *14*, 15.
Aminoarsanes	*Synthesis* **1982**, 173.
3-Amino-2-azetidinones	*Tetrahedron* **1991**, *47*, 7503.
3-Amino-2*H*-azirines	*Angew. Chem. Int. Ed. Engl.* **1991**, *30*, 238.
2-Aminobenzophenones	*Synthesis* **1980**, 677.
2-Aminobenzimidazoles	*Synthesis* **1983**, 861.
5-Amino-2,4-pentadienals	*Synthesis* **1980**, 589.
Aminosugars	*Pure & Appl. Chem.* **1989**, *61*, 1217.
Aminyl free radicals	*Angew. Chem. Int. Ed. Engl.* **1975**, *14*, 783.
Ammonium compounds	
Alkylations	*Org. Reactions* **1953**, *7*, 99.
Rearrangement reactions	*Org. Reactions* **1970**, *18*, 403.
Ammonium formate	
Catalytic hydrogen transfer	*Synthesis* **1988**, 91.
Ammoxidation	
Propene	*Angew. Chem. Int. Ed. Engl.* **1966**, *5*, 642.
Anchimeric assistance	
Bond homolyses	
Reaction mechanisms	*Angew. Chem. Int. Ed. Engl.* **1979**, *18*, 173.

Anchored complexes	
Catalysis	*Pure & Appl. Chem,* **1980**, *52*, 2075.
Anh-Felkin addition	
Alkoxycarbonyl compounds	*Angew. Chem. Int. Ed. Engl.* **1984**, *23*, 556.
Animal defense	*Tetrahedron* **1982**, *38*, 1853ff.
Anion hosts	
Host-guest chemistry	*Pure & Appl. Chem.* **1989**, *61*, 1535.
Anionic reactions	
Ab initio calculations	*Chem. Rev.* **1987**, *87*, 535.
Cyclizations	
Phenols	*Chem. Soc. Rev.* **1983**, *12*, 213.
Cycloadditions	*Angew. Chem. Int. Ed. Engl.* **1974**, *13*, 627.
Cyclopentanoids	*Chem. Rev.* **1989**, *89*, 1467.
Hydrocarbons	*Chem. Rev.* **1987**, *87*, 557.
Organometallic chemistry	*Acc. Chem. Res.* **1988**, *21*, 147.
Photochemistry	*Chem. Rev.* **1979**, *79*, 253.
Anisotropic media	*Tetrahedron* **1987**, *43*, 1197.
Annelated arenes	*Angew. Chem. Int. Ed. Engl.* **1980**, *19*, 243.
Annelated indoles	*Chem. Rev.* **1989**, *89*, 1681.
Annelations	*Chem. Rev.* **1986**, *86*, 831.
	SYNLETT **1990**, 433.
	Tetrahedron **1976**, *32*, 3.
Heteroannulations	*Tetrahedron* **1980**, *36*, 2359.
Annulenes	
Dehydro[8]annulenes	*Acc. Chem. Res.* **1982**, *15*, 96.
Heteroannulenes	*Acc. Chem. Res.* **1976**, *9*, 453.
Oxidations	*Chem. Rev.* **1984**, *84*, 603.
Reductions	*Chem. Rev.* **1984**, *84*, 603.
Annulenoannulenes	*Angew. Chem. Int. Ed. Engl.* **1979**, *18*, 202.
Anomeric effect	
Organosulfur chemistry	
Heterocycles	*Acc. Chem. Res.* **1989**, *22*, 357.
Reaction mechanisms	*Tetrahedron* **1992**, *48*, 5019.
Anomerization	*Acc. Chem. Res.* **1986**, *19*, 307.
Carbohydrates	*Acc. Chem. Res.* **1978**, *11*, 136.
Anthocyanidins	*Tetrahedron* **1983**, *39*, 3005.
Anthocyanins	*Angew. Chem. Int. Ed. Engl.* **1991**, *30*, 17.
	Tetrahedron **1983**, *39*, 3005.
Anthracenes	
Photochemistry	*Pure & Appl. Chem,* **1980**, *52*, 2633.
	Pure & Appl. Chem. **1982**, *54*, 1589.
Anthracyclinones	*Tetrahedron* **1984**, *40*, 4537ff.
	Tetrahedron **1990**, *46*, 291.
Synthesis	*Angew. Chem. Int. Ed. Engl.* **1986**, *25*, 790.

Arenes	*SYNLETT* **1990**, 564.
Additions	*Synthesis* **1980**, 165.
	Synthesis **1980**, 769.
Arylations	*Acc. Chem. Res.* **1989**, *22*, 275.
Cycloadditions	
Photochemistry	*Pure & Appl. Chem.* **1990**, *62*, 1597.
Epoxides	*Acc. Chem. Res.* **1974**, *7*, 85.
Organometallic complexes	*Pure & Appl. Chem.* **1981**, *53*, 2379.
Arenesulfonylhydrazones	
Lithioalkenes	*Org. Reactions* **1990**, *39*, 1.
Arenesulfonyloxy groups	
Electrophiles	*Tetrahedron* **1991**, *47*, 1109.
Arginase	
Nutritional control	*Acc. Chem. Res.* **1970**, *3*, 113.
Aromatic amino acid hydrolase	
Reaction mechanisms	*Acc. Chem. Res.* **1988**, *21*, 101.
Aromatic compounds	*Angew. Chem. Int. Ed. Engl.* **1980**, *19*, 243.
	Angew. Chem. Int. Ed. Engl. **1991**, *30*, 1598.
	Chem. Soc. Rev. **1986**, *15*, 261.
	SYNLETT **1991**, 369.
Alkenes	
Palladium chemistry	*Synthesis* **1973**, 524.
Alkylations	*Chem. Rev.* **1987**, *87*, 1277.
Amides	
Orthometallations	*Chem. Rev.* **1990**, *90*, 879.
Benzyne	*Acc. Chem. Res.* **1969**, *2*, 273.
Biosynthesis	*Tetrahedron* **1978**, *34*, 3353.
Bis-Wittig reactions	*Synthesis* **1975**, 765.
Carbenes	*Acc. Chem. Res.* **1988**, *21*, 236.
Carboxymethylations	*Synthesis* **1970**, 628.
Chloromethylation	*Org. Reactions* **1942**, *1*, 63.
Complexes	
Organoiron chemistry	*Tetrahedron* **1983**, *39*, 4027.
Coupling reactions	*Chem. Rev.* **1987**, *87*, 357.
Cycloaddition reactions	*Chem. Rev.* **1989**, *89*, 827.
Photochemistry	*Pure & Appl. Chem*, **1980**, *52*, 2669.
Dianions	*Chem. Rev.* **1974**, *74*, 243.
Diazoacetic esters	*Org. Reactions* **1970**, *18*, 217.
Halides	
Cyanations	*Chem. Rev.* **1987**, *87*, 779.
Halogenations	*Pure & Appl. Chem.* **1963**, *7*, 193.
Reaction mechanisms	*Acc. Chem. Res.* **1974**, *7*, 361.
Hydrogenation	*Acc. Chem. Res.* **1979**, *12*, 324.

Aromatic compounds, cont'd.	
Hydroxylation	*Acc. Chem. Res.* **1971**, *4*, 337.
Photochemical	*Synthesis* **1974**, 173.
Ipso substitution	*Acc. Chem. Res.* **1980**, *13*, 51.
Radical reactions	*Pure & Appl. Chem.* **1981**, *53*, 239.
Ketones	
Nucleophilic addition	*Acc. Chem. Res.* **1988**, *21*, 414.
Polymerizations	*Angew. Chem. Int. Ed. Engl.* **1972**, *11*, 974.
Nitrations	*Acc. Chem. Res.* **1971**, *4*, 240.
	Acc. Chem. Res. **1971**, *4*, 248.
	Synthesis **1977**, 217.
Ipso substitution	*Acc. Chem. Res.* **1976**, *9*, 287.
Reaction mechanisms	*Acc. Chem. Res.* **1987**, *20*, 53.
Side reactions	*Synthesis* **1977**, 217.
Nitrenes	*Acc. Chem. Res.* **1972**, *5*, 303.
Nitro compounds	
Nucleophilic displacement	*Tetrahedron* **1978**, *34*, 2057.
Reductions	*Synthesis* **1969**, 9.
Nucleic acids	
Binding	*Acc. Chem. Res.* **1988**, *21*, 66.
Organolithium compounds	*Synthesis* **1983**, 957.
Amines	*Acc. Chem. Res.* **1982**, *15*, 306.
Photoadditions	*Chem. Rev.* **1987**, *87*, 811.
Photochemistry	*Chem. Rev.* **1974**, *74*, 29.
	Pure & Appl. Chem. **1982**, *54*, 1633.
Photosubstitution	*Chem. Rev.* **1975**, *75*, 353.
	Pure & Appl. Chem. **1975**, *41*, 601.
	Pure & Appl. Chem. **1983**, *55*, 331.
Polyaromatic compounds	
Arynes	*Acc. Chem. Res.* **1978**, *11*, 283.
Polycarbonates	*Angew. Chem. Int. Ed. Engl.* **1991**, *30*, 1598.
Polyketides	
Biosynthesis	*Tetrahedron* **1977**, *33*, 2159.
Protecting groups	*Synthesis* **1979**, 921.
Radical anions	*Chem. Rev.* **1974**, *74*, 243.
Reductions	*Tetrahedron* **1989**, *45*, 1579.
Metal ammonia reductions	*Synthesis* **1970**, 161.
Sulfonations	*Org. Reactions* **1946**, *3*, 141.
Trifluoromethyl compounds	
Nucleophiles	*Acc. Chem. Res.* **1978**, *11*, 197.
Wittig reaction	*Synthesis* **1975**, 765.
Aromatic diazonium ions	*Angew. Chem. Int. Ed. Engl.* **1978**, *17*, 141.
Aromatic indices	*Pure & Appl. Chem.* **1982**, *54*, 1097.

Aromatic systems	
Hyperconjugation	*Chem. Rev.* **1971**, *71*, 139.
Aromaticity	*Angew. Chem. Int. Ed. Engl.* **1971**, *10*, 761.
	Pure & Appl. Chem. **1990**, *62*, 417.
Annulenoannulenes	*Angew. Chem. Int. Ed. Engl.* **1979**, *18*, 202.
Aromatic indices	*Pure & Appl. Chem.* **1982**, *54*, 1097.
Bent benzene	*Pure & Appl. Chem.* **1990**, *62*, 373.
Benzenoid polycyclic compounds	*Pure & Appl. Chem.* **1982**, *54*, 1075.
Benzvalene	*Angew. Chem. Int. Ed. Engl.* **1981**, *20*, 529.
Bond order	*Pure & Appl. Chem.* **1983**, *55*, 269.
Bridged annulenes	*Pure & Appl. Chem.* **1982**, *54*, 1015.
Conductivity	*Pure & Appl. Chem.* **1982**, *54*, 1051.
Cyclohepta[*b*][1,4]benzoxazine	*Pure & Appl. Chem.* **1982**, *54*, 975.
Cyclooctatetraene	*Tetrahedron* **1975**, *31*, 2855.
	Pure & Appl. Chem. **1982**, *54*, 987.
Cyclophanes	*Pure & Appl. Chem.* **1990**, *62*, 373.
Cyclopropenylidene compounds	*Pure & Appl. Chem.* **1982**, *54*, 1059.
Diatropicity	*Pure & Appl. Chem.* **1982**, *54*, 1115.
4n π-Electron molecules	*Pure & Appl. Chem.* **1982**, *54*, 927.
8π-Electron heterocycles	*Angew. Chem. Int. Ed. Engl.* **1975**, *14*, 581.
Fullerenes	*Pure & Appl. Chem.* **1990**, *62*, 407.
Heterocycles	*Acc. Chem. Res.* **1972**, *5*, 281.
Homoaromaticity	*Angew. Chem. Int. Ed. Engl.* **1978**, *17*, 106.
Magnetic circular dichroism	*Tetrahedron* **1984**, *40*, 3845.
Nonbenzenoid phanes	*Pure & Appl. Chem.* **1982**, *54*, 957.
Organic metals	*Angew. Chem. Int. Ed. Engl.* **1977**, *16*, 519.
Organometallic compounds	*Pure & Appl. Chem.* **1990**, *62*, 383.
Polycyclic aromatic ions	*Pure & Appl. Chem.* **1982**, *54*, 1005.
Polycyclic conjugated systems	*Pure & Appl. Chem.* **1982**, *54*, 939.
Polysilanes	*Pure & Appl. Chem.* **1982**, *54*, 1041.
Porphyrins	
Theoretical aspects	*Chem. Rev.* **1975**, *75*, 85.
Principle of areno-analogy	
Heterocyclopolyaromatic compounds	*Angew. Chem. Int. Ed. Engl.* **1979**, *18*, 1.
Quinoid systems	*Pure & Appl. Chem.* **1990**, *62*, 395.
Synthesis	*Pure & Appl. Chem.* **1986**, *58*, 1ff.
Synthetic methods	*Pure & Appl. Chem.* **1980**, *52*, 1397ff.
Theoretical aspects	*Pure & Appl. Chem.* **1980**, *52*, 1397ff.
	Pure & Appl. Chem. **1982**, *54*, 1097.
	Pure & Appl. Chem. **1982**, *54*, 1129.
	Pure & Appl. Chem. **1986**, *58*, 1ff.
Troponoids	*Pure & Appl. Chem.* **1982**, *54*, 975.

Arsabenzene	*Acc. Chem. Res.* **1978**, *11*, 153.
Arsenides	*Synthesis* **1974**, 328.
Aryl cations	
Reaction mechanisms	*Chem. Soc. Rev.* **1979**, *8*, 353.
Aryl compounds	
Coupling reactions	*Tetrahedron* **1980**, *36*, 3327.
Aryl ethers	
Oxidations	*Chem. Rev.* **1969**, *69*, 499.
Aryl ethylenes	
Photochemistry	*Pure & Appl. Chem.* **1984**, *56*, 1225.
Aryl halides	
Dehydrogenations	*Acc. Chem. Res.* **1972**, *5*, 139.
Nucleophilic substitution	
Copper catalysis	*Tetrahedron* **1984**, *40*, 1433.
Photocyclizations	*Chem. Soc. Rev.* **1981**, *10*, 181.
Aryl α-methylene groups	
Autoxidation	*Chem. Rev.* **1970**, *70*, 944.
Aryl radicals	
Arenediazomium ions	*Chem. Rev.* **1988**, *88*, 765.
Aryl synthons	*SYNLETT* **1991**, 134.
Aryl triflates	
Coupling reactions	*Acc. Chem. Res.* **1988**, *21*, 47.
2-Aryl(aralkyl)-4-arylidene (alkylidene)-5(4*H*)oxazolones	*Synthesis* **1975**, 749.
α-Arylalkanoic acids	*Angew. Chem. Int. Ed. Engl.* **1984**, *23*, 413.
Arylations	*Chem. Soc. Rev.* **1986**, *15*, 261.
	Tetrahedron **1988**, *44*, 3039.
Arenes	*Acc. Chem. Res.* **1989**, *22*, 275.
Azonium salts	
Unsaturated compounds	*Org. Reactions* **1976**, *24*, 225.
Diazonium salts	*Org. Reactions* **1960**, *11*, 189.
	Org. Reactions **1976**, *24*, 225.
Homolytic	*Chem. Rev.* **1957**, *57*, 123.
Meerwein arylation reaction	*Org. Reactions* **1976**, *24*, 225.
Organobismuth reagents	*Chem. Rev.* **1989**, *89*, 1487.
Unsaturated compounds	
Azonium salts	*Org. Reactions* **1976**, *24*, 225.
Aryldiazonium cations	
Co-ordination chemistry	*Chem. Soc. Rev.* **1975**, *4*, 443.
4-Aryldihydropyridines	*Angew. Chem. Int. Ed. Engl.* **1981**, *20*, 762.
Arylhydrazines	*Tetrahedron* **1980**, *36*, 161.
Arylhydrazones	*Tetrahedron* **1980**, *36*, 161.
Aryliodine(III) dicarboxylates	*Chem. Soc. Rev.* **1981**, *10*, 377.
Arylnitrenes	*Angew. Chem. Int. Ed. Engl.* **1979**, *18*, 900.

2-Arylpropionic acids *Tetrahedron* **1986**, *42*, 4095.
Asymmetric synthesis *Tetrahedron: Asymmetry* **1992**, *3*, 163.
Aryltrichloromethylcarbinols
Nucleophilic addition *Synthesis* **1971**, 131.
Aryltrimethylsilanes *Synthesis* **1979**, 841.
Arynes *Acc. Chem. Res.* **1978**, *11*, 283.
Aromatic substitution *Angew. Chem.* **1960**, *72*, 91.
Asymmetric induction *Acc. Chem. Res.* **1985**, *18*, 280.
Asymmetric mechanisms
Biological specificity *Pure & Appl. Chem.* **1982**, *54*, 1819.
Reaction mechanisms *Pure & Appl. Chem.* **1982**, *54*, 1819.
Asymmetric membranes *Angew. Chem. Int. Ed. Engl.* **1975**, *14*, 457.
Asymmetric methods *Acc. Chem. Res.* **1984**, *17*, 338.
Acc. Chem. Res. **1987**, *20*, 72.
Angew. Chem. Int. Ed. Engl. **1991**, *30*, 1193.
Pure & Appl. Chem. **1990**, *62*, 1209.
Pure & Appl. Chem. **1991**, *63*, 1591.
See also alkylations, chiral auxiliaries, cycloaddition reactions, enzymes, heterogenous catalysis, insect pheromones, organometallic chemistry, and reductions.
Synthesis **1978**, 329.
Tetrahedron **1979**, *35*, 2797.
Tetrahedron **1984**, *40*, 1213ff.
Tetrahedron **1986**, *42*, 5157.
Addition reactions
Nucleophilic additions *Chem. Rev.* **1975**, *75*, 521.
Anh-Felkin addition *Angew. Chem. Int. Ed. Engl.* **1984**, *23*, 556.
Chelation controlled addition *Angew. Chem. Int. Ed. Engl.* **1984**, *23*, 556.
Circular dichroism
Absolute configurations *Angew. Chem. Int. Ed. Engl.* **1979**, *18*, 363.
Liquid crystals *Angew. Chem. Int. Ed. Engl.* **1984**, *23*, 348.
Radical chemistry *Tetrahedron* **1981**, *37*, 3073.
Stereochemistry
Crystal structures *Angew. Chem. Int. Ed. Engl.* **1975**, *14*, 460.
ATP *Pure & Appl. Chem.* **1984**, *56*, 1025.
Analogues *Pure & Appl. Chem.* **1984**, *56*, 1025.
Nucleotides
Phosphorylations *Pure & Appl. Chem.* **1980**, *52*, 2213.
Phosphorothioate analogues *Acc. Chem. Res.* **1979**, *12*, 204.
Phosphorylations *Pure & Appl. Chem.* **1980**, *52*, 2213.

Autoxidation	
Aryl α-methylene groups	*Chem. Rev.* **1970**, *39*, 944.
Free radical	*Acc. Chem. Res.* **1968**, *1*, 193.
Inhibition	*Chem. Rev.* **1961**, *61*, 563.
Organometallic compounds	*Angew. Chem.* **1959**, *71*, 541.
Avermectins	*Pure & Appl. Chem.* **1987**, *59*, 299.
	Pure & Appl. Chem. **1990**, *62*, 1231.
	Chem. Soc. Rev. **1991**, *20*, 211.
	Chem. Soc. Rev. **1991**, *20*, 271.
Aza-arenes	*Angew. Chem. Int. Ed. Engl.* **1979**, *18*, 707.
3-Azabicyclo[3.3.1]nonanes	*Chem. Rev.* **1981**, *81*, 149.
Aza-crown ethers	*Chem. Rev.* **1989**, *89*, 929.
	Pure & Appl. Chem. **1989**, *61*, 1619.
	Synthesis **1982**, 997.
Azacyclophanes	
Inclusion complexes	*Pure & Appl. Chem.* **1988**, *60*, 549.
Azadienes	
2-Aza-1,3-dienes	*SYNLETT* **1990**, 129.
Diels-Alder reaction	*Chem. Rev.* **1986**, *86*, 781.
	Tetrahedron **1983**, *39*, 2869.
Azapeptides	*Synthesis* **1989**, 405.
Azaquinones	*Angew. Chem. Int. Ed. Engl.* **1973**, *12*, 139.
Azastilbenes	
Photochemistry	*Pure & Appl. Chem.* **1982**, *54*, 1705.
Azetidines	*Chem. Rev.* **1979**, *79*, 331.
Azides	*Acc. Chem. Res.* **1971**, *4*, 9.
	Chem. Rev. **1988**, *88*, 297.
Addition reactions	*Chem. Rev.* **1969**, *69*, 353.
Asymmetric synthesis	*Acc. Chem. Res.* **1971**, *4*, 9.
Pyrolysis	*Angew. Chem. Int. Ed. Engl.* **1987**, *26*, 504.
Azidotrimethylsilane	*Synthesis* **1980**, 861.
Azines	*Synthesis* **1976**, 349.
	Tetrahedron **1985**, *41*, 237.
	Tetrahedron **1988**, *44*, 1.
Aziridines	
Asymmetric synthesis	*Acc. Chem. Res.* **1973**, *6*, 341.
Isomerization	*Angew. Chem. Int. Ed. Engl.* **1962**, *1*, 528.
Azirines	
Cycloadditions	*Synthesis* **1975**, 483.
Photochemical	*Acc. Chem. Res.* **1976**, *9*, 317.
Azlactones	*Org. Reactions* **1946**, *3*, 198.
Azo compounds	
Elimination reactions	*Angew. Chem. Int. Ed. Engl.* **1977**, *16*, 835.
Isomerization	*Acc. Chem. Res.* **1973**, *6*, 275.

Spectroscopic properties — *Angew. Chem. Int. Ed. Engl.* **1973**, *12*, 224.
Thermolysis — *Angew. Chem. Int. Ed. Engl.* **1977**, *16*, 835.
Azo dyes
Oxidative couplings — *Angew. Chem. Int. Ed. Engl.* **1962**, *1*, 640.
Azoalkanes — *Angew. Chem. Int. Ed. Engl.* **1980**, *19*, 762.
Decomposition
Reaction mechanisms — *Chem. Rev.* **1980**, *80*, 99.
Photochemistry — *Angew. Chem. Int. Ed. Engl.* **1986**, *25*, 661.
Azobenzene
Photochemistry — *Chem. Rev.* **1989**, *89*, 1915.
Chem. Soc. Rev. **1972**, *1*, 481.
Azomethine — *Tetrahedron* **1976**, *32*, 2165.
Azomethine imines — *Synthesis* **1973**, 123.
Azomethine ylides — *Synthesis* **1973**, 123.
Azomethines
Lead tetraacetate — *Chem. Rev.* **1973**, *73*, 93.
Azonium salts
Arylation
Unsaturated compounds — *Org. Reactions* **1976**, *24*, 225.
Azoxyalkanes — *Acc. Chem. Res.* **1974**, *7*, 421.
Azulene
Quinones — *Pure & Appl. Chem.* **1983**, *55*, 363.
Bacteria methanogenic coenzyme F420 — *Acc. Chem. Res.* **1986**, *19*, 216.
Bacterial chemotaxis — *Chem. Rev.* **1987**, *87*, 997.
Bacterial flagella — *Angew. Chem. Int. Ed. Engl.* **1973**, *12*, 683.
Bacterial membranes — *Angew. Chem. Int. Ed. Engl.* **1979**, *18*, 337.
Peptidoglycans
Biosynthesis — *Acc. Chem. Res.* **1971**, *4*, 297.
Bacterial ribosome — *Angew. Chem. Int. Ed. Engl.* **1982**, *21*, 23.
Bacterial toxins — *Pure & Appl. Chem.* **1987**, *59*, 1477.
Bacteriorhodopsin — *Acc. Chem. Res.* **1985**, *18*, 331.
Pure & Appl. Chem. **1986**, *58*, 719.
Photochemistry — *Angew. Chem. Int. Ed. Engl.* **1976**, *15*, 17.
Baker's yeast
Reductions
Asymmetric synthesis — *Chem. Rev.* **1991**, *91*, 49.
Basic research
Economics — *Angew. Chem. Int. Ed. Engl.* **1982**, *21*, 109.
Beech lignin — *Angew. Chem. Int. Ed. Engl.* **1974**, *13*, 313.
Benzene
Photochemistry — *Tetrahedron* **1976**, *32*, 1309.
Tetrahedron **1977**, *33*, 2459.

Benzenoid polycyclic compounds	
Aromaticity	*Pure & Appl. Chem.* **1982**, *54*, 1075.
Benzilic acid	
Rearrangements	*Quart. Rev.* **1960**, *14*, 221.
	Chem. Rev. **1974**, *74*, 279.
Benzimidazoles	*Chem. Rev.* **1973**, *73*, 113.
Benzocyclopropenes	*Angew. Chem. Int. Ed. Engl.* **1971**, *10*, 34.
1,4-Benzodiazepines	*Synthesis* **1975**, 415.
Benzofuroxan	*Org. Reactions* **1948**, *4*, 269.
Benzoins	*Chem. Rev.* **1977**, *77*, 1.
Benzomorphans	
Benzoquinones	
Oxidations	*Org. Reactions* **1948**, *4*, 305.
Benzotriazole	*Tetrahedron* **1991**, *47*, 2683.
Benzoylformates	
Reductions	*Tetrahedron: Asymmetry* **1991**, *2*, 299.
Benzyl group	
Hydrogenolysis	*Org. Reactions* **1953**, *7*, 263.
Benzylic compounds	
Polymetalations	*Tetrahedron* **1983**, *39*, 2733.
Benzyne	
Addition reactions	*Acc. Chem. Res.* **1974**, *7*, 6.
	Acc. Chem. Res. **1974**, *7*, 301.
Aromatic compounds	*Acc. Chem. Res.* **1969**, *2*, 273.
Organometallic chemistry	*Chem. Rev.* **1988**, *88*, 1047.
Berry pseudorotation	*Angew. Chem. Int. Ed. Engl.* **1971**, *10*, 687.
Betaines	
Heterocycles	*Tetrahedron* **1985**, *41*, 2239.
Betweenanenes	*Acc. Chem. Res.* **1980**, *13*, 213.
Biaryl compounds	*Angew. Chem. Int. Ed. Engl.* **1990**, *29*, 977.
Bicyclic amide acetals	*Synthesis* **1971**, 16.
Bicyclic endoperoxides	*Chem. Rev.* **1981**, *81*, 91.
Bicyclic ketones	*Tetrahedron* **1981**, *37*, 1238.
Oxygen insertion	*Tetrahedron* **1981**, *37*, 2697.
Bicyclo[n.1.0]alkenes	*Chem. Rev.* **1989**, *89*, 1147.
Bicycloamidines	*Synthesis* **1972**, 591.
Bicyclo[3.2.0]heptanones	*Synthesis* **1977**, 155.
Bicyclomycin	*Chem. Rev.* **1988**, *88*, 511.
Bicyclo[2.1.0]pentane derivatives	*Tetrahedron* **1976**, *32*, 641.
Bifunctional anions	*Angew. Chem. Int. Ed. Engl.* **1976**, *15*, 321.
	Chem. Soc. Rev. **1990**, *19*, 1.
Bifunctional reagents	*Pure & Appl. Chem.* **1988**, *60*, 107.
Bile pigments	*Chem. Soc. Rev.* **1975**, *4*, 363.
Photochemistry	*Angew. Chem. Int. Ed. Engl.* **1983**, *22*, 656.

Biliproteins	*Angew. Chem. Int. Ed. Engl.* **1981**, *20*, 241.
Bilirubin	
Jaundice	*Acc. Chem. Res.* **1984**, *17*, 417.
Photochemistry	*Angew. Chem. Int. Ed. Engl.* **1983**, *22*, 656.
Biliverdin	
Photochemistry	*Angew. Chem. Int. Ed. Engl.* **1983**, *22*, 656.
BINAP	
Asymmetric synthesis	*Acc. Chem. Res.* **1990**, *23*, 345.
Binding	*Tetrahedron* **1986**, *42*, 1917.
BINOL compounds	
Asymmetric synthesis	*Pure & Appl. Chem.* **1981**, *53*, 2315.
Biocatalytic reactions	*Angew. Chem. Int. Ed. Engl.* **1985**, *24*, 539.
Biochemical	
Excited species	*Pure & Appl. Chem.* **1984**, *56*, 1179.
Methylations	*Chem. Rev.* **1945**, *36*, 315.
	Quart. Rev. **1955**, *9*, 255.
Photochemistry	*Pure & Appl. Chem.* **1984**, *56*, 1179.
Structure activity relationships	*Acc. Chem. Res.* **1969**, *2*, 232.
Biological chemistry	*Chem. Rev.* **1987**, *87*, 861.
Bioinorganic chemistry	*Pure & Appl. Chem.* **1983**, *55*, 1089.
Organometallic compounds	*J. Organometal. Chem.* **1974**, *76*, 265.
Biological membranes	*See membranes*
Biological specificity	*Pure & Appl. Chem.* **1982**, *54*, 1819.
Biomaterials	*Angew. Chem. Int. Ed. Engl.* **1982**, *21*, 837.
Biomembranes	*Angew. Chem. Int. Ed. Engl.* **1972**, *11*, 551.
	Angew. Chem. Int. Ed. Engl. **1990**, *29*, 1269.
Biomimetic approaches	*Chem. Soc. Rev.* **1972**, *1*, 553.
Bioorganic chemistry	*Tetrahedron* **1991**, *47*, 2351ff.
Natural products	*Pure & Appl. Chem.* **1981**, *53*, 1259.
Peroxides	*Angew. Chem. Int. Ed. Engl.* **1983**, *22*, 529.
Biopolymers	*Angew. Chem. Int. Ed. Engl.* **1974**, *13*, 121.
	Pure & Appl. Chem. **1987**, *59*, 437ff.
Bio-sensors	*Angew. Chem. Int. Ed. Engl.* **1991**, *30*, 516.
Biosynthesis	*See under specific compound or class of compounds.*
NMR	*Chem. Soc. Rev.* **1975**, *4*, 497.
	Chem. Soc. Rev. **1979**, *8*, 539.
	Pure & Appl. Chem. **1981**, *53*, 1241.
	Pure & Appl. Chem. **1986**, *58*, 753.
Spectroscopic methods	*Tetrahedron* **1983**, *39*, 3441ff.
Biotechnology	*Angew. Chem. Int. Ed. Engl.* **1991**, *30*, 677.
	Pure & Appl. Chem. **1988**, *60*, 821ff.
Biradical reactions	*Chem. Rev.* **1989**, *89*, 521.
Birth control	*Angew. Chem. Int. Ed. Engl.* **1977**, *16*, 506.

Bis(4,5-dihydrooxazolyl) derivatives	*Angew. Chem. Int. Ed. Engl.* **1991**, *30*, 542.
Bis(diphenylphosphino)methane	*Chem. Soc. Rev.* **1983**, *12*, 99.
Bishomocubanes	*Chem. Rev.* **1989**, *89*, 1011.
Bis-indole alkaloids	*Tetrahedron* **1982**, *38*, 223.
Bismuth chemistry	*Acc. Chem. Res.* **1978**, *11*, 363.
	Pure & Appl. Chem. **1987**, *59*, 937.
	Synthesis **1974**, 328.
Bis(trimethylsilyl)diimine	*Angew. Chem. Int. Ed. Engl.* **1971**, *10*, 374.
Bleomycin	
DNA	
Binding	*Acc. Chem. Res.* **1986**, *19*, 383.
Degradation	*Chem. Rev.* **1987**, *87*, 1107.
Blood groups	
Carbohydrates	*Chem. Soc. Rev.* **1978**, *7*, 423.
	Pure & Appl. Chem. **1981**, *53*, 89.
	Pure & Appl. Chem. **1984**, *56*, 807.
Blood substitutes	*Angew. Chem. Int. Ed. Engl.* **1978**, *17*, 621.
Bond homolyses	
Anchimeric assistance	
Reaction mechanisms	*Angew. Chem. Int. Ed. Engl.* **1979**, *18*, 173.
Bone	
Hydroxyapatitie	*Acc. Chem. Res.* **1975**, *8*, 273.
Boraheterocycles	*Tetrahedron* **1977**, *33*, 2331.
Boron hydrides	
Theoretical aspects	*Pure & Appl. Chem.* **1982**, *54*, 1143.
Boronates	
Fragmentations	*Synthesis* **1971**, 229.
Substitution reactions	*Acc. Chem. Res.* **1970**, *3*, 186.
Boronic esters	*Pure & Appl. Chem.* **1985**, *57*, 1741.
	Synthesis **1975**, 147.
	Tetrahedron **1989**, *45*, 1859.
Synthetic methods	*Acc. Chem. Res.* **1988**, *21*, 294.
Branched chain sugars	*Angew. Chem. Int. Ed. Engl.* **1972**, *11*, 159.
Synthesis	*Pure & Appl. Chem.* **1981**, *53*, 113.
Branched tannins	*Pure & Appl. Chem.* **1982**, *54*, 2465.
Bredt's rule	*Angew. Chem. Int. Ed. Engl.* **1973**, *12*, 464.
	Chem. Soc. Rev. **1974**, *3*, 41.
Brefeldin A	
Biosynthesis	*Acc. Chem. Res.* **1983**, *16*, 7.
Bridged annulenes	
Aromaticity	*Pure & Appl. Chem.* **1982**, *54*, 1015.
Bridged bicyclic ketones	
Ring expansion	*Tetrahedron* **1987**, *43*, 3.

Bridgehead alkenes *Angew. Chem. Int. Ed. Engl.* **1975**, *14*, 528.
Pure & Appl. Chem. **1987**, *59*, 1613.
Tetrahedron **1980**, *36*, 1683.
Bridgehead cage alcohols *Chem. Rev.* **1989**, *89*, 1035.
Bridgehead substitution *Angew. Chem.* **1960**, *72*, 147.
Chem. Rev. **1954**, *54*, 1065.
Bromination
Alkenes *Acc. Chem. Res.* **1990**, *23*, 87.
Alkyl bromides *Acc. Chem. Res.* **1984**, *17*, 160.
N-Bromosuccinimide
Reaction mechanisms *Acc. Chem. Res.* **1978**, *11*, 381.
Bromine azide
Alkenes *Acc. Chem. Res.* **1971**, *4*, 9.
Bromo compounds *Synthesis* **1977**, 145.
1-Bromo-19-methylbiladienes-*ac* *SYNLETT* **1990**, 235.
Bryophytes *Angew. Chem. Int. Ed. Engl.* **1991**, *30*, 130.
Bryostatin 1 *Pure & Appl. Chem.* **1988**, *60*, 1587.
Buckminsterfullerene *Angew. Chem. Int. Ed. Engl.* **1991**, *30*, 678.
Chem. Rev. **1991**, *91*, 1213.
Building blocks *Angew. Chem. Int. Ed. Engl.* **1979**, *18*, 429.
Organic metals *Angew. Chem. Int. Ed. Engl.* **1991**, *30*, 1121.
Butadiene
Addition reactions
Palladium chemistry *Acc. Chem. Res.* **1973**, *6*, 8.
Cooligomerization *Chem. Soc. Rev.* **1977**, *6*, 235.
Unsaturated compounds *Angew. Chem. Int. Ed. Engl.* **1965**, *4*, 330.
Dimerization *Angew. Chem. Int. Ed. Engl.* **1965**, *4*, 327.
Polymerization *Chem. Soc. Rev.* **1977**, *6*, 235.
Stereoselection *Acc. Chem. Res.* **1980**, *13*, 1.
Butadiene-Iron tricarbonyl complexes *Synthesis* **1989**, 341.
Cadaverine *Acc. Chem. Res.* **1982**, *15*, 290.
Cage compounds *Chem. Rev.* **1985**, *85*, 341.
SYNLETT **1991**, 73.
Tetrahedron **1988**, *44*, 2377.
Cage effects
Micelles
Reaction mechanisms *Pure & Appl. Chem.* **1981**, *53*, 259.
Calciferols *Acc. Chem. Res.* **1983**, *16*, 81.
Chem. Soc. Rev. **1977**, *6*, 83.
Calicheamicin
DNA
Antitumor compounds *Acc. Chem. Res.* **1991**, *24*, 235.

Calixarenes	*Acc. Chem. Res.* **1983**, *16*, 161.
	Pure & Appl. Chem. **1988**, *60*, 483.
Cation binding	*Pure & Appl. Chem.* **1989**, *61*, 1597.
Host-guest chemistry	*Pure & Appl. Chem.* **1986**, *58*, 1523.
Camphor	
Chiral auxiliaries	*Tetrahedron* **1987**, *43*, 1969.
Synthetic methods	*Pure & Appl. Chem.* **1990**, *62*, 1241.
Camptothecin	*Chem. Rev.* **1973**, *73*, 385.
	Tetrahedron **1981**, *37*, 1047.
Cannabis	*Chem. Rev.* **1976**, *76*, 75.
	Pure & Appl. Chem. **1986**, *58*, 693.
Carbalkoxycarbenes	*Acc. Chem. Res.* **1974**, *7*, 415.
	Chem. Rev. **1974**, *74*, 431.
Carbanions	*Acc. Chem. Res.* **1980**, *13*, 161.
	Chem. Soc. Rev. **1973**, *2*, 397.
Aldehydes	*Tetrahedron* **1988**, *44*, 4653.
Alkylations	*Chem. and Ind. (London)* **1973**, 937.
Allylic carbanions	*Org. Reactions* **1982**, *27*, 1.
Aminations	*Chem. Rev.* **1989**, *89*, 1947.
Benzylic carbanions	*Org. Reactions* **1982**, *27*, 1.
Delocalized carbanions	*SYNLETT* **1991**, 207.
Dianions	*Tetrahedron* **1991**, *47*, 4223.
Dipole stabilized carbanions	*Chem. Rev.* **1978**, *78*, 275.
Electron transfers	*Acc. Chem. Res.* **1972**, *5*, 169.
Elimination reactions	
Alkenes	*Acc. Chem. Res.* **1970**, *3*, 281.
Functionalized carbanions	*Org. Reactions* **1982**, *27*, 1.
Meisenheimer-Jackson complexes	*Acc. Chem. Res.* **1974**, *7*, 181.
Nitriles	*Org. Reactions* **1984**, *31*, 1.
NMR	*Pure & Appl. Chem.* **1989**, *61*, 709.
Spirocyclic compounds	*Acc. Chem. Res.* **1972**, *5*, 354.
Vinyl carbanions	*Tetrahedron* **1988**, *44*, 4653.
Carbene anion radicals	*Tetrahedron* **1989**, *45*, 3993.
Carbenes	*Angew. Chem. Int. Ed. Engl.* **1971**, *10*, 529.
	Angew. Chem. Int. Ed. Engl. **1971**, *10*, 537.
	Angew. Chem. Int. Ed. Engl. **1987**, *26*, 275.
	Angew. Chem. Int. Ed. Engl. **1991**, *30*, 674.
	Chem. Rev. **1988**, *88*, 1293.
	Chem. Rev. **1991**, *91*, 197.
	Chem. Rev. **1991**, *91*, 263.
	Tetrahedron **1985**, *41*, 1423ff.

Alkenes	
Reaction mechanisms	*Angew. Chem. Int. Ed. Engl.* **1990**, *29*, 1371.
Alkylations	*Chem. and Ind. (London)* **1973**, 881.
Alkylidenecarbenes	*Acc. Chem. Res.* **1982**, *15*, 348.
Aromatic compounds	*Acc. Chem. Res.* **1988**, *21*, 236.
Carbalkoxycarbenes	*Acc. Chem. Res.* **1974**, *7*, 415.
Carbene anion radicals	*Tetrahedron* **1989**, *45*, 3993.
Carbene complexes	*Angew. Chem. Int. Ed. Engl.* **1984**, *23*, 587.
	Angew. Chem. Int. Ed. Engl. **1988**, *27*, 1456.
	SYNLETT **1991**, 381.
Organometallic chemistry	*Pure & Appl. Chem.* **1990**, *62*, 691.
Carbenoids	*Tetrahedron* **1982**, *38*, 2751.
Intramolecular	*Tetrahedron* **1992**, *48*, 5385.
Organolithiums	*Angew. Chem. Int. Ed. Engl.* **1972**, *11*, 473.
Carboalkoxycarbenes	*Chem. Rev.* **1974**, *74*, 431.
Carbonyl-carbene complexes	*Pure & Appl. Chem.* **1983**, *55*, 1689.
Chromium carbenes	*Pure & Appl. Chem.* **1983**, *55*, 1745.
Dihalocarbenes	*Acc. Chem. Res.* **1972**, *5*, 65.
	Chem. Rev. **1988**, *88*, 1293.
Halocarbenes	*Org. Reactions* **1963**, *13*, 55.
Metal carbene complexes	*Chem. Rev.* **1986**, *86*, 919.
	Chem. Rev. **1987**, *87*, 411.
	Chem. Rev. **1988**, *88*, 1293.
	Chem. Soc. Rev. **1973**, *2*, 99.
Cycloadditions	*Pure & Appl. Chem.* **1988**, *60*, 137.
Organometallic chemistry	*Pure & Appl. Chem.* **1990**, *62*, 1021.
	Acc. Chem. Res. **1986**, *19*, 348.
	SYNLETT **1990**, 441.
Photochemistry	*Acc. Chem. Res.* **1977**, *10*, 85.
Radical ions	*Acc. Chem. Res.* **1988**, *21*, 400.
Reactivity	*Acc. Chem. Res.* **1989**, *22*, 15.
Rearrangements	*Topics Curr. Chem.* **1976**, *62*, 173.
Spectroscopy	*Acc. Chem. Res.* **1984**, *17*, 283.
Stable carbenes	*Angew. Chem. Int. Ed. Engl.* **1991**, *30*, 674.
Unsaturated carbenes	*Chem. Rev.* **1978**, *78*, 383.
	Chem. Rev. **1991**, *91*, 197.
	Pure & Appl. Chem. **1983**, *55*, 369.
Transition metal complexes	*Acc. Chem. Res.* **1980**, *13*, 327.
Vinylidene carbenes	*Chem. Rev.* **1991**, *91*, 197.
Carbenium ions	*Pure & Appl. Chem.* **1984**, *56*, 1819.
	Chem. Soc. Rev. **1987**, *16*, 75.
α-Acylcarbenium ions	*Acc. Chem. Res.* **1980**, *13*, 207.

Carbides	
Organometallic chemistry	*Pure & Appl. Chem.* **1990**, *62*, 1021.
Carbocations	*Acc. Chem. Res.* **1980**, *13*, 161.
	Chem. Rev. **1991**, *91*, 375.
	Chem. Rev. **1991**, *91*, 1625.
	Chem. Soc. Rev. **1973**, *2*, 397.
	Pure & Appl. Chem. **1991**, *63*, 231.
	See also carbenium and carbonium ions.
Cyclopropyl groups	*Chem. Rev.* **1974**, *74*, 315.
Destabilized carbocations	*Angew. Chem. Int. Ed. Engl.* **1984**, *23*, 20.
Electrophilic reactions	*Angew. Chem. Int. Ed. Engl.* **1973**, *12*, 173.
Multiply charged carbocations	*Tetrahedron* **1984**, *40*, 4161.
Organosilicon chemistry	*Tetrahedron* **1990**, *46*, 2677.
Pyramidal carbocations	*Angew. Chem. Int. Ed. Engl.* **1981**, *20*, 991.
Solvent effects	*Pure & Appl. Chem.* **1984**, *56*, 1797.
Carbodiimides	*Chem. Rev.* **1981**, *81*, 589.
	Tetrahedron **1981**, *37*, 233.
Carbohydrates	*Angew. Chem. Int. Ed. Engl.* **1973**, *12*, 721.
	Angew. Chem. Int. Ed. Engl. **1988**, *27*, 1267.
	Chem. Soc. Rev. **1973**, *2*, 355.
	Chem. Soc. Rev. **1979**, *8*, 85.
	Chem. Soc. Rev. **1987**, *16*, 161.
	Pure & Appl. Chem. **1981**, *53*, 15.
	Pure & Appl. Chem. **1984**, *56*, 779.
	Pure & Appl. Chem. **1984**, *56*, 859.
	Pure & Appl. Chem. **1987**, *59*, 1477.
	Pure & Appl. Chem. **1989**, *61*, 1283.
	See also blood types, oligosaccharides, and polysaccharides
	Tetrahedron **1987**, *43*, 2389.
	Tetrahedron **1991**, *47*, 6079.
	Tetrahedron **1992**, *48*, 2803.
Acetal formation	*Chem. Rev.* **1979**, *79*, 491.
Acetolysis	*Adv. Carbohydr. Chem.* **1967**, *22*, 11.
Anomeric center	*Pure & Appl. Chem.* **1991**, *63*, 507.
Anomeric effect	
Reaction mechanisms	*Tetrahedron* **1992**, *48*, 5019.
Antibiotic sugars	*Angew. Chem. Int. Ed. Engl.* **1971**, *10*, 236.
Antibody binding sites	*Pure & Appl. Chem.* **1989**, *61*, 1171.
Antigens	*Pure & Appl. Chem.* **1991**, *63*, 561.
Asymmetric synthesis	*SYNLETT* **1990**, 173.
Bacteria	
Enzymes	*Pure & Appl. Chem.* **1983**, *55*, 637.

Blood groups	*Pure & Appl. Chem.* **1981**, *53*, 89.
	Pure & Appl. Chem. **1984**, *56*, 807.
Branched sugars	*Angew. Chem. Int. Ed. Engl.* **1972**, *11*, 159.
	SYNLETT **1990**, 715.
Synthesis	*Pure & Appl. Chem.* **1981**, *53*, 113.
Chirons	*Acc. Chem. Res.* **1979**, *12*, 159.
	Pure & Appl. Chem. **1983**, *55*, 565.
	SYNLETT **1990**, 173.
	Tetrahedron **1984**, *40*, 3161.
Cloning	*Pure & Appl. Chem.* **1987**, *59*, 1489.
Complexes	*Chem. Soc. Rev.* **1980**, *9*, 415.
Deoxy sugars	*Chem. Rev.* **1991**, *91*, 25.
Deoxyfluoro sugars	*Chem. Rev.* **1991**, *91*, 25.
Ezomycins	*Pure & Appl. Chem.* **1981**, *53*, 129.
Fungal degradation	*Pure & Appl. Chem.* **1981**, *53*, 33.
Glycoconjugates	*Pure & Appl. Chem.* **1989**, *61*, 1257.
Glycoproteins	*Pure & Appl. Chem.* **1981**, *53*, 45.
	Pure & Appl. Chem. **1987**, *59*, 1457.
Glycosidation	*Angew. Chem. Int. Ed. Engl.* **1974**, *13*, 157.
	Pure & Appl. Chem. **1991**, *63*, 519.
	Synthesis **1991**, 583.
	Tetrahedron **1990**, *46*, 5835.
Glycosphingolipids	*Pure & Appl. Chem.* **1989**, *61*, 1307.
Hexenuloses	*Chem. Rev.* **1982**, *82*, 287.
Higher monosaccharides	*Angew. Chem. Int. Ed. Engl.* **1987**, *26*, 15.
Industrial chemistry	*Pure & Appl. Chem.* **1989**, *61*, 1313.
Interaction regulation	*Pure & Appl. Chem.* **1981**, *53*, 79.
Ketoses	
Structures	*Acc. Chem. Res.* **1976**, *9*, 418.
Lectins	*Pure & Appl. Chem.* **1991**, *63*, 499.
Lipopolysaccharides	*Pure & Appl. Chem.* **1989**, *61*, 1271.
Liquid crystals	*Acc. Chem. Res.* **1986**, *19*, 168.
Metal complexes	*Chem. Soc. Rev.* **1979**, *8*, 221.
N-Acetylneuraminic acid	*Synthesis* **1991**, 583.
NMR	*Chem. Rev.* **1973**, *73*, 669.
	Chem. Soc. Rev. **1975**, *4*, 401.
	Pure & Appl. Chem. **1981**, *53*, 45.
	Pure & Appl. Chem. **1991**, *63*, 529.
Oxidations	*Synthesis* **1971**, 70.
Phosphates	
Anomerization	*Acc. Chem. Res.* **1978**, *11*, 136.
Polyamino monosaccharides	*Synthesis* **1971**, 359.
Polymethylpolysaccharides	*Pure & Appl. Chem.* **1981**, *53*, 107.

Carbohydrates, cont'd.	
Polysaccharides	
Enzymes	*Tetrahedron* **1985**, *41*, 2957.
Microbial polysaccharides	*Pure & Appl. Chem.* **1984**, *56*, 879.
Protein-carbohydrate	*Pure & Appl. Chem.* **1987**, *59*, 1447.
interactions	*Pure & Appl. Chem.* **1989**, *61*, 1293.
Pyranose derivatives	*Pure & Appl. Chem.* **1989**, *61*, 1235.
Radical chemistry	*Pure & Appl. Chem.* **1988**, *60*, 1655.
Ribose	*Acc. Chem. Res.* **1988**, *21*, 294.
Spin-labeled	*Chem. Rev.* **1986**, *86*, 203.
Stereochemistry	*Pure & Appl. Chem.* **1987**, *59*, 1521.
Stereoselective reactions	*Pure & Appl. Chem.* **1984**, *56*, 845.
Steroidal glycosides	*Pure & Appl. Chem.* **1982**, *54*, 1935.
Structural chemistry	*Angew. Chem. Int. Ed. Engl.* **1990**, *29*, 823.
	Chem. Rev. **1986**, *86*, 203.
	Chem. Soc. Rev. **1981**, *10*, 409.
	Pure & Appl. Chem. **1984**, *56*, 1031.
	Pure & Appl. Chem. **1989**, *61*, 1193.
Synthesis	*Chem. Rev.* **1986**, *86*, 35.
	Pure & Appl. Chem. **1987**, *59*, 1509.
	Pure & Appl. Chem. **1989**, *61*, 1217.
	Pure & Appl. Chem. **1989**, *61*, 1235.
	Pure & Appl. Chem. **1989**, *61*, 1257.
	Synthesis **1991**, 499.
	Synthesis **1991**, 583.
	Tetrahedron **1990**, *46*, 1ff.
Enzymes	*Synthesis* **1991**, 499.
	Tetrahedron **1989**, *45*, 5365.
Synthetic methods	*Pure & Appl. Chem.* **1989**, *61*, 1243.
	Pure & Appl. Chem. **1991**, *63*, 507.
	Pure & Appl. Chem. **1991**, *63*, 545.
2,3,6-Trideoxy-3-aminohexoses	*Chem. Rev.* **1986**, *86*, 35.
2,3,6-Trideoxy-3-nitrohexoses	*Chem. Rev.* **1986**, *86*, 35.
Unsaturated carbohydrates	
2,3-Unsaturated sugars	*Acc. Chem. Res.* **1975**, *8*, 192.
Synthons	*Acc. Chem. Res.* **1985**, *18*, 347.
Wittig reaction	*Adv. Carbohydr. Chem.* **1972**, *27*, 227.
Yeast usage	*Pure & Appl. Chem.* **1987**, *59*, 1501.
Carbometallations	*Acc. Chem. Res.* **1987**, *20*, 65.
	Synthesis **1981**, 841.
Carbon dioxide	*Angew. Chem. Int. Ed. Engl.* **1988**, *27*, 661.
Organometallic complexes	*Chem. Rev.* **1988**, *88*, 747.
Carbon disulfide	*Synthesis* **1984**, 797.

Carbon glycosides	*Acc. Chem. Res.* **1990**, *23*, 201.
Carbon monosulfide	*Chem. Rev.* **1988**, *88*, 391.
Carbon monoxide	*Angew. Chem. Int. Ed. Engl.* **1966**, *5*, 435.
Insertion reactions	*Angew. Chem. Int. Ed. Engl.* **1977**, *16*, 299.
Organoboranes	*Acc. Chem. Res.* **1969**, *2*, 65.
Carbon suboxide	*Angew. Chem. Int. Ed. Engl.* **1974**, *13*, 491.
Carbonic acid derivatives	*Angew. Chem. Int. Ed. Engl.* **1973**, *12*, 630.
Carbonic anhydrase	*Acc. Chem. Res.* **1983**, *16*, 272.
	Acc. Chem. Res. **1988**, *21*, 30.
Metal ion function	*Angew. Chem. Int. Ed. Engl.* **1972**, *11*, 408.
Carbon-sulfur bond cleavage	*Synthesis* **1990**, 89.
Carbonyl compounds	*Synthesis* **1981**, 501.
	SYNLETT **1990**, 67.
Alkenes	*Chem. Rev.* **1989**, *89*, 1513.
Alkylations	*Angew. Chem. Int. Ed. Engl.* **1982**, *21*, 96.
	Chem. Rev. **1987**, *87*, 1277.
Asymmetric synthesis	*Pure & Appl. Chem.* **1988**, *60*, 1597.
	Tetrahedron: Asymmetry **1991**, *2*, 1.
Allylations	
Palladium catalysts	*Acc. Chem. Res.* **1987**, *20*, 140.
Asymmetric synthesis	*Tetrahedron: Asymmetry* **1990**, *1*, 477.
Chiroptical properties	*Tetrahedron* **1986**, *42*, 777.
Cluster compounds	*Pure & Appl. Chem.* **1982**, *54*, 97.
Condensation reactions	*Pure & Appl. Chem.* **1964**, *9*, 337.
Coupling reactions	*Chem. Rev.* **1988**, *88*, 733.
Cycloadditions	*Pure & Appl. Chem.* **1971**, *27*, 679.
Diazomethane	*Org. Reactions* **1954**, *8*, 364.
Nucleophilic acylation	
Umpolung	*Tetrahedron* **1976**, *32*, 1943.
Nucleophilic additions	
Organometallic reagents	
Asymmetric synthesis	*Angew. Chem. Int. Ed. Engl.* **1991**, *30*, 49.
Reaction mechanisms	*Pure & Appl. Chem.* **1980**, *52*, 545.
Organometallic chemistry	
Nucleophilic additions	*Angew. Chem. Int. Ed. Engl.* **1991**, *30*, 49.
Organophosphorus compounds	*Org. Reactions* **1988**, *36*, 175.
Oxidations	*Org. Reactions* **1990**, *39*, 297.
Oxidative cyclizations	*Synthesis* **1970**, 279.
Oxoalkylations	
Nitroalkenes	*Acc. Chem. Res.* **1985**, *18*, 284.
Photochemistry	*Chem. Soc. Rev.* **1972**, *1*, 465.
	Pure & Appl. Chem. **1982**, *54*, 1579.
	Pure & Appl. Chem. **1982**, *54*, 1723.

Carbonyl compounds, cont'd.
Reductions — *Angew. Chem.* **1956**, *68*, 601.
Asymmetric synthesis — *Pure & Appl. Chem.* **1981**, *53*, 2315.
Unsaturated carbonyl compounds — *Pure & Appl. Chem.* **1984**, *56*, 91.
Vilsmeier reagents — *Tetrahedron* **1992**, *48*, 3659.

Carbonyl equivalents
Umpolung — *Chem. Soc. Rev.* **1982**, *11*, 493.
Synthesis **1977**, 357.

Carbonyl insertions — *Acc. Chem. Res.* **1984**, *17*, 67.

Carbonyl oxides — *Chem. Rev.* **1991**, *91*, 335.

1,2-Carbonyl transpositions — *Tetrahedron* **1983**, *39*, 345.

Carbonyl ylides — *Tetrahedron* **1976**, *32*, 2165.

Carbonylations — *Chem. Rev.* **1988**, *88*, 1011.
Pure & Appl. Chem. **1985**, *57*, 1875.
Pure & Appl. Chem. **1988**, *60*, 35.
Carboxylic acids — *Synthesis* **1973**, 509.
Esters — *Synthesis* **1973**, 509.
Organoboranes — *Acc. Chem. Res.* **1969**, *2*, 65.
Organometallic reagents — *Synthesis* **1985**, 253.
Transition metal catalysis — *Synthesis* **1973**, 509.

Carboxyl group
Protecting groups — *Tetrahedron* **1980**, *36*, 2409.
Oxazoles — *Chem. Rev.* **1986**, *86*, 845.

Carboxylic acid derivatives — *See also esters and orthoesters*
Acid chlorides
Nitriles — *Synthesis* **1973**, 189.
Acid halides
Organometallic compounds — *Org. Reactions* **1954**, *8*, 28.
Addition reactions — *Synthesis* **1970**, 99.
Photochemistry — *Chem. Rev.* **1978**, *78*, 97.
Radical reactions — *Synthesis* **1970**, 99.
Reductions — *Org. Reactions* **1988**, *36*, 249.
Tetrahedral intermediates — *Acc. Chem. Res.* **1981**, *14*, 306.

Carboxylic acids
Aldehydes — *Org. Reactions* **1954**, *8*, 218.
Angew. Chem. Int. Ed. Engl. **1980**, *19*, 171.
Asymmetric synthesis — *Tetrahedron: Asymmetry* **1992**, *3*, 163.
Carbonylations — *Synthesis* **1973**, 509.
Coupling reactions — *Synthesis* **1979**, 633.
Decarboxylations
Lead tetraacetate — *Org. Reactions* **1972**, *19*, 279.
Dianions — *Synthesis* **1982**, 521.

Ketones	*Org. Reactions* **1970**, *18*, 1.
	Angew. Chem. Int. Ed. Engl. **1980**, *19*, 171.
Orthoesters	*Synthesis* **1974**, 153.
Silver salts	*Org. Reactions* **1957**, *9*, 332.
Carboxylic carbonic anhydrides	*Acc. Chem. Res.* **1969**, *2*, 296.
Carboxymethylation	*Synthesis* **1970**, 628.
Carboxypeptidase A	*Acc. Chem. Res.* **1989**, *22*, 62.
	Acc. Chem. Res. **1972**, *5*, 219.
Structure	*Acc. Chem. Res.* **1982**, *15*, 232.
Carbynes	
Complexes	*Adv. Organomet. Chem.* **1976**, *14*, 1.
	J. Organometal. Chem. **1975**, *100*, 59.
Carcinogens	*Chem. Soc. Rev.* **1980**, *9*, 241.
Hydrocarbons	
Metabolites	*Acc. Chem. Res.* **1981**, *14*, 218.
Carotenoids	*Pure & Appl. Chem.* **1985**, *57*, 639ff.
	Pure & Appl. Chem. **1991**, *63*, 1ff.
Carpyrinic acid	*Synthesis* **1972**, 464.
Castanospermine	*Tetrahedron* **1992**, *48*, 4045.
Catalysis	*Acc. Chem. Res.* **1986**, *19*, 121.
	See also organometallic chemistry
Anchored complexes	*Pure & Appl. Chem,* **1980**, *52*, 2075.
Alloys	*Chem. Rev.* **1975**, *75*, 650.
Asymmetric synthesis	*Tetrahedron* **1989**, *45*, 6901.
Bis(4,5-dihydrooxazolyl) derivatives	*Angew. Chem. Int. Ed. Engl.* **1991**, *30*, 542.
Hydrogen transfer	*Chem. Rev.* **1974**, *74*, 567.
	Chem. Rev. **1985**, *85*, 129.
Ammonium formate	*Synthesis* **1988**, 91.
Hydrogenation	*Acc. Chem. Res.* **1977**, *10*, 15.
Palladium compounds	*Pure & Appl. Chem.* **1982**, *54*, 197.
Shape selectivity	*Pure & Appl. Chem,* **1980**, *52*, 2091.
Surface chemistry	*Acc. Chem. Res.* **1986**, *19*, 24.
Catalytic antibodies	*Acc. Chem. Res.* **1989**, *22*, 287.
	Angew. Chem. Int. Ed. Engl. **1990**, *29*, 1296.
Catechol derived macrocycles	*Pure & Appl. Chem.* **1988**, *60*, 545.
Catenones	
Nucleic acids	*Acc. Chem. Res.* **1973**, *6*, 252.
Cation complexation	
Macromolecules	*Pure & Appl. Chem.* **1988**, *60*, 461.
Carbohydrates	*Chem. Soc. Rev.* **1980**, *9*, 415.

Cation separations	
Macrocycles	*Pure & Appl. Chem.* **1988**, *60*, 453.
Cationic species	*Chem. Soc. Rev.* **1987**, *16*, 1.
CC1065	
DNA binding	*Acc. Chem. Res.* **1986**, *19*, 230.
Cell cultures	*SYNLETT* **1991**, 11.
Cells with manipulated functions	*Angew. Chem. Int. Ed. Engl.* **1981**, *20*, 325.
Cembranolides	*Chem. Rev.* **1988**, *88*, 719.
Cephalosporin	*Acc. Chem. Res.* **1973**, *6*, 32.
***Cephalotaxaus* alkaloids**	*Acc. Chem. Res.* **1975**, *8*, 158.
Ceric ion oxidations	*Synthesis* **1973**, 347.
Chaos	*Acc. Chem. Res.* **1987**, *20*, 436.
Charge transfers	*Chem. Soc. Rev.* **1986**, *15*, 475.
Organometallic compounds	
Reaction mechanisms	*Pure & Appl. Chem.* **1991**, *63*, 255.
Photochemistry	*Pure & Appl. Chem.* **1984**, *56*, 1255.
	Pure & Appl. Chem. **1986**, *58*, 1285.
Chelation controlled addition	*Pure & Appl. Chem.* **1991**, *63*, 1591.
Alkoxycarbonyl compounds	*Angew. Chem. Int. Ed. Engl.* **1984**, *23*, 556.
Chemical defense	*Angew. Chem. Int. Ed. Engl.* **1981**, *20*, 164.
	Angew. Chem. Int. Ed. Engl. **1978**, *17*, 635.
Sponges	*Pure & Appl. Chem.* **1986**, *58*, 357.
Chemical ecology	*Angew. Chem. Int. Ed. Engl.* **1976**, *15*, 214.
Chemical mutagenesis	*Angew. Chem. Int. Ed. Engl.* **1971**, *10*, 302.
Chemical oscillators	*Acc. Chem. Res.* **1977**, *10*, 273.
	Acc. Chem. Res. **1976**, *9*, 438.
	Acc. Chem. Res. **1977**, *10*, 214.
Reaction mechanisms	*Chem. Rev.* **1973**, *73*, 365.
Chemical plant protection	
Historical perspective	*Angew. Chem. Int. Ed. Engl.* **1972**, *11*, 260.
Chemical selectivity	
Isoinversion principle	*Angew. Chem. Int. Ed. Engl.* **1991**, *30*, 477.
Chemical sensors	*Angew. Chem. Int. Ed. Engl.* **1991**, *30*, 516.
	Pure & Appl. Chem. **1989**, *61*, 1605.
	Pure & Appl. Chem. **1989**, *61*, 1613.
Chemiluminescence	*Angew. Chem. Int. Ed. Engl.* **1990**, *29*, 362.
Luciferins	*Acc. Chem. Res.* **1974**, *7*, 135.
Chemotaxis	
Bacterial chemotaxis	*Chem. Rev.* **1987**, *87*, 997.
Chimeras	*Angew. Chem. Int. Ed. Engl.* **1976**, *15*, 181.
Chiral auxiliaries	
Camphor	*Tetrahedron* **1987**, *43*, 1969.
Chiral bases	
Carbonyl compounds	*Tetrahedron: Asymmetry* **1991**, *2*, 1.

Chiral complexes	
Organometallic chemistry	*Chem. Rev.* **1987**, *87*, 761.
Chiral crystals	
Asymmetric synthesis	*Acc. Chem. Res.* **1979**, *12*, 191.
Chiral Lewis acids	*Synthesis* **1991**, 1.
Chiral separations	
Cyclodextrins	*Angew. Chem. Int. Ed. Engl.* **1990**, *29*, 939.
Chiral sulfoxides	*Pure & Appl. Chem.* **1988**, *60*, 1699.
Chirality	*Chem. Rev.* **1979**, *79*, 17.
	Chem. Soc. Rev. **1973**, *2*, 397.
	Chem. Soc. Rev. **1986**, *15*, 189.
	Tetrahedron **1981**, *37*, 4123.
	Tetrahedron **1982**, *38*, 3.
Crop protection	*Angew. Chem. Int. Ed. Engl.* **1991**, *30*, 1193.
Theoretical aspects	*Chem. Soc. Rev.* **1989**, *18*, 187.
Transfer	*Angew. Chem. Int. Ed. Engl.* **1991**, *30*, 49.
Chirons	*Pure & Appl. Chem.* **1987**, *59*, 299.
	SYNLETT **1990**, 173.
Amino acids	*Angew. Chem. Int. Ed. Engl.* **1991**, *30*, 1531.
α-Amino acids	*Tetrahedron* **1991**, *47*, 6079.
Bicyclic lactams	*Tetrahedron* **1991**, *47*, 9503.
Carbohydrates	*Acc. Chem. Res.* **1979**, *12*, 159.
	Pure & Appl. Chem. **1983**, *55*, 565.
	Tetrahedron **1984**, *40*, 3161.
	Tetrahedron **1991**, *47*, 6079.
2,3-*O*-Isopropylideneglyceraldehyde	*Tetrahedron* **1986**, *42*, 447.
Retrosynthetic analysis	*Pure & Appl. Chem.* **1990**, 62, 1887.
Chloral	*Chem. Rev.* **1975**, *75*, 259.
Chloramine-T	*Chem. Rev.* **1978**, *78*, 65.
3-Chloro-2-aza-2-propeniminium compounds	*Synthesis* **1988**, 655.
Chlorinated heterocycles	
Dehalogenation reactions	*Synthesis* **1985**, 586.
Chlorinations	*Angew. Chem. Int. Ed. Engl.* **1975**, *14*, 801.
Alkanes	
Photochemistry	*Acc. Chem. Res.* **1990**, *23*, 219.
Amines	*Angew. Chem.* **1962**, *74*, 848.
Hydrocarbons	*Chem. Rev.* **1963**, *63*, 355.
Steroids	*Acc. Chem. Res.* **1980**, *13*, 170.
Chloromethylation	
Aromatic compounds	*Org. Reactions* **1942**, *1*, 63.

Chloromethyl sulfones	*Acc. Chem. Res.* **1987**, *20*, 282.
Chloromethyltrimethylsilane	*Synthesis* **1985**, 717.
Chloroperoxidase	
Reaction mechanisms	*Chem. Rev.* **1987**, *87*, 1255.
Chlorophyll	
Biosynthesis	*Chem. Soc. Rev.* **1977**, 6, 467.
NMR	*Chem. Soc. Rev.* **1977**, 6, 467.
Protein	*Acc. Chem. Res.* **1980**, *13*, 309.
3-Chloro-2-propeniminium salts	*Synthesis* **1979**, 241.
3-Chloropropyltrialkoxysilanes	*Angew. Chem. Int. Ed. Engl.* **1986**, *25*, 236.
Chlorosulfenyl isocyanate	*Chem. Rev.* **1976**, *76*, 389.
	Synthesis **1986**, 437.
α-Chlorosulfides	*Tetrahedron* **1986**, *42*, 3731.
Choleic acids	
Photochemistry	*Pure & Appl. Chem.* **1980**, *52*, 2693.
Cholinergic excitation	*Angew. Chem. Int. Ed. Engl.* **1984**, *23*, 195.
Chorismate metabolism	*Chem. Rev.* **1990**, *90*, 1105.
4-Chromanones	*Angew. Chem. Int. Ed. Engl.* **1982**, *21*, 247.
Chromatin	
DNA	
Histones	*Acc. Chem. Res.* **1975**, *8*, 327.
Chromium chemistry	*SYNLETT* **1991**, 381.
	Synthesis **1974**, 1.
Arene complexes	*Pure & Appl. Chem.* **1985**, *57*, 1855.
Carbene complexes	*Pure & Appl. Chem.* **1990**, *62*, 691.
Catalysis	
Polymerization	*Angew. Chem. Int. Ed. Engl.* **1971**, *10*, 776.
Chromosomal nucleohistones	*Angew. Chem. Int. Ed. Engl.* **1972**, *11*, 496.
Chymotrypsin	
Models	*Acc. Chem. Res.* **1987**, *20*, 146.
CIDEP	*Chem. Soc. Rev.* **1979**, *8*, 29.
Photochemistry	*Acc. Chem. Res.* **1977**, *10*, 161.
CIDNP	*Acc. Chem. Res.* **1972**, *5*, 18.
	Acc. Chem. Res. **1972**, *5*, 25.
	Acc. Chem. Res. **1985**, *18*, 196.
Carbenes	
Photochemistry	*Acc. Chem. Res.* **1977**, *10*, 85.
Circular dichroism	*Chem. Rev.* **1975**, *75*, 323.
	Tetrahedron **1981**, *37*, 4123.
Absolute configurations	*Angew. Chem. Int. Ed. Engl.* **1979**, *18*, 363.
Alkenes	*Tetrahedron* **1976**, *32*, 2475.
Amines	*Chem. Rev.* **1983**, *83*, 359.
Carbohydrates	*Pure & Appl. Chem.* **1984**, *56*, 1031.
Cyclohexenones	*Tetrahedron* **1982**, *38*, 3.

Oligosaccharides	*Pure & Appl. Chem.* **1984**, *56*, 1031.
Salicylidenamino chirality rule	*Chem. Rev.* **1983**, *83*, 359.
Symmetry rules	*Acc. Chem. Res.* **1974**, *7*, 258.
Clathrates	*Acc. Chem. Res.* **1969**, *2*, 344.
	Angew. Chem. Int. Ed. Engl. **1985**, *24*, 727.
Host-guest chemistry	*Acc. Chem. Res.* **1978**, *11*, 81.
	Chem. Soc. Rev. **1978**, *7*, 65.
Clay supported reagents	*Synthesis* **1985**, 909.
Cleavage reactions	*Adv. Org. Chem.* **1972**, *8*, 181.
Alkenes	*Angew. Chem. Int. Ed. Engl.* **1978**, *17*, 150.
Disulfides	*Chem. Rev.* **1959**, *59*, 583.
Esters	*Org. Reactions* **1976**, *24*, 187.
Ethers	*Chem. Rev.* **1954**, *54*, 615.
	Synthesis **1983**, 249.
Ketones	*Adv. Org. Chem.* **1972**, *8*, 209.
	Org. Reactions **1957**, *9*, 1.
Organosilanes	*J. Organometal. Chem.* **1975**, *100*, 43.
Organosulfur compounds	*Chem. Rev.* **1951**, *49*, 1.
Polyenes	*Angew. Chem. Int. Ed. Engl.* **1978**, *17*, 150.
Sulfonamides	*Chem. Rev.* **1959**, *59*, 1077.
Cloning	*Pure & Appl. Chem.* **1987**, *59*, 1489.
Cluster compounds	*Pure & Appl. Chem.* **1982**, *54*, 113.
	Pure & Appl. Chem. **1982**, *54*, 131.
C-Nitroso compounds	
Electrophiles	*Chem. Soc. Rev.* **1977**, 6, 1.
Coal hydrogenation	
Oil production	*Angew. Chem.. Int. Ed. Engl.* **1976**, *15*, 341.
Cobalt chemistry	*Chem. Rev.* **1983**, *83*, 203.
Catalysis	*Pure & Appl. Chem.* **1985**, *57*, 1819.
[2+2+2] Cycloadditions	*Angew. Chem. Int. Ed. Engl.* **1984**, *23*, 539.
Oxo reaction	*Acc. Chem. Res.* **1981**, *14*, 259.
Propargyl compounds	*Acc. Chem. Res.* **1987**, *20*, 207.
Pyridines	*Angew. Chem. Int. Ed. Engl.* **1978**, *17*, 505.
Radical reactions	*Chem. Soc. Rev.* **1988**, *17*, 361.
Coelenterates	*Pure & Appl. Chem.* **1982**, *54*, 1981.
Coenzyme B_{12}	
Mechanistic studies	*Pure & Appl. Chem.* **1983**, *55*, 1059.
Coenzyme Q	*Angew. Chem. Int. Ed. Engl.* **1974**, *13*, 559.
Coenzymes	
Enzyme-coenzyme binding sites	*Acc. Chem. Res.* **1982**, *15*, 128.
Flavins	
Redox chemistry	*Acc. Chem. Res.* **1980**, *13*, 148.
	Acc. Chem. Res. **1980**, *13*, 256.

Collodial semiconductors
Photochemistry — *Pure & Appl. Chem.* **1984**, *56*, 1215.
Colloidal polymers
Fluorescence — *Pure & Appl. Chem.* **1984**, *56*, 1281.
Color dyes
Photography — *Angew. Chem. Int. Ed. Engl.* **1983**, *22*, 191.
Complex reducing agents — *Angew. Chem. Int. Ed. Engl.* **1983**, *22*, 599.
Complexes — *Chem. Rev.* **1975**, *75*, 439.
Topics Curr. Chem. **1976**, *65*, 105.
Ligands
Stereochemical interactions — *Pure & Appl. Chem.* **1983**, *55*, 65.
Polyethers — *Angew. Chem. Int. Ed. Engl.* **1972**, *11*, 16.
Computer applications
Drug design — *Acc. Chem. Res.* **1987**, *20*, 322.
Molecular design — *Angew. Chem. Int. Ed. Engl.* **1971**, *4*, 393.
Molecular dynamics — *Angew. Chem. Int. Ed. Engl.* **1990**, *29*, 992.
Predicton of reactions — *Pure & Appl. Chem.* **1990**, *62*, 1921.
Synthesis design — *Angew. Chem. Int. Ed. Engl.* **1990**, *29*, 1286.
Pure & Appl. Chem. **1988**, *60*, 1563.
Pure & Appl. Chem. **1988**, *60*, 1573.
Condensation reactions
Carbonyl compounds — *Pure & Appl. Chem.* **1964**, *9*, 337.
Oxidation-reduction condensation — *Angew. Chem. Int. Ed. Engl.* **1976**, *15*, 94.
Polymerizations — *Chem. Rev.* **1946**, *39*, 137.
Conduritols — *Tetrahedron* **1990**, *46*, 3715.
Conformational analysis — *Chem. Rev.* **1974**, *74*, 519.
Pure & Appl. Chem. **1987**, *59*, 1661.
Heterocycles
Pentamethylene heterocycles — *Chem. Rev.* **1975**, *75*, 611.
Chem. Soc. Rev. **1978**, *7*, 399.
Alcohols — *Chem. Soc. Rev.* **1976**, *5*, 411.
Amines — *Chem. Soc. Rev.* **1976**, *5*, 411.
Lactones — *Pure & Appl. Chem.* **1982**, *54*, 2515.
Ring compounds — *Chem. Soc. Rev.* **1974**, *3*, 1.
Chem. Rev. **1980**, *80*, 231.
Small molecules — *Chem. Soc. Rev.* **1972**, *1*, 293.
Conjugate additions
Alkylations — *Org. Reactions* **1990**, *38*, 225.
Organocopper reagents — *Org. Reactions* **1972**, *19*, 1.
Unsaturated carbonyl compounds — *Org. Reactions* **1990**, *38*, 225.
Vinyl sulfones — *Chem. Rev.* **1986**, *86*, 903.
Conjugated enamines — *Tetrahedron* **1984**, *40*, 2989.

Conjugated esters	*Pure & Appl. Chem.* **1986**, *58*, 1257.
Conjugated ketenes	*Chem. Rev.* **1986**, *86*, 821.
Conjugated nitroalkenes	*Chem. Rev.* **1986**, *86*, 751.
Conjugated systems	
Addition reactions	*Chem. Rev.* **1959**, *59*, 329.
Photoisomerization	*Pure & Appl. Chem.* **1984**, *56*, 1241.
Conjugation reactions	*Angew. Chem. Int. Ed. Engl.* **1981**, *20*, 135.
Cooligomerization	
Butadiene	
Unsaturated compounds	*Angew. Chem. Int. Ed. Engl.* **1965**, *4*, 330.
Coordinated π-hydrocarbon complexes	
Nucleophilic additions	*Chem. Rev.* **1984**, *84*, 525.
Coordination compounds	
Aryldiazonium cations	*Chem. Soc. Rev.* **1975**, *4*, 443.
Photocatalysis	*Pure & Appl. Chem.* **1990**, *62*, 1489.
Polyazacycloalkanes	*Pure & Appl. Chem.* **1988**, *60*, 525.
Copolymerization	*Chem. Rev.* **1950**, *46*, 191.
Butadiene	*Chem. Soc. Rev.* **1977**, *6*, 235.
Zwitterions	*Angew. Chem. Int. Ed. Engl.* **1977**, *16*, 826.
Copper chemistry	
Carbenoids	*Pure & Appl. Chem.* **1985**, *57*, 1839.
Catalysis	*Pure & Appl. Chem.* **1985**, *57*, 1827.
Grignard reagents	*Tetrahedron* **1984**, *40*, 641.
Isonitrile complexes	*Synthesis* **1975**, 291.
Nucleophilic substitution	
Aryl halides	*Tetrahedron* **1984**, *40*, 1433.
Organolithium reagents	*Tetrahedron* **1984**, *40*, 641.
Thiacrown ethers	*Pure & Appl. Chem.* **1988**, *60*, 501.
Corey's reagent	*Tetrahedron* **1987**, *43*, 2609.
Corrins	*Angew. Chem. Int. Ed. Engl.* **1977**, *16*, 233.
	Chem. Soc. Rev. **1976**, *5*, 377.
	Tetrahedron **1976**, *32*, 1599.
Biosynthesis	*Acc. Chem. Res.* **1978**, *11*, 29.
NMR	*Pure & Appl. Chem.* **1981**, *53*, 1215.
Coumarins	
3:4-Fused ring systems	*Synthesis* **1982**, 337.
"Counterattack" reagents	*Tetrahedron* **1989**, *45*, 1233.
Coupling reactions	
Acetals	*Synthesis* **1987**, 1043.
Aldehydes	*Synthesis* **1979**, 633.
Alkenes	*Chem. Rev.* **1989**, *89*, 1513.
	Org. Reactions **1984**, *32*, 375.
	Pure & Appl. Chem. **1980**, *52*, 2417.

Coupling reactions, cont'd.	
Alkyl halides	*Pure & Appl. Chem.* **1980**, *52*, 669.
Alkynes	*Org. Reactions* **1984**, *32*, 375.
Aromatic compounds	*Chem. Rev.* **1987**, *87*, 357.
Aryl triflates	*Acc. Chem. Res.* **1988**, *21*, 47.
Arylations	*Tetrahedron* **1980**, *36*, 3327.
Diazonium salts	*Org. Reactions* **1960**, *11*, 189.
Carbonyl compounds	*Chem. Rev.* **1988**, *88*, 733.
Low-valent titanium	*Chem. Rev.* **1989**, *89*, 1513.
Carboxylic acids	*Synthesis* **1979**, 633.
Diazonium compounds	*Org. Reactions* **1959**, *10*, 1.
Enol triflates	
Organotin chemistry	*Acc. Chem. Res.* **1988**, *21*, 47.
Imines	*Pure & Appl. Chem.* **1980**, *52*, 2417.
Ketones	*Synthesis* **1979**, 633.
Nickel chemistry	*Acc. Chem. Res.* **1982**, *15*, 340.
Organoaluminum chemistry	*Org. Reactions* **1984**, *32*, 375.
Organocuprates	
Enol triflates	*Acc. Chem. Res.* **1988**, *21*, 47.
Organotin chemistry	*Pure & Appl. Chem.* **1985**, *57*, 1771.
Palladium catalysis	*Angew. Chem. Int. Ed. Engl.* **1986**, *25*, 508.
Oxidative	*Acc. Chem. Res.* **1982**, *15*, 340.
	Pure & Appl. Chem. **1985**, *57*, 1771.
	SYNLETT **1991**, 845.
Peptides	*Synthesis* **1972**, 453.
Polyenes	*Pure & Appl. Chem.* **1981**, *53*, 2323.
Cram's rule	*Pure & Appl. Chem.* **1991**, *63*, 1591.
	See also Anh-Felkin addition
Crop protection	*Angew. Chem. Int. Ed. Engl.* **1991**, *30*, 1193.
Cross-coupling reactions	
Nucleic acids and Proteins	*Pure & Appl. Chem.* **1980**, *52*, 2717.
Crown ethers	*Tetrahedron* **1980**, *36*, 461.
Aza-crown ethers	*Chem. Rev.* **1989**, *89*, 929.
	Pure & Appl. Chem. **1989**, *61*, 1619.
Functionalization	*Pure & Appl. Chem.* **1988**, *60*, 539.
Historical perspective	*Angew. Chem. Int. Ed. Engl.* **1988**, *27*, 1021.
	Pure & Appl. Chem. **1988**, *60*, 445.
	Pure & Appl. Chem. **1988**, *60*, 467.
Host-guest chemistry	*Acc. Chem. Res.* **1978**, *11*, 8.
Immobilized	*Pure & Appl. Chem.* **1989**, *61*, 1619.
Perfluorocrown ethers	*Pure & Appl. Chem.* **1988**, *60*, 473.
Sensors	*Pure & Appl. Chem.* **1989**, *61*, 1605.

Synthesis	*Synthesis* **1976**, 168.
Thiocrown ethers	*Acc. Chem. Res.* **1988**, *21*, 141.
Cryoenzymology	*Acc. Chem. Res.* **1977**, *10*, 233.
Cryptates	*Acc. Chem. Res.* **1978**, *11*, 49.
Inclusion complexes	*Pure & Appl. Chem.* **1980**, *52*, 2303.
	Pure & Appl. Chem. **1980**, *52*, 2441.
Cryptophanes	*Tetrahedron* **1987**, *43*, 5725.
Cubanes	*Chem. Rev.* **1989**, *89*, 997.
Cumulenes	*Chem. Rev.* **1989**, *89*, 1111.
Curtin-Hammett Winstein-Holness Kinetics	*Chem. Rev.* **1983**, *83*, 73.
Cyanation	*Chem. Rev.* **1987**, *87*, 779.
Cyanic esters	*Angew. Chem. Int. Ed. Engl.* **1972**, *11*, 949.
α-Cyano esters	
Dealkoxycarbonylations	*Synthesis* **1982**, 805.
Cyanoethylation	*Org. Reactions* **1949**, *5*, 79.
Cyanohydrins	*Acc. Chem. Res.* **1986**, *19*, 307.
Acyl anion equivalents	*Tetrahedron* **1983**, *39*, 3207.
Cyanoketenes	
Cycloadditions	*Acc. Chem. Res.* **1979**, *12*, 125.
	Chem. Soc. Rev. **1981**, *10*, 289.
Cyanotrimethylsilane	*Synthesis* **1980**, 861.
Cyclic compounds	
NMR	*Chem. Rev.* **1977**, *77*, 599.
Cyclic enediol phosphoryl (CEP) method	
Phosphodiesters	*Synthesis* **1985**, 449.
Cyclization reactions	*Chem. Rev.* **1986**, *86*, 857.
	Pure & Appl. Chem. **1990**, *62*, 1941.
	See also radical reactions
	Tetrahedron **1990**, *46*, 1385.
Alkenes	*Acc. Chem. Res.* **1968**, *1*, 1.
Alkynes	*Angew. Chem. Int. Ed. Engl.* **1969**, *8*, 727.
Transition metal catalysis	*Acc. Chem. Res.* **1977**, *10*, 1.
Allylic pyrophosphates	*Acc. Chem. Res.* **1985**, *18*, 220.
Carbometallations	*Acc. Chem. Res.* **1987**, *20*, 65.
Carbon monoxide	*Angew. Chem. Int. Ed. Engl.* **1966**, *5*, 435.
Conformations	
Reaction mechanisms	*Chem. Rev.* **1981**, *81*, 491.
Dialdehydes	*Angew. Chem. Int. Ed. Engl.* **1964**, *3*, 211.
Dienes	*Tetrahedron* **1974**, *30*, 1651.
2,2′-Disubstituted biphenyls	*Chem. Rev.* **1968**, *68*, 209.
Enones	*Synthesis* **1975**, 1.
Free radical	*Acc. Chem. Res.* **1971**, *4*, 386.

Cyclization reactions, cont'd.	
Heteroaromatic compounds	*Topics Curr. Chem.* **1978**, *73*, 125.
3-Methylcyclohex-2-enone derivatives	*Chem. Soc. Rev.* **1980**, *9*, 265.
N-Chloroalkenylamines	*Angew. Chem. Int. Ed. Engl.* **1983**, *22*, 337.
N-Haloamines	*Chem. Rev.* **1963**, *63*, 55.
Photochemical	*Chem. Rev.* **1966**, *66*, 373.
Polyenes	
Biomimetic	*Angew. Chem. Int. Ed. Engl.* **1976**, *15*, 9.
Polyphosphoric acid	*Angew. Chem.* **1954**, *66*, 435.
Radical cyclizations	
Stereochemistry	*Acc. Chem. Res.* **1991**, *24*, 139.
Reaction mechanisms	*Chem. Rev.* **1981**, *81*, 491.
Unsaturated acids	*Quart. Rev.* **1964**, *18*, 211.
Unsaturated carbonyl compounds	*Synthesis* **1975**, 1.
Ylide malonodinitriles	*Synthesis* **1976**, 705.
Ylides	*Acc. Chem. Res.* **1991**, *24*, 22.
Cycloaddition reactions	*Chem. Rev.* **1976**, *76*, 187.
	Pure & Appl. Chem. **1971**, *27*, 635.
	SYNLETT **1990**, 433.
Acyl isocyanates	*Synthesis* **1974**, 461.
Acyliron complexes	*SYNLETT* **1991**, 1.
Alkenes	*Chem. Rev.* **1988**, *88*, 793.
	Pure & Appl. Chem. **1980**, *52*, 2669.
Cyclopropanes	*Acc. Chem. Res.* **1984**, *17*, 56.
Organosulfur chemistry	*Tetrahedron* **1988**, *44*, 6755.
Alkynes	*Tetrahedron* **1981**, *37*, 3765.
Alkyne equivalents	*Tetrahedron* **1984**, *40*, 2585.
Transition metal catalysis	*Chem. Rev.* **1988**, *88*, 1081.
Allyl cations	*Angew. Chem. Int. Ed. Engl.* **1973**, *12*, 819.
Dienes	*Angew. Chem. Int. Ed. Engl.* **1984**, *23*, 1.
Allylstannanes	*SYNLETT* **1991**, 1.
Anionic	
Organolithium compounds	*Angew. Chem. Int. Ed. Engl.* **1974**, *13*, 627.
Aromatic compounds	*Chem. Rev.* **1989**, *89*, 827.
	Pure & Appl. Chem, **1980**, *52*, 2669.
1-Azirines	*Synthesis* **1975**, 483.
Carbonyl compounds	*Pure & Appl. Chem.* **1971**, *27*, 679.
Cation radical cycloadditions	*Tetrahedron* **1989**, *45*, 5307.
Cyanoketenes	*Acc. Chem. Res.* **1979**, *12*, 125.
	Chem. Soc. Rev. **1981**, *10*, 289.
[2+2] Cycloadditions	
Photochemical	*Acc. Chem. Res.* **1980**, *13*, 45.

[2+2+2] Cycloadditions	
Cobalt catalysis	*Angew. Chem. Int. Ed. Engl.* **1984**, *23*, 539.
[3+2] Cycloadditions	*Angew. Chem. Int. Ed. Engl.* **1977**, *16*, 10.
Trimethylenemethane equivalents	*Angew. Chem. Int. Ed. Engl.* **1986**, *25*, 1.
[4+2] Cycloadditions	*Angew. Chem. Int. Ed. Engl.* **1977**, *16*, 10.
Cyclobutanes	*Org. Reactions* **1962**, *12*, 1.
Cyclopentenones	*Org. Reactions* **1991**, *40*, 1.
Cyclophanes	*Pure & Appl. Chem.* **1987**, *59*, 1627.
Cyclopropanes	
Alkenes	*Acc. Chem. Res.* **1984**, *17*, 56.
Cyclopropanones	*Acc. Chem. Res.* **1969**, *2*, 25.
Cyclopropenes	*Synthesis* **1972**, 675.
Dienes	
Allyl cations	*Angew. Chem. Int. Ed. Engl.* **1984**, *23*, 1.
1,3-Dipolar cycloadditions	*Angew. Chem. Int. Ed. Engl.* **1963**, *2*, 565.
	Angew. Chem. Int. Ed. Engl. **1963**, *2*, 633.
	Tetrahedron **1977**, *33*, 3009.
Azides	*Chem. Rev.* **1969**, *69*, 353.
Diazoalkanes	*Tetrahedron* **1991**, *47*, 2925.
Hetarynes	*Angew. Chem. Int. Ed. Engl.* **1965**, *4*, 553.
Intramolecular	*Angew. Chem., Int. Ed. Engl.* **1976**, *15*, 123.
Nitrones	*Synthesis* **1975**, 205.
Reaction mechanisms	*Tetrahedron* **1977**, *33*, 3009.
Solvent effects	*Synthesis* **1973**, 71.
1,5-Dipolar cycloadditions	*Chem. Rev.* **1979**, *79*, 181.
Fluoroallenes	*Acc. Chem. Res.* **1991**, *24*, 63.
Hetero-1,3,5-hexatrienes	*Angew. Chem. Int. Ed. Engl.* **1980**, *19*, 973.
Heteroaromatic compounds	*Chem. Rev.* **1989**, *89*, 827.
Heterocycles	*Synthesis* **1975**, 483.
Heterodienes	
N-Acyl imines	*Chem. Rev.* **1989**, *89*, 1525.
Intramolecular	*Angew. Chem. Int. Ed. Engl.* **1976**, *15*, 123.
	Angew. Chem. Int. Ed. Engl. **1977**, *16*, 10.
	Synthesis **1978**, 793.
	Chem. Rev. **1988**, *88*, 793.
Isocyanates	*Acc. Chem. Res.* **1969**, *2*, 186.
	Synthesis **1974**, 461.
Ketenes	*Chem. Rev.* **1988**, *88*, 793.
Keteniminium salts	*Chem. Rev.* **1988**, *88*, 793.
Metal carbene complexes	*Pure & Appl. Chem.* **1988**, *60*, 137.
Metal catalysis	*Pure & Appl. Chem.* **1988**, *60*, 1615.
Metal-carbon bonds	*Ann. N.Y. Acad. Sci.* **1974**, *239*, 100.

Cycloaddition reactions, cont'd.

N-Acyl imines	
Heterodienes	*Chem. Rev.* **1989**, *89*, 1525.
Orbital symmetry	*Acc. Chem. Res.* **1969**, *1*, 20.
Organolithium compounds	*Angew. Chem. Int. Ed. Engl.* **1974**, *13*, 627.
Organosulfur chemistry	*Tetrahedron* **1988**, *44*, 6755.
Oxyallyl cations	*Tetrahedron* **1986**, *42*, 4611.
Pauson-Khand cycloaddition	*Org. Reactions* **1991**, *40*, 1.
Photochemical	*Acc. Chem. Res.* **1968**, *1*, 50.
	Acc. Chem. Res. **1968**, *1*, 257.
	Acc. Chem. Res. **1982**, *15*, 80.
	Adv. Photochem. **1968**, *6*, 301.
	Chem. Rev. **1969**, *69*, 845.
	Pure & Appl. Chem. **1982**, *54*, 1633.
Alkenes	*Pure & Appl. Chem.* **1990**, *62*, 1597.
Arenes	*Pure & Appl. Chem.* **1990**, *62*, 1597.
Azirines	*Acc. Chem. Res.* **1976**, *9*, 317.
Enones	
Alkenes	*Synthesis* **1970**, 287.
Polar cycloadditions	*Angew. Chem. Int. Ed. Engl.* **1973**, *12*, 212.
Quinodimethanes	*Synthesis* **1978**, 793.
Reaction mechanisms	*Pure & Appl. Chem.* **1980**, *52*, 2283.
	Pure & Appl. Chem. **1981**, *53*, 171.
Solvent effects	*Pure & Appl. Chem.* **1980**, *52*, 2283.
Spiro compounds	*Synthesis* **1978**, 77.
Steroids	
Intramolecular cycloadditions	*Tetrahedron* **1981**, *37*, 3.
Stilbenes	*Pure & Appl. Chem.* **1989**, *61*, 635.
Strained molecules	*Acc. Chem. Res.* **1971**, *4*, 128.
Sulfinyl dienophiles	*Acc. Chem. Res.* **1988**, *21*, 313.
Tetracyanoethylene	*Synthesis* **1987**, 749.
Enol ethers	*Acc. Chem. Res.* **1977**, *10*, 117.
	Acc. Chem. Res. **1977**, *10*, 199.
Theoretical aspects	*Chem. Rev.* **1972**, *72*, 157.
Unsaturated compounds	
Photochemical	*Chem. Rev.* **1983**, *83*, 1.
Zwitterionic intermediates	*Acc. Chem. Res.* **1977**, *10*, 117.
	Acc. Chem. Res. **1977**, *10*, 199.

Cyclobutadiene	*Angew. Chem. Int. Ed. Engl.* **1974**, *13*, 425.
	Angew. Chem. Int. Ed. Engl. **1988**, *27*, 309.
Metal complexes	*Chem. Rev.* **1977**, *77*, 691.
Theoretical aspects	*Tetrahedron* **1980**, *36*, 343.

Cyclobutanes	*Angew. Chem. Int. Ed. Engl.* **1975**, *14*, 473.
	Org. Reactions **1962**, *12*, 1.
	Synthesis **1974**, 539.
Cycloadditions	*Org. Reactions* **1962**, *12*, 1.
Photochemistry	
Reaction mechanisms	*Pure & Appl. Chem.* **1986**, *58*, 1249.
Ring contractions	*Acc. Chem. Res.* **1972**, *5*, 33.
Ring opening	*Acc. Chem. Res.* **1982**, *15*, 135.
Cyclobutanones	*Acc. Chem. Res.* **1974**, *7*, 85.
	Angew. Chem. Int. Ed. Engl. **1988**, *27*, 797.
Cyclobutendione	*Synthesis* **1978**, 1.
	Synthesis **1978**, 649.
	Synthesis **1978**, 869.
	Angew. Chem. Int. Ed. Engl. **1988**, *27*, 797.
Cyclodextrins	*Pure & Appl. Chem.* **1986**, *58*, 1529.
	Tetrahedron **1983**, *39*, 1417.
Chiral separations	*Angew. Chem. Int. Ed. Engl.* **1990**, *29*, 939.
Enzyme models	*Acc. Chem. Res.* **1987**, *20*, 146.
Host-guest chemistry	*Acc. Chem. Res.* **1982**, *15*, 66.
	Angew. Chem. Int. Ed. Engl. **1980**, *19*, 344.
Photochemistry	*Acc. Chem. Res.* **1988**, *21*, 300.
Cyclofunctionalization	*Tetrahedron* **1990**, *46*, 3321.
Cycloheptatrienes	
Rearrangements	*Acc. Chem. Res.* **1968**, *1*, 152.
Cyclohepta[*b*][1,4]benzoxazine	
Aromaticity	*Pure & Appl. Chem.* **1982**, *54*, 975.
Cyclohexadiene derivatives	*SYNLETT* **1991**, 369.
Cyclohexa-3,5-diene-1,2-diols	
Asymmetric synthesis	*Tetrahedron: Asymmetry* **1992**, *3*, 795.
Halogenations	*Acc. Chem. Res.* **1974**, *7*, 361.
Cyclohexadienones	*Acc. Chem. Res.* **1971**, *4*, 337.
Photochemistry	*Acc. Chem. Res.* **1978**, *11*, 65.
	Pure & Appl. Chem. **1988**, *60*, 981.
Rearrangements	*Acc. Chem. Res.* **1975**, *8*, 245.
Cyclohexanes	
Asymmetric synthesis	*Acc. Chem. Res.* **1990**, *23*, 207.
Cyclohexene	
Addition reactions	
Photochemistry	*Acc. Chem. Res.* **1969**, *2*, 33.
Cyclohexenones	
Circular dichroism	*Tetrahedron* **1982**, *38*, 3.
Cycloimmonium ylides	*Tetrahedron* **1976**, *32*, 2647.

Cycloisomerization	
Enynes	
Palladium catalysis	*Acc. Chem. Res.* **1990**, *23*, 34.
Cyclometalation reactions	*Angew. Chem. Int. Ed. Engl.* **1977**, *16*, 73.
Platinum chemistry	*Chem. Rev.* **1986**, *86*, 451.
Cyclooctane derivatives	*Tetrahedron* **1992**, *48*, 5757.
Cyclooctatetraenes	*Tetrahedron* **1975**, *31*, 2855.
Aromaticity	*Pure & Appl. Chem.* **1982**, *54*, 987.
Ring inversions	*Pure & Appl. Chem.* **1982**, *54*, 987.
Cyclooligomerization	
Transition metal catalysis	*Angew. Chem. Int. Ed. Engl.* **1973**, *12*, 975.
Cyclopalladation	
Asymmetric synthesis	*Pure & Appl. Chem.* **1983**, *55*, 1837.
Cyclopentaannellations	*Synthesis* **1984**, 529.
Cyclopentadienyl compounds	
Platinum compounds	*Chem. Soc. Rev.* **1981**, *10*, 1.
Cyclopentane derivatives	*Pure & Appl. Chem.* **1988**, *60*, 27.
Cyclopentanoids	*Chem. Soc. Rev.* **1982**, *11*, 141.
Anions	*Chem. Rev.* **1989**, *89*, 1467.
Photochemistry	*Angew. Chem. Int. Ed. Engl.* **1982**, *21*, 820.
Cyclopentene	
Homopolymerization	*Angew. Chem. Int. Ed. Engl.* **1964**, *3*, 723.
Polymerization	*Angew. Chem. Int. Ed. Engl.* **1964**, *3*, 723.
Cyclopentenes	
Vinylcyclopropanes	*Org. Reactions* **1985**, *33*, 247.
Cyclopentenoid compounds	*Pure & Appl. Chem.* **1986**, *58*, 781.
Cyclophanes	*Chem. Rev.* **1986**, *86*, 957.
	Pure & Appl. Chem. **1987**, *59*, 1627.
	Pure & Appl. Chem. **1989**, *61*, 1523.
	Pure & Appl. Chem. **1990**, *62*, 373.
Host-guest chemistry	*Angew. Chem. Int. Ed. Engl.* **1988**, *27*, 362.
Multibridged [2_n]cyclophanes	*Angew. Chem. Int. Ed. Engl.* **1982**, *21*, 469.
Multilayered cyclophanes	*Acc. Chem. Res.* **1978**, *11*, 251.
Strain	*Acc. Chem. Res.* **1971**, *4*, 204.
Cyclophosphazenes	
Substituton reactions	*Chem. Rev.* **1991**, *91*, 119.
Cyclopolymerization	*Chem. Soc. Rev.* **1972**, *1*, 523.
	Pure Appl. Chem. **1970**, *23*, 255.
Cyclopropanations	*Acc. Chem. Res.* **1986**, *19*, 348.
Metal carbene complexes	*Pure & Appl. Chem.* **1988**, *60*, 137.
Cyclopropane	*Angew. Chem. Int. Ed. Engl.* **1979**, *18*, 809.
Cyclopropane amino acids	*Tetrahedron* **1990**, *46*, 2231.

Cyclopropanes	*Chem. Rev.* **1989**, *89*, 165.
	Org. Reactions **1973**, *20*, 1.
Alkenes	*Chem. Rev.* **1987**, *87*, 411.
Cycloadditions	*Acc. Chem. Res.* **1984**, *17*, 56.
Asymmetric synthesis	*Chem. Rev.* **1989**, *89*, 1247.
	Pure & Appl. Chem. **1985**, *57*, 1839.
Biosynthesis	*Acc. Chem. Res.* **1971**, *4*, 199.
Carbenes	
Halocarbenes	*Org. Reactions* **1963**, *13*, 55.
Cycloadditions	
Alkenes	*Acc. Chem. Res.* **1984**, *17*, 56.
Enzymes	*Angew. Chem. Int. Ed. Engl.* **1988**, *27*, 537.
Photochemistry	*Acc. Chem. Res.* **1987**, *20*, 107.
Silyl-substituted cyclopropanes	*Chem. Rev.* **1986**, *86*, 755.
Cyclopropanols	*Chem. Rev.* **1974**, *74*, 605.
	Acc. Chem. Res. **1968**, *1*, 33.
Cyclopropanones	*Acc. Chem. Res.* **1969**, *2*, 25.
Hemiacetals	*Chem. Rev.* **1983**, *83*, 619.
Cycloproparenes	*Chem. Rev.* **1989**, *89*, 1161.
	Tetrahedron **1988**, *44*, 1305.
Cyclopropenes	
[2+2]Cycloadditions	*Synthesis* **1982**, 701.
1,3-Dipolar additions	*Synthesis* **1982**, 701.
Photochemistry	*Acc. Chem. Res.* **1979**, *12*, 310.
Cyclopropenones	*Chem. Rev.* **1974**, *74*, 189.
Cyclopropenylidene compounds	
Aromaticity	*Pure & Appl. Chem.* **1982**, *54*, 1059.
Cyclopropyl alkenes	
Rearrangements	*Acc. Chem. Res.* **1978**, *11*, 204.
Cyclopropyl groups	
Carbocations	*Chem. Rev.* **1974**, *74*, 315.
Cyclopropyl radical	*Tetrahedron* **1981**, *37*, 1625.
Cyclopropyl tosylates	
Solvolysis	*Acc. Chem. Res.* **1968**, *1*, 33.
Cyclopropylmethyl compounds	*Angew. Chem. Int. Ed. Engl.* **1975**, *14*, 473.
Cycloreversions	
[2+2] Cycloreversions	*Angew. Chem. Int. Ed. Engl.* **1982**, *21*, 225.
1,3-Dipolar cycloreversions	*Angew. Chem. Int. Ed. Engl.* **1979**, *18*, 721.
Cyclosporine	*Angew. Chem. Int. Ed. Engl.* **1985**, *24*, 77.
Cyclothiazenes	*Acc. Chem. Res.* **1984**, *17*, 166.
Cyclotrimerization	
Alkynes	*Pure Appl. Chem.* **1973**, *33*, 489.
Cyclotriveratrylenes	*Tetrahedron* **1987**, *43*, 5725.

Cysteine proteases	*Tetrahedron* **1976**, *32*, 291.
Cytochalasins	*Angew. Chem. Int. Ed. Engl.* **1973**, *12*, 370
Diels Alder reaction	*Acc. Chem. Res.* **1991**, *24*, 229.
Cytochrome c	
Electrochemistry	*Acc. Chem. Res.* **1988**, *21*, 407.
Cytochrome c oxidase	*Angew. Chem. Int. Ed. Engl.* **1983**, *22*, 275.
	Chem. Rev. **1990**, *90*, 1247.
Cytochrome c_3	*Acc. Chem. Res.* **1983**, *16*, 2.
Cytochrome oxidase	*Pure & Appl. Chem.* **1987**, *59*, 743.
Cytochrome P-450	*Pure & Appl. Chem.* **1987**, *59*, 759.
Reaction mechanisms	*Chem. Rev.* **1987**, *87*, 1255.
Suicide destruction	*Acc. Chem. Res.* **1984**, *17*, 9.
***Daphniphyllum* alkaloids**	
Biomimetic synthesis	*Pure & Appl. Chem.* **1990**, *62*, 1911.
Biosynthesis	*Pure & Appl. Chem.* **1989**, *61*, 289.
Dealkoxycarbonylations	*Synthesis* **1982**, 805.
Dealkylations	*Synthesis* **1988**, 749.
Amine	*Synthesis* **1989**, 1.
Ester cleavage	*Org. Reactions* **1976**, *24*, 187.
Deaminations	*Quart. Rev.* **1961**, *15*, 418.
Deazaflavin	*Acc. Chem. Res.* **1986**, *19*, 216.
Decarbonylations	*Synthesis* **1969**, 157.
Transition metal catalysed	*Synthesis* **1969**, 157.
Decarboxylations	
Enzymes	*Chem. Rev.* **1987**, *87*, 863.
Lead tetraacetate	*Org. Reactions* **1972**, *19*, 279.
Dehalogenation reactions	
Chlorinated heterocycles	*Synthesis* **1985**, 586.
Dehydrations	*Angew. Chem. Int. Ed. Engl.* **1975**, *14*, 801.
Carbohydrates	*Adv. Carbohydrate Chem.* **1973**, *28*, 161.
Dehydroannulenes	*Acc. Chem. Res.* **1972**, *5*, 81.
Dehydrocyclizations	*Chem. Rev.* **1953**, *53*, 353.
Dehydrogenase	*Acc. Chem. Res.* **1974**, *7*, 40.
Dehydrogenations	
Aryl halides	*Acc. Chem. Res.* **1972**, *5*, 139.
Ionic	*Angew. Chem. Int. Ed. Engl.* **1962**, *1*, 613.
N-Bromosuccinimide	*Chem. Rev.* **1963**, *63*, 21.
Polycyclic hydroaromatic compounds	*Chem. Rev.* **1978**, *78*, 317.
Dehydro[8]annulenes	*Acc. Chem. Res.* **1982**, *15*, 96.
Delocalized carbanions	*SYNLETT* **1991**, 207.
DENDRAL	
Structure determination	*Pure & Appl. Chem.* **1982**, *54*, 2425.

Diazoalkanes	
1,3-Dipolar cycloadditions	*Tetrahedron* **1991**, *47*, 2925.
Electrophilic reactions	*Synthesis* **1985**, 569.
Organometallic compounds	*Angew. Chem. Int. Ed. Engl.* **1978**, *17*, 800.
Diazocarbonyl compounds	
Intramolecular reactions	*Org. Reactions* **1979**, *26*, 361.
Rhodium catalysis	*Tetrahedron* **1991**, *47*, 1765.
α-Diazocarbonyl compounds	
Decomposition	*Tetrahedron* **1981**, *37*, 2407.
Wolff rearrangement	*Angew. Chem. Int. Ed. Engl.* **1975**, *14*, 32.
Diazomalonic esters	*Synthesis* **1973**, 137.
Diazomethane	
Carbonyl compounds	*Org. Reactions* **1954**, *8*, 364.
Methylations	*Adv. Heterocycl. Chem.* **1963**, *2*, 245.
Diazonium compounds	
Arylations	
Coupling reactions	*Org. Reactions* **1960**, *11*, 189.
Meerwein arylation reaction	*Org. Reactions* **1976**, *24*, 225.
Coupling reactions	*Org. Reactions* **1959**, *10*, 1.
Diazotizations	*Quart. Rev.* **1961**, *15*, 418.
Amines	*Chem. Rev.* **1975**, *75*, 241.
Heterocycles	*Chem. Rev.* **1975**, *75*, 241.
DIBAL	
Reductions	*Synthesis* **1975**, 617.
Dibenzo[*b,f*]azepines	*Chem. Rev.* **1974**, *74*, 101.
Dibenzo-18-crown-6	*Pure & Appl. Chem.* **1988**, *60*, 467.
Diboraheterocycles	*Pure & Appl. Chem.* **1987**, *59*, 947.
Diboranes	
Reductions	*Chem. Rev.* **1976**, *76*, 773.
β-Dicarbonyl compounds	
Acylations	*Org. Reactions* **1954**, *8*, 59.
Alkylations	*Org. Reactions* **1969**, *17*, 155
Tautomerism	*Chem. Soc. Rev.* **1984**, *13*, 69.
Dicarboxylic acid anhydrides	
Friedal Crafts reaction	*Org. Reactions* **1949**, *5*, 229.
Dichloromethylenammonium salts	*Angew. Chem. Int. Ed. Engl.* **1973**, *12*, 806.
Dichloro(methyl)phosphane	*Angew. Chem. Int. Ed. Engl.* **1981**, *20*, 223.
Didehydroamino acids	*Angew. Chem. Int. Ed. Engl.* **1991**, *30*, 1051.
	Synthesis **1988**, 159.
Didehydropeptides	*Synthesis* **1988**, 159.

Dienals	
Organocuprates	
Addition reactions	*Pure & Appl. Chem.* **1983**, *55*, 1759.
Dienes	*Acc. Chem. Res.* **1988**, *21*, 47.
Additions	*Synthesis* **1980**, 165.
	Synthesis **1980**, 769.
Photochemical	*Pure & Appl. Chem.* **1988**, *60*, 1009.
Allyl cations	
Cycloadditions	*Angew. Chem. Int. Ed. Engl.* **1984**, *23*, 1.
Boronate fragmentation	*Synthesis* **1971**, 229.
Cyclizations	*Tetrahedron* **1974**, *30*, 1651.
Cyclopolymerization	*Pure & Appl. Chem.* **1970**, *23*, 255.
Hydroboration	*Org Reactions* **1963**, *13*, 18.
Organoiron chemistry	*Acc. Chem. Res.* **1980**, *13*, 463.
Organometallic complexes	*Pure & Appl. Chem.* **1981**, *53*, 2379.
Photooxidation	*Tetrahedron* **1991**, *47*, 1343.
Siloxy dienes	*Acc. Chem. Res.* **1981**, *14*, 400.
Dienones	
Organocuprates	
Addition reactions	*Pure & Appl. Chem.* **1983**, *55*, 1759.
Diepoxides	
Rearrangements	*Quart. Rev.* **1967**, *21*, 456.
β-Diesters	
Dealkyoxycarbonylations	*Synthesis* **1982**, 805.
1,2-Diethynylcyclopropanes	
Rearrangements	*Acc. Chem. Res.* **1973**, *6*, 25.
1,2-Difunctionality	*Tetrahedron* **1992**, *48*, 2803.
1,3-Difunctionality	*Tetrahedron* **1992**, *48*, 2803.
1,1-Dihaloalkyl heterocumulenes	*Tetrahedron* **1991**, *47*, 1563.
Dihalocarbenes	*Acc. Chem. Res.* **1972**, *5*, 65.
	Chem. Rev. **1988**, *88*, 1293.
Dihydroanthrocenes	*Acc. Chem. Res.* **1971**, *4*, 393.
1,4-Dihydroaromatic compounds	*Acc. Chem. Res.* **1973**, *6*, 25.
Dihydroerythronolide A	*Pure & Appl. Chem.* **1987**, *59*, 345.
Dihydrofolated reductase	
Inhibitors	*Chem. Rev.* **1984**, *84*, 333.
	Acc. Chem. Res. **1969**, *2*, 129.
1,2-Dihydro-3*H*-indazol-3-ones	*Synthesis* **1978**, 633.
2,3-Dihydroisoxazoles	*Chem. Rev.* **1983**, *83*, 241.
5,10-Dihydrophenophosphazine derivatives	*Chem. Rev.* **1987**, *87*, 289.
7,12-Dihydropleiadenes	
Transannular reactions	*Acc. Chem. Res.* **1969**, *2*, 210.

Dihydropyridines — *Angew. Chem. Int. Ed. Engl.* **1991**, *30*, 1559.
Chem. Rev. **1972**, *72*, 1.
Chem. Rev. **1982**, *82*, 223.

Diimides
- Monosubstitution — *Acc. Chem. Res.* **1971**, *4*, 193.
- Reductions — *Org. Reactions* **1991**, *40*, 91.

α,ω-Diisocyanatocarbodiimides — *Angew. Chem. Int. Ed. Engl.* **1981**, *20*, 819.

Diketene — *Acc. Chem. Res.* **1974**, *7*, 265.
Chem. Rev. **1986**, *86*, 241.
- Heterocycles — *Acc. Chem. Res.* **1974**, *7*, 265.

β-Diketones — *Angew. Chem. Int. Ed. Engl.* **1971**, *10*, 225.
- Metalla-β-diketones — *Acc. Chem. Res.* **1981**, *14*, 109.

2,5-Diketopiperazines — *Pure & Appl. Chem.* **1983**, *55*, 1799.

Dimerization
- Alkenes — *Synthesis* **1974**, 539.
  - *Transition metal catalysis* — *Chem. Rev.* **1986**, *86*, 353.
- Butadiene — *Angew. Chem. Int. Ed. Engl.* **1965**, *4*, 327.
- Electrochemical
  - *Cathodic* — *Angew. Chem. Int. Ed. Engl.* **1972**, *11*, 760.
- Photochemical — *Chem. Rev.* **1952**, *51*, 1.

Di-π-methane
- Rearrangement — *Chem. Rev.* **1973**, *73*, 531.
  - *Photochemical* — *Acc. Chem. Res.* **1982**, *15*, 312.

Dimethyl sulfoxide
- Oxidations — *Chem. Rev.* **1967**, *67*, 247.
  Synthesis **1981**, 165.
  Synthesis **1990**, 87.

Diols — *Chem. Rev.* **1980**, *80*, 187.
- Alkenes — *Org. Reactions* **1984**, *30*, 457.

1,3-Dioxanes — *Acc. Chem. Res.* **1970**, *3*, 1.

1,2-Dioxetane
- Photochemistry — *Acc. Chem. Res.* **1974**, *7*, 97.

Dioxirane
- Oxidations — *Acc. Chem. Res.* **1989**, *22*, 205.

Dioxiranes — *Chem. Rev.* **1989**, *89*, 1359.

1,3-Dioxolan-2-ylium cations
- Heterocycles — *Chem. Rev.* **1972**, *72*, 357.

Dipolar cycloadditions — *See also cycloadditions*
Tetrahedron **1985**, *41*, 3447ff.
- Alkenes — *Org. Reactions* **1988**, *36*, 1.
- Isoxazolines — *Acc. Chem. Res.* **1984**, *17*, 410.
- Nitrones — *Org. Reactions* **1988**, *36*, 1.

2,2′-Disubstituted biphenyls — *Chem. Rev.* **1968**, *68*, 209.

2,6-Disubstituted oxanes — *Chem. Rev.* **1983**, *83*, 379.

2,6-Disubstituted thianes	*Chem. Rev.* **1983**, *83*, 379.
Disulfides	
Cleavage reactions	*Chem. Rev.* **1959**, *59*, 583.
***vic*-Disulfoxides**	*Chem. Rev.* **1984**, *84*, 117.
Diterpene alkaloids	*Pure & Appl. Chem.* **1986**, *58*, 711.
Dithianes	*Acc. Chem. Res.* **1970**, *3*, 1.
	Tetrahedron **1989**, *45*, 7643.
Umpolung reagents	*Synthesis* **1969**, 17.
Dithioacetals	*Acc. Chem. Res.* **1991**, *24*, 257.
Dithiocarboxylic acid derivatives	*Synthesis* **1983**, 605.
Dithiolene dyes	*Tetrahedron* **1991**, *47*, 909.
Diyl trapping	
Intramolecular	*Chem. Rev.* **1986**, *86*, 875.
Diyne reaction	*Synthesis* **1974**, 761.
Diynes	*Synthesis* **1974**, 761.
DMAP catalysis	*Chem. Soc. Rev.* **1983**, *12*, 129.
DNA	*Pure & Appl. Chem.* **1987**, *59*, 325.
Acridine dye binding	*Angew. Chem. Int. Ed. Engl.* **1986**, *25*, 115.
Biosynthesis	*Acc. Chem. Res.* **1977**, *10*, 418.
Calicheamicin	*Acc. Chem. Res.* **1991**, *24*, 235.
Conformations	*Acc. Chem. Res.* **1986**, *19*, 180.
Degradation	
Bleomycin	*Chem. Rev.* **1987**, *87*, 1107.
DNA-protein interactions	*Acc. Chem. Res.* **1980**, *13*, 155.
Drug interactions	*Acc. Chem. Res.* **1979**, *12*, 118.
Ethidium dye binding	*Angew. Chem. Int. Ed. Engl.* **1986**, *25*, 115.
Extrachromosomal DNA in bacteria	*Angew. Chem. Int. Ed. Engl.* **1973**, *12*, 517.
Histones	*Acc. Chem. Res.* **1975**, *8*, 327.
Intercalation	*Acc. Chem. Res.* **1990**, *23*, 271.
Interlocking rings	*Acc. Chem. Res.* **1973**, *6*, 252.
Isomerizations	*Chem. Soc. Rev.* **1989**, *18*, 53.
Metal complexes	*Chem. Rev.* **1971**, *71*, 439.
Methylation	*Angew. Chem. Int. Ed. Engl.* **1984**, *23*, 919.
NMR	
Solid state	*Chem. Rev.* **1991**, *91*, 1545.
Protein	
Recognition	*Acc. Chem. Res.* **1980**, *13*, 155.
Repair processes	*Angew. Chem. Int. Ed. Engl.* **1973**, *12*, 264.
Structures	*Acc. Chem. Res.* **1987**, *20*, 276.
	Pure & Appl. Chem. **1984**, *56*, 989.
Synthesis	*Acc. Chem. Res.* **1991**, *24*, 278.
	Tetrahedron **1983**, *39*, 3.
Viruses	*Acc. Chem. Res.* **1974**, *7*, 169.

DNA polymerase *Chem. Rev.* **1990**, *90*, 1291.
Dodecahedrane *Chem. Rev.* **1989**, *89*, 1051.
Double asymmetric induction *Angew. Chem. Int. Ed. Engl.* **1985**, *24*, 1.
Drimanes *Acc. Chem. Res.* **1983**, *16*, 81.
Drug design *Angew. Chem. Int. Ed. Engl.* **1977**, *16*, 766.
Angew. Chem. Int. Ed. Engl. **1981**, *20*, 217.
Chem. Soc. Rev. **1974**, *3*, 273.
Chem. Soc. Rev. **1979**, *8*, 563.
Chem. Soc. Rev. **1984**, *13*, 279.
Tetrahedron **1989**, *45*, 4327.
Computer applications *Acc. Chem. Res.* **1987**, *20*, 322.
Protein crystallography *Angew. Chem. Int. Ed. Engl.* **1986**, *25*, 767.
Drug inhibition
Enzymes *Acc. Chem. Res.* **1988**, *21*, 348.
Drug receptors *Angew. Chem. Int. Ed. Engl.* **1977**, *16*, 766.
Dyeing *Chem. Soc. Rev.* **1972**, *1*, 145.
Dyes
Optical sensors *Angew. Chem. Int. Ed. Engl.* **1991**, *30*, 958.
Semi-conductor properties *Angew. Chem. Int. Ed. Engl.* **1972**, *11*, 1051.
Xanthene dye *Acc. Chem. Res.* **1989**, *22*, 171.
Echinoderms *Chem. Soc. Rev.* **1972**, *1*, 1.
Pure & Appl. Chem. **1986**, *58*, 423.
Economics
Basic research *Angew. Chem. Int. Ed. Engl.* **1982**, *21*, 109.
Effective charge
Reaction mechanisms *Chem. Soc. Rev.* **1986**, *15*, 125.
Eicosanoids *Angew. Chem. Int. Ed. Engl.* **1991**, *30*, 1100.
Biosynthesis *Pure & Appl. Chem.* **1987**, *59*, 269.
Electro-optical Kerr effect *Angew. Chem. Int. Ed. Engl.* **1977**, *16*, 663.
Electrochemical *Angew. Chem. Int. Ed. Engl.* **1986**, *25*, 683.
Aniodic
Oxidations *Tetrahedron* **1991**, *47*, 531ff.
Substitution reactions *Tetrahedron* **1976**, *32*, 2185.
Black carbons *Angew. Chem. Int. Ed. Engl.* **1983**, *22*, 950.
Cyctochrome c *Acc. Chem. Res.* **1988**, *21*, 407.
Dimerization
Cathodic *Angew. Chem. Int. Ed. Engl.* **1972**, *11*, 760.
Electron transfers *Chem. Rev.* **1983**, *83*, 425.
Glucose
Oxidations *Pure & Appl. Chem.* **1991**, *63*, 1599.
Organic halides *Synthesis* **1990**, 369.

Organometallic chemistry	*Acc. Chem. Res.* **1972**, *5*, 415.
	Synthesis **1973**, 377.
Oxidations	*Chem. Rev.* **1968**, *68*, 449.
	SYNLETT **1990**, 301.
Aromatic compounds	
Amines	*Acc. Chem. Res.* **1969**, *2*, 175.
Indoles	*Chem. Rev.* **1990**, *90*, 795.
Photochemistry	*Angew. Chem. Int. Ed. Engl.* **1988**, *27*, 63.
	Pure & Appl. Chem, **1980**, *52*, 2649.
Protecting group removal	*Angew. Chem. Int. Ed. Engl.* **1976**, *15*, 281.
Radical chemistry	
Synthetic methods	*Angew. Chem. Int. Ed. Engl.* **1982**, *21*, 256.
Reaction mechanisms	*Chem. Soc. Rev.* **1975**, *4*, 471.
Reductions	
Enzymatic reactions	*Angew. Chem. Int. Ed. Engl.* **1985**, *24*, 539.
Substitution reactions	
Aromatic compounds	*Acc. Chem. Res.* **1973**, *6*, 106.
Synthetic methods	*Angew. Chem. Int. Ed. Engl.* **1981**, *20*, 911.
	Synthesis **1971**, 285.
	Tetrahedron **1984**, *40*, 811.
	Tetrahedron **1984**, *40*, 935.
	Tetrahedron **1991**, *47*, 531ff.
Ultramicroelectrodes	*Angew. Chem. Int. Ed. Engl.* **1991**, *30*, 170.
Vitamin B_{12}	*Acc. Chem. Res.* **1983**, *16*, 235.
	Pure & Appl. Chem. **1987**, *59*, 363.
Electrocyclic reactions	*Acc. Chem. Res.* **1968**, *1*, 17.
1,5-Electrocyclic reactions	*Angew. Chem. Int. Ed. Engl.* **1980**, *19*, 947.
1,7-Electrocyclic reactions	*Synthesis* **1991**, 181.
Photochemistry	*Acc. Chem. Res.* **1978**, *11*, 334.
Stereochemistry	*Acc. Chem. Res.* **1969**, *2*, 152.
Electron acceptors	*SYNLETT* **1991**, 301.
Electron loss	
Photochemistry	*Pure & Appl. Chem.* **1981**, *53*, 223.
Electron spin alignment	*Pure & Appl. Chem.* **1987**, *59*, 1595.
Electron spin echo spectroscopy	
Triplets	*Chem. Rev.* **1984**, *84*, 1.
Electron transfers	*Pure & Appl. Chem.* **1988**, *60*, 1025.
Carbanions	*Acc. Chem. Res.* **1972**, *5*, 169.
Marcus-Hush theory	*Acc. Chem. Res.* **1978**, *11*, 94.
Membrane systems	*Pure & Appl. Chem.* **1982**, *54*, 1693.
Metalloproteins	*Pure & Appl. Chem.* **1983**, *55*, 1049.
Organocuprates	*Acc. Chem. Res.* **1976**, *9*, 59.
Organometallic chemistry	
Reaction mechanisms	*Pure & Appl. Chem,* **1980**, *52*, 571.

Electron transfers, cont'd.	
Photochemistry	*Pure & Appl. Chem.* **1982**, *54*, 1605.
	Pure & Appl. Chem. **1986**, *58*, 1279.
	Synthesis **1989**, 233.
Reaction mechanisms	*Pure & Appl. Chem.* **1982**, *54*, 1885.
Polymer systems	*Pure & Appl. Chem.* **1982**, *54*, 1693.
Reaction mechanisms	*Acc. Chem. Res.* **1987**, *20*, 53.
	Angew. Chem. Int. Ed. Engl. **1988**, *27*, 1227.
	Tetrahedron **1982**, *38*, 1025ff.
Electrophilic reactions	*Angew. Chem. Int. Ed. Engl.* **1973**, *12*, 173.
Additions	
Alkenes	*Acc. Chem. Res.* **1981**, *14*, 227.
Alkynes	*Acc. Chem. Res.* **1981**, *14*, 227.
Allenes	
Reaction mechanisms	*Chem. Rev.* **1983**, *83*, 263.
Anthraquinones	*Tetrahedron* **1990**, *46*, 291.
Heterocycles	*Pure & Appl. Chem.* **1991**, *63*, 243.
Norbornene	*Acc. Chem. Res.* **1969**, *2*, 152.
Strained alkanes	*Acc. Chem. Res.* **1969**, *2*, 152.
Alkenes	*Acc. Chem. Res.* **1981**, *14*, 227.
Arenesulfonyloxy groups	*Tetrahedron* **1991**, *47*, 1109.
C-Nitroso compounds	*Chem. Soc. Rev.* **1977**, 6, 1.
Diazoalkanes	*Synthesis* **1985**, 569.
Reaction mechanisms	*Angew. Chem. Int. Ed. Engl.* **1980**, *19*, 151.
Substitution	
Allylsilanes	*Org. Reactions* **1989**, *37*, 57.
Amines	
α-Metalloamine equivalents	*Chem. Rev.* **1984**, *84*, 471.
Aromatic compounds	*Acc. Chem. Res.* **1971**, *4*, 240.
	Acc. Chem. Res. **1971**, *4*, 248.
	Acc. Chem. Res. **1974**, *7*, 361.
	Chem. Soc. Rev. **1974**, *3*, 167.
	Synthesis **1971**, 455.
Nitration	*Acc. Chem. Res.* **1987**, *20*, 53.
Reaction mechanisms	*Acc. Chem. Res.* **1974**, *7*, 361.
Heterocycles	*Angew. Chem. Int. Ed. Engl.* **1967**, 6, 608.
Organosilicon chemistry	*Synthesis* **1979**, 761.
Trimethylsilyl groups	*Acc. Chem. Res.* **1977**, *10*, 1.
Saturated compounds	*Angew. Chem. Int. Ed. Engl.* **1962**, *1*, 382.
Trisdialkylaminobenzenes	*Acc. Chem. Res.* **1989**, *22*, 27.
Vinylsilanes	*Org. Reactions* **1989**, *37*, 57.
Electrophoresis	
Historical perspective	*Chem. Soc. Rev.* **1985**, *14*, 225.

Electrophotography	*Angew. Chem. Int. Ed. Engl.* **1977**, *16*, 374.
Electroreductions	*Synthesis* **1986**, 873.
Elimination addition reactions	
Acyl transfers	*Chem. Rev.* **1975**, *75*, 627.
Elimination reactions	
Acetals	*Pure & Appl. Chem.* **1988**, *60*, 49.
	Angew. Chem. Int. Ed. Engl. **1967**, *6*, 534.
Alkenes	*Angew. Chem. Int. Ed. Engl.* **1965**, *4*, 49.
Carbanions	*Acc. Chem. Res.* **1970**, *3*, 281.
E_2	*Chem. Rev.* **1980**, *80*, 453.
Nucleofugality	*Acc. Chem. Res.* **1979**, *12*, 198.
Reaction mechanisms	*Acc. Chem. Res.* **1972**, *5*, 374.
	Angew. Chem. Int. Ed. Engl. **1972**, *11*, 200.
Allyl oxo carboxylates	*Acc. Chem. Res.* **1987**, *20*, 140.
Azo compounds	*Angew. Chem. Int. Ed. Engl.* **1977**, *16*, 835.
Base catalysed	*Acc. Chem. Res.* **1973**, *6*, 410.
Bimolecular	
Carbanion mechanism	*Quart. Rev.* **1957**, *21*, 490.
Reaction mechanisms	*Angew. Chem.* **1962**, *74*, 731.
	Angew. Chem. Int. Ed. Engl. **1972**, *11*, 200.
	Pure & Appl. Chem. **1971**, *25*, 655.
Concerted reactions	*Acc. Chem. Res.* **1972**, *5*, 374.
	Acc. Chem. Res. **1976**, *9*, 19.
1,4-Conjugate reactions	*Acc. Chem. Res.* **1971**, *4*, 393.
Cycloeliminations	*Angew. Chem. Int. Ed. Engl.* **1971**, *10*, 529.
	Angew. Chem. Int. Ed. Engl. **1971**, *10*, 537.
β-Eliminations	*Tetrahedron* **1975**, *31*, 2999.
Haloalkenes	*Acc. Chem. Res.* **1984**, *17*, 137.
Reaction mechanisms	*Acc. Chem. Res.* **1972**, *5*, 374.
	Acc. Chem. Res. **1989**, *22*, 211.
	Chem. Soc. Rev. **1972**, *1*, 163.
Sulfilimines	*Tetrahedron* **1977**, *33*, 2359.
Thiol esters	*Acc. Chem. Res.* **1986**, *19*, 186.
Ellipticines	*Tetrahedron* **1986**, *42*, 2389.
6*H*-Pyrido[4,3-*b*]carbazole	*Synthesis* **1984**, 289.
Enamides	*Synthesis* **1987**, 421.
Photochemistry	*Synthesis* **1978**, 489.
Enamines	*Synthesis* **1970**, 510.
Conjugated enamines	*Tetrahedron* **1984**, *40*, 2989.
Reaction mechanisms	*Tetrahedron* **1982**, *38*, 1975.
	Tetrahedron **1982**, *38*, 3363.
Enaminones	*Chem. Soc. Rev.* **1977**, *6*, 277.
Endoperoxides	*Angew. Chem. Int. Ed. Engl.* **1983**, *22*, 805.

ENDOR spectroscopy — *Angew. Chem. Int. Ed. Engl.* **1984**, *23*, 173.
Acc. Chem. Res. **1991**, *24*, 164.

Enediol phosphates
Phosphorylation — *Acc. Chem. Res.* **1978**, *11*, 239.
Enediynes — *Angew. Chem. Int. Ed. Engl.* **1991**, *30*, 1387.
Energy metabolism
Mitochondria — *Angew. Chem. Int. Ed. Engl.* **1972**, *11*, 792.
Enol ethers
Nitroalkenes — *Acc. Chem. Res.* **1985**, *18*, 284.
Silyl enol ethers
Enolates — *Acc. Chem. Res.* **1985**, *18*, 181.
Nitroalkenes — *Acc. Chem. Res.* **1985**, *18*, 284.
Tetracyanoethylene
Cycloaddition reactions — *Acc. Chem. Res.* **1977**, *10*, 117.
Acc. Chem. Res. **1977**, *10*, 199.

Enol triflates
Coupling reactions — *Acc. Chem. Res.* **1988**, *21*, 47.
Enolates — *Pure & Appl. Chem.* **1983**, *55*, 1749.
Pure & Appl. Chem. **1985**, *57*, 1789.

Alkylations
Asymmetric synthesis — *Pure & Appl. Chem.* **1981**, *53*, 1109.
Peptides — *Angew. Chem. Int. Ed. Engl.* **1988**, *27*, 1624.
Asymmetric synthesis — *Tetrahedron* **1989**, *45*, 4913.
Chiral enolates
Aldol reaction — *Angew. Chem. Int. Ed. Engl.* **1987**, *26*, 24.
Ketones — *Tetrahedron* **1976**, *32*, 2979.
Organometallic chemistry — *Pure & Appl. Chem.* **1988**, *60*, 1.
Peptides
Alkylations — *Angew. Chem. Int. Ed. Engl.* **1988**, *27*, 1624.
Reaction mechanisms — *Tetrahedron* **1977**, *33*, 2737.
Enols — *Acc. Chem. Res.* **1988**, *21*, 135.
Chem. Rev. **1979**, *79*, 515.

Asymmetric synthesis — *Tetrahedron* **1989**, *45*, 4913.
Conformations — *Acc. Chem. Res.* **1988**, *21*, 442.
Reaction mechanisms — *Pure & Appl. Chem.* **1991**, *63*, 213.
Tautomerism — *Acc. Chem. Res.* **1988**, *21*, 42.
Enones — *Acc. Chem. Res.* **1988**, *21*, 47.
Chem. Rev. **1989**, *89*, 1591.

Alkenes
Cycloadditions
Photochemistry — *Synthesis* **1970**, 287.
Cyclizations — *Synthesis* **1975**, 1.
Heterodienes — *Chem. Rev.* **1975**, *75*, 651.
Reductions — *Tetrahedron* **1986**, *42*, 6351.

Enterobactin biosynthetic pathway	*Chem. Rev.* **1990**, *90*, 1105.
Enynes	*Acc. Chem. Res.* **1988**, *21*, 47.
Cycloisomerization	*Acc. Chem. Res.* **1990**, *23*, 34.
Enzymatic cleavage	
Peptides	*Acc. Chem. Res.* **1970**, *3*, 81.
	Acc. Chem. Res. **1970**, *3*, 249.
Enzymes	*Acc. Chem. Res.* **1975**, *8*, 1.
	Acc. Chem. Res. **1981**, *14*, 284.
	Acc. Chem. Res. **1985**, *18*, 128.
	Angew. Chem. Int. Ed. Engl. **1971**, *10*, 548.
	Angew. Chem. Int. Ed. Engl. **1977**, *16*, 137.
	Angew. Chem. Int. Ed. Engl. **1977**, *16*, 205.
	Angew. Chem. Int. Ed. Engl. **1977**, *16*, 285.
	Angew. Chem. Int. Ed. Engl. **1985**, *24*, 85.
	Angew. Chem. Int. Ed. Engl. **1991**, *30*, 931.
	Chem. Rev. **1990**, *90*, 1327.
	Chem. Soc. Rev. **1973**, 2, 1.
	Chem. Soc. Rev. **1974**, *3*, 387.
	Chem. Soc. Rev. **1979**, *8*, 85.
	Chem. Soc. Rev. **1979**, *8*, 447.
	Chem. Soc. Rev. **1988**, *17*, 383.
	Compare with non-enzymatic reactions.
	Pure & Appl. Chem. **1983**, *55*, 637.
	Pure & Appl. Chem. **1984**, *56*, 979.
	Pure & Appl. Chem. **1990**, *62*, 1859.
	See also Baker's yeast, and specific enzyme names.
	Tetrahedron **1984**, *40*, 269.
Active site manipulations	*Angew. Chem. Int. Ed. Engl.* **1988**, *27*, 913.
Alkanes	*Chem. Rev.* **1990**, *90*, 1343.
Amino acids	*Acc. Chem. Res.* **1973**, 6, 361.
Analogues	*Chem. Soc. Rev.* **1979**, *8*, 85.
Apolar solvents	
Reverse micelles	*Angew. Chem. Int. Ed. Engl.* **1985**, *24*, 439.
Artifical enzymes	*Tetrahedron* **1984**, *40*, 269.
Asymmetric synthesis	*Acc. Chem. Res.* **1990**, *23*, 114.
	Tetrahedron **1986**, *42*, 3351.
Bacterial ribosome	*Angew. Chem. Int. Ed. Engl.* **1982**, *21*, 23.
Biomimetic models	*Pure & Appl. Chem.* **1991**, *63*, 265.
Carbohydrates	
Synthesis	*Synthesis* **1991**, 499.
	Tetrahedron **1989**, *45*, 5365.
Chemical modifications	*Acc. Chem. Res.* **1971**, *4*, 353.

Enzymes, cont'd.	
Chorismate metabolism	*Chem. Rev.* **1990**, *90*, 1105.
Cyclopropane derivatives	*Angew. Chem. Int. Ed. Engl.* **1988**, *27*, 537.
Decarboxylation	*Chem. Rev.* **1987**, *87*, 863.
Discrimination	*Acc. Chem. Res.* **1991**, *24*, 209.
Drug inhibition	*Acc. Chem. Res.* **1988**, *21*, 348.
Electron transfers	*Chem. Rev.* **1990**, *90*, 1359.
Enterobactin biosynthetic pathway	*Chem. Rev.* **1990**, *90*, 1105.
Enzyme-coenzyme binding sites	*Acc. Chem. Res.* **1982**, *15*, 128.
Erythromycin	
Biosynthesis	*Angew. Chem. Int. Ed. Engl.* **1991**, *30*, 1302.
Esterolytic enzymes	*Synthesis* **1991**, 1049.
Esters	*Org. Reactions* **1989**, *37*, 1.
Fatty acids	
Unsaturated fatty acids	*Acc. Chem. Res.* **1969**, *2*, 196.
Glycosyl transfer	
Reaction mechanisms	*Chem. Rev.* **1990**, *90*, 1171.
Histidine	*Angew. Chem. Int. Ed. Engl.* **1978**, *17*, 583.
Hydrolysis	*Angew. Chem. Int. Ed. Engl.* **1964**, *3*, 1.
Hydroxylations	
Molecular oxygen	*Angew. Chem. Int. Ed. Engl.* **1972**, *11*, 701.
Immobilized enzymes	*Acc. Chem. Res.* **1979**, *12*, 344.
	Chem. Soc. Rev. **1977**, 6, 215.
Inhibitors	*Chem. Rev.* **1987**, *87*, 1183.
Nitro compounds	*Acc. Chem. Res.* **1983**, *16*, 418.
Ionic intermediates	
Reaction mechanisms	*Chem. Rev.* **1990**, *90*, 1151.
Kinetics	*Chem. Rev.* **1989**, *89*, 789.
Lipolytic enzymes	*Synthesis* **1991**, 1049.
Marine microorganisms	*Pure & Appl. Chem.* **1982**, *54*, 1951.
Metalloenzymes	*Pure & Appl. Chem.* **1991**, *63*, 265.
ENDOR	*Acc. Chem. Res.* **1991**, *24*, 164.
Methylations	
Reaction mechanisms	*Chem. Rev.* **1990**, *90*, 1275.
Mimics	*Pure & Appl. Chem.* **1990**, *62*, 1859.
Models	*Acc. Chem. Res.* **1987**, *20*, 146.
Molecular recognition	*Acc. Chem. Res.* **1987**, *20*, 79.
Molybdo-enzymes	*Pure & Appl. Chem.* **1984**, *56*, 1645.
Natural selection	*Acc. Chem. Res.* **1975**, *8*, 1.
Nitrenes	
Labelling	*Acc. Chem. Res.* **1972**, *5*, 155.

NMR	*Acc. Chem. Res.* **1977**, *10*, 246.
Nucleoside phosphorothioates	*Angew. Chem. Int. Ed. Engl.* **1975**, *14*, 160.
Organometallic chemistry	*Angew. Chem. Int. Ed. Engl.* **1991**, *30*, 931.
Peptide synthesis	*Angew. Chem. Int. Ed. Engl.* **1991**, *30*, 1437.
Pesticides	*Acc. Chem. Res.* **1988**, *21*, 348.
Phosphates	*Tetrahedron* **1982**, *38*, 1541.
Phosphoryl transfer enzymes	*Acc. Chem. Res.* **1971**, *4*, 214.
Polysaccharides	*Tetrahedron* **1985**, *41*, 2957.
Protein engineering	*Angew. Chem. Int. Ed. Engl.* **1984**, *23*, 467.
Proteinase inhibitors	*Angew. Chem. Int. Ed. Engl.* **1974**, *13*, 10.
Purines	
Binding sites	*Acc. Chem. Res.* **1982**, *15*, 128.
Pyridine nucleotides	*Acc. Chem. Res.* **1973**, *6*, 289.
Pyridoxal phosphate	*Acc. Chem. Res.* **1980**, *13*, 455.
	Angew. Chem. Int. Ed. Engl. **1980**, *19*, 441.
Reaction mechanisms	*Acc. Chem. Res.* **1982**, *15*, 232.
	Angew. Chem. Int. Ed. Engl. **1977**, *16*, 449.
	Chem. Soc. Rev. **1972**, *1*, 319.
	Tetrahedron **1978**, *34*, 813.
Reactive intermediates	*Angew. Chem. Int. Ed. Engl.* **1990**, *29*, 355.
	Chem. Soc. Rev. **1984**, *13*, 97.
Rearrangements	
Allylic rearrangements	*Chem. Rev.* **1990**, *90*, 1203.
Reductions	
Deoxyribonucleotides	*Angew. Chem. Int. Ed. Engl.* **1974**, *13*, 569.
Electrochemistry	*Angew. Chem. Int. Ed. Engl.* **1985**, *24*, 539.
Ribonucleotides	*Angew. Chem. Int. Ed. Engl.* **1974**, *13*, 569.
RNA cleavage	*Angew. Chem. Int. Ed. Engl.* **1990**, *29*, 749.
Sesquiterpenes	
Biosynthesis	*Chem. Rev.* **1990**, *90*, 1089.
Site-directed mutagenesis	*Chem. Rev.* **1987**, *87*, 1079.
Stereochemistry	*Pure & Appl. Chem.* **1984**, *56*, 1005.
Structures	*Angew. Chem. Int. Ed. Engl.* **1985**, *24*, 639.
	Chem. Soc. Rev. **1972**, *1*, 319.
Gene sequences	*Acc. Chem. Res.* **1989**, *22*, 232.
Subunit exchange	
Chimeras	*Angew. Chem. Int. Ed. Engl.* **1976**, *15*, 181.
Suicide inactivators	*Acc. Chem. Res.* **1976**, *9*, 313.
	Pure & Appl. Chem. **1981**, *53*, 149.
Suicide inhibitors	*Acc. Chem. Res.* **1983**, *16*, 418.
Suicide substrates	*Tetrahedron* **1982**, *38*, 871.
Sulfides	*Chem. Rev.* **1988**, *88*, 473.

Enzymes, cont'd.
Sulfoxides — *Chem. Rev.* **1988**, *88*, 473.
Synthesis — *Acc. Chem. Res.* **1989**, *22*, 47.
Synthetic methods — *Angew. Chem. Int. Ed. Engl.* **1988**, *27*, 622.
Toxins — *Acc. Chem. Res.* **1975**, *8*, 281.
Transition states — *Acc. Chem. Res.* **1972**, *5*, 10.
Vitamin B_{12} — *Chem. Soc. Rev.* **1985**, *14*, 161.
Enzymology — *Tetrahedron* **1991**, *47*, 5919ff.
Epidermal growth factor — *Angew. Chem. Int. Ed. Engl.* **1987**, *26*, 717.
Epimerization — *Acc. Chem. Res.* **1986**, *19*, 307.
Episulfones — *Acc. Chem. Res.* **1968**, *1*, 209.
Synthesis **1970**, 393.
Epoxidations
Alkenes
Asymmetric synthesis — *Angew. Chem. Int. Ed. Engl.* **1991**, *30*, 403.
Peroxy acids — *Org. Reactions* **1953**, *7*, 378.
Tetrahedron **1976**, *32*, 2855.
Asymmetric synthesis — *Tetrahedron: Asymmetry* **1991**, *2*, 481ff.
Peroxy acids — *Org. Reactions* **1953**, *7*, 378.
Transition metal catalysis — *Chem. Rev.* **1989**, *89*, 431.
Epoxides — *Synthesis* **1984**, 629.
Arenes — *Acc. Chem. Res.* **1974**, *7*, 85.
Halide ions — *Angew. Chem. Int. Ed. Engl.* **1972**, *11*, 1041.
Isomerizations — *Org. Reactions* **1983**, *29*, 345.
Organofluorine compounds — *Angew. Chem. Int. Ed. Engl.* **1985**, *24*, 161.
Polymerization — *Acc. Chem. Res.* **1974**, *7*, 294.
Ring opening reactions — *Angew. Chem. Int. Ed. Engl.* **1977**, *16*, 572.
Epoxyalcohols
Asymmetric synthesis — *Chem. Rev.* **1991**, *91*, 437.
Kinetic resolutions — *Pure & Appl. Chem.* **1983**, *55*, 589.
EPSP synthase — *Chem. Rev.* **1990**, *90*, 1131.
Ergot alkaloids — *Tetrahedron* **1980**, *36*, 3123.
Biosynthesis — *Tetrahedron* **1976**, *32*, 873.
Erythrocytes — *Pure & Appl. Chem.* **1984**, *56*, 907.
Erythromycin — *Angew. Chem. Int. Ed. Engl.* **1991**, *30*, 1452.
Biosynthesis — *Angew. Chem. Int. Ed. Engl.* **1991**, *30*, 1302.
Erythronolide — *Pure & Appl. Chem.* **1989**, *61*, 1235.
ESR
Reaction mechanisms — *Chem. Soc. Rev.* **1979**, *8*, 1.
Esterases — *Angew. Chem. Int. Ed. Engl.* **1985**, *24*, 617.

Eukaryotes
Fatty acids — *Acc. Chem. Res.* **1990**, *23*, 363.
Transcription — *Angew. Chem. Int. Ed. Engl.* **1987**, *26*, 218.
Euphorbia lathyris
Hydrocarbons — *Pure & Appl. Chem.* **1981**, *53*, 1101.
Evolution — *Angew. Chem. Int. Ed. Engl.* **1974**, *13*, 186.
Angew. Chem. Int. Ed. Engl. **1981**, *20*, 143.
Angew. Chem. Int. Ed. Engl. **1981**, *20*, 233.
Angew. Chem. Int. Ed. Engl. **1977**, *16*, 285.
Chem. Rev. **1989**, *89*, 789.
Tetrahedron **1984**, *40*, 1093.
Exchange reactions
Organolithium compounds — *Acc. Chem. Res.* **1968**, *1*, 23.
Exciplexs — *Pure & Appl. Chem,* **1982**, *54*, 1885.
Excited molecules — *Pure & Appl. Chem,* **1980**, *52*, 2591.
Excited species
Biochemical systems — *Pure & Appl. Chem.* **1984**, *56*, 1179.
Extrusion reactions — *Tetrahedron* **1988**, *44*, 6241.
Ezomycins
Synthesis — *Pure & Appl. Chem.* **1981**, *53*, 129.
Far Infra-red
Fourier transform IR — *Angew. Chem. Int. Ed. Engl.* **1976**, *15*, 25.
Farnesyl pyrophosphate synthetase — *Acc. Chem. Res.* **1978**, *11*, 307.
Fats and oils — *Angew. Chem. Int. Ed. Engl.* **1988**, *27*, 41.
Fatty acids — *Acc. Chem. Res.* **1976**, *9*, 34.
Angew. Chem. Int. Ed. Engl. **1976**, *15*, 61.
Enzymes — *Acc. Chem. Res.* **1969**, *2*, 196.
Eukaryotes — *Acc. Chem. Res.* **1990**, *23*, 363.
Fenestranes — *Chem. Rev.* **1987**, *87*, 399.
Fenton's reagent — *Acc. Chem. Res.* **1975**, *8*, 125.
Ferricyanide
Oxidations — *Chem. Rev.* **1958**, *58*, 439.
Ferritin — *Angew. Chem. Int. Ed. Engl.* **1973**, *12*, 57.
Ferrocenes — *Acc. Chem. Res.* **1973**, 6, 1.
Org. Reactions **1969**, *17*, 1.
Films
Photochemistry — *Tetrahedron* **1982**, *38*, 2455.
Five-membered ring compounds
Conformations — *Chem. Rev.* **1980**, *80*, 231.
Electrophilic substitution — *Adv. Heterocycl. Chem.* **1971**, *13*, 235.
Flash thermolysis — *Angew. Chem. Int. Ed. Engl.* **1977**, *16*, 365.
Flash vacuum pyrolysis — *Acc. Chem. Res.* **1969**, *2*, 367.

Flavanoids	
Oligomeric flavanoids	*Tetrahedron* **1992**, *48*, 1743.
Flavins	
Photochemistry	*Chem. Soc. Rev.* **1982**, *11*, 15.
Flavor chemistry	*Chem. Soc. Rev.* **1978**, *7*, 185.
Consumer acceptance	*Chem. Soc. Rev.* **1978**, *7*, 212.
Development	*Chem. Soc. Rev.* **1978**, *7*, 177.
Legislation	*Chem. Soc. Rev.* **1978**, *7*, 195.
Organoleptic properties	*Chem. Soc. Rev.* **1978**, *7*, 167.
Potable spirits	*Chem. Soc. Rev.* **1978**, *7*, 201.
Structural considerations	*Chem. Soc. Rev.* **1978**, *7*, 167.
Flavylium salts	*Tetrahedron* **1983**, *39*, 3005.
Fluorenylmethoxycarbonyl	
Protecting groups	
Amines	*Acc. Chem. Res.* **1987**, *20*, 401.
Fluoride ion	
Synthetic methods	*Chem. Rev.* **1980**, *80*, 429.
Fluorinated organometallic reagents	*Tetrahedron* **1992**, *48*, 189.
Fluorinated β-sultones	*Angew. Chem. Int. Ed. Engl.* **1972**, *11*, 583.
Fluorinated ketones	*Tetrahedron* **1991**, *47*, 3207.
Fluorinations	*Angew. Chem. Int. Ed. Engl.* **1981**, *20*, 647.
	Tetrahedron **1978**, *34*, 3.
Aminofluorosulfuranes	*Org. Reactions* **1988**, *35*, 513.
Monofluoroaliphatic compounds	*Org. Reactions* **1974**, *21*, 125.
Sulfur tetrafluoride	*Org. Reactions* **1974**, *21*, 1.
	Org. Reactions **1985**, *34*, 319.
Fluoro compounds	*Chem. Soc. Rev.* **1987**, *16*, 381,
	Org. Reactions **1944**, *2*, 49.
	Tetrahedron **1991**, *47*, 3207.
Hydrogen fluoride	*Tetrahedron* **1991**, *47*, 5329.
Hyperconjugation	*Chem. Rev.* **1971**, *71*, 139.
Organometallic compounds	*Chem. Rev.* **1991**, *91*, 553.
Fluoroallenes	
Cycloadditions	*Acc. Chem. Res.* **1991**, *24*, 63.
Fluorocarbons	
Metal atoms	*Angew. Chem. Int. Ed. Engl.* **1975**, *14*, 287.
α-Fluorocarbonyl compounds	*Tetrahedron* **1985**, *41*, 1111.
Fluoronated isocyanates	*Angew. Chem. Int. Ed. Engl.* **1977**, *16*, 735.
Food systems	
Maillard reaction	*Angew. Chem. Int. Ed. Engl.* **1990**, *29*, 565.
Formamide acetals	*Tetrahedron* **1979**, *35*, 1675.
Formylating agents	*Chem. Rev.* **1987**, *87*, 671.

Forskolin	*Tetrahedron* **1992**, *48*, 963.
Fossils	*Angew. Chem. Int. Ed. Engl.* **1971**, *10*, 209.
Fourier transforms	*Chem. Soc. Rev.* **1975**, *4*, 569.
Four-membered ring compounds	
Conformations	*Chem. Rev.* **1980**, *80*, 231.
Isocyanides	*Synthesis* **1985**, 1083.
Reaction mechanisms	*Chem. Soc. Rev.* **1976**, *5*, 149.
Fragmentation reactions	
Heterolytic fragmentations	*Angew. Chem. Int. Ed. Engl.* **1967**, 6, 1.
	Angew. Chem. Int. Ed. Engl. **1969**, *8*, 535.
Free radical reactions	*Synthesis* **1973**, 1.
Addition reactions	
Unsaturated compounds	*Chem. Rev.* **1962**, *62*, 599.
	Org. Reactions **1963**, *13*, 91.
Alkenes	*Acc. Chem. Res.* **1975**, *8*, 165.
Autoxidation	*Acc. Chem. Res.* **1968**, *1*, 193.
Cyclizations	*Acc. Chem. Res.* **1971**, *4*, 386.
Organoboranes	
Substitution reactions	*Angew. Chem. Int. Ed. Engl.* **1972**, *11*, 692.
Structure activity relationships	*Angew. Chem. Int. Ed. Engl.* **1970**, *9*, 830.
Substitution reactions	
Organoboranes	*Angew. Chem. Int. Ed. Engl.* **1972**, *11*, 692.
Sulfur compounds	*Angew. Chem. Int. Ed. Engl.* **1964**, *3*, 602.
Fremy's radical	*Chem. Rev.* **1971**, *71*, 229.
FTIR	*Acc. Chem. Res.* **1981**, *14*, 171.
Fulvalenes	
Organometallic derivatives	*SYNLETT* **1990**, 493.
Fulvenes	
Organometallic chemistry	*Synthesis* **1970**, 449.
Functionalized polymers	*Chem. Rev.* **1981**, *81*, 557.
Fungicides	
3-Phenylpropylamines	*Angew. Chem. Int. Ed. Engl.* **1980**, *19*, 184.
Furans	
Macrocycles	*Chem. Rev.* **1977**, *77*, 513.
Gabriel reagent	*Acc. Chem. Res.* **1991**, *24*, 285.
Gallium compounds	*Angew. Chem. Int. Ed. Engl.* **1985**, *24*, 893.
Gas chromatography	*Acc. Chem. Res.* **1970**, *3*, 33.
Chiral separations	*Angew. Chem. Int. Ed. Engl.* **1984**, *23*, 747.
	Angew. Chem. Int. Ed. Engl. **1990**, *29*, 939.
Gas phase flow thermolysis	*Angew. Chem. Int. Ed. Engl.* **1986**, *25*, 414.
Gas phase reactions	
Gas phase ions	*Pure & Appl. Chem.* **1984**, *56*, 1831.
Organic ions	*Pure & Appl. Chem.* **1984**, *56*, 1809.

Gel permeation chromatography	*Acc. Chem. Res.* **1973**, *6*, 348.
Gels	*Angew. Chem. Int. Ed. Engl.* **1979**, *18*, 599.
Gene expression	*Angew. Chem. Int. Ed. Engl.* **1988**, *27*, 1040.
Gene sequences	
Enzymes	
Structures	*Acc. Chem. Res.* **1989**, *22*, 232.
Genetic engineering	
Plants	*Angew. Chem. Int. Ed. Engl.* **1987**, *26*, 382.
Genetic information	*Angew. Chem. Int. Ed. Engl.* **1971**, *10*, 638.
	Angew. Chem. Int. Ed. Engl. **1972**, *11*, 883.
	Angew. Chem. Int. Ed. Engl. **1985**, *24*, 1015.
Germylenes	*Chem. Rev.* **1991**, *91*, 311.
Ghosts	*Acc. Chem. Res.* **1975**, *8*, 81.
Gibberellins	*Acc. Chem. Res.* **1983**, *16*, 48.
Globular proteins	*Angew. Chem. Int. Ed. Engl.* **1977**, *16*, 23.
Glucose	
Oxidations	
Electrochemistry	*Pure & Appl. Chem.* **1991**, *63*, 1599.
α-Glucosidase inhibtiors	*Angew. Chem. Int. Ed. Engl.* **1981**, *20*, 744.
Glucosides	
Hydrolysis	*Acc. Chem. Res.* **1984**, *17*, 305.
Glutaconaldehyde	*Synthesis* **1980**, 589.
Glutamate dehydrogenase	*Acc. Chem. Res.* **1971**, *4*, 379.
Glutamate receptor inhibitor	*Pure & Appl. Chem.* **1990**, *62*, 1223.
γ-Glutamyl transpeptidase	*Pure & Appl. Chem.* **1984**, *56*, 821.
Glutaraldehyde monoacetal	*Synthesis* **1985**, 592.
Glutathione xenobiotic metabolites	*Acc. Chem. Res.* **1991**, *24*, 264.
Glycals	
Glycosidation	*Acc. Chem. Res.* **1990**, *23*, 201.
Glycans	*Pure & Appl. Chem.* **1984**, *56*, 779.
	Pure & Appl. Chem. **1984**, *56*, 859.
Glyceraldehyde	*Acc. Chem. Res.* **1988**, *21*, 294.
Glycidic esters	
Darzens condensation	*Org. Reactions* **1949**, *5*, 413.
Glycidols	
Asymmetric synthesis	*Chem. Rev.* **1991**, *91*, 437.
Glycoconjugates	*Pure & Appl. Chem.* **1989**, *61*, 1257.
Glycogen phosphorylase	*Acc. Chem. Res.* **1978**, *11*, 239.
Glycol derivatives	*Angew. Chem. Int. Ed. Engl.* **1980**, *19*, 441.
Asymmetric synthesis	*Pure & Appl. Chem.* **1987**, *59*, 353.

Glycolipids	*Angew. Chem. Int. Ed. Engl.* **1991**, *30*, 1611.
	Pure & Appl. Chem. **1987**, *59*, 1477.
	Pure & Appl. Chem. **1991**, *63*, 529.
Immunomodulators	*Angew. Chem. Int. Ed. Engl.* **1991**, *30*, 1611.
Glycopeptides	*Angew. Chem. Int. Ed. Engl.* **1987**, *26*, 294.
Glycoproteins	*Chem. Soc. Rev.* **1973**, *2*, 355.
	Chem. Soc. Rev. **1984**, *13*, 15.
	Pure & Appl. Chem. **1984**, *56*, 779.
	Pure & Appl. Chem. **1984**, *56*, 859.
	Pure & Appl. Chem. **1987**, *59*, 1457.
	Pure & Appl. Chem. **1989**, *61*, 1283.
Human serum glycoproteins	*Angew. Chem. Int. Ed. Engl.* **1973**, *12*, 721.
NMR	*Pure & Appl. Chem.* **1981**, *53*, 45.
Structures	*Angew. Chem. Int. Ed. Engl.* **1990**, *29*, 823.
Glycopyranans	*Acc. Chem. Res.* **1973**, *6*, 184.
Glycosaminoglycans	*Pure & Appl. Chem.* **1991**, *63*, 545.
Glycosidation	*Synthesis* **1991**, 583.
	Tetrahedron **1990**, *46*, 5835.
Glycals	*Acc. Chem. Res.* **1990**, *23*, 201.
Glycosides	*Angew. Chem. Int. Ed. Engl.* **1974**, *13*, 157.
Carbon glycosides	*Acc. Chem. Res.* **1990**, *23*, 201.
Koenigs-Knorr method	*Angew. Chem. Int. Ed. Engl.* **1986**, *25*, 212.
Steroidal	*Pure & Appl. Chem.* **1986**, *58*, 799.
Glycosphingolipids	*Pure & Appl. Chem.* **1989**, *61*, 1307.
Glycosyl transfer	
Enzymes	*Chem. Rev.* **1990**, *90*, 1171.
Glycosylations	*Pure & Appl. Chem.* **1991**, *63*, 519.
Palladium catalysis	*Acc. Chem. Res.* **1990**, *23*, 201.
Glyoxalase	
Reaction mechanisms	*Angew. Chem. Int. Ed. Engl.* **1985**, *24*, 31.
Gold compounds	*Chem. Soc. Rev.* **1980**, *9*, 217.
Green bacteriochlorophyll	*Acc. Chem. Res.* **1980**, *13*, 309.
Grignard reagents	
Activated magnesium	*Synthesis* **1981**, 585.
Addition reactions	
Ketones	
Reaction mechanisms	*Acc. Chem. Res.* **1974**, *7*, 272.
Nitroalkenes	*Acc. Chem. Res.* **1984**, *17*, 109.
Allyl Grignard reagents	*Synthesis* **1971**, 347.
Copper catalysis	*Tetrahedron* **1984**, *40*, 641.
Crotyl Grignard reagents	*Synthesis* **1971**, 347.
Di-Grignard reagents	*Angew. Chem. Int. Ed. Engl.* **1987**, *26*, 990.

Ketones	
Addition reactions	
Reaction mechanisms	*Acc. Chem. Res.* **1974**, *7*, 272.
Nitroalkenes	
Addition reactions	*Acc. Chem. Res.* **1984**, *17*, 109.
Nucleophilic additions	
Carbonyl compounds	
Reaction mechanisms	*Pure & Appl. Chem.* **1980**, *52*, 545.
Organometallic compounds	*Tetrahedron* **1975**, *31*, 2735.
Radical reactions	*Acc. Chem. Res.* **1991**, *24*, 255.
Reaction mechanisms	*Acc. Chem. Res.* **1990**, *23*, 286.
	Acc. Chem. Res. **1991**, *24*, 95.
Griseoviridin	
Synthesis	*Pure & Appl. Chem.* **1982**, *54*, 2537.
Guanidine derivatives	*Chem. Soc. Rev.* **1985**, *14*, 375.
Haemoglobin	*Chem. Soc. Rev.* **1983**, *12*, 387.
Halichondrins	*Pure & Appl. Chem.* **1986**, *58*, 701.
Halide ions	
Epoxides	*Angew. Chem. Int. Ed. Engl.* **1972**, *11*, 1041.
Hallucinogens	*Chem. Rev.* **1983**, *83*, 633.
Haloalkenes	
Elimination reactions	*Acc. Chem. Res.* **1984**, *17*, 137.
Oxidation	*Chem. Rev.* **1976**, *76*, 801.
Halobacteria	
Biosynthesis	*Angew. Chem. Int. Ed. Engl.* **1976**, *15*, 187.
α-Haloboronic esters	*Chem. Rev.* **1989**, *89*, 1535.
Halocarbenes	*Org. Reactions* **1963**, *13*, 55.
Haloformates	*Pure & Appl. Chem.* **1988**, *60*, 1715.
Halogenated hydrocarbons	
Microbial degradation	*Angew. Chem. Int. Ed. Engl.* **1986**, *25*, 779.
Halogenated ketones	*Synthesis* **1971**, 415.
	Tetrahedron **1981**, *37*, 2949.
Halogenations	*Synthesis* **1970**, 7.
Aromatic compounds	
Reaction mechanisms	*Acc. Chem. Res.* **1974**, *7*, 361.
	Pure & Appl. Chem. **1963**, *7*, 193.
Cyclohexadienes	*Acc. Chem. Res.* **1974**, *7*, 361.
Halogen-metal exchange	
Organolithiums	*Org. Reactions* **1951**, *6*, 339.
α-Haloisocyanates	*Synthesis* **1980**, 85.
Halolactones	*Chem. Soc. Rev.* **1979**, *8*, 171.
Halomalondialdehydes	*Angew. Chem. Int. Ed. Engl.* **1975**, *14*, 86.
α-Halosulfones	*Acc. Chem. Res.* **1968**, *1*, 209.

Halosulfonium salts	*Tetrahedron* **1982**, *38*, 2597.
Halovinylene carbonates	*Angew. Chem. Int. Ed. Engl.* **1974**, *13*, 520.
Hammett constants	*Chem. Rev.* **1991**, *91*, 165.
Helicenes	*Angew. Chem. Int. Ed. Engl.* **1974**, *13*, 649.
Heterohelicenes	*Acc. Chem. Res.* **1971**, *4*, 65.
Helices	
Self-assembly	*Angew. Chem. Int. Ed. Engl.* **1991**, *30*, 407.
Heme	*Acc. Chem. Res.* **1987**, *20*, 289.
Hemerythrin	
Oxygen carrier	*Acc. Chem. Res.* **1984**, *17*, 16.
Hemiterpenes	
Biosynthesis	*Tetrahedron* **1978**, *34*, 143.
Hemoglobin	*Acc. Chem. Res.* **1980**, *13*, 353.
	Angew. Chem. Int. Ed. Engl. **1971**, *10*, 663.
	Angew. Chem. Int. Ed. Engl. **1972**, *11*, 576.
	Angew. Chem. Int. Ed. Engl. **1978**, *17*, 407.
Oxygen binding	*Acc. Chem. Res.* **1977**, *10*, 265.
Hemoproteins	*Acc. Chem. Res.* **1987**, *20*, 289.
	Chem. Rev. **1981**, *81*, 543.
Cytochrome c_3	*Acc. Chem. Res.* **1983**, *16*, 2.
Porphyrins	*Acc. Chem. Res.* **1981**, *14*, 102.
Herbicides	*Angew. Chem. Int. Ed. Engl.* **1991**, *30*, 1621.
Hetarynes	*Angew. Chem. Int. Ed. Engl.* **1971**, *10*, 20.
	Tetrahedron **1982**, *38*, 427.
1,3-Dipolar cycloadditions	*Angew. Chem. Int. Ed. Engl.* **1965**, *4*, 553.
Heteroallenes	*Chem. Rev.* **1978**, *78*, 569.
Heteroannulenes	*Acc. Chem. Res.* **1976**, *9*, 453.
Theoretical aspects	*Pure & Appl. Chem.* **1982**, *54*, 1129.
Heteroaromatic amines	
Mutagens	*Acc. Chem. Res.* **1984**, *17*, 403.
Heteroaromatic betaines	*Angew. Chem. Int. Ed. Engl.* **1976**, *15*, 1.
Heteroaromatic compounds	*Tetrahedron* **1983**, *39*, 3845.
Cycloaddition reactions	*Chem. Rev.* **1989**, *89*, 827.
Rearrangements	
N-Oxides	*Acc. Chem. Res.* **1976**, *9*, 192.
Substitution reactions	
N-Oxides	*Acc. Chem. Res.* **1976**, *9*, 192.
Heteroaromatic *N*-oxides	*Acc. Chem. Res.* **1976**, *9*, 192.
Heteroaryltrimethylsilanes	*Synthesis* **1979**, 841.
Heterocumulenes	
1,1-Dihaloalkyl heterocumulenes	*Tetrahedron* **1991**, *47*, 1563.

Heterocycles	*Angew. Chem. Int. Ed. Engl.* **1971**, *10*, 450.
	Angew. Chem. Int. Ed. Engl. **1975**, *14*, 665.
	Angew. Chem. Int. Ed. Engl. **1984**, *23*, 420.
	Chem. Rev. **1971**, *71*, 315.
	See also under specific classes of compounds
	SYNLETT **1990**, 557.
	Synthesis **1972**, 151.
	Synthesis **1974**, 775.
	Synthesis **1987**, 525.
Acylations	*Angew. Chem. Int. Ed. Engl.* **1973**, *12*, 119.
Alkylations	*Acc. Chem. Res.* **1977**, *10*, 186.
Aminations	*Org. Reactions* **1942**, *1*, 91.
3-Amino-2*H*-azirines	*Angew. Chem. Int. Ed. Engl.* **1991**, *30*, 238.
Aromaticity	*Acc. Chem. Res.* **1972**, *5*, 281.
Betaines	*Tetrahedron* **1985**, *41*, 2239.
Conformational analysis	*Angew. Chem. Int. Ed. Engl.* **1972**, *11*, 739.
	Chem. Rev. **1975**, *75*, 611.
Cycloadditions	*Synthesis* **1975**, 483.
Diazoacetic esters	*Org. Reactions* **1970**, *18*, 217.
Diazotization	*Chem. Rev.* **1975**, *75*, 241.
Diketene	*Acc. Chem. Res.* **1974**, *7*, 265.
1,3-Dioxolan-2-ylium cations	*Chem. Rev.* **1972**, *72*, 357.
Dipoles	*Chem. Soc. Rev.* **1987**, *16*, 89.
1,5-Electrocyclizations	*Angew. Chem. Int. Ed. Engl.* **1980**, *19*, 947.
8π-Electron heterocycles	*Angew. Chem. Int. Ed. Engl.* **1975**, *14*, 581.
Electrophilic additions	*Pure & Appl. Chem.* **1991**, *63*, 243.
Electrophilic substitutions	*Angew. Chem. Int. Ed. Engl.* **1967**, 6, 608.
Five-membered rings	*Chem. Rev.* **1986**, *86*, 795.
Isothiocyanates	*Chem. Rev.* **1991**, *91*, 1.
Isoxazoles	*Synthesis* **1975**, 20.
Macrocycles	*Pure & Appl. Chem.* **1986**, *58*, 1437.
Meso-ionic heterocycles	*Tetrahedron* **1982**, *38*, 2965.
Metallations	*Tetrahedron* **1983**, *39*, 1955ff.
Methylenecyclopropane derivatives	*Angew. Chem. Int. Ed. Engl.* **1980**, *19*, 276.
Natural products	*SYNLETT* **1990**, 557.
Nitrile oxides	*Synthesis* **1970**, 344.
Nitro compounds	*Synthesis* **1974**, 613.
Nitrogen heterocycles	*Acc. Chem. Res.* **1972**, *5*, 303.
Medium ring nitrogen heterocycles	*Tetrahedron* **1991**, *47*, 9131.
Photooxidation	*Chem. Rev.* **1979**, *79*, 447.

Heterocycles, cont'd.	
Nitrogen inversion barrier	*Angew. Chem. Int. Ed. Engl.* **1981**, *20*, 521.
Nucleophilic substitution	*Synthesis* **1991**, 103.
Organolithium compounds	*Synthesis* **1983**, 957.
Organophosphorus chemistry	*Chem. Rev.* **1977**, *77*, 121.
	Acc. Chem. Res. **1972**, *5*, 204.
Organosulfur chemistry	
Anomeric effect	*Acc. Chem. Res.* **1989**, *22*, 357.
Oxidations	
Nitrogen aromatic heterocycles	*Pure & Appl. Chem.* **1971**, *25*, 365.
Oxygen heterocycles	*Pure & Appl. Chem.* **1987**, *59*, 969.
Palladium catalysis	*Pure & Appl. Chem.* **1983**, *55*, 1845.
Phosphorus-nitrogen heterocycles	*Synthesis* **1978**, 557.
Photochemistry	*Chem. Rev.* **1977**, *77*, 473.
Decomposition	
Gas phase	*Chem. Rev.* **1977**, *77*, 473.
Polyaromatic compounds	*Angew. Chem. Int. Ed. Engl.* **1979**, *18*, 1.
Polymerization	*Pure & Appl. Chem.* **1976**, *48*, 335.
	Pure & Appl. Chem. **1981**, *53*, 1645ff.
"Rabbit ear" effect	*Acc. Chem. Res.* **1970**, *3*, 1.
Ring transformations	*Synthesis* **1975**, 20.
Ring closures	
Reaction mechanisms	*Tetrahedron* **1987**, *43*, 5171.
Ring-transformations	*Tetrahedron* **1982**, *38*, 3537.
Seven-membered heterocycles	*Chem. Rev.* **1973**, *73*, 293.
	Synthesis **1988**, 569.
	Synthesis **1991**, 181.
Tautomerism	*Acc. Chem. Res.* **1977**, *10*, 186.
Reaction mechanisms	*Pure & Appl. Chem.* **1987**, *59*, 1577.
Thermochemistry	
Decomposition	
Gas phase	*Chem. Rev.* **1977**, *77*, 473.
Valence isomerism	*Angew. Chem. Int. Ed. Engl.* **1971**, *10*, 11.
Ylides	*Chem. Soc. Rev.* **1987**, *16*, 89.
Heterocyclic bases	
Amination	*Org. Reactions* **1942**, *1*, 91.
Heterocyclic intermediates	*Pure & Appl. Chem.* **1988**, *60*, 1679.
Alkenes	*Tetrahedron* **1990**, *46*, 3321.
Heterodienes	
Enones	*Chem. Rev.* **1975**, *75*, 651.
N-Acyl imines	
Cycloaddition reactions	*Chem. Rev.* **1989**, *89*, 1525.

Holographic methods
  Photochemistry — *Angew. Chem. Int. Ed. Engl.* **1983**, *22*, 582.
Homoaromaticity — *Acc. Chem. Res.* **1988**, *21*, 243.
  Angew. Chem. Int. Ed. Engl. **1978**, *17*, 106.
  Homotropylium cation — *Acc. Chem. Res.* **1984**, *17*, 347.
Homocubanes — *Chem. Rev.* **1989**, *89*, 1011.
Homoenolate anions
  1-Alkoxy-1-siloxycyclopropanes — *Pure & Appl. Chem.* **1988**, *60*, 115.
  Homoenolate equivalents — *Tetrahedron* **1983**, *39*, 205.
Homogeneous catalysis — *Angew. Chem. Int. Ed. Engl.* **1973**, *12*, 954.
  Chem. Rev. **1973**, *73*, 21.
  Chem. Soc. Rev. **1972**, *1*, 337.
  J. Organomet. Chem. **1980**, *200*, 349.
  Org. Reactions **1976**, *24*, 1.
  Alkanes
    Activation — *Acc. Chem. Res.* **1975**, *8*, 113.
  Asymmetric synthesis — *Chem. Rev.* **1989**, *89*, 257.
  Hydrogenation — *Chem. Rev.* **1973**, *73*, 213.
  Org. Reactions **1976**, *24*, 1.
    Historical perspective — *J. Organomet. Chem.* **1980**, *200*, 133.
  Iridium complexes — *Acc. Chem. Res.* **1978**, *11*, 301.
  Acc. Chem. Res. **1979**, *12*, 331.
  Iridium hydrides — *Acc. Chem. Res.* **1984**, *17*, 221.
  Nickel chemistry — *Angew. Chem. Int. Ed. Engl.* **1990**, *29*, 235.
  Organometallic compounds — *Acc. Chem. Res.* **1968**, *1*, 23.
  Phosphorus ligands
    Steric effects — *Chem. Rev.* **1977**, *77*, 313.
  Rhodium complexes — *Acc. Chem. Res.* **1978**, *11*, 301.
    Rhodium hydrides — *Acc. Chem. Res.* **1984**, *17*, 221.
Homologation
  Three carbon homologation agents — *Chem. Rev.* **1984**, *84*, 409.
Homolytic
  Arylation — *Chem. Rev.* **1957**, *57*, 123.
Homopolymerization
  Cyclopentene — *Angew. Chem. Int. Ed. Engl.* **1964**, *3*, 723.
Homotropylium cation
  Homoaromaticity — *Acc. Chem. Res.* **1984**, *17*, 347.
Hops — *Angew. Chem. Int. Ed. Engl.* **1978**, *17*, 710.
Host-guest chemistry — *Angew. Chem. Int. Ed. Engl.* **1988**, *27*, 1009.
  Angew. Chem. Int. Ed. Engl. **1991**, *30*, 1417.
  Chem. Rev. **1991**, *91*, 1721.
  Pure & Appl. Chem. **1988**, *60*, 477.

Alkenes	*Tetrahedron* **1987**, *43*, 3839.
Anion hosts	*Pure & Appl. Chem.* **1989**, *61*, 1535.
Calixarenes	*Pure & Appl. Chem.* **1986**, *58*, 1523.
Cation-macrocyle complexes	*Chem. Rev.* **1985**, *85*, 271.
Chiral recognition	*Acc. Chem. Res.* **1978**, *11*, 8.
Clathrates	*Acc. Chem. Res.* **1978**, *11*, 81.
	Angew. Chem. Int. Ed. Engl. **1985**, *24*, 727.
	Chem. Soc. Rev. **1978**, *7*, 65.
Cooperation	*Angew. Chem. Int. Ed. Engl.* **1991**, *30*, 1086.
Crown ethers	*Acc. Chem. Res.* **1978**, *11*, 8.
Cyclodextrins	*Acc. Chem. Res.* **1982**, *15*, 66.
	Angew. Chem. Int. Ed. Engl. **1980**, *19*, 344.
Photochemistry	*Acc. Chem. Res.* **1988**, *21*, 300.
Cyclophanes	*Angew. Chem. Int. Ed. Engl.* **1988**, *27*, 362.
Metal ion complexes	*Chem. Rev.* **1974**, *74*, 351.
Molecular inclusion compounds	*Angew. Chem. Int. Ed. Engl.* **1985**, *24*, 727.
Polyaza compounds	*Tetrahedron* **1991**, *47*, 6851.
HPLC	
Proteins	*Angew. Chem. Int. Ed. Engl.* **1986**, *25*, 535.
HSAB principle	*Chem. Rev.* **1975**, *75*, 1.
Organometallic reactions	*Tetrahedron* **1985**, *41*, 3.
Human plasma proteins	*Angew. Chem. Int. Ed. Engl.* **1980**, *19*, 87.
Human serum glycoproteins	*Angew. Chem. Int. Ed. Engl.* **1973**, *12*, 721.
Hun's rule	*Pure & Appl. Chem.* **1987**, *59*, 1571.
Hydrations	*Angew. Chem. Int. Ed. Engl.* **1966**, *5*, 646.
Nucleic acids	*Chem. Rev.* **1987**, *87*, 589.
Hydrazidinyl radicals	*Angew. Chem. Int. Ed. Engl.* **1973**, *12*, 455.
Hydrazine	*Tetrahedron* **1977**, *33*, 3183.
Hydride reagents	
Ketones	*Tetrahedron* **1979**, *35*, 449.
Reductions	
Historical perspective	*Tetrahedron* **1979**, *35*, 567.
Hydride transfers	
Organometallic chemistry	
Reaction mechanisms	*Pure & Appl. Chem.* **1988**, *60*, 131.
Reaction mechanisms	*Tetrahedron* **1986**, *42*, 941ff.
Hydroboration	*Pure & Appl. Chem.* **1983**, *55*, 1387.
	Tetrahedron **1961**, *12*, 117.
	Tetrahedron **1977**, *33*, 2331.
Alkenes	*Org. Reactions* **1963**, *13*, 4.
Alkynes	*Org. Reactions* **1963**, *13*, 20.
Catecholborane	*Tetrahedron* **1976**, *32*, 981.
Dienes	*Org Reactions* **1963**, *13*, 18.
Thexylborane	*Synthesis* **1974**, 77.
Transition metal catalysis	*Chem. Rev.* **1991**, *91*, 1179.

Hydrocarbons	*See also alkanes, and alkenes.*
Activation	*Acc. Chem. Res.* **1985**, *18*, 302.
	Acc. Chem. Res. **1989**, *22*, 91.
Addition reactions	*Synthesis* **1974**, 309.
Base catalysed	*Acc. Chem. Res.* **1974**, *7*, 155.
Anions	*Chem. Rev.* **1987**, *87*, 557.
Carcinogens	
Metabolites	*Acc. Chem. Res.* **1981**, *14*, 218.
Chlorinations	*Chem. Rev.* **1963**, *63*, 355.
Combustion	
Reactions mechanisms	*Acc. Chem. Res.* **1979**, *12*, 223.
(from) *Euphorbia lathyris*	*Pure & Appl. Chem.* **1981**, *53*, 1101.
Photonitrosation	*Angew. Chem.* **1959**, *71*, 229.
Rearrangement reactions	*Chem. Rev.* **1972**, *72*, 181.
Strain	*Chem. Rev.* **1989**, *89*, 1125.
Synthesis	
Microorganisms	*Chem. Soc. Rev.* **1974**, *3*, 309.
Hydrocyanation	
Unsaturated carbonyl compounds	*Org. Reactions* **1977**, *25*, 255.
Hydroformylation	
Alkenes	*Acc. Chem. Res.* **1981**, *14*, 259.
Rhodium catalysis	*Angew. Chem. Int. Ed. Engl.* **1980**, *19*, 178.
Hydrogen abstraction	*Chem. Rev.* **1971**, *71*, 247.
Hydrogen cyanide	
Natural products	
Evolution	*Tetrahedron* **1984**, *40*, 1093.
Hydrogen fluoride	*Tetrahedron* **1991**, *47*, 5329.
Hydrogen migrations	
Allylamines	
Asymmetric synthesis	*Synthesis* **1991**, 665.
Hydrogenations	*Angew. Chem. Int. Ed. Engl.* **1988**, *27*, 679.
Alkenes	*Chem. Rev.* **1954**, *54*, 575.
	Synthesis **1973**, 457.
Reaction mechanisms	*Acc. Chem. Res.* **1969**, *2*, 289.
Alkynes	*Synthesis* **1973**, 457.
Aromatic compounds	*Acc. Chem. Res.* **1979**, *12*, 324.
Transition metal complexes	*Acc. Chem. Res.* **1979**, *12*, 324.
Asymmetric synthesis	*Adv. Catalysis* **1976**, *25*, 81.
Catalytic	*Acc. Chem. Res.* **1977**, *10*, 15.
	Chem. Rev. **1974**, *74*, 567.
Esters	*Org. Reactions* **1954**, *8*, 1.

Homogeneous	*Chem. Rev.* **1973**, *73*, 21.
	Org. Reactions **1976**, *24*, 1.
	Synthesis **1981**, 85.
Historical perspective	*J. Organomet. Chem.* **1980**, *200*, 133.
Ionic	*Angew. Chem. Int. Ed. Engl.* **1962**, *1*, 613.
	Synthesis **1974**, 633.
Nickel-copper alloy	*Chem. Rev.* **1980**, *80*, 417.
Reductions	
Asymmetric synthesis	*Angew. Chem. Int. Ed. Engl.* **1971**, *10*, 871.
Rhodium compounds	
Asymmetric synthesis	*Acc. Chem. Res.* **1983**, *16*, 106.
Transition metal complexes	*Adv. Organomet. Chem.* **1979**, *17*, 319.
Asymmetric synthesis	*Acc. Chem. Res.* **1982**, *15*, 395.
Hydrogenolysis	
Benzyl group removal	*Org. Reactions* **1953**, *7*, 263.
Organic halides	*Synthesis* **1980**, 425.
Small ring compounds	*Chem. Rev.* **1963**, *63*, 123.
Hydrolysis	
Acetals	*Chem. Rev.* **1974**, *74*, 581.
Amides	*Quart. Rev.* **1970**, *24*, 553.
	Tetrahedron **1975**, *31*, 2463.
Enzymes	*Angew. Chem. Int. Ed. Engl.* **1964**, *3*, 1.
Esters	*Tetrahedron* **1975**, *31*, 2463.
Ketals	*Chem. Rev.* **1974**, *74*, 581.
Ketene dithioacetal	*Acc. Chem. Res.* **1986**, *19*, 370.
Ortho esters	*Chem. Rev.* **1974**, *74*, 581.
Orthophosphate esters	*Acc. Chem. Res.* **1970**, *3*, 257.
Phosphate esters	*Acc. Chem. Res.* **1968**, *1*, 70.
Thioacetals	*Chem. Soc. Rev.* **1990**, *19*, 55.
Hydroquinones	*Chem. Rev.* **1986**, *86*, 821.
Hydrosilylations	*Angew. Chem. Int. Ed. Engl.* **1979**, *18*, 837.
Homogeneous catalysis	*Adv. Organomet. Chem.* **1979**, *17*, 407.
Rhodium catalysts	
Asymmetric synthesis	*Angew. Chem. Int. Ed. Engl.* **1983**, *22*, 897.
Hydroxamic acid derivatives	*Chem. Rev.* **1989**, *89*, 1563.
Photochemistry	*Chem. Rev.* **1991**, *91*, 477.
Hydroxamic acids	*Angew. Chem. Int. Ed. Engl.* **1974**, *13*, 376.
Hydroxyapatitie	*Acc. Chem. Res.* **1975**, *8*, 273.
5-Hydroxyindoles	
Nenitzescu reaction	*Org. Reactions* **1973**, *20*, 337.
Hydroxyl radical	
Gas phase	
Reaction mechanisms	*Chem. Rev.* **1986**, *86*, 69.

Hydroxylamine derivatives	
Conformations	*Tetrahedron* **1981**, *37*, 849.
Hydroxylamines	*Acc. Chem. Res.* **1973**, *6*, 191.
	Tetrahedron **1977**, *33*, 3183.
Inversion	*Tetrahedron* **1984**, *40*, 3345.
α-Hydroxylaminooximes	*Synthesis* **1986**, 704.
Hydroxylations	
Alkenes	
Peroxy acids	*Org. Reactions,* **1953**, *7*, 378.
Aromatic compounds	*Acc. Chem. Res.* **1971**, *4*, 337.
Photochemistry	*Synthesis* **1974**, 173.
Enzymes	
Molecular oxygen	*Angew. Chem. Int. Ed. Engl.* **1972**, *11*, 701.
Molecular oxygen	
Enzymes	*Angew. Chem. Int. Ed. Engl.* **1972**, *11*, 701.
Peroxy acids	*Org. Reactions* **1953**, *7*, 378.
Photochemical	
Aromatic compounds	*Synthesis* **1974**, 173.
Steroids	
Biosynthesis	*Chem. Soc. Rev.* **1982**, *11*, 371.
Terpenes	
Biosynthesis	*Tetrahedron* **1984**, *40*, 3597.
Hydroxypyridine	
Tautomerism	*Acc. Chem. Res.* **1977**, *10*, 186.
Hydrozirconation	*Angew. Chem. Int. Ed. Engl.* **1976**, *15*, 333.
Hydrozulenes	*Synthesis* **1972**, 517.
Hypercarbon carbocations	
Reaction mechanisms	*Pure & Appl. Chem.* **1981**, *53*, 201.
Hyperconjugation	
Transition states	*Pure & Appl. Chem.* **1984**, *56*, 1755.
Hypervalent iodine	*Acc. Chem. Res.* **1986**, *19*, 244.
	Synthesis **1984**, 709.
Aryliodine(III) dicarboxylates	*Chem. Soc. Rev.* **1981**, *10*, 377.
[Hydroxy(organosulfonyloxy) iodo]arenes	*SYNLETT* **1990**, 365.
Oxidations	*Synthesis* **1990**, 431.
Hypobromites	*Tetrahedron* **1976**, *32*, 517.
Hypoiodite reaction	*Synthesis* **1971**, 501.
Imidazole derivatives	
Metal complexes	*Chem. Rev.* **1974**, *74*, 471.
Imides	
Alkenylations	*Synthesis* **1975**, 685.
Cyclic imides	
Photochemistry	*Acc. Chem. Res.* **1978**, *11*, 407.
Photochemistry	*Pure & Appl. Chem.* **1988**, *60*, 941.

Imines	*Synthesis* **1976**, 349.
Carbenes	*Pure & Appl. Chem.* **1983**, *55*, 1745.
Coupling reactions	*Pure & Appl. Chem.* **1980**, *52*, 2417.
Ester enolates	*Chem. Rev.* **1989**, *89*, 1447.
Michael addition	*Tetrahedron: Asymmetry* **1992**, *3*, 459.
Organometallic compounds	*Pure & Appl. Chem.* **1990**, *62*, 605.
Photochemistry	*Chem. Rev.* **1977**, *77*, 37.
	Chem. Soc. Rev. **1977**, *6*, 63.
Reaction mechanisms	*Acc. Chem. Res.* **1989**, *22*, 211.
Iminium salts	
Photochemistry	*Tetrahedron* **1983**, *39*, 3845.
η^2-Iminoacyl group	*Chem. Rev.* **1988**, *88*, 1059.
Iminophosphanes	*Angew. Chem. Int. Ed. Engl.* **1991**, *30*, 217.
Iminophosphoranes	*Synthesis* **1973**, 123.
Immobilization	
Biomaterials	*Angew. Chem. Int. Ed. Engl.* **1982**, *21*, 837.
Enzymes	*Chem. Soc. Rev.* **1977**, *6*, 215.
	Pure & Appl. Chem. **1987**, *59*, 1501.
Inclusion complexes	
Azacyclophanes	*Pure & Appl. Chem.* **1988**, *60*, 549.
Cryptates	*Pure & Appl. Chem.* **1980**, *52*, 2303.
	Pure & Appl. Chem. **1980**, *52*, 2441.
Metallomacrocycles	*Pure & Appl. Chem.* **1988**, *60*, 555.
1*H*-Indazol-3-ols	*Synthesis* **1978**, 633.
Indium compounds	*Angew. Chem. Int. Ed. Engl.* **1985**, *24*, 893.
Indole alkaloids	*Acc. Chem. Res.* **1970**, *3*, 151.
	Acc. Chem. Res. **1984**, *17*, 35.
	Tetrahedron **1983**, *39*, 3627ff.
Indole-2,3-quinodimethanes	*Chem. Rev.* **1989**, *89*, 1681.
Indoles	
Organometallic chemistry	*Angew. Chem. Int. Ed. Engl.* **1988**, *27*, 113.
Oxidations	*Heterocycl.* **1977**, *8*, 743.
Electrochemistry	*Chem. Rev.* **1990**, *90*, 795.
Indolino-spiropyrans	
Photochemistry	*Pure & Appl. Chem.* **1990**, 62, 1483.
Indolizines	*Synthesis* **1976**, 209.
Industrial chemistry	*Angew. Chem. Int. Ed. Engl.* **1990**, *29*, 1228.
Asymmetric synthesis	*Tetrahedron* **1991**, *47*, 4789.
Carbohydrates	*Pure & Appl. Chem.* **1989**, *61*, 1313.
Dyes	
Semi-conductor properties	*Angew. Chem. Int. Ed. Engl.* **1972**, *11*, 1051.
Host-guest chemistry	
Cyclodextrins	*Angew. Chem. Int. Ed. Engl.* **1980**, *19*, 344.

Industrial chemistry, cont'd.	
Maleic anhydride	*Chem. Rev.* **1988**, *88*, 55.
Oil production	
Coal hydrogenation	*Angew. Chem. Int. Ed. Engl.* **1976**, *15*, 341.
Organometallic reagents	*Pure & Appl. Chem.* **1985**, *57*, 1809.
Oxidations	
Methanol	*Chem. Rev.* **1985**, *85*, 235.
Palladium catalysis	*Synthesis* **1990**, 739.
Photochemistry	*Angew. Chem. Int. Ed. Engl.* **1978**, *17*, 16.
	Pure & Appl. Chem. **1986**, *58*, 1267.
Polysaccharides	*Pure & Appl. Chem.* **1989**, *61*, 1315.
Process design	
Thermodynamics	
Reaction mechanisms	*Angew. Chem. Int. Ed. Engl.* **1990**, *29*, 1246.
Risk management	*Chem. Soc. Rev.* **1979**, *8*, 419.
Supercritical extractions	*Angew. Chem. Int. Ed. Engl.* **1978**, *17*, 751.
Wittig reaction	*Angew. Chem. Int. Ed. Engl.* **1977**, *16*, 423.
Inhibition	
Autoxidation	*Chem. Rev.* **1961**, *61*, 563.
	Angew. Chem. Int. Ed. Engl. **1985**, *24*, 466.
	Angew. Chem. Int. Ed. Engl. **1991**, *30*, 613.
Dihydrofolated reductase	*Chem. Rev.* **1984**, *84*, 333.
Enzymes	*Chem. Rev.* **1987**, *87*, 1183.
Oligonucleotides	*Angew. Chem. Int. Ed. Engl.* **1991**, *30*, 613.
Radical reactions	*Chem. Rev.* **1987**, *87*, 1313.
Inositol derivatives	*Chem. Soc. Rev.* **1989**, *18*, 83.
Insecticides	*Chem. Soc. Rev.* **1978**, *7*, 473.
Insects	*See also natural products*
Attractants	*Chem. Soc. Rev.* **1973**, *2*, 75.
Juvenile hormones	*Acc. Chem. Res.* **1980**, *13*, 297.
Natural products	*Chem. Soc. Rev.* **1975**, *4*, 263.
Pheromones	*Angew. Chem. Int. Ed. Engl.* **1973**, *12*, 644.
	Pure & Appl. Chem. **1982**, *54*, 2479.
	Synthesis **1977**, 817.
	Synthesis **1978**, 413.
	Tetrahedron **1977**, *33*, 1845.
	Tetrahedron **1989**, *45*, 3233.
Insertion reactions	
Carbon monoxide	*Angew. Chem. Int. Ed. Engl.* **1977**, *16*, 299.
Carbonyl insertions	*Acc. Chem. Res.* **1984**, *17*, 67.
Nickel chemistry	*Acc. Chem. Res.* **1973**, 6, 422.
Interaction regulation	
Carbohydrates	*Pure & Appl. Chem.* **1981**, *53*, 79.

Intersystem crossing	
Photochemistry	*Pure & Appl. Chem.* **1984**, *56*, 1167.
Intrabridgehead chemistry	*Acc. Chem. Res.* **1983**, *16*, 321.
	Tetrahedron **1990**, *46*, 683.
Intramolecular reactions	
Acylations	*Angew. Chem. Int. Ed. Engl.* **1973**, *12*, 119.
Additions	
Allylsilanes	*Synthesis* **1988**, 263.
Propargylsilanes	*Synthesis* **1988**, 263.
Cycloadditions	*Angew. Chem. Int. Ed. Engl.* **1976**, *15*, 123.
	Angew. Chem. Int. Ed. Engl. **1977**, *16*, 10.
Diels Alder reaction	*Chem. Soc. Rev.* **1987**, *16*, 187.
	Pure & Appl. Chem. **1981**, *53*, 1181.
	Tetrahedron: Asymmetry **1991**, *2*, 1173ff.
o-Xylylenes	*Chem. Soc. Rev.* **1980**, *9*, 41.
Ene reaction	*Angew. Chem. Int. Ed. Engl.* **1978**, *17*, 476.
Reaction mechanisms	*Angew. Chem. Int. Ed. Engl.* **1981**, *20*, 487.
Iodine azide	*Angew. Chem. Int. Ed. Engl.* **1979**, *18*, 507.
Alkenes	*Acc. Chem. Res.* **1971**, *4*, 9.
Iodoaromatic compounds	*Synthesis* **1988**, 923.
Iodotrimethylsilane	*Synthesis* **1980**, 861.
	Tetrahedron **1982**, *38*, 2225.
Ion cyclotron resonance	*Acc. Chem. Res.* **1971**, *4*, 114.
Ion selectivity	*Pure & Appl. Chem.* **1986**, *58*, 1445.
Ionophores	*Pure & Appl. Chem.* **1988**, *60*, 567.
Ion transport	*Angew. Chem. Int. Ed. Engl.* **1985**, *24*, 905.
	Chem. Soc. Rev. **1980**, *9*, 281.
Ionophores	*Acc. Chem. Res.* **1983**, *16*, 161.
Ion selectivity	*Pure & Appl. Chem.* **1988**, *60*, 567.
Iridium complexes	*Pure & Appl. Chem.* **1984**, *56*, 13.
Homogeneous catalysis	*Acc. Chem. Res.* **1978**, *11*, 301.
	Acc. Chem. Res. **1979**, *12*, 331.
Iridium hydrides	*Acc. Chem. Res.* **1984**, *17*, 221.
Iridoid glycosides	*Angew. Chem. Int. Ed. Engl.* **1983**, *22*, 828.
Iron chemistry	*Chem. Rev.* **1983**, *83*, 203.
Ketones	
Cyclic ketones	*Acc. Chem. Res.* **1976**, *9*, 209.
Macromolecules	*Pure & Appl. Chem.* **1988**, *60*, 495.
Iron(II) macrocycles	
Biomimetic applications	*Pure & Appl. Chem.* **1988**, *60*, 561.
Iron-porphyrins	*Pure & Appl. Chem.* **1986**, *58*, 1493.
Iron proteins	*Chem. Rev.* **1990**, *90*, 1447.
Isatoic anhydride	*Synthesis* **1980**, 505.
Isobenzofurans	*Tetrahedron* **1988**, *44*, 2093.

Isocyanates	*Chem. Soc. Rev.* **1975**, *4*, 231.
	Chem. Rev. **1972**, *72*, 457.
	Chem. Soc. Rev. **1974**, *3*, 209.
	Synthesis **1987**, 525.
Acylations	*Chem. Soc. Rev.* **1975**, *4*, 231.
Addition reactions	*Acc. Chem. Res.* **1969**, *2*, 186.
Cycloadditions	*Acc. Chem. Res.* **1969**, *2*, 186.
	Synthesis **1974**, 461.
Metal catalysis	*Chem. Rev.* **1989**, *89*, 1927.
Isocyanide-cyanide rearrangement	*Angew. Chem. Int. Ed. Engl.* **1991**, *30*, 893.
Isocyanides	*Angew. Chem. Int. Ed. Engl.* **1982**, *21*, 810.
	Angew. Chem. Int. Ed. Engl. **1988**, *27*, 1456.
Addition reactions	*Synthesis* **1969**, 65.
Four-membered rings	*Synthesis* **1985**, 1083.
Metallations	*SYNLETT* **1990**, 245.
α-Metalloisocyanides	*Angew. Chem. Int. Ed. Engl.* **1974**, *13*, 789.
	Angew. Chem. Int. Ed. Engl. **1977**, *16*, 339.
Polymerization	*Chem. Rev.* **1972**, *72*, 101.
Isocyanophosphine derivatives	*Synthesis* **1986**, 793.
Isoindenes	
Reactive intermediates	*Acc. Chem. Res.* **1980**, *13*, 270.
Isoinversion principle	*Angew. Chem. Int. Ed. Engl.* **1991**, *30*, 477.
Isoleucyl-tRNA-synthetase	*Acc. Chem. Res.* **1987**, *20*, 79.
	Angew. Chem. Int. Ed. Engl. **1988**, *27*, 773.
Isomerase enzymes	
Vitamin B_{12}	*Chem. Soc. Rev.* **1985**, *14*, 161.
Isomerization	*Angew. Chem. Int. Ed. Engl.* **1972**, *11*, 1072.
Acylcyclopentenes	*Tetrahedron* **1976**, *32*, 641.
Alicyclic-aromatic	*Chem. Rev.* **1943**, *33*, 89.
Alkenes	*Acc. Chem. Res.* **1986**, *19*, 78.
	Chem. Rev. **1955**, *55*, 625.
	Synthesis **1969**, 97.
	Synthesis **1970**, 405.
	Tetrahedron **1974**, *30*, 1861.
Alkynes	*Quart. Rev.* **1970**, *24*, 585.
Allylic compounds	*Pure & Appl. Chem.* **1985**, *57*, 1845.
Aziridine derivatives	*Angew. Chem. Int. Ed. Engl.* **1962**, *1*, 528.
Azocompounds	*Acc. Chem. Res.* **1973**, *6*, 275.
DNA	*Chem. Soc. Rev.* **1989**, *18*, 53.
Double bonds	*Tetrahedron* **1974**, *30*, 1861.
Epoxides	*Org. Reactions* **1983**, *29*, 345.

Photochemistry	
Cyclopentene derivatives	*Tetrahedron* **1976**, *32*, 641.
Strained ring compounds	*Angew. Chem. Int. Ed. Engl.* **1963**, *2*, 1.
Isomers	
Structural chemistry	*Chem. Soc. Rev.* **1974**, *3*, 355.
Isonitrin B	*SYNLETT* **1989**, 9.
Isopenicillin N synthase	*Chem. Rev.* **1990**, *90*, 1079.
Isopeptide crosslinks	*Angew. Chem. Int. Ed. Engl.* **1974**, *13*, 514.
Isoprenoids	
Biosynthesis	*Acc. Chem. Res.* **1985**, *18*, 230.
	Pure & Appl. Chem. **1990**, *62*, 1259.
Isoquinoline alkaloids	*Pure & Appl. Chem.* **1986**, *58*, 685.
Isoquinolines	
Photochemistry	*Acc. Chem. Res.* **1972**, *5*, 212.
Isosterism	
Drug design	*Chem. Soc. Rev.* **1979**, *8*, 563.
Isothiocyanates	*Synthesis* **1987**, 525.
Heterocycles	*Chem. Rev.* **1991**, *91*, 1.
Isotope effects	
Elimination reactions	*Chem. Soc. Rev.* **1972**, *1*, 163.
Reaction mechanisms	*Chem. Soc. Rev.* **1974**, *3*, 513.
	Chem. Soc. Rev. **1989**, *18*, 209.
Isoureas	*Synthesis* **1979**, 561.
Isoxazoles	*Synthesis* **1975**, 20.
	Synthesis **1987**, 857.
Isoxazolines	*Acc. Chem. Res.* **1984**, *17*, 410.
Jaundice	*Acc. Chem. Res.* **1984**, *17*, 417.
Juvenile hormones	*Acc. Chem. Res.* **1970**, *3*, 120.
	Acc. Chem. Res. **1980**, *13*, 297.
Kekulé	
Historical perspective	*Angew. Chem. Int. Ed. Engl.* **1979**, *18*, 641.
Kekulé molecules	*Pure & Appl. Chem.* **1987**, *59*, 1571.
Ketals	
Hydrolysis	*Acc. Chem. Res.* **1972**, *5*, 264.
Reaction mechanisms	*Chem. Rev.* **1974**, *74*, 581.
Ketene dithioacetals	*Synthesis* **1990**, 171.
Hydrolysis	*Acc. Chem. Res.* **1986**, *19*, 370.
Ketene equivalents	*Synthesis* **1977**, 289.
Ketenes	*Acc. Chem. Res.* **1990**, *23*, 273.
	Chem. Soc. Rev. **1975**, *4*, 231.
	Org. Reactions **1946**, *3*, 108.
Acylations	*Chem. Soc. Rev.* **1975**, *4*, 231.
Addition reactions	*Tetrahedron* **1986**, *42*, 2587.
Cycloaddition reactions	*Chem. Rev.* **1988**, *88*, 793.

Ketenimines	*Angew. Chem. Int. Ed. Engl.* **1971**, *10*, 435.
	Angew. Chem. Int. Ed. Engl. **1988**, *27*, 1456.
Keteniminium salts	
Cycloaddition reactions	*Chem. Rev.* **1988**, *88*, 793.
α-**Ketoalkylsulfonates**	
Solvolysis	*Acc. Chem. Res.* **1985**, *18*, 3.
Ketocarbenes	
Addition reactions	*Angew. Chem.* **1961**, *73*, 368.
β-**Ketoesters**	
Dealkoxycarbonylations	*Synthesis* **1982**, 805.
Ketones	
Acid halides	
Organometallic compounds	*Org. Reactions* **1954**, *8*, 28.
Acylations	*Org. Reactions* **1954**, *8*, 59.
Intramolecular acylations	*Org. Reactions* **1944**, *2*, 114.
Addition reactions	
Grignard reagents	
Reaction mechanisms	*Acc. Chem. Res.* **1974**, *7*, 272.
	Chem. Rev. **1975**, *75*, 521.
Aldehydes	*Angew. Chem. Int. Ed. Engl.* **1980**, *19*, 171.
Alkylations	
Nitrogen derivatives	*Synthesis* **1983**, 517.
Aromatic ketones	
Nucleophilic addition	*Acc. Chem. Res.* **1988**, *21*, 414.
Asymmetric synthesis	*Tetrahedron* **1989**, *45*, 4913.
Carboxylic acids	*Angew. Chem. Int. Ed. Engl.* **1980**, *19*, 171.
Organolithiums	*Org. Reactions* **1970**, *18*, 1.
Cleavage reactions	*Org. Reactions* **1957**, *9*, 1.
Clemmensen reduction	*Org. Reactions* **1975**, *22*, 401.
Coupling reactions	*Synthesis* **1979**, 633.
Cyclic ketones	
Iron chemistry	*Acc. Chem. Res.* **1976**, *9*, 209.
Enolates	*Tetrahedron* **1976**, *32*, 2979.
Grignard reagents	
Addition reactions	
Reaction mechanisms	*Acc. Chem. Res.* **1974**, *7*, 272.
α-Hydrogen exchange	*Acc. Chem. Res.* **1978**, *11*, 1.
Nucleophilic additions	
Organometallic compounds	
Reaction mechanisms	
Stereochemistry	*Chem. Rev.* **1975**, *75*, 521.
Organoboranes	*Acc. Chem. Res.* **1969**, *2*, 65.

Organometallic compounds	
Nucleophilic additions	
Reaction mechanisms	
Stereochemistry	*Chem. Rev.* **1975**, *75*, 521.
Oxacarbenes	*Acc. Chem. Res.* **1975**, *8*, 209.
Oxidations	*Synthesis* **1984**, 369.
Oxy-Cope rearrangement	*Angew. Chem. Int. Ed. Engl.* **1990**, *29*, 609.
Photochemistry	*Tetrahedron* **1974**, *30*, 1891.
NMR	
Reaction mechanisms	*Pure & Appl. Chem.* **1988**, *60*, 933.
Non-enolizable	*Chem. Rev.* **1947**, *40*, 181.
Norrish type II reactions	*Pure & Appl. Chem.* **1988**, *60*, 999.
Photocyclizations	*Acc. Chem. Res.* **1971**, *4*, 168.
Photoelimination reactions	*Acc. Chem. Res.* **1971**, *4*, 168.
Reductions	*Acc. Chem. Res.* **1983**, *16*, 399.
	Acc. Chem. Res. **1984**, *17*, 338.
	Angew. Chem. Int. Ed. Engl. **1984**, *23*, 570.
	Pure & Appl. Chem. **1977**, *49*, 1049.
	Tetrahedron **1986**, *42*, 6351.
Hydride reagents	*Tetrahedron* **1979**, *35*, 449.
Stereoselectivity	*Acc. Chem. Res.* **1984**, *17*, 338.
Steroidal ketones	*Synthesis* **1972**, 526.
Titanium coupling	
Alkenes	*Acc. Chem. Res.* **1983**, *16*, 405.
Ketoses	
Structures	*Acc. Chem. Res.* **1976**, *9*, 418.
Kinetic control	
Reaction mechanisms	*Chem. Soc. Rev.* **1987**, *16*, 313.
Kinetic resolutions	
Epoxyalcohols	*Pure & Appl. Chem.* **1983**, *55*, 589.
Kinetic studies	
Radical reactions	*Pure & Appl. Chem.* **1984**, *56*, 1767.
Labelled compounds	
Organoboranes	*Acc. Chem. Res.* **1984**, *17*, 215.
Lac repressor	*Angew. Chem. Int. Ed. Engl.* **1971**, *10*, 160.
Lactam acetals	*Tetrahedron* **1988**, *44*, 5975.
β-Lactamase	*Acc. Chem. Res.* **1985**, *18*, 97.
	Pure & Appl. Chem. **1987**, *59*, 279.
Lactams	*Synthesis* **1972**, 151.
	Tetrahedron **1981**, *37*, 1238.
Isomerization	*Angew. Chem. Int. Ed. Engl.* **1975**, *14*, 43.
Polymerization	*Angew. Chem. Int. Ed. Engl.* **1975**, *14*, 43.

β-Lactams *Acc. Chem. Res.* **1973**, 6, 32.
Acc. Chem. Res. **1986**, *19*, 49.
Angew. Chem., Int. Ed. Engl. **1985**, *24*, 180.
Chem. Rev. **1976**, *76*, 113.
Chem. Rev. **1989**, *89*, 1447.
Chem. Soc. Rev. **1976**, *5*, 181.
Pure & Appl. Chem. **1987**, *59*, 455ff.
Pure & Appl. Chem. **1989**, *61*, 325.
Synthesis **1973**, 327.
Synthesis **1975**, 547.
Tetrahedron **1978**, *34*, 1731.
Tetrahedron **1983**, *39*, 2445ff.
Tetrahedron **1988**, *44*, 5615.
Aminolysis *Acc. Chem. Res.* **1984**, *17*, 144.
Biosynthesis *Tetrahedron* **1977**, *33*, 1545.
Carbenes
Imines *Pure & Appl. Chem.* **1983**, *55*, 1745.
Hydrolysis
Reaction mechanisms *Acc. Chem. Res.* **1984**, *17*, 144.
Polymerizations *Angew. Chem. Int. Ed. Engl.* **1962**, *1*, 481.
Synthesis *Org. Reactions* **1957**, *9*, 388.
Lactones *Synthesis* **1972**, 151.
β-Lactones *Org. Reactions* **1954**, *8*, 305.
Conformations *Pure & Appl. Chem.* **1982**, *54*, 2515.
Radical chemistry *Pure & Appl. Chem.* **1988**, *60*, 1659.
Unsaturated lactones *Chem. Rev.* **1976**, *76*, 625.
Lactose
Biosynthesis *Acc. Chem. Res.* **1970**, *3*, 41.
Lanosterol
Biosynthesis *Acc. Chem. Res.* **1968**, *1*, 1.
Acc. Chem. Res. **1968**, *1*, 111.
Lanthanide chemistry *Pure & Appl. Chem.* **1988**, *60*, 1725.
Tetrahedron **1986**, *42*, 6573.
Chem. Soc. Rev. **1988**, *17*, 69.
Lanthanide shift reagents *Chem. Rev.* **1973**, *73*, 553.
Chem. Soc. Rev. **1973**, *2*, 49.
Lantibiotics *Angew. Chem. Int. Ed. Engl.* **1991**, *30*, 1051.
Lascaloid A
Biosynthesis *Acc. Chem. Res.* **1983**, *16*, 7.
Lawesson's reagents *Tetrahedron* **1985**, *41*, 5061.
Lead(IV) acetate azides *Synthesis* **1972**, 285.

Lead tetraacetate	
Azomethines	*Chem. Rev.* **1973**, *73*, 93.
Decarboxylation	*Org. Reactions* **1972**, *19*, 279.
Oxidations	*Angew. Chem.* **1958**, *70*, 173.
Alcohols	*Synthesis* **1970**, 209.
Lectins	*Pure & Appl. Chem.* **1989**, *61*, 1283.
Binding sites	*Pure & Appl. Chem.* **1991**, *63*, 499.
Leuckart reaction	*Org. Reactions* **1949**, *5*, 301.
Leukotrienes	*Acc. Chem. Res.* **1985**, *18*, 87.
	Angew. Chem. Int. Ed. Engl. **1982**, *21*, 902.
	Angew. Chem. Int. Ed. Engl. **1983**, *22*, 805.
	Chem. Soc. Rev. **1982**, *11*, 321.
	Pure & Appl. Chem. **1981**, *53*, 1203.
	Tetrahedron **1983**, *39*, 1687.
Lewis acids	*SYNLETT* **1990**, 74.
Chiral Lewis acids	*Synthesis* **1991**, 1.
Ene reaction	*Acc. Chem. Res.* **1980**, *13*, 419.
	Acc. Chem. Res. **1980**, *13*, 426.
Organocopper reagents	*Angew. Chem. Int. Ed. Engl.* **1986**, *25*, 947.
Titanium tetrachloride	*Angew. Chem. Int. Ed. Engl.* **1977**, *16*, 817.
Lewis bases	
Metallomacrocycles	*Pure & Appl. Chem.* **1988**, *60*, 555.
Ligands	*Angew. Chem. Int. Ed. Engl.* **1991**, *30*, 296.
	See also chiral ligands
Complexes	
Stereochemical interactions	*Pure & Appl. Chem.* **1983**, *55*, 65.
Random screening	*Angew. Chem. Int. Ed. Engl.* **1991**, *30*, 296.
Lignans	*Chem. Soc. Rev.* **1982**, *11*, 75.
	Tetrahedron **1990**, *46*, 5029.
Limonene	*Pure & Appl. Chem.* **1990**, *62*, 1263
Lincomycins	
7-Deoxy-7-substituted alkylthiolincomycins	*Tetrahedron* **1984**, *40*, 1633.
Lipases	*Angew. Chem. Int. Ed. Engl.* **1985**, *24*, 617.
Lipid membranes	
Photochemistry	*Pure & Appl. Chem.* **1988**, *60*, 1039.
Lipids	
Biosynthesis	*Acc. Chem. Res.* **1972**, *5*, 361.
Membranes	*Acc. Chem. Res.* **1975**, *8*, 321.
Protein-lipid interactions	*Angew. Chem. Int. Ed. Engl.* **1990**, *29*, 1269.
Lipolytic enzymes	*Synthesis* **1991**, 1049.
Lipopeptides	*Tetrahedron* **1989**, *45*, 6331.
Lipopolysaccharides	*Pure & Appl. Chem.* **1989**, *61*, 1271.

Lipoproteins	*Acc. Chem. Res.* **1973**, *6*, 398.
Lipoxins	*Angew. Chem. Int. Ed. Engl.* **1991**, *30*, 1100.
Lipoxygenase	*Pure & Appl. Chem.* **1987**, *59*, 269.
Lipoxygenase products	*Acc. Chem. Res.* **1985**, *18*, 87.
Liquid crystals	*Angew. Chem. Int. Ed. Engl.* **1973**, *12*, 617.
	Angew. Chem. Int. Ed. Engl. **1984**, *23*, 348.
	Tetrahedron **1988**, *44*, 3413.
Carbohydrates	*Acc. Chem. Res.* **1986**, *19*, 168.
NMR	*Acc. Chem. Res.* **1971**, *4*, 81.
Polymeric liquid crystals	*Angew. Chem. Int. Ed. Engl.* **1990**, *29*, 1256.
Liquid membranes	*Pure & Appl. Chem.* **1986**, *58*, 1453.
Lithioalkenes	
Arenesulfonylhydrazones	*Org. Reactions* **1990**, *39*, 1.
Lithium aluminum hydride	
Rearrangements	
Reductions	*Synthesis* **1974**, 691.
Reductions	*Org. Reactions* **1951**, *6*, 469.
	Synthesis **1974**, 691.
Lithium perchlorate in ether	*Angew. Chem. Int. Ed. Engl.* **1991**, *30*, 1306.
Longifolene	
Rearrangement reactions	*Acc. Chem. Res.* **1981**, *14*, 82.
Low energy electron diffraction	*Acc. Chem. Res.* **1976**, *9*, 392.
Low-valent metals	
Reductions	*Synthesis* **1979**, 1.
Low-valent titanium	*Synthesis* **1989**, 883.
Coupling reactions	*Chem. Rev.* **1989**, *89*, 1513.
Luciferins	
Chemiluminescence	*Acc. Chem. Res.* **1974**, *7*, 135.
Lysosyme	
Denaturation	*Acc. Chem. Res.* **1970**, *3*, 62.
Lysozyme	*Acc. Chem. Res.* **1984**, *17*, 305.
Macrocycles	
Actinide complexes	*Chem. Soc. Rev.* **1988**, *17*, 69.
Alkali metal complexes	*Acc. Chem. Res.* **1978**, *11*, 49.
Azacrown compounds	*Synthesis* **1982**, 997.
Catechol derived macrocycles	*Pure & Appl. Chem.* **1988**, *60*, 545.
Cation complexes	*Chem. Rev.* **1982**, *82*, 209.
	Chem. Rev. **1985**, *85*, 271.
Cation separations	*Pure & Appl. Chem.* **1988**, *60*, 453.
Chiral macrocycles	*Angew. Chem. Int. Ed. Engl.* **1984**, *23*, 782.
Esters	*Chem. Rev.* **1979**, *79*, 37.

Furans	*Chem. Rev.* **1977**, *77*, 513.
Heterocycles	*Pure & Appl. Chem.* **1986**, *58*, 1437.
Ion selectivity	*Pure & Appl. Chem.* **1986**, *58*, 1445.
Iron(II) macrocycles	
Biomimetic applications	*Pure & Appl. Chem.* **1988**, *60*, 561.
Lactones	*Tetrahedron* **1977**, *33*, 3041.
Lanthanide complexes	*Chem. Soc. Rev.* **1988**, *17*, 69.
Lipophilic cage ligands	*Pure & Appl. Chem.* **1986**, *58*, 1503.
Liquid membranes	*Pure & Appl. Chem.* **1986**, *58*, 1453.
Metal complexes	*Chem. Soc. Rev.* **1975**, *4*, 421.
	Pure & Appl. Chem. **1989**, *61*, 1575.
	Pure & Appl. Chem. **1989**, *61*, 1587.
	Pure & Appl. Chem. **1989**, *61*, 1593.
Molecular "belts"	*Pure & Appl. Chem.* **1989**, *61*, 1581.
Platinum complexes	*Pure & Appl. Chem.* **1988**, *60*, 517.
Polyamines	*Pure & Appl. Chem.* **1986**, *58*, 1461.
	Tetrahedron **1992**, *48*, 6175.
Polyarsanes	*Pure & Appl. Chem.* **1986**, *58*, 1429.
Polyphosphanes	*Pure & Appl. Chem.* **1986**, *58*, 1429.
Pyridines	*Chem. Rev.* **1977**, *77*, 513.
Receptors	*Chem. Soc. Rev.* **1989**, *18*, 409.
Redox centers	*Chem. Soc. Rev.* **1989**, *18*, 409.
Redox processes	*Chem. Soc. Rev.* **1990**, *19*, 239.
Reductions	*Pure & Appl. Chem.* **1989**, *61*, 1555.
Ring enlargement	*Tetrahedron* **1988**, *44*, 1573.
Schiff bases	*Pure & Appl. Chem.* **1986**, *58*, 1437.
Thiophenes	*Chem. Rev.* **1977**, *77*, 513.
Macrolactam alkaloids	*Pure & Appl. Chem.* **1981**, *53*, 1141.
Macrolides	*Acc. Chem. Res.* **1988**, *21*, 294.
	Angew. Chem. Int. Ed. Engl. **1977**, *16*, 585.
	Tetrahedron **1977**, *33*, 3041.
	Tetrahedron **1985**, *41*, 3569.
Biosynthesis	*Acc. Chem. Res.* **1983**, *16*, 7.
Synthetic methods	*Tetrahedron* **1977**, *33*, 683.
Macromolecules	*See also crown ethers and polymerization*
Allosteric effects	*Pure & Appl. Chem.* **1988**, *60*, 581.
Cation complexation	*Pure & Appl. Chem.* **1988**, *60*, 461.
Chemical sensors	*Pure & Appl. Chem.* **1989**, *61*, 1613.
Copper complexes	
Hydrolysis	*Pure & Appl. Chem.* **1988**, *60*, 489.
Cross-linking	*Angew. Chem. Int. Ed. Engl.* **1976**, *15*, 451.
High pressure	
Reaction mechanisms	*Angew. Chem. Int. Ed. Engl.* **1975**, *14*, 137.
Historical perspective	*Angew. Chem. Int. Ed. Engl.* **1981**, *20*, 303.

Macromolecules, cont'd.	
Immobilization	
Biomaterials	*Angew. Chem. Int. Ed. Engl.* **1982**, *21*, 837.
Iron complexes	*Pure & Appl. Chem.* **1988**, *60*, 495.
Metal complexes	*Chem. Soc. Rev.* **1979**, *8*, 221.
Triazacyclononane derivatives	*Pure & Appl. Chem.* **1988**, *60*, 509.
Molecular assemblies	*Angew. Chem. Int. Ed. Engl.* **1983**, *22*, 565.
NMR	*Pure & Appl. Chem.* **1982**, *54*, 559ff.
Osmium complexes	*Pure & Appl. Chem.* **1988**, *60*, 495.
Polyethylene fibrids	*Angew. Chem. Int. Ed. Engl.* **1978**, *17*, 820.
Polystyrene	
Historical perspective	*Angew. Chem. Int. Ed. Engl.* **1981**, *20*, 344.
Receptors	*Chem. Soc. Rev.* **1984**, *13*, 279.
Ruthenium complexes	*Pure & Appl. Chem.* **1988**, *60*, 495.
Self-assembly	*SYNLETT* **1991**, 445.
Thiacrown ethers	
Copper complexes	*Pure & Appl. Chem.* **1988**, *60*, 501.
Macrophages	*Pure & Appl. Chem.* **1984**, *56*, 907.
Madumycin	*Pure & Appl. Chem.* **1982**, *54*, 2537.
Magnesium chemistry	
Amino acids	*Angew. Chem. Int. Ed. Engl.* **1990**, *29*, 1090.
Phase transfer catalysis	*Acc. Chem. Res.* **1988**, *21*, 261.
Reductions	*Chem. Rev.* **1957**, *57*, 417.
Magnetic circular dichroism	*Tetrahedron* **1984**, *40*, 3845.
Maleic anhydride	*Chem. Rev.* **1988**, *88*, 55.
Malonaldehyde monoacetal	*Synthesis* **1985**, 592.
Malononitrile	*Synthesis* **1978**, 165.
	Synthesis **1978**, 241.
Malononitrile derivatives	*Synthesis* **1981**, 925.
Mandelates	
Asymmetric synthesis	*Tetrahedron: Asymmetry* **1991**, 2, 299.
Manganese chemistry	
Photosynthesis	*Acc. Chem. Res.* **1980**, *13*, 249.
Manganese dioxide	
Oxidations	*Synthesis* **1976**, 65.
	Synthesis **1976**, 133.
Mannich bases	*Synthesis* **1973**, 703.
	Tetrahedron **1990**, *46*, 1791.
Marine natural products	*See natural products*
Mass spectometry	
Field ionization/field desorption mass spectrometry	*Tetrahedron* **1982**, *38*, 1125.
	Tetrahedron **1980**, *36*, 2687.

Mathematic models	
Reaction mechanisms	*Angew. Chem. Int. Ed. Engl.* **1990**, *29*, 1235.
Mechanism	*See under specific reaction*
Medicinal chemistry	*Angew. Chem. Int. Ed. Engl.* **1991**, *30*, 1278.
	Chem. Rev. **1989**, *89*, 1765.
	Chem. Soc. Rev. **1987**, *16*, 437.
	See also antibiotics, antibodies, and natural products
Abiotic receptors	*Chem. Soc. Rev.* **1983**, *12*, 285.
Adrenoceptor agonists	*Tetrahedron* **1991**, *47*, 9953.
Adrenoceptor antagonists	*Tetrahedron* **1991**, *47*, 9953.
Aminoacyl-tRNA synthetases	*Angew. Chem. Int. Ed. Engl.* **1981**, *20*, 217.
Antibiotics	*Angew. Chem. Int. Ed. Engl.* **1991**, *30*, 1051.
	Angew. Chem. Int. Ed. Engl. **1991**, *30*, 1387.
	Angew. Chem. Int. Ed. Engl. **1977**, *16*, 687.
Anticancer compounds	*Acc. Chem. Res.* **1986**, *19*, 293.
	Angew. Chem. Int. Ed. Engl. **1985**, *24*, 31.
	Angew. Chem. Int. Ed. Engl. **1991**, *30*, 1387.
	Pure & Appl. Chem. **1982**, *54*, 1919.
Enediynes	*Angew. Chem. Int. Ed. Engl.* **1991**, *30*, 1387.
Antifungal compounds	*Angew. Chem. Int. Ed. Engl.* **1987**, *26*, 320.
Antiinflammatory compounds	*Angew. Chem. Int. Ed. Engl.* **1984**, *23*, 413.
	Tetrahedron **1986**, *42*, 4095.
Anti-leprosy	*Chem. Soc. Rev.* **1987**, *16*, 437.
Antimalarials	*Angew. Chem. Int. Ed. Engl.* **1974**, *13*, 559.
Antimetabolites	*Acc. Chem. Res.* **1969**, *2*, 202.
	Angew. Chem. Int. Ed. Engl. **1974**, *13*, 559.
Antitumor agents	*Acc. Chem. Res.* **1982**, *15*, 308.
	Angew. Chem. Int. Ed. Engl. **1981**, *20*, 305.
	Pure & Appl. Chem. **1982**, *54*, 2523.
	Pure & Appl. Chem. **1984**, *56*, 1011.
	Pure & Appl. Chem. **1986**, *58*, 701.
Calicheamicin	*Acc. Chem. Res.* **1991**, *24*, 235.
Antiviral agents	*Pure & Appl. Chem.* **1985**, *57*, 423.
Marine natural products	*Pure & Appl. Chem.* **1981**, *53*, 795.
5-Substituted pyrimidine nucleoside analogues	*Pure & Appl. Chem.* **1983**, *55*, 623.
Blood coagulation	*Angew. Chem. Int. Ed. Engl.* **1971**, *10*, 85.

Medicinal chemistry, cont'd.

Blood substitutes	
Perfluoro compounds	*Angew. Chem. Int. Ed. Engl.* **1978**, *17*, 621.
Calcium agonists	*Angew. Chem. Int. Ed. Engl.* **1991**, *30*, 1559.
Calcium antagonists	*Angew. Chem. Int. Ed. Engl.* **1981**, *20*, 762.
	Angew. Chem. Int. Ed. Engl. **1991**, *30*, 1559.
Cancer	*Chem. Soc. Rev.* **1975**, *4*, 289.
Metals	*Chem. Rev.* **1972**, *72*, 203.
Chemotherapy	*Acc. Chem. Res.* **1969**, *2*, 202.
	Angew. Chem. Int. Ed. Engl. **1976**, *15*, 410.
Structure-activity relationships	*Acc. Chem. Res.* **1969**, *2*, 232.
Cholesterol homeostasis	*Angew. Chem. Int. Ed. Engl.* **1986**, *25*, 583.
CNS	*Angew. Chem. Int. Ed. Engl.* **1971**, *10*, 51.
	Chem. Rev. **1990**, *90*, 795.
Drugs	*Chem. Rev.* **1989**, *89*, 1765.
Myelin membrane	*Angew. Chem. Int. Ed. Engl.* **1990**, *29*, 958.
Coenzyme Q	*Angew. Chem. Int. Ed. Engl.* **1974**, *13*, 559.
Fibrosis	*Angew. Chem. Int. Ed. Engl.* **1971**, *10*, 85.
Glycine receptor	*Angew. Chem. Int. Ed. Engl.* **1985**, *24*, 365.
Hallucinogens	*Chem. Rev.* **1983**, *83*, 633.
Hereditary diseases	*Angew. Chem. Int. Ed. Engl.* **1973**, *12*, 1.
Histaminergic receptors	*Chem. Soc. Rev.* **1985**, *14*, 375.
Histocompatability antigens	*Angew. Chem. Int. Ed. Engl.* **1978**, *17*, 342.
Hormonal control	*Acc. Chem. Res.* **1970**, *3*, 113.
Hormone receptors	*Angew. Chem. Int. Ed. Engl.* **1976**, *15*, 741.
	Acc. Chem. Res. **1973**, *6*, 169.
Hormones	*Angew. Chem. Int. Ed. Engl.* **1972**, *11*, 7.
Peptides	*Acc. Chem. Res.* **1990**, *23*, 338.
Immune chemistry	*Angew. Chem. Int. Ed. Engl.* **1988**, *27*, 1028.
Immune regulatory agents	*Tetrahedron* **1989**, *45*, 4327.
Immunoactive peptides	*Pure & Appl. Chem.* **1987**, *59*, 317.
Immunology	*Angew. Chem. Int. Ed. Engl.* **1985**, *24*, 810.
Immunomodulators	*Angew. Chem. Int. Ed. Engl.* **1991**, *30*, 1611.
Immunostimulants	*Pure & Appl. Chem.* **1990**, *62*, 1217.
	Tetrahedron **1989**, *45*, 6331.
Immunosuppresants	*Acc. Chem. Res.* **1969**, *2*, 202.
	Angew. Chem. Int. Ed. Engl. **1991**, *30*, 954.
	Angew. Chem. Int. Ed. Engl. **1985**, *24*, 77.
Thiopurines	*Acc. Chem. Res.* **1969**, *2*, 202.

Influenza viruses	*Angew. Chem. Int. Ed. Engl.* **1986**, *25*, 47.
Inorganic medicinal chemistry	*Top. Curr. Chem.* **1976**, *64*, 181.
Local anesthetics	*Chem. Rev.* **1991**, *91*, 1109.
Methadone	
Opiod receptors	*Acc. Chem. Res.* **1978**, *11*, 21.
Mitochondrial adenine nucleotide translocation	*Angew. Chem. Int. Ed. Engl.* **1977**, *16*, 695.
Muscular dystrophy	
ESR	*Acc. Chem. Res.* **1977**, *10*, 111.
Nerve growth factor	*Angew. Chem. Int. Ed. Engl.* **1987**, *26*, 707.
Neurohormones	*Angew. Chem. Int. Ed. Engl.* **1973**, *12*, 255.
Neurophyrins	*Angew. Chem. Int. Ed. Engl.* **1979**, *18*, 846.
Neurotoxins	*Pure & Appl. Chem.* **1984**, *56*, 1049.
Nonsteroidal antiinflammatory agents	*Angew. Chem. Int. Ed. Engl.* **1972**, *11*, 460.
Nuclear medicine	*Angew. Chem. Int. Ed. Engl.* **1973**, *12*, 712.
Oligonucleotide-drug complexes	*Angew. Chem. Int. Ed. Engl.* **1991**, *30*, 1254.
Opioid receptors	*Acc. Chem. Res.* **1978**, *11*, 21.
Peptides	*Angew. Chem. Int. Ed. Engl.* **1991**, *30*, 1278.
Sickle cell disease	*Angew. Chem. Int. Ed. Engl.* **1976**, *15*, 749.
Silaoligosaccharide receptor	*Pure & Appl. Chem.* **1984**, *56*, 797.
Sphingolipid storage diseases	*Angew. Chem. Int. Ed. Engl.* **1977**, *16*, 273.
Steroids	*Angew. Chem. Int. Ed. Engl.* **1991**, *30*, 1278.
Suicide substrates	*Tetrahedron* **1982**, *38*, 871.
Toxication	*Angew. Chem. Int. Ed. Engl.* **1981**, *20*, 135.
Viroids	*Angew. Chem. Int. Ed. Engl.* **1980**, *19*, 231.
Virus diseases	*Angew. Chem. Int. Ed. Engl.* **1976**, *15*, 410.
Viruses	*Angew. Chem. Int. Ed. Engl.* **1976**, *15*, 410.
	Pure & Appl. Chem. **1983**, *55*, 637.
	Pure & Appl. Chem. **1987**, *59*, 1477.
DNA	*Acc. Chem. Res.* **1974**, *7*, 169.
Influenza virus	*Pure & Appl. Chem.* **1984**, *56*, 797.
RNA	*Acc. Chem. Res.* **1974**, *7*, 169.
Meldrum's acid	*Chem. Soc. Rev.* **1978**, *7*, 345.
Membranes	*Angew. Chem. Int. Ed. Engl.* **1971**, *10*, 388.
	Chem. Rev. **1990**, *90*, 355.
	Chem. Soc. Rev. **1977**, 6, 325.
	Pure & Appl. Chem. **1991**, *63*, 529.
Artifical nerves	*Angew. Chem. Int. Ed. Engl.* **1990**, 29, 20.
Artifical skin	*Angew. Chem. Int. Ed. Engl.* **1990**, 29, 20.
Bacterial membranes	*Angew. Chem. Int. Ed. Engl.* **1979**, *18*, 337.

Membranes, cont'd.
Electron transport — *Pure & Appl. Chem.* **1989**, *61*, 1569.
Ion carriers — *Angew. Chem. Int. Ed. Engl.* **1985**, *24*, 905.
Ion channels — *Angew. Chem. Int. Ed. Engl.* **1985**, *24*, 905.
Ion pumps — *Angew. Chem. Int. Ed. Engl.* **1985**, *24*, 905.
Ion transport — *Angew. Chem. Int. Ed. Engl.* **1985**, *24*, 905.
Chem. Rev. **1991**, *91*, 793.
Lipids — *Acc. Chem. Res.* **1975**, *8*, 321.
Mimetics
Reaction mechanisms — *Pure & Appl. Chem.* **1982**, *54*, 1809.
Models — *Chem. Rev.* **1988**, *88*, 455.
NMR — *Acc. Chem. Res.* **1983**, *16*, 266.
Phospholipids — *Angew. Chem. Int. Ed. Engl.* **1984**, *23*, 257.
Pure & Appl. Chem. **1982**, *54*, 2443.
NMR — *Acc. Chem. Res.* **1987**, *20*, 221.
Photochemistry — *Pure & Appl. Chem.* **1982**, *54*, 1693.
Photosynthesis — *Acc. Chem. Res.* **1978**, *11*, 257.
Potassium transport — *Angew. Chem. Int. Ed. Engl.* **1971**, *10*, 882.
Proteins
Cross-linking — *Acc. Chem. Res.* **1988**, *21*, 435.
NMR — *Acc. Chem. Res.* **1987**, *20*, 221.
Sodium transport — *Angew. Chem. Int. Ed. Engl.* **1971**, *10*, 882.
Synthetic membranes — *Angew. Chem. Int. Ed. Engl.* **1982**, *21*, 660.
Vesicle membranes — *Angew. Chem. Int. Ed. Engl.* **1984**, *23*, 100.
Vesicles — *Acc. Chem. Res.* **1980**, *13*, 7.
Mercury chemistry — *SYNLETT* **1991**, 597.
Catalysis
[3,3]-Sigmatropic rearrangements — *Angew. Chem. Int. Ed. Engl.* **1984**, *23*, 579.
Salts
Alkenes — *Synthesis* **1971**, 527.
Mesitylenesulfonylhydroxylamine
Aminations — *Synthesis* **1977**, 1.
Metabolism — *Angew. Chem. Int. Ed. Engl.* **1985**, *24*, 451.
[2.2]Metacyclophanes — *Angew. Chem. Int. Ed. Engl.* **1972**, *11*, 73.
Metal aluminide catalysis — *Chem. Rev.* **1986**, *86*, 763.
Metal bases — *Angew. Chem. Int. Ed. Engl.* **1983**, *22*, 927.
Metal binding
Proteins — *Chem. Soc. Rev.* **1977**, 6, 139.
Metal boride catalysis — *Chem. Rev.* **1986**, *86*, 763.
Metal carbene complexes — *See carbenes*
Metal carbonyls — *Pure & Appl. Chem.* **1980**, *52*, 607.
Synthesis **1971**, 55.

Photochemical	*Chem. Rev.* **1974**, *74*, 401.
Substitution reactions	*Angew. Chem. Int. Ed. Engl.* **1964**, *3*, 730.
Metal clusters	*Pure & Appl. Chem.* **1986**, *58*, 529.
Biological aspects	*Chem. Soc. Rev.* **1981**, *10*, 455.
Metal complexes	*Pure & Appl. Chem.* **1986**, *58*, 1429.
	Pure & Appl. Chem. **1989**, *61*, 1575.
	Pure & Appl. Chem. **1989**, *61*, 1587.
	Pure & Appl. Chem. **1989**, *61*, 1593.
Amides	*Chem. Rev.* **1982**, *82*, 385.
Asymmetric synthesis	*Pure & Appl. Chem.* **1983**, *55*, 99.
Carbohydrates	*Chem. Soc. Rev.* **1979**, *8*, 221.
Imidazole derivatives	*Chem. Rev.* **1974**, *74*, 471.
Macrocycles	*Chem. Soc. Rev.* **1975**, *4*, 421.
Macromolecules	*Chem. Soc. Rev.* **1979**, *8*, 221.
Peptides	*Chem. Rev.* **1982**, *82*, 385.
	Pure & Appl. Chem. **1983**, *55*, 23.
Phosphorus NMR	*Chem. Rev.* **1981**, *81*, 229.
Photochemistry	*Pure & Appl. Chem.* **1986**, *58*, 1193.
RNA	*Chem. Rev.* **1971**, *71*, 439.
Triazacyclononane derivatives	*Pure & Appl. Chem.* **1988**, *60*, 509.
Metal hydrides	
Reductions	*Org. Reactions* **1988**, *36*, 249.
Transition metal chemistry	*Angew. Chem. Int. Ed. Engl.* **1991**, *30*, 759.
Metal ion complexes	
DNA	*Chem. Rev.* **1971**, *71*, 439.
Host-guest chemistry	*Chem. Rev.* **1974**, *74*, 351.
Nucleotides	*Chem. Rev.* **1980**, *80*, 365.
Oligopeptides	*Pure & Appl. Chem.* **1983**, *55*, 107.
Proteins	*Pure & Appl. Chem.* **1983**, *55*, 35.
Metal surfaces	
Organometallic chemistry	*Pure & Appl. Chem.* **1982**, *54*, 83.
Metal-halogen exchange	
Organolithiums	*Org. Reactions* **1951**, *6*, 339.
Metallations	
Heterocycles	*Tetrahedron* **1983**, *39*, 1955ff.
Isocyanides	*SYNLETT* **1990**, 245.
Organolithiums	*Chem. Rev.* **1969**, *69*, 693.
	Org. Reactions **1954**, *8*, 258.
Thioethers	*Acc. Chem. Res.* **1989**, *22*, 152.
Metalloboranes	*Pure & Appl. Chem.* **1983**, *55*, 1415.
Metallocarboranes	*Pure & Appl. Chem.* **1982**, *54*, 43.
Metalloenzymes	*Acc. Chem. Res.* **1986**, *19*, 363.
	Pure & Appl. Chem. **1991**, *63*, 265.
ENDOR	*Acc. Chem. Res.* **1991**, *24*, 164.

α-Metalloisocyanides	*Angew. Chem. Int. Ed. Engl.* **1974**, *13*, 789.
	Angew. Chem. Int. Ed. Engl. **1977**, *16*, 339.
Metallomacrocycles	*Pure & Appl. Chem.* **1988**, *60*, 555.
Metalloporphyrins	
Organic metals	*Acc. Chem. Res.* **1983**, *16*, 15.
Metalloproteins	
Electron transfers	*Pure & Appl. Chem.* **1983**, *55*, 1049.
Metathesis	*Acc. Chem. Res.* **1972**, *5*, 127.
	Angew. Chem. Int. Ed. Engl. **1976**, *15*, 401.
	Chem. Soc. Rev. **1975**, *4*, 155.
Methane	
Activation	*Angew. Chem. Int. Ed. Engl.* **1991**, *30*, 820.
Oxidative coupling	*Chem. Soc. Rev.* **1989**, *18*, 251.
Superacid	*Acc. Chem. Res.* **1987**, *20*, 422.
Methanol	
Methane	
Oxidation	*Chem. Rev.* **1985**, *85*, 235.
Methionine	
Biosynthesis	*Acc. Chem. Res.* **1968**, *1*, 97.
Methyl radicals	
Hydrogen abstraction	
Reaction mechanisms	*Chem. Rev.* **1971**, *71*, 247.
Methylchlorophosphanes	*Angew. Chem. Int. Ed. Engl.* **1973**, *12*, 877.
3-Methylcyclohex-2-enone derivatives	*Chem. Soc. Rev.* **1980**, *9*, 265.
Methylene bridges	
Organometallic chemistry	*Pure & Appl. Chem.* **1982**, *54*, 65.
	Pure & Appl. Chem. **1984**, *56*, 59.
Methylene groups	
Substitution reactions	
Ethers	*Tetrahedron* **1974**, *30*, 1683.
α-Methylene lactones	*Synthesis* **1975**, 67.
	Synthesis **1986**, 157.
α-Methylene buytrolactones	*Angew. Chem., Int. Ed. Engl.* **1985**, *24*, 94.
Methyleneamino phosphine derivatives	*Synthesis* **1986**, 793.
Methylenecyclopropane derivatives	*Angew. Chem. Int. Ed. Engl.* **1980**, *19*, 276.
Methylenetetrahydrofolate reductase	*Chem. Rev.* **1990**, *90*, 1275.
Methyltetrahydrofolatehomocysteine methyltransferase	*Chem. Rev.* **1990**, *90*, 1275.
Methylthiotrimethylsilane	*Synthesis* **1980**, 861.
Mevalonic acid	
Biosynthesis	*Tetrahedron* **1986**, *42*, 3.

Mevinic acids	*Tetrahedron* **1986**, *42*, 4909.
Mevinoids	*Pure & Appl. Chem.* **1988**, *60*, 1555.
Micelles	*Acc. Chem. Res.* **1979**, *12*, 111.
	Acc. Chem. Res. **1983**, *16*, 251.
Cage effects	
Reaction mechanisms	*Pure & Appl. Chem.* **1981**, *53*, 259.
Phospholipids	*Acc. Chem. Res.* **1983**, *16*, 251.
Photochemistry	*Angew. Chem. Int. Ed. Engl.* **1980**, *19*, 675.
	Tetrahedron **1982**, *38*, 2455.
Reaction mechanisms	*Pure & Appl. Chem.* **1981**, *53*, 259.
Surfactants	*Acc. Chem. Res.* **1979**, *12*, 111.
	Angew. Chem. Int. Ed. Engl. **1988**, *27*, 902.
Microencapsulation	*Angew. Chem. Int. Ed. Engl.* **1975**, *14*, 539.
Micro-organisms	*Angew. Chem. Int. Ed. Engl.* **1977**, *16*, 205.
	Chem. Soc. Rev. **1988**, *17*, 383.
Degradations	
Halogenated hydrocarbons	*Angew. Chem. Int. Ed. Engl.* **1986**, *25*, 779.
Hydrocarbon synthesis	*Chem. Soc. Rev.* **1974**, *3*, 309.
Metabolites	*Pure & Appl. Chem.* **1986**, *58*, 773.
Polysaccharides	*Chem. Soc. Rev.* **1990**, *19*, 29.
	Pure & Appl. Chem. **1984**, *56*, 879.
Proteins	*Chem. Soc. Rev.* **1979**, *8*, 143.
Reactions	*Synthesis* **1969**, 120.
	Synthesis **1969**, 147.
Reductions	
Ketones	*Angew. Chem. Int. Ed. Engl.* **1984**, *23*, 570.
Migrations	
Phosphine oxides	*Acc. Chem. Res.* **1978**, *11*, 401.
Milbemycins	*Chem. Soc. Rev.* **1991**, *20*, 211.
	Chem. Soc. Rev. **1991**, *20*, 271.
Mitochondria	
Electron transport	*Acc. Chem. Res.* **1972**, *5*, 234.
Energy converting system	*Angew. Chem. Int. Ed. Engl.* **1980**, *19*, 659.
Mitochondrial adenine nucleotide translocation	*Angew. Chem. Int. Ed. Engl.* **1977**, *16*, 695.
Möbius-Hückel concept	*Acc. Chem. Res.* **1971**, *4*, 272.
Mold metabolites	*Pure & Appl. Chem.* **1989**, *61*, 281.
Molecular assemblies	*Angew. Chem. Int. Ed. Engl.* **1983**, *22*, 565.
Molecular "belts"	*Pure & Appl. Chem.* **1989**, *61*, 1581.
Molecular biology	*Angew. Chem. Int. Ed. Engl.* **1981**, *20*, 233.
Gene fragments	*Angew. Chem. Int. Ed. Engl.* **1983**, *22*, 13.
Signal transducers	*Angew. Chem. Int. Ed. Engl.* **1972**, *11*, 1.
Molecular dynamics	*Angew. Chem. Int. Ed. Engl.* **1987**, *26*, 1221.
	Angew. Chem. Int. Ed. Engl. **1990**, *29*, 992.

Molecular engineering	*Pure & Appl. Chem.* **1984**, *56*, 967.
Molecular genetics	*Tetrahedron* **1991**, *47*, 5919ff.
Molecular hysteresis	
Cybernetic significance	*Angew. Chem. Int. Ed. Engl.* **1973**, *12*, 356.
Molecular mechanics calculations	*Angew. Chem. Int. Ed. Engl.* **1983**, *22*, 1.
Molecular organization	
Spherands	*Angew. Chem. Int. Ed. Engl.* **1986**, *25*, 1039.
Molecular recognition	*Acc. Chem. Res.* **1991**, *24*, 332.
	Angew. Chem. Int. Ed. Engl. **1988**, *27*, 113.
	Angew. Chem. Int. Ed. Engl. **1990**, *29*, 245.
	Angew. Chem. Int. Ed. Engl. **1990**, *29*, 1304.
	Angew. Chem. Int. Ed. Engl. **1991**, *30*, 1417.
	Chem. Soc. Rev. **1986**, *15*, 63.
	Pure & Appl. Chem. **1989**, *61*, 1517.
	Pure & Appl. Chem. **1989**, *61*, 1547.
	Pure & Appl. Chem. **1989**, *61*, 1563.
	Pure & Appl. Chem. **1990**, *62*, 1251.
Chiral recognition	*Acc. Chem. Res.* **1978**, *11*, 8.
Cyclophanes	*Pure & Appl. Chem.* **1989**, *61*, 1523.
Enzymes	*Acc. Chem. Res.* **1987**, *20*, 79.
Host-guest chemistry	*Angew. Chem. Int. Ed. Engl.* **1991**, *30*, 1417.
Sandwich compounds	*Angew. Chem. Int. Ed. Engl.* **1991**, *30*, 407.
Molecular self-organization	
Origin of Life	*Angew. Chem. Int. Ed. Engl.* **1981**, *20*, 500.
Molecular shapes	*Chem. Soc. Rev.* **1978**, *7*, 507.
Molybdenum compounds	*Pure & Appl. Chem.* **1985**, *57*, 1789.
Molybdo-enzymes	*Pure & Appl. Chem.* **1984**, *56*, 1645.
Monoclonal antibodies	*Angew. Chem. Int. Ed. Engl.* **1985**, *24*, 139.
	Angew. Chem. Int. Ed. Engl. **1985**, *24*, 827.
Monolayers	
Photochemistry	*Angew. Chem. Int. Ed. Engl.* **1979**, *18*, 440.
Monooxygenases	*Angew. Chem. Int. Ed. Engl.* **1988**, *27*, 333.
Reaction mechanisms	*Acc. Chem. Res.* **1988**, *21*, 101.
Monoterpenes	
Biosynthesis	*Chem. Rev.* **1972**, *72*, 115.
	Chem. Rev. **1987**, *87*, 929.
Morphinandienone alkaloids	*Chem. Rev.* **1971**, *71*, 47.
Mössbauer spectroscopy	*Acc. Chem. Res.* **1977**, *10*, 153.
Multilayered cyclophanes	*Acc. Chem. Res.* **1978**, *11*, 251.

Muramylpeptides	*Tetrahedron* **1989**, *45*, 6331.
Mushroom metabolites	*Tetrahedron* **1981**, *37*, 2199.
Mutagens	*Acc. Chem. Res.* **1984**, *17*, 403.
Mycotoxins	*Angew. Chem. Int. Ed. Engl.* **1984**, *23*, 493.
	Pure & Appl. Chem. **1980**, *52*, 165ff.
	Pure & Appl. Chem. **1986**, *58*, 211ff.
	Pure & Appl. Chem. **1989**, *61*, 7.
	Tetrahedron **1989**, *45*, 2237ff.
Myelin membrane	*Angew. Chem. Int. Ed. Engl.* **1990**, *29*, 958.
Myoglobin	*Chem. Soc. Rev.* **1983**, *12*, 387.
Model systems	*Acc. Chem. Res.* **1970**, *3*, 90.
Oxygen binding	*Acc. Chem. Res.* **1977**, *10*, 265.
***N*-Acetylneuraminic acid**	*Synthesis* **1991**, 583.
***N*-Acyliiminium intermediates**	
Intramolecular reactions	*Tetrahedron* **1985**, *41*, 4367.
***N*-Acylimines**	
Cycloaddition reactions	*Chem. Rev.* **1989**, *89*, 1525.
NAD	
Proteins	*Acc. Chem. Res.* **1977**, *10*, 92.
NADH mimics	
Reductions	*Tetrahedron: Asymmetry* **1991**, *2*, 299.
Nafion-H	*Synthesis* **1986**, 513.
Naphthalene	
Thermal rearrangement	*Acc. Chem. Res.* **1982**, *15*, 52.
Naphthalene nitriles	
Addition reactions	
Photochemistry	*Pure & Appl. Chem.* **1988**, *60*, 1009.
Naphthalene radical anions	
Alkyl halides	*Acc. Chem. Res.* **1971**, *4*, 400.
Narcotics	
Nutmeg	*Angew. Chem. Int. Ed. Engl.* **1971**, *10*, 370.
Natural products	
Ants	*Chem. Soc. Rev.* **1984**, *13*, 245.
Aphids	*Chem. Soc. Rev.* **1975**, *4*, 263.
Bio-active compounds	*Chem. Soc. Rev.* **1984**, *13*, 131.
Biomimetic approaches	*Chem. Soc. Rev.* **1972**, *1*, 553.
Biosynthesis	*Tetrahedron* **1991**, *47*, 5919ff.
Building blocks	*Angew. Chem. Int. Ed. Engl.* **1979**, *18*, 429.
Carrier-bound substances	*Angew. Chem. Int. Ed. Engl.* **1972**, *11*, 249.
Chemical ecology	*Angew. Chem. Int. Ed. Engl.* **1976**, *15*, 214.
Diamines	*Acc. Chem. Res.* **1982**, *15*, 290.
Diels Alder reaction	*Acc. Chem. Res.* **1986**, *19*, 250.
Hydrogen cyanide	
Evolution	*Tetrahedron* **1984**, *40*, 1093.

Natural products, cont'd.

Marine natural products	*Pure & Appl. Chem.* **1982**, *54*, 1907.
	Pure & Appl. Chem. **1982**, *54*, 1963.
	Pure & Appl. Chem. **1982**, *54*, 1995.
	Pure & Appl. Chem. **1986**, *58*, 395.
	Pure & Appl. Chem. **1989**, *61*, 293.
	Tetrahedron **1977**, *33*, 1421.
	Tetrahedron **1984**, *40*, 979ff.
Algae	
Invertebrates	*Acc. Chem. Res.* **1977**, *10*, 40.
Metabolites	*Pure & Appl. Chem.* **1981**, *53*, 853.
Signal substances	*Angew. Chem. Int. Ed. Engl.* **1982**, *21*, 643.
Antibiotics	*Pure & Appl. Chem.* **1982**, *54*, 1951.
Antimicrobial agents	*Pure & Appl. Chem.* **1981**, *53*, 795.
Antineoplastic agents	*Pure & Appl. Chem.* **1981**, *53*, 795.
Antiviral agents	*Pure & Appl. Chem.* **1981**, *53*, 795.
Biosynthesis	*Pure & Appl. Chem.* **1990**, *62*, 1259.
Enzymes	*Pure & Appl. Chem.* **1982**, *54*, 1951.
Mediterranean invertibrates	*Pure & Appl. Chem.* **1986**, *58*, 375.
Prokaryotes	
Toxins	*Pure & Appl. Chem.* **1982**, *54*, 1919.
Tumor promoters	*Pure & Appl. Chem.* **1982**, *54*, 1919.
Sterols	*Pure & Appl. Chem.* **1981**, *53*, 873.
Toxins	*Acc. Chem. Res.* **1977**, *10*, 33.
	Acc. Chem. Res. **1977**, *10*, 40.
	Pure & Appl. Chem. **1982**, *54*, 1973.
Wax esters	*Pure & Appl. Chem.* **1981**, *53*, 967.
Shellfish poisons	*Pure & Appl. Chem.* **1989**, *61*, 7.
Signal substances	
Marine algae	*Angew. Chem. Int. Ed. Engl.* **1982**, *21*, 643.
Metabolites	
Polycyclic aromatic metabolites	*Synthesis* **1986**, 605.
Oxygen transport	*Chem. Rev.* **1979**, *79*, 139.
Pacifigorgiol	*Pure & Appl. Chem.* **1982**, *54*, 1915.
Pakistani medicinal plants	*Pure & Appl. Chem.* **1986**, *58*, 663.
Photochemistry	*Quart. Rev.* **1970**, *24*, 37.
Polyamines	*Acc. Chem. Res.* **1982**, *15*, 290.
Quinghaosu	*Pure & Appl. Chem.* **1986**, *58*, 817.
Screening	*Angew. Chem. Int. Ed. Engl.* **1991**, *30*, 296.
Seaweed	*Acc. Chem. Res.* **1977**, *10*, 40.
Snake venom	*Pure & Appl. Chem.* **1986**, *58*, 407.
Stereoelectronic effects	*Chem. Rev.* **1987**, *87*, 1047.
Structure determination	*Tetrahedron* **1991**, *47*, 3511ff.
DENDRAL	*Pure & Appl. Chem.* **1982**, *54*, 2425.

Synthesis	*Chem. Rev.* **1991**, *91*, 1237.
	Pure & Appl. Chem. **1987**, *59*, 373ff.
Synthetic methods	*Pure & Appl. Chem.* **1981**, *53*, 1081.
	Pure & Appl. Chem. **1986**, *58*, 675.
	Pure & Appl. Chem. **1988**, *60*, 1549.
Tropical plants	*Tetrahedron* **1980**, *36*, 1465.
West African medicinal plants	*Pure & Appl. Chem.* **1986**, *58*, 653.
***N*-Bromoimides**	
Reaction mechanisms	*Pure & Appl. Chem.* **1991**, *63*, 205.
***N*-Bromosuccinimide**	
Brominations	*Acc. Chem. Res.* **1978**, *11*, 381.
Dehydrogenations	*Chem. Rev.* **1963**, *63*, 21.
***N*-Chloroalkenylamines**	
Cyclizations	*Angew. Chem. Int. Ed. Engl.* **1983**, *22*, 337.
***N*-Chlorocarbonylisocyanate**	*Angew. Chem. Int. Ed. Engl.* **1977**, *16*, 743
Neighboring group effects	
Nitrobenzene	*Chem. Rev.* **1972**, *72*, 627.
Nitrobenzene derivatives	*Chem. Rev.* **1973**, *73*, 190.
Reaction mechanisms	*Chem. Soc. Rev.* **1973**, *2*, 295.
Ring compounds	*Chem. Soc. Rev.* **1974**, *3*, 1.
Neocarzinostatin	*Acc. Chem. Res.* **1991**, *24*, 191.
Neolignans	*Chem. Soc. Rev.* **1982**, *11*, 75.
***N*-Haloamides**	
Rearrangement reactions	*Synthesis* **1971**, 1.
***N*-Haloamines**	
Cyclizations	*Chem. Rev.* **1963**, *63*, 55.
***N*-Hydroxy-α-amino acids**	*Chem. Rev.* **1986**, *86*, 697.
***N*-Hydroxyimides**	*Angew. Chem. Int. Ed. Engl.* **1974**, *13*, 376.
Nickel chemistry	*Chem. Rev.* **1983**, *83*, 203.
	Pure & Appl. Chem. **1984**, *56*, 1621.
	Pure & Appl. Chem. **1990**, *62*, 1941.
π-Allyl complexes	*Chem. Soc. Rev.* **1985**, *14*, 93.
	Org. Reactions **1972**, *19*, 115.
Catalysis	
Coupling reactions	*Pure & Appl. Chem.* **1980**, *52*, 2417.
Oligomerization	*Adv. Organomet. Chem.* **1979**, *17*, 105.
Complexes	*Pure & Appl. Chem.* **1980**, *52*, 669.
Coupling reactions	*Acc. Chem. Res.* **1982**, *15*, 340.
Homogeneous catalysis	*Angew. Chem. Int. Ed. Engl.* **1990**, *29*, 235.
Insertion reactions	*Acc. Chem. Res.* **1973**, *6*, 422.
Nickel peroxide	
Oxidations	*Chem. Rev.* **1975**, *75*, 491.
Ninhydrin	
Ninhydrin derivatives	*Tetrahedron* **1991**, *47*, 8791.
1,2,3-Tricarbonyl compounds	*Tetrahedron* **1978**, *34*, 1285.

Nitrations	
Aromatic compounds	*Acc. Chem. Res.* **1971**, *4*, 240.
	Acc. Chem. Res. **1971**, *4*, 248.
	Synthesis **1977**, 217.
Ipso substitution	*Acc. Chem. Res.* **1976**, *9*, 287.
Reaction mechanisms	*Acc. Chem. Res.* **1987**, *20*, 53.
Side reactions	*Synthesis* **1977**, 217.
Nitrenes	
Enzymes	
Labelling	*Acc. Chem. Res.* **1972**, *5*, 155.
Rearrangements	*Topics Curr. Chem.* **1976**, *62*, 173.
Aromatic compounds	*Acc. Chem. Res.* **1972**, *5*, 303.
Nitrile oxides	
Heterocycles	*Synthesis* **1970**, 344.
Nitriles	
Acid chlorides	*Synthesis* **1973**, 189.
Alkylations	*Org. Reactions* **1957**, *9*, 107.
Alkynes	*Angew. Chem. Int. Ed. Engl.* **1978**, *17*, 505.
Carbanions	*Org. Reactions* **1984**, *31*, 1.
Nitrilium ions	
1,3-Dipoles	*Acc. Chem. Res.* **1980**, *13*, 448.
Nitro compounds	*Org. Reactions* **1962**, *12*, 101.
Enzymes	
Inhibitors	*Acc. Chem. Res.* **1983**, *16*, 418.
Heterocycles	*Synthesis* **1974**, 613.
Nucleophilic substitution	
Reaction mechanisms	*Chem. Soc. Rev.* **1988**, *17*, 285.
Reductions	*Acc. Chem. Res.* **1974**, *7*, 281.
Nitroacetic acid	*Synthesis* **1979**, 666.
Nitroalkanes	*Tetrahedron* **1990**, *46*, 7313ff.
Nucleophilic substitution	*Synthesis* **1991**, 423.
Reductive cleavage	*Synthesis* **1986**, 693.
Nitroalkenes	*Chem. Rev.* **1986**, *86*, 751.
	Chem. Soc. Rev. **1991**, *20*, 95.
	Tetrahedron **1990**, *46*, 7313ff.
Addition reactions	
Grignard reagents	*Acc. Chem. Res.* **1984**, *17*, 109.
Alkyl anion synthons	*Synthesis* **1988**, 833.
Asymmetric synthesis	*SYNLETT* **1991**, 603.
Carbonyl compounds	
Oxoalkylations	*Acc. Chem. Res.* **1985**, *18*, 284.
Enol ethers	*Acc. Chem. Res.* **1985**, *18*, 284.
Grignard reagents	
Addition reactions	*Acc. Chem. Res.* **1984**, *17*, 109.
Silyl enol ethers	*Acc. Chem. Res.* **1985**, *18*, 284.

Nitroarenes	
Chloromethyl sulfones	*Acc. Chem. Res.* **1987**, *20*, 282.
Zinin reduction	*Org. Reactions* **1973**, *20*, 455.
Nitrobenzenes	
Neighboring group effects	*Chem. Rev.* **1972**, *72*, 627.
	Chem. Rev. **1973**, *73*, 190.
Nitroenamines	*Tetrahedron* **1981**, *37*, 1453.
α-Nitroketones	*Synthesis* **1980**, 264.
Nitrogen bridgehead compounds	
Spectroscopic techniques	*Chem. Rev.* **1971**, *71*, 109.
Nitrogen compounds	
Addition reactions	
Polar additions	*Acc. Chem. Res.* **1969**, *2*, 186.
Oxidations	*Quart. Rev.* **1971**, *25*, 407.
Reductions	*Org. Reactions* **1988**, *36*, 249.
Nitrogen elimination	*Angew. Chem. Int. Ed. Engl.* **1977**, *16*, 835.
Nitrogen fixation	*Angew. Chem. Int. Ed. Engl.* **1975**, *14*, 80.
	Angew. Chem. Int. Ed. Engl. **1975**, *14*, 514.
	Chem. Rev. **1973**, *73*, 247.
	Chem. Rev. **1978**, *78*, 589.
	Chem. Soc. Rev. **1972**, *1*, 121.
Organotitanium compounds	*Acc. Chem. Res.* **1970**, *3*, 361.
Nitrogen heterocycles	*Acc. Chem. Res.* **1972**, *5*, 303.
	Angew. Chem. Int. Ed. Engl. **1988**, *27*, 1456.
Medium ring nitrogen hetereocycles	*Tetrahedron* **1991**, *47*, 9131.
Nitrogen ylides	
Cycloimmonium ylides	*Tetrahedron* **1976**, *32*, 2647.
Nitrogenase	*Acc. Chem. Res.* **1981**, *14*, 201.
	Acc. Chem. Res. **1991**, *24*, 1.
	Angew. Chem. Int. Ed. Engl. **1975**, *14*, 514.
Iron-molybdenum cofactor	*Chem. Rev.* **1990**, *90*, 1377.
Nitrones	
Alkaloids	*Acc. Chem. Res.* **1979**, *12*, 396.
Dipolar cycloadditions	*Org. Reactions* **1988**, *36*, 1.
1,3-Dipolar cycloadditions	*Synthesis* **1975**, 205.
Photochemistry	*Chem. Rev.* **1970**, *70*, 236.
Nitrosamines	
Organometallic reagents	*Angew. Chem. Int. Ed. Engl.* **1975**, *14*, 15.
Photochemistry	*Acc. Chem. Res.* **1973**, *6*, 354.
Tobacco	*Acc. Chem. Res.* **1979**, *12*, 92.
Nitrosations	*Org. Reactions* **1953**, *7*, 327.
	Chem. Rev. **1959**, *59*, 497.
	Quart Rev. **1961**, *15*, 418.
Aliphatic compounds	*Org. Reactions* **1953**, *7*, 327.

Nitroso compounds	*Chem. Soc. Rev.* **1990**, *19*, 355.
Nucleophilic substitution	
Reaction mechanisms	*Chem. Soc. Rev.* **1988**, *17*, 285.
Nitrosoacetylamine reaction	
Biaryl compounds	*Org. Reactions* **1944**, *2*, 224.
Nitrosoalkenes	*Chem. Soc. Rev.* **1983**, *12*, 53.
Nitrosoalkynes	*Chem. Soc. Rev.* **1983**, *12*, 53.
Nitrosonium ions	
Synthetic methods	*Acc. Chem. Res.* **1980**, *13*, 330.
Nitrous acid	
Amines	*Acc. Chem. Res.* **1971**, *4*, 315.
Nitroxyl radicals	*Synthesis* **1971**, 190.
	Synthesis **1971**, 401.
NMR	*Angew. Chem. Int. Ed. Engl.* **1980**, *19*, 411.
	Chem. Soc. Rev. **1976**, *5*, 1.
	Chem. Soc. Rev. **1983**, *12*, 1.
Alkaloids	*Acc. Chem. Res.* **1974**, *7*, 46.
Allylic compounds	
Spin-spin coupling	*Chem. Rev.* **1976**, *76*, 593.
Anti-aromaticity	*Acc. Chem. Res.* **1973**, *6*, 393.
Binding phenomena	*Tetrahedron* **1989**, *45*, 581.
Biosynthesis	*Chem. Soc. Rev.* **1975**, *4*, 497.
	Chem. Soc. Rev. **1979**, *8*, 539.
	Pure & Appl. Chem. **1981**, *53*, 1241.
	Pure & Appl. Chem. **1986**, *58*, 753.
Carbanions	*Pure & Appl. Chem.* **1989**, *61*, 709.
Carbohydrates	*Chem. Rev.* **1973**, *73*, 669.
	Chem. Soc. Rev. **1975**, *4*, 401.
	Pure & Appl. Chem. **1981**, *53*, 45.
Carbon-13 NMR	*Acc. Chem. Res.* **1978**, *11*, 469.
Fourier transforms	*Angew. Chem. Int. Ed. Engl.* **1971**, *10*, 673.
CIDNP	*Acc. Chem. Res.* **1985**, *18*, 196.
Computers	*Angew. Chem. Int. Ed. Engl.* **1972**, *11*, 375.
Conformations	*Acc. Chem. Res.* **1975**, *8*, 306.
Corrins	*Pure & Appl. Chem.* **1981**, *53*, 1215.
Coupling	*Acc. Chem. Res.* **1974**, *7*, 333.
Coupling constants	*Acc. Chem. Res.* **1980**, *13*, 400.
	Angew. Chem. Int. Ed. Engl. **1972**, *11*, 861.
Solvent effects	*Chem. Rev.* **1973**, *73*, 53.
Cyclic compounds	*Chem. Rev.* **1977**, *77*, 599.
Deuterium NMR	*Acc. Chem. Res.* **1987**, *20*, 221.
Membranes	*Acc. Chem. Res.* **1983**, *16*, 266.
Dianions	*Tetrahedron* **1988**, *44*, 6957.

2D-NMR	*Angew. Chem. Int. Ed. Engl.* **1988**, *27*, 490.
	Angew. Chem. Int. Ed. Engl. **1988**, *27*, 1655.
	Chem. Rev. **1990**, *90*, 935.
	Chem. Soc. Rev. **1990**, *19*, 381.
Double resonance	*Angew. Chem. Int. Ed. Engl.* **1971**, *10*, 472.
Enantiomeric purity	*Chem. Rev.* **1991**, *91*, 1441.
Enzyme-bound substrates	*Acc. Chem. Res.* **1977**, *10*, 246.
Fluorine NMR	
Solid state nmr	*Chem. Rev.* **1991**, *91*, 1427.
Gas phase	*Chem. Rev.* **1991**, *91*, 1375.
Glycoproteins	*Pure & Appl. Chem.* **1981**, *53*, 45.
High pressure	*Chem. Rev.* **1991**, *91*, 1339.
High temperature	*Chem. Rev.* **1991**, *91*, 1353.
Imaging	*Acc. Chem. Res.* **1983**, *16*, 114.
INADEQUATE	*Angew. Chem. Int. Ed. Engl.* **1987**, *26*, 625.
Lanthanide shift reagents	*Chem. Rev.* **1973**, *73*, 553.
	Chem. Soc. Rev. **1973**, *2*, 49.
Linear prediction theory	*Chem. Rev.* **1991**, *91*, 1413.
Liquid crystals	*Acc. Chem. Res.* **1971**, *4*, 81.
Liquids	
Partially aligned liquids	*Acc. Chem. Res.* **1984**, *17*, 172.
Long range hyperfine coupling	*Chem. Rev.* **1976**, *76*, 157.
Macromolecules	*Pure & Appl. Chem.* **1982**, *54*, 559ff.
Membranes	*Acc. Chem. Res.* **1983**, *16*, 266.
	Acc. Chem. Res. **1987**, *20*, 221.
Molecular structure	*Tetrahedron* **1989**, *45*, 581.
Multidimensional	*Chem. Rev.* **1991**, *91*, 1507.
Nitrogen NMR	*Angew. Chem. Int. Ed. Engl.* **1986**, *25*, 383.
	Chem. Rev. **1981**, *81*, 205.
	Tetrahedron **1989**, *45*, 581.
Nucleic acids	*Acc. Chem. Res.* **1983**, *16*, 35.
NMR microscopy	*Angew. Chem. Int. Ed. Engl.* **1990**, *29*, 1.
Nuclear Overhauser effect	*See also nuclear Overhauser effect*
Intramolecular	*Chem. Rev.* **1971**, *71*, 167.
	Chem. Rev. **1971**, *71*, 617.
Oligosaccharides	*Pure & Appl. Chem.* **1983**, *55*, 605.
Organolithium compounds	*Angew. Chem. Int. Ed. Engl.* **1987**, *26*, 1212.
Organometallic chemistry	*Pure & Appl. Chem.* **1986**, *58*, 513.
	Angew. Chem. Int. Ed. Engl. **1986**, *25*, 861.
Reaction mechanisms	*Pure & Appl. Chem.* **1989**, *61*, 699.
Organotin chemistry	*Pure & Appl. Chem.* **1982**, *54*, 29.
Oxygen-17 NMR	*Angew. Chem. Int. Ed. Engl.* **1978**, *17*, 246
Steric pertubation	*Tetrahedron* **1989**, *45*, 3613.

NMR, cont'd.	
Parahydrogen induced polarization	*Acc. Chem. Res.* **1991**, *24*, 110.
Phospholipids	*Acc. Chem. Res.* **1987**, *20*, 221.
Phosphorus NMR	
Membranes	*Acc. Chem. Res.* **1983**, *16*, 266.
Metal complexes	*Chem. Rev.* **1981**, *81*, 229.
Organometallic compounds	*Acc. Chem. Res.* **1981**, *14*, 266.
Polycrystalline solids	*Chem. Rev.* **1975**, *75*, 203.
Polymers	*Angew. Chem. Int. Ed. Engl.* **1988**, *27*, 1468.
Polynucleotides	*Acc. Chem. Res.* **1974**, *7*, 33.
Porphyrins	*Pure & Appl. Chem.* **1981**, *53*, 1215.
Prochiral groups	*Chem. Rev.* **1975**, *75*, 307.
Proteins	*Acc. Chem. Res.* **1978**, *11*, 469.
	Acc. Chem. Res. **1981**, *14*, 291.
	Acc. Chem. Res. **1987**, *20*, 221.
Conformations	*Acc. Chem. Res.* **1989**, *22*, 36.
Histidine residues	*Acc. Chem. Res.* **1975**, *8*, 70.
Pulse methods	*Angew. Chem. Int. Ed. Engl.* **1983**, *22*, 350.
R values	
Ring conformations	*Acc. Chem. Res.* **1971**, *4*, 87.
Reaction mechanisms	*Acc. Chem. Res.* **1978**, *11*, 277.
RNA	*Acc. Chem. Res.* **1974**, *7*, 33.
	Acc. Chem. Res. **1977**, *10*, 396.
	Acc. Chem. Res. **1991**, *24*, 152.
Base pairing	*Acc. Chem. Res.* **1977**, *10*, 396.
Rotational barriers	*Acc. Chem. Res.* **1981**, *14*, 253.
Selective excitation	*Chem. Rev.* **1991**, *91*, 1397.
Sesquiterpenes	*Pure & Appl. Chem.* **1981**, *53*, 1241.
Shift reagents	*Angew. Chem. Int. Ed. Engl.* **1972**, *11*, 675.
Lanthanide shift reagents	*Chem. Rev.* **1973**, *73*, 553.
Sigmatropic shifts	*Acc. Chem. Res.* **1982**, *15*, 2.
Sodium 23 NMR	*Angew. Chem. Int. Ed. Engl.* **1978**, *17*, 254.
Solid state	*Angew. Chem. Int. Ed. Engl.* **1972**, *11*, 607.
	Chem. Rev. **1991**, *91*, 1307.
	Chem. Rev. **1991**, *91*, 1427.
	Chem. Rev. **1991**, *91*, 1545.
	Chem. Soc. Rev. **1986**, *15*, 225.
Reactive intermediates	*Acc. Chem. Res.* **1982**, *15*, 208.
Spin lattice relaxation	*Angew. Chem. Int. Ed. Engl.* **1975**, *14*, 144.
	Chem. Rev: **1991**, *91*, 1591.
Carbohydrates	*Chem. Soc. Rev.* **1975**, *4*, 401.

Spin-spin couplings	
Solvent dependence	*Chem. Rev.* **1973**, *73*, 53.
Spin systems	*Angew. Chem. Int. Ed. Engl.* **1990**, *29*, 374.
Transition metal dihydrogen complexes	*Acc. Chem. Res.* **1990**, *23*, 95.
Zero field	*Acc. Chem. Res.* **1987**, *20*, 47.
Nomenclature	*Angew. Chem. Int. Ed. Engl.* **1988**, *27*, 1050.
Cyclic conjugated polyolefins	*Angew. Chem. Int. Ed. Engl.* **1972**, *11*, 404.
Nodal nomenclature	*Angew. Chem. Int. Ed. Engl.* **1984**, *23*, 33.
Stereochemical	*Angew. Chem. Int. Ed. Engl.* **1982**, *21*, 567.
	Angew. Chem. Int. Ed. Engl. **1982**, *21*, 654.
	Angew. Chem. Int. Ed. Engl. **1986**, *25*, 882.
Non-enzymatic reactions	*Chem. Soc. Rev.* **1980**, 9, 63.
Non-classical carbocations	*Acc. Chem. Res.* **1983**, *16*, 432.
	Acc. Chem. Res. **1983**, *16*, 440.
	Acc. Chem. Res. **1983**, *16*, 448.
	See also carbocations, carbonium ions, and carbenium ions
	Tetrahedron **1976**, *32*, 179.
Norbornyl carbocation	*Acc. Chem. Res.* **1986**, *19*, 34.
	Acc. Chem. Res. **1986**, *19*, 36.
	Acc. Chem. Res. **1973**, 6, 377.
	Acc. Chem. Res. **1976**, 9, 41.
Reaction mechanisms	*Pure & Appl. Chem.* **1982**, *54*, 1783.
Nonbenzenoid phanes	
Aromaticity	*Pure & Appl. Chem.* **1982**, *54*, 957.
Norbornadienes	
Thermolysis	*Acc. Chem. Res.* **1985**, *18*, 248.
Norbornenes	
Electrophilic additions	*Acc. Chem. Res.* **1969**, *2*, 152.
Hydration	*Acc. Chem. Res.* **1985**, *18*, 254.
7-Norbornenyl systems	
Solvolysis	*Acc. Chem. Res.* **1968**, *1*, 239.
19-Norsteroids	*Acc. Chem. Res.* **1976**, 9, 412.
***N*-Oxides**	
Heteroaromatic compounds	*Acc. Chem. Res.* **1976**, 9, 192.
Photochemistry	*Chem. Rev.* **1984**, *84*, 43.
Nuclear Overhauser effect	*Chem. Rev.* **1971**, *71*, 167.
Intramolecular nuclear Overhauser effect	*Chem. Rev.* **1971**, *71*, 617.
Nucleic acid polymerase	
Nucleotides	*Acc. Chem. Res.* **1969**, *2*, 338.

Nucleic acids	*Acc. Chem. Res.* **1979**, *12*, 423.
	Angew. Chem. Int. Ed. Engl. **1973**, *12*, 264.
	Chem. Soc. Rev. **1980**, *9*, 241.
	Tetrahedron **1984**, *40*, 1ff.
Aromatic compounds	
Binding	*Acc. Chem. Res.* **1988**, *21*, 66.
Catenones	*Acc. Chem. Res.* **1973**, *6*, 252.
Conformations	*Acc. Chem. Res.* **1969**, *2*, 257.
Hydrations	*Chem. Rev.* **1987**, *87*, 589.
NMR	*Acc. Chem. Res.* **1983**, *16*, 35.
Organometallic chemistry	*Pure & Appl. Chem.* **1990**, *62*, 613.
Photochemistry	*Acc. Chem. Res.* **1985**, *18*, 134.
	Pure & Appl. Chem. **1973**, *34*, 281.
	Pure & Appl. Chem, **1980**, *52*, 2705.
	Pure & Appl. Chem. **1980**, *52*, 2717.
Protein-nucleic acid interactions	*Chem. Rev.* **1987**, *87*, 981.
Proteins	
Cross-coupling reactions	*Pure & Appl. Chem.* **1980**, *52*, 2717.
Nucleofugal anions	*Acc. Chem. Res.* **1985**, *18*, 154
Nucleofugality	*Acc. Chem. Res.* **1979**, *12*, 198.
Nucleophilic reactions	
Acylations	*Angew. Chem. Int. Ed. Engl.* **1969**, *8*, 639.
	Tetrahedron **1976**, *32*, 1943.
Umpolung	*Synthesis* **1969**, 17.
	Tetrahedron **1976**, *32*, 1943.
Additions	*Org. Reactions* **1991**, *40*, 407.
	SYNLETT **1990**, 564.
Alkenes	
Reaction mechanisms	*Tetrahedron* **1989**, *45*, 4017.
Alkynes	
Organoiron chemistry	*Acc. Chem. Res.* **1988**, *21*, 229.
Anthraquinones	*Tetrahedron* **1990**, *46*, 291.
Aryltrichloromethylcarbinols	*Synthesis* **1971**, 131.
Carbonyl compounds	
Organometallic chemistry	*Angew. Chem. Int. Ed. Engl.* **1991**, *30*, 49.
Reaction mechanisms	*Pure & Appl. Chem.* **1980**, *52*, 545.
Coordinated π-hydrocarbon complexes	
Reaction mechanisms	*Chem. Rev.* **1984**, *84*, 525.
Ketones	
Aromatic ketones	*Acc. Chem. Res.* **1988**, *21*, 414.
Organometallic compounds	
Reaction mechanisms	
Stereochemistry	*Chem. Rev.* **1975**, *75*, 521.

Organometallic chemistry	*Tetrahedron* **1978**, *34*, 3047.
Arenes	*Pure & Appl. Chem.* **1981**, *53*, 2379.
Carbonyl compounds	*Angew. Chem. Int. Ed. Engl.* **1991**, *30*, 49.
Dienes	*Pure & Appl. Chem.* **1981**, *53*, 2379.
Ketones	
Reaction mechanisms	
Stereochemistry	*Chem. Rev.* **1975**, *75*, 521.
Quinones	*Tetrahedron* **1991**, *47*, 8043.
Tetrahydropyridinium salts	*Acc. Chem. Res.* **1984**, *17*, 289.
Alkyl halides	*Acc. Chem. Res.* **1988**, *21*, 414.
Aromatic compounds	
Trifluoromethyl compounds	*Acc. Chem. Res.* **1978**, *11*, 197.
Aryl trichloromethyl carbinols	*Synthesis* **1971**, 131.
Asymmetric synthesis	
Chiral sulfoxides	*Synthesis* **1981**, 185.
Catalytic	
Organosilicon chemistry	*Tetrahedron* **1988**, *44*, 2675.
Displacements	
Aromatic nitro compounds	*Tetrahedron* **1978**, *34*, 2057.
Stereochemistry	*Acc. Chem. Res.* **1970**, *3*, 321.
Eliminations	*Chem. Rev.* **1978**, *78*, 517.
Organotitanium reagents	*Angew. Chem. Int. Ed. Engl.* **1983**, *22*, 31.
Organozirconium reagents	*Angew. Chem. Int. Ed. Engl.* **1983**, *22*, 31.
Pyridines	*Tetrahedron* **1981**, *37*, 3423.
Substitution	*Acc. Chem. Res.* **1987**, *20*, 282.
	Tetrahedron **1988**, *44*, 1.
Alcohols	*Org. Reactions* **1983**, *29*, 1.
Aliphatic compounds	
Reaction mechanisms	*Chem. Soc. Rev.* **1990**, *19*, 83.
Alkynes	*Acc. Chem. Res.* **1976**, *9*, 358.
Aromatic substitution	*Acc. Chem. Res.* **1977**, *10*, 125.
Aryl halides	
Copper catalysis	*Tetrahedron* **1984**, *40*, 1433.
Heterocyclic compounds	*Synthesis* **1991**, 103.
Nitro compounds	*Chem. Soc. Rev.* **1988**, *17*, 285.
Nitroalkanes	*Synthesis* **1991**, 423.
Nitroso compounds	*Chem. Soc. Rev.* **1988**, *17*, 285.
Organometallic chemistry	
Allyl compounds	*Tetrahedron* **1980**, *36*, 1901.
Organosilicon chemistry	*Chem. Rev.* **1990**, *90*, 17.
Organosulfur chemistry	*Chem. Rev.* **1976**, *76*, 747.
Pyridines	
Reaction mechanisms	*Chem. Soc. Rev.* **1984**, *13*, 47.
S_N2	*Acc. Chem. Res.* **1983**, *16*, 363.
Nucleoside phosphorothioates	*Angew. Chem. Int. Ed. Engl.* **1975**, *14*, 160.

Nucleosides	*Angew. Chem. Int. Ed. Engl.* **1973**, *12*, 591.
	Chem. Rev. **1989**, *89*, 503.
	Chem. Soc. Rev. **1977**, 6, 43.
	Tetrahedron **1984**, *40*, 1ff.
Carbocyclic nucleosides	*Tetrahedron* **1992**, *48*, 571.
Structure determination	*Acc. Chem. Res.* **1991**, *24*, 81.
Transition metal complexes	*Acc. Chem. Res.* **1985**, *18*, 32.
Nucleotides	*Angew. Chem. Int. Ed. Engl.* **1973**, *12*, 591.
	Chem. Rev. **1989**, *89*, 503.
	Chem. Soc. Rev. **1977**, 6, 43.
	Tetrahedron **1984**, *40*, 1ff.
Analogues	*Angew. Chem. Int. Ed. Engl.* **1977**, *16*, 695.
Dephosphorylations	*Pure & Appl. Chem.* **1983**, *55*, 137.
Metal ion complexes	*Chem. Rev.* **1980**, *80*, 365.
Nucleic acid polymerase	*Acc. Chem. Res.* **1969**, *2*, 338.
Phosphorothioate analogues	*Acc. Chem. Res.* **1979**, *12*, 204.
	Angew. Chem. Int. Ed. Engl. **1983**, *22*, 423.
Phosphorylations	*Pure & Appl. Chem.* **1980**, *52*, 2213.
Replication	*Angew. Chem. Int. Ed. Engl.* **1990**, *29*, 36.
Nutmeg	*Angew. Chem. Int. Ed. Engl.* **1971**, *10*, 370.
Nutritional control	*Acc. Chem. Res.* **1970**, *3*, 113.
***N*-Vinylcarbazole**	
Polymerization	*Pure & Appl. Chem.* **1973**, *34*, 329.
Nyctinastenes	*Pure & Appl. Chem.* **1982**, *54*, 2501.
***O-D*-alcohols**	*Synthesis* **1972**, 254.
Oil production	
Coal hydrogenation	*Angew. Chem. Int. Ed. Engl.* **1976**, *15*, 341.
2,3-*O*-Isopropylideneglyceraldehyde	
Chirons	*Tetrahedron* **1986**, *42*, 447.
Oligomerization	
Alkenes	*Chem. Rev.* **1991**, *91*, 613.
Alkynes	*Russ. Chem. Rev.* **1974**, *43*, 48.
Palladium catalysis	*Acc. Chem. Res.* **1976**, *9*, 93.
	Acc. Chem. Res. **1976**, *9*, 93.
Oligonucleotides	*Angew. Chem. Int. Ed. Engl.* **1991**, *30*, 296.
	Angew. Chem. Int. Ed. Engl. **1991**, *30*, 613.
	Angew. Chem. Int. Ed. Engl. **1991**, *30*, 822.
	Tetrahedron **1978**, *34*, 3143.
Antisense oligonucleotides	*Chem. Rev.* **1990**, *90*, 543.
Inhibitors	*Angew. Chem. Int. Ed. Engl.* **1991**, *30*, 613.
Ligands	
Random screeening	*Angew. Chem. Int. Ed. Engl.* **1991**, *30*, 296.
Oligonucleotide-drug complexes	*Angew. Chem. Int. Ed. Engl.* **1991**, *30*, 1254.

Phosphoramidite approach	*Tetrahedron* **1992**, *48*, 2223.
Photochemistry	*Angew. Chem. Int. Ed. Engl.* **1991**, *30*, 822.
Probes	*Angew. Chem. Int. Ed. Engl.* **1991**, *30*, 613.
Structural chemistry	*Angew. Chem. Int. Ed. Engl.* **1991**, *30*, 1254.
Synthesis	*Chem. Rev.* **1977**, *77*, 183.
	Synthesis **1975**, 222.
Oligopeptides	*Angew. Chem. Int. Ed. Engl.* **1991**, *30*, 822.
Metal ions	*Pure & Appl. Chem.* **1983**, *55*, 107.
Photochemistry	*Angew. Chem. Int. Ed. Engl.* **1991**, *30*, 822.
Oligosaccharides	*Chem. Soc. Rev.* **1984**, *13*, 15.
	Pure & Appl. Chem. **1987**, *59*, 1447.
	Pure & Appl. Chem. **1987**, *59*, 1465.
Antibody binding sites	*Pure & Appl. Chem.* **1989**, *61*, 1171.
Conformations	*Acc. Chem. Res.* **1986**, *19*, 307.
Glycosides	
Koenigs-Knorr method	*Angew. Chem. Int. Ed. Engl.* **1986**, *25*, 212.
NMR	*Pure & Appl. Chem.* **1983**, *55*, 605.
Oligosaccharide-protein interactions	*Chem. Soc. Rev.* **1989**, *18*, 347.
	Pure & Appl. Chem. **1983**, *55*, 577.
Structure determination	*Pure & Appl. Chem.* **1989**, *61*, 1193.
Structures	*Pure & Appl. Chem.* **1984**, *56*, 1031.
Synthesis	*Angew. Chem. Int. Ed. Engl.* **1982**, *21*, 155.
	Pure & Appl. Chem. **1991**, *63*, 555.
Oncogenes	*Angew. Chem. Int. Ed. Engl.* **1990**, *29*, 707.
	Angew. Chem. Int. Ed. Engl. **1990**, *29*, 716.
Oncology	*Angew. Chem. Int. Ed. Engl.* **1988**, *27*, 1267.
Onium salts	*Angew. Chem. Int. Ed. Engl.* **1979**, *18*, 707.
Optical brighteners	*Angew. Chem. Int. Ed. Engl.* **1975**, *14*, 665.
Optical sensors	
Dyes	*Angew. Chem. Int. Ed. Engl.* **1991**, *30*, 958.
Organic chemistry	*Angew. Chem. Int. Ed. Engl.* **1990**, *29*, 1209.
Technological perspectives	*Pure & Appl. Chem.* **1986**, *58*, 825ff.
Organic metals	*Acc. Chem. Res.* **1974**, *7*, 232.
	Acc. Chem. Res. **1979**, *12*, 79.
	Acc. Chem. Res. **1984**, *17*, 227.
	Acc. Chem. Res. **1985**, *18*, 261.
	Pure & Appl. Chem. **1987**, *59*, 975.
Aromaticity	*Angew. Chem. Int. Ed. Engl.* **1977**, *16*, 519.
Building blocks	*Angew. Chem. Int. Ed. Engl.* **1991**, *30*, 1121.
Metalloporphyrins	*Acc. Chem. Res.* **1983**, *16*, 15.
Organic superconductors	*Acc. Chem. Res.* **1984**, *17*, 227.
	Acc. Chem. Res. **1985**, *18*, 261.

Organic synthesis	
Design	*Acc. Chem. Res.* **1987**, *20*, 237.
Overview	*Angew. Chem. Int. Ed. Engl.* **1990**, *29*, 1320.
Organoactinides	*Acc. Chem. Res.* **1980**, *13*, 276.
Organoalkali metal compounds	
Ether cleavage	*Angew. Chem. Int. Ed. Engl.* **1987**, *26*, 972.
Rearrangements	*Angew. Chem. Int. Ed. Engl.* **1978**, *17*, 313.
Organoaluminum chemistry	*Angew. Chem. Int. Ed. Engl.* **1978**, *17*, 169.
	Angew. Chem. Int. Ed. Engl. **1985**, *24*, 668.
	Org. Reactions **1984**, *32*, 375.
	Pure & Appl. Chem. **1983**, *55*, 1853.
	Pure & Appl. Chem. **1988**, *60*, 21.
	Tetrahedron **1988**, *44*, 5001.
Organoarsinic chemistry	
Ylides	*Chem. Soc. Rev.* **1987**, *16*, 45.
Organobismuth compounds	*Chem. Rev.* **1982**, *82*, 15.
Arylation reactions	*Chem. Rev.* **1989**, *89*, 1487.
Organoboranes	*Chem. Soc. Rev.* **1974**, *3*, 443.
	Pure & Appl. Chem. **1987**, *59*, 907.
	Synthesis **1974**, 77.
	Synthesis **1976**, 633.
	Synthesis **1986**, 973.
	Tetrahedron **1981**, *37*, 3547.
Alcohols	*Acc. Chem. Res.* **1969**, *2*, 65.
Aldehydes	*Acc. Chem. Res.* **1969**, *2*, 65.
Alkenes	
Alcohols	*Org. Reactions* **1963**, *13*, 1.
Allylboranes	*Pure & Appl. Chem.* **1987**, *59*, 895.
Asymmetric synthesis	*Acc. Chem. Res.* **1988**, *21*, 287.
Carbon monoxide	*Acc. Chem. Res.* **1969**, *2*, 65.
Carbonylations	*Acc. Chem. Res.* **1969**, *2*, 65.
Free radicals	
Substitution reactions	*Angew. Chem. Int. Ed. Engl.* **1972**, *11*, 692.
Ketones	*Acc. Chem. Res.* **1969**, *2*, 65.
Labelled compounds	*Acc. Chem. Res.* **1984**, *17*, 215.
Oxidations	*Org. Reactions* **1953**, *13*, 22.
Radical reactions	*Angew. Chem. Int. Ed. Engl.* **1972**, *11*, 692.
Reductions	*Chem. Rev.* **1989**, *89*, 1553.
Retention (of configuration)	*Pure & Appl. Chem.* **1987**, *59*, 879.
Trialkylboranes	*Acc. Chem. Res.* **1976**, *9*, 26.
Organoborates	*Acc. Chem. Res.* **1982**, *15*, 178.
	Chem. Soc. Rev. **1977**, *6*, 393.
	Org. Reactions **1985**, *33*, 1.
	Pure & Appl. Chem. **1991**, *63*, 339.
Photochemistry	*Pure & Appl. Chem.* **1990**, *62*, 1565.

Organoboron chemistry	*Chem. Rev.* **1973**, *73*, 465.
	Chem. Soc. Rev. **1982**, *11*, 191.
	J. Organometal. Chem. **1975**, *100*, 3.
	Org. Reactions **1985**, *33*, 1.
	Pure & Appl. Chem. **1985**, *57*, 1759.
	Pure & Appl. Chem. **1986**, *58*, 629.
	Pure & Appl. Chem. **1987**, *59*, 837.
	Pure & Appl. Chem. **1988**, *60*, 123.
	Pure & Appl. Chem. **1988**, *60*, 1705.
	Pure & Appl. Chem. **1991**, *63*, 307.
	Synthesis **1973**, 635.
	Tetrahedron **1977**, *33*, 2331.
Boraheterocycles	*Tetrahedron* **1977**, *33*, 2331.
Catecholborane	*Tetrahedron* **1976**, *32*, 981.
Diboraheterocycles	*Pure & Appl. Chem.* **1987**, *59*, 947.
Organometallic complexe	*Pure & Appl. Chem.* **1987**, *59*, 847.
Rearrangement reactions	*Synthesis* **1973**, 635.
Organocadmium reagents	*Chem. Rev.* **1978**, *78*, 491.
Organocobaloximes	*Acc. Chem. Res.* **1983**, *16*, 343.
Pyridines	*Angew. Chem. Int. Ed. Engl.* **1985**, *24*, 248.
Organocopper chemistry	*Org. Reactions* **1972**, *19*, 1.
	SYNLETT **1991**, 539.
	Synthesis **1972**, 63.
	Tetrahedron **1989**, *45*, 349ff.
Addition reactions	
Conjugate additions	*Org. Reactions* **1972**, *19*, 1.
Lewis acids	*Angew. Chem. Int. Ed. Engl.* **1986**, *25*, 947.
Organocuprates	*Synthesis* **1972**, 63.
Addition reactions	*Pure & Appl. Chem.* **1983**, *55*, 1759.
Electron transfer reactions	*Acc. Chem. Res.* **1976**, *9*, 59.
Enol triflates	*Acc. Chem. Res.* **1988**, *21*, 47.
Higher order cuprates	*SYNLETT* **1990**, 119.
	Synthesis **1987**, 325.
	Tetrahedron **1984**, *40*, 5005.
Unsaturated compounds	*Pure & Appl. Chem.* **1988**, *60*, 57.
Oxidative couplings	*Angew. Chem. Int. Ed. Engl.* **1974**, *13*, 291.
Substitution reactions	*Org. Reactions* **1975**, *22*, 253.
Unsaturated carbonyl compounds	*Pure & Appl. Chem.* **1984**, *56*, 91.
Vinyloxiranes	*Chem. Rev.* **1989**, *89*, 1503.
Organofluorine chemistry	*Acc. Chem. Res.* **1988**, *21*, 307.
	Angew. Chem. Int. Ed. Engl. **1991**, *30*, 361.
	Chem. Rev. **1986**, *86*, 997.
	Tetrahedron **1987**, *43*, 3123.

Organofluorine chemistry, cont'd.	
Asymmetric synthesis	*Tetrahedron: Asymmetry* **1990**, *1*, 661.
Epoxides	*Angew. Chem. Int. Ed. Engl.* **1985**, *24*, 161.
Thermolytic reactions	*Synthesis* **1976**, 374.
Organogermanium chemistry	*Pure & Appl. Chem.* **1984**, *56*, 137.
	Pure & Appl. Chem. **1991**, *63*, 231.
Germylenes	*Chem. Rev.* **1991**, *91*, 311.
Multiple bonds	*Angew. Chem. Int. Ed. Engl.* **1991**, *30*, 902.
Strained molecules	*Angew. Chem. Int. Ed. Engl.* **1991**, *30*, 958.
Organoiron chemistry	*Acc. Chem. Res.* **1974**, *7*, 122.
	Acc. Chem. Res. **1974**, *7*, 122.
	Pure & Appl. Chem. **1983**, *55*, 1767.
	Pure & Appl. Chem. **1984**, *56*, 129.
Acyliron complexes	
Cycloaddition reactions	*SYNLETT* **1991**, 1.
Alkenes	*Pure & Appl. Chem.* **1982**, *54*, 145.
Alkynes	
Nucleophilic addition	*Acc. Chem. Res.* **1988**, *21*, 229.
Aromatic complexes	*Tetrahedron* **1983**, *39*, 4027.
Asymmetric synthesis	*Pure & Appl. Chem.* **1988**, *60*, 13.
Butadiene-iron tricarbonyl complexes	*Synthesis* **1989**, 341.
Dienes	*Acc. Chem. Res.* **1980**, *13*, 463.
Ferrocenylphosphine metal complexes	
Asymmetric synthesis	*Pure & Appl. Chem.* **1988**, *60*, 7.
Polybromoketones	*Acc. Chem. Res.* **1979**, *12*, 61.
Tetracarbonylhydridoferrates	*Chem. Rev.* **1990**, *90*, 1041.
Tricarbonyl(diene)iron complexes	*Acc. Chem. Res.* **1980**, *13*, 463.
Organolanthanide chemistry	*Acc. Chem. Res.* **1980**, *13*, 276.
	Angew. Chem. Int. Ed. Engl. **1984**, *23*, 474.
Alkenes	
Polymerization	*Acc. Chem. Res.* **1985**, *18*, 51.
Organolithium chemistry	*Acc. Chem. Res.* **1982**, *15*, 300.
	Acc. Chem. Res. **1986**, *19*, 356.
	Chem. Rev. **1990**, *90*, 1061.
	Chem. Rev. **1991**, *91*, 137.
	Org. Reactions **1954**, *8*, 258.
	Synthesis **1975**, 83.
1,3-Anionic cycloadditions	*Angew. Chem. Int. Ed. Engl.* **1974**, *13*, 627.
Aromatic compounds	*Synthesis* **1983**, 957.
Amines	
Tertiary amines	*Acc. Chem. Res.* **1982**, *15*, 306.

Carbenoids	*Angew. Chem. Int. Ed. Engl.* **1972**, *11*, 473.
Carboxylic acids	*Org. Reactions* **1970**, *18*, 1.
Copper catalysis	*Tetrahedron* **1984**, *40*, 641.
Cycloadditions	*Angew. Chem. Int. Ed. Engl.* **1974**, *13*, 627.
Exchange reactions	*Acc. Chem. Res.* **1968**, *1*, 23.
Functionalized organolithiums	*Org. Reactions* **1979**, 26, 1.
Halogen-lithium exchange	*Org. Reactions* **1951**, 6, 339.
Alkyl halides	*Acc. Chem. Res.* **1982**, *15*, 300.
Heterocycles	*Synthesis* **1983**, 957.
Metallations	*Chem. Rev.* **1969**, 69, 693.
	Org. Reactions **1954**, *8*, 258.
NMR	*Angew. Chem. Int. Ed. Engl.* **1987**, 26, 1212.
Reaction mechanisms	*Chem. Soc. Rev.* **1991**, *20*, 167.
Structures	*Pure & Appl. Chem.* **1983**, *55*, 355.
	Pure & Appl. Chem. **1984**, *56*, 151.
Organomanganese chemistry	*SYNLETT* **1990**, 564.
Organomercury chemistry	*Angew. Chem. Int. Ed. Engl.* **1978**, *17*, 27.
	Angew. Chem. Int. Ed. Engl. **1985**, *24*, 553.
	Tetrahedron **1982**, *38*, 1713.
Organomercury lyase	*Pure & Appl. Chem.* **1987**, *59*, 295.
Organometallic chemistry	*Acc. Chem. Res.* **1977**, *10*, 301.
	Angew. Chem. Int. Ed. Engl. **1974**, *13*, 701.
	Angew. Chem. Int. Ed. Engl. **1987**, *26*, 723.
	Angew. Chem. Int. Ed. Engl. **1991**, *30*, 931.
	Angew. Chem. Int. Ed. Engl. **1991**, *30*, 1119.
	Chem. Rev. **1972**, *72*, 545.
	Chem. Rev. **1987**, *87*, 319.
	Chem. Rev. **1988**, *88*, 1031.
	Chem. Soc. Rev. **1973**, *2*, 271.
	Chem. Soc. Rev. **1980**, *9*, 25.
	Pure & Appl. Chem. **1981**, *53*, 2333.
	Pure & Appl. Chem. **1981**, *53*, 2357.
	Pure & Appl. Chem. **1982**, *54*, 59.
	Pure & Appl. Chem. **1982**, *54*, 177.
	Pure & Appl. Chem. **1985**, *57*, 1809.
	Pure & Appl. Chem. **1986**, *58*, 481.
	Pure & Appl. Chem. **1987**, *59*, 847.
	Pure & Appl. Chem. **1989**, *61*, 1665.
	Pure & Appl. Chem. **1989**, *61*, 1681.
	Pure & Appl. Chem. **1990**, *62*, 731.
	See also transition metal chemistry
	SYNLETT **1990**, 10.
	SYNLETT **1990**, 441.

(Cont'd.)

Organometallic chemistry, cont'd.

	SYNLETT **1990**, 493.
	Tetrahedron **1975**, *31*, 2735.
	Tetrahedron **1978**, *34*, 2827.
	Top. Curr. Chem. **1980**, *92*, 109.
Acyl complexes	*Chem. Soc. Rev.* **1988**, *17*, 147.
Addition reactions	
Ketones	*Chem. Rev.* **1975**, *75*, 521.
Alkanes	*Chem. Rev.* **1985**, *85*, 245.
	Chem. Soc. Rev. **1982**, *11*, 283.
	Pure & Appl. Chem, **1980**, *52*, 649.
Activation	*Chem. Rev.* **1990**, *90*, 403.
Functionalization	*Pure & Appl. Chem.* **1990**, *62*, 1539.
Alkenes	*Chem. Rev.* **1973**, *73*, 163.
	Chem. Rev. **1988**, *88*, 1047.
	Pure & Appl. Chem. **1986**, *58*, 495.
	Pure & Appl. Chem. **1988**, *60*, 65.
Intramolecular complexes	*Angew. Chem. Int. Ed. Engl.* **1982**, *21*, 889.
Alkyl transfers	*Acc. Chem. Res.* **1974**, *7*, 351.
Alkylations	*Pure & Appl. Chem.* **1982**, *54*, 189.
Alkyltransition metal complexes	
Photochemistry	*Angew. Chem. Int. Ed. Engl.* **1984**, *23*, 766.
Alkyne clusters	*Chem. Rev.* **1983**, *83*, 203.
Alkynes	*Acc. Chem. Res.* **1987**, *20*, 65.
	Chem. Rev. **1973**, *73*, 163.
	Chem. Rev. **1988**, *88*, 1047.
	Pure & Appl. Chem. **1982**, *54*, 113.
	Pure & Appl. Chem. **1990**, *62*, 1021.
π-Allyl complexes	*Chem. Soc. Rev.* **1985**, *14*, 93.
	Pure & Appl. Chem. **1982**, *54*, 197.
Allylic alkylations	*Pure & Appl. Chem.* **1982**, *54*, 189.
Anions	*Acc. Chem. Res.* **1988**, *21*, 147.
Arbuzov-like dealkylations	*Chem. Rev.* **1984**, *84*, 215.
	Chem. Rev. **1984**, *84*, 577.
Arene complexes	*Angew. Chem. Int. Ed. Engl.* **1985**, *24*, 893.
	Chem. Rev. **1982**, *82*, 499.
Arenes	*Pure & Appl. Chem.* **1981**, *53*, 2379.
Aromaticity	*Pure & Appl. Chem.* **1990**, *62*, 383.
Asymmetric synthesis	*Angew. Chem. Int. Ed. Engl.* **1971**, *10*, 249.
	Pure & Appl. Chem. **1983**, *55*, 1781.
	Pure & Appl. Chem. **1985**, *57*, 1883.
	Pure & Appl. Chem. **1988**, *60*, 1607.
Benzynes	*Chem. Rev.* **1988**, *88*, 1047.
Biological aspects	*Chem. Soc. Rev.* **1981**, *10*, 455.
C_1 ligands	*Angew. Chem. Int. Ed. Engl.* **1991**, *30*, 1119.

Carbene complexes	*Pure & Appl. Chem.* **1990**, *62*, 691.
	Chem. Rev. **1991**, *91*, 197.
	Pure & Appl. Chem. **1990**, *62*, 1021.
Carbides	*Pure & Appl. Chem.* **1990**, *62*, 1021.
Carbon dioxide	*Chem. Rev.* **1988**, *88*, 747.
Carbonyl cluster compounds	*Pure & Appl. Chem.* **1982**, *54*, 97.
Carbonyl compounds	
Nucleophilic additions	*Angew. Chem. Int. Ed. Engl.* **1991**, *30*, 49.
Carbonylation	*Synthesis* **1985**, 253.
	Chem. Rev. **1988**, *88*, 1011.
Catalysis	*Pure & Appl. Chem.* **1980**, *52*, 607.
Charge transfer complexes	
Reaction mechanisms	*Pure & Appl. Chem.* **1991**, *63*, 255.
Chiral complexes	*Chem. Rev.* **1987**, *87*, 761.
Cluster compounds	*Pure & Appl. Chem.* **1982**, *54*, 113.
	Pure & Appl. Chem. **1982**, *54*, 131.
Theoretical chemistry	*Pure & Appl. Chem,* **1980**, *52*, 705.
Cycloalkynes	*Chem. Rev.* **1988**, *88*, 1047.
Cyclobutadiene metal complexes	*Chem. Rev.* **1977**, *77*, 691.
Diazoalkanes	*Angew. Chem. Int. Ed. Engl.* **1978**, *17*, 800.
Dienes	*Pure & Appl. Chem.* **1981**, *53*, 2379.
π-Donors	*Angew. Chem. Int. Ed. Engl.* **1990**, *29*, 627.
Electrochemistry	*Acc. Chem. Res.* **1972**, *5*, 415.
	Synthesis **1973**, 377.
Electron transfer reactions	*Acc. Chem. Res.* **1985**, *18*, 160.
Reaction mechanisms	*Pure & Appl. Chem,* **1980**, *52*, 571.
Encapsulated metal ions	*Pure & Appl. Chem.* **1986**, *58*, 1511.
Enolates	*Pure & Appl. Chem.* **1985**, *57*, 1789.
	Pure & Appl. Chem. **1988**, *60*, 1.
Enzymes	
Proteins	*Angew. Chem. Int. Ed. Engl.* **1991**, *30*, 931.
Ferrocenes	*Org. Reactions* **1969**, *17*, 1.
Fischer-Tropsch synthesis	*Angew. Chem. Int. Ed. Engl.* **1982**, *21*, 117.
Fluorinated organometallic reagents	*Tetrahedron* **1992**, *48*, 189.
Fluorine compounds	*Chem. Rev.* **1991**, *91*, 553.
Free radical reactions	*Pure & Appl. Chem.* **1986**, *58*, 575.
Fulvenes	*Synthesis* **1970**, 449.
Gas phase	*Chem. Rev.* **1991**, *91*, 1121.
Gold compounds	*Chem. Soc. Rev.* **1980**, *9*, 217.
Group 4-7 transition metals	*Chem. Rev.* **1976**, *76*, 243.
	Pure & Appl. Chem. **1984**, *56*, 47.
Complexes	*Angew. Chem. Int. Ed. Engl.* **1983**, *22*, 927.
Group 6 metals	*Pure & Appl. Chem.* **1988**, *60*, 1017.

Organometallic chemistry, cont'd.	
Group 8 metals	*Pure & Appl. Chem.* **1980**, *52*, 635.
Oxidations	*Pure & Appl. Chem.* **1981**, *53*, 2389.
Half-sandwich compounds	*Angew. Chem. Int. Ed. Engl.* **1983**, *22*, 927.
Heavy main group elements	*Angew. Chem. Int. Ed. Engl.* **1982**, *21*, 410.
Heterocycles	*Chem. Rev.* **1973**, *73*, 293.
α-Heterosubstituted reagents	*Tetrahedron* **1980**, *36*, 2531.
HSAB	*Tetrahedron* **1985**, *41*, 3.
Hydride transfers	
Reaction mechanisms	*Pure & Appl. Chem.* **1988**, *60*, 131.
Imines	*Pure & Appl. Chem.* **1990**, *62*, 605.
Indoles	*Angew. Chem. Int. Ed. Engl.* **1988**, *27*, 113.
Ketones	
Nucleophilic additions	
Reaction mechanisms	
Stereochemistry	*Chem. Rev.* **1975**, *75*, 521.
β-Lactams	*Tetrahedron* **1988**, *44*, 5615.
Metal alkenes	*Angew. Chem. Int. Ed. Engl.* **1972**, *11*, 596.
Metal alkynes	*Angew. Chem. Int. Ed. Engl.* **1972**, *11*, 596.
Metal bases	*Angew. Chem. Int. Ed. Engl.* **1983**, *22*, 927.
Metal carbonyls	*Pure & Appl. Chem.* **1980**, *52*, 607.
Metal catalysis	*Pure & Appl. Chem.* **1981**, *53*, 2419.
Metal clusters	*Pure & Appl. Chem.* **1986**, *58*, 529.
Metal surfaces	*Pure & Appl. Chem.* **1982**, *54*, 83.
Metallacycles	*Angew. Chem. Int. Ed. Engl.* **1987**, *26*, 990.
Metalla-β-diketones	*Acc. Chem. Res.* **1981**, *14*, 109.
Metalla-enes	*Angew. Chem. Int. Ed. Engl.* **1991**, *30*, 673.
Metallaoxetanes	
Oxygen transfer	*Chem. Rev.* **1990**, *90*, 1483.
Metalla-ynes	*Angew. Chem. Int. Ed. Engl.* **1991**, *30*, 673.
Metallocarboranes	*Pure & Appl. Chem.* **1982**, *54*, 43.
Metallocenes	*Acc. Chem. Res.* **1979**, *12*, 415.
	Acc. Chem. Res. **1980**, *13*, 276.
	Chem. Soc. Rev. **1988**, *17*, 453.
	Pure & Appl. Chem. **1984**, *56*, 63.
	Pure & Appl. Chem. **1991**, *63*, 813.
[m,m]Metallocenophanes	*Angew. Chem. Int. Ed. Engl.* **1986**, *25*, 702.
Methylene bridges	*Pure & Appl. Chem.* **1982**, *54*, 65.
	Pure & Appl. Chem. **1984**, *56*, 59.
Michael additions	
Vinyl sulfoxides	*Pure & Appl. Chem.* **1981**, *53*, 2307.
Molecular orbitals	*Tetrahedron* **1982**, *38*, 1339.
Multidentate ligands	*Angew. Chem. Int. Ed. Engl.* **1979**, *18*, 753.
Natural products	*Tetrahedron* **1985**, *41*, 5741ff.
Nitrosamines	*Angew. Chem. Int. Ed. Engl.* **1975**, *14*, 15.

NMR	*Angew. Chem. Int. Ed. Engl.* **1986**, *25*, 861.
	Chem. Rev. **1981**, *81*, 205.
	Pure & Appl. Chem. **1986**, *58*, 513.
Reaction mechanisms	*Pure & Appl. Chem.* **1989**, *61*, 699.
Nucleic acids	*Pure & Appl. Chem.* **1990**, *62*, 613.
Nucleophilic additions	*Tetrahedron* **1978**, *34*, 3047.
Carbonyl compounds	
Asymmetric synthesis	*Angew. Chem. Int. Ed. Engl.* **1991**, *30*, 49.
Ketones	
Reaction mechanisms	
Stereochemistry	*Chem. Rev.* **1975**, *75*, 521.
Nucleophilic substitution	
Allyl compounds	*Tetrahedron* **1980**, *36*, 1901.
Oxidations	*Organometal. Chem. Rev. (A)* **1970**, 6, 209.
	Russ. Chem. Rev. **1966**, *35*, 613.
Group IV compounds	*Organometal. Chem. Rev. (A)* **1970**, 6, 209.
Reaction mechanisms	*Pure & Appl. Chem.* **1984**, *56*, 35.
Oxidative cleavage	*Progr. Inorg. Chem.* **1977**, 22, 409.
Phosphazenes	*SYNLETT* **1990**, 651.
Phosphorus ligands	
NMR	*Acc. Chem. Res.* **1981**, *14*, 266.
Steric effects	*Chem. Rev.* **1977**, *77*, 313.
Photolytic reactions	*Pure & Appl. Chem.* **1982**, *54*, 23.
Porphyrins	*Acc. Chem. Res.* **1986**, *19*, 209.
	Chem. Rev. **1988**, *88*, 1121.
	Pure & Appl. Chem, **1980**, *52*, 681.
Radiolabeled compounds	*Tetrahedron* **1989**, *45*, 6601.
Reaction mechanisms	*Pure & Appl. Chem,* **1980**, *52*, 741.
	Pure & Appl. Chem. **1981**, *53*, 161.
Photochemistry	*Pure & Appl. Chem.* **1982**, *54*, 161.
Rearrangements	*Acc. Chem. Res.* **1968**, *1*, 257.
	Pure & Appl. Chem. **1990**, *62*, 1933.
Remote functionalization	*Acc. Chem. Res.* **1989**, *22*, 282.
Sandwich compounds	*Angew. Chem. Int. Ed. Engl.* **1977**, *16*, 1.
	Angew. Chem. Int. Ed. Engl. **1991**, *30*, 407.
	Chem. Rev. **1986**, *86*, 957.
	J. Organomet. Chem. **1980**, *200*, 335.
	SYNLETT **1991**, 369.
Historical perspectives	*J. Organometal. Chem.* **1975**, *100*, 273.
Steroidal hormones	*Pure & Appl. Chem.* **1986**, *58*, 597.
Structural chemistry	*Angew. Chem. Int. Ed. Engl.* **1990**, *29*, 256.
Substitution reactions	*Acc. Chem. Res.* **1983**, *16*, 350.
Surface chemistry	*Acc. Chem. Res.* **1977**, *10*, 287.
Tetracyanoethylene	*Synthesis* **1987**, 959.
Transition metal compounds	*Acc. Chem. Res.* **1987**, *20*, 379.

Organomethyl compounds	*Chem. Soc. Rev.* **1980**, *9*, 25.
Organonickel chemistry	*Angew. Chem. Int. Ed. Engl.* **1988**, *27*, 185.
Organopalladium chemistry	*Pure & Appl. Chem.* **1990**, *62*, 1867.
	Acc. Chem. Res. **1983**, *16*, 335.
	Synthesis **1970**, 225.
	Synthesis **1985**, 233.
	Tetrahedron **1977**, *33*, 2615.
Alkenes	
Addition reactions	*Chem. Rev.* **1989**, *89*, 1433.
π-Allyl complexes	*Pure & Appl. Chem.* **1989**, *61*, 1673.
Organophosphorus chemistry	*Acc. Chem. Res.* **1969**, *2*, 353.
	Acc. Chem. Res. **1969**, *2*, 373.
	Angew. Chem. Int. Ed. Engl. **1976**, *15*, 468.
	Angew. Chem. Int. Ed. Engl. **1982**, *21*, 492.
	Chem. Soc. Rev. **1974**, *3*, 87.
Alkenes	*Org. Reactions* **1977**, *25*, 73.
	Org. Reactions **1988**, *36*, 175.
Arbuzov-like dealkylations	*Chem. Rev.* **1984**, *84*, 215.
Asymmetric synthesis	*Pure & Appl. Chem.* **1980**, *52*, 843.
Reductions	*Pure & Appl. Chem.* **1980**, *52*, 843.
Carbanions	
Wittig reaction	*Chem. Rev.* **1974**, *74*, 87.
Carbonyl compounds	*Org. Reactions* **1988**, *36*, 175.
Cyclic phosphorus esters	
Asymmetric synthesis	*Tetrahedron* **1980**, *36*, 2059.
Dichloro(methyl)phosphane	*Angew. Chem. Int. Ed. Engl.* **1981**, *20*, 223.
Heterocycles	*Acc. Chem. Res.* **1972**, *5*, 204.
	Chem. Rev. **1971**, *71*, 315.
	Chem. Rev. **1977**, *77*, 121.
Peroxides	*Chem. Rev.* **1981**, *81*, 49.
Phosphaallene ylides	*Angew. Chem. Int. Ed. Engl.* **1977**, *16*, 349.
Phosphacumulene ylides	*Angew. Chem. Int. Ed. Engl.* **1977**, *16*, 349.
Phosphanediyls	*Angew. Chem. Int. Ed. Engl.* **1975**, *14*, 523.
Phosphinidenes	*Angew. Chem. Int. Ed. Engl.* **1975**, *14*, 523.
Phosphorus-bridging carbonyl derivatives	*SYNLETT* **1991**, 671.
Phosphorus-carbon bond cleavage	*Chem. Rev.* **1985**, *85*, 171.
Phosphorus-carbon unsaturation	*Pure & Appl. Chem.* **1987**, *59*, 977.
Spin-labelled compounds	*Synthesis* **1981**, 682.
Substitution reactions	*Angew. Chem. Int. Ed. Engl.* **1973**, *12*, 91.
Three-membered phosphorus heterocycles	*Chem. Rev.* **1990**, *90*, 997.
Ylides	*Acc. Chem. Res.* **1974**, *7*, 6.
	Acc. Chem. Res. **1975**, *8*, 62.

Organoplatinum chemistry	
Cyclopentadienyl platinum compounds	*Chem. Soc. Rev.* **1981**, *10*, 1.
Organorhodium chemistry	
Alkanes	
Activation	*Acc. Chem. Res.* **1985**, *18*, 302.
Organoscandium chemistry	
Lewis acids	*SYNLETT* **1990**, 74.
Organoselenium chemistry	*Acc. Chem. Res.* **1968**, *1*, 202.
	Acc. Chem. Res. **1979**, *12*, 22.
	Acc. Chem. Res. **1984**, *17*, 28.
	Tetrahedron **1978**, *34*, 1049.
	Tetrahedron **1985**, *41*, 4727ff.
Alkenes	*Acc. Chem. Res.* **1984**, *17*, 28.
Extrusion reactions	*Tetrahedron* **1988**, *44*, 6241.
Organosilicon chemistry	*Chem. Soc. Rev.* **1978**, *7*, 15
	Chem. Soc. Rev. **1981**, *10*, 83.
	Pure & Appl. Chem. **1983**, *55*, 1707.
	Pure & Appl. Chem. **1988**, *60*, 71.
	Pure & Appl. Chem. **1991**, *63*, 231.
	See also under specific compounds
	Tetrahedron **1983**, *39*, 839ff.
	Tetrahedron **1988**, *44*, 281.
	Tetrahedron **1988**, *44*, 3761ff.
Alkenes	*Acc. Chem. Res.* **1977**, *10*, 442.
Asymmetric synthesis	*Pure & Appl. Chem.* **1990**, *62*, 1879.
Bioactivity	*Top. Curr. Chem.* **1979**, *84*, 1.
Brook rearrangement	*Acc. Chem. Res.* **1974**, *7*, *77*.
Carbocations	*Tetrahedron* **1990**, *46*, 2677.
Carbosilanes	*Angew. Chem. Int. Ed. Engl.* **1987**, *26*, 1111.
3-Chloropropyltrialkoxysilanes	*Angew. Chem. Int. Ed. Engl.* **1986**, *25*, 236.
Coupling reactions	*SYNLETT* **1991**, 845.
Cyclic organosilicon compounds	
Reaction mechanisms	*Pure & Appl. Chem,* **1980**, *52*, 615.
Disilenes	*Pure & Appl. Chem.* **1984**, *56*, 163.
Double bonds	*Acc. Chem. Res.* **1975**, *8*, 18.
	Acc. Chem. Res. **1982**, *15*, 283.
Electrophilic substitition	*Acc. Chem. Res.* **1977**, *10*, 1.
	Org. Reactions **1989**, *37*, 57.
	Synthesis **1979**, 761.
Heteroaryltrimethylsilanes	*Synthesis* **1979**, 841.
Hindered compounds	*Pure & Appl. Chem.* **1986**, *58*, 623.
Historical perspective	*J. Organomet. Chem.* **1980**, *200*, 11.
Hypervalent silicon compounds	*Pure & Appl. Chem.* **1988**, *60*, 99.

Organosilicon chemistry, cont'd.	
Multiple bonds	*Angew. Chem. Int. Ed. Engl.* **1991**, *30*, 902.
	Chem. Rev. **1985**, *85*, 419.
Nucleophilic catalysis	*Tetrahedron* **1988**, *44*, 2675.
Nucleophilic substitution	
Stereochemistry	*Chem. Rev.* **1990**, *90*, 17.
Peterson reaction	*Acc. Chem. Res.* **1977**, *10*, 442.
	Org. Reactions **1990**, *38*, 1.
	SYNLETT **1991**, 764.
Alkenes	*Synthesis* **1984**, 384.
Protecting groups	*Synthesis* **1985**, 817.
Radical cations	*Acc. Chem. Res.* **1982**, *15*, 9.
Rearrangement reactions	*Acc. Chem. Res.* **1974**, *7*, 77.
Silacyclopropanes	*J. Organometal. Chem.* **1975**, *100*, 237.
Silenes	
Photochemistry	*Acc. Chem. Res.* **1987**, *20*, 329.
Silicon multiple bonds	*Angew. Chem. Int. Ed. Engl.* **1987**, *26*, 1201.
Silicon-carbon bond cleavage	*Synthesis* **1984**, 991.
Silicon-proton exchange	*Acc. Chem. Res.* **1970**, *3*, 299.
Silyl anions	*Acc. Chem. Res.* **1987**, *20*, 127.
Silylations	*Acc. Chem. Res.* **1970**, *3*, 299.
Silyl-proton exchange	*Acc. Chem. Res.* **1970**, *3*, 299.
Silyl-substituted cyclopropanes	*Chem. Rev.* **1986**, *86*, 755.
Steric effects	*Chem. Rev.* **1989**, *89*, 1599.
Strained molecules	*Angew. Chem. Int. Ed. Engl.* **1991**, *30*, 958.
Synthons	*Top. Curr. Chem.* **1980**, *88*, 33.
Umpolung	
Carbonyl equivalents	*Chem. Soc. Rev.* **1982**, *11*, 493.
Vinylsilanes	*Chem. Rev.* **1986**, *86*, 857.
	Org. Reactions **1989**, *37*, 57.
Ylides	*Acc. Chem. Res.* **1975**, *8*, 62.
Organostibines	
Coordination chemistry	*Acc. Chem. Res.* **1978**, *11*, 363.
Organosulfur chemistry	*Acc. Chem. Res.* **1978**, *11*, 453.
	Chem. Soc. Rev. **1989**, *18*, 123.
	Pure & Appl. Chem. **1990**, *62*, 1949.
	See also under specific compounds
	Synthesis **1972**, 101.
	Synthesis **1978**, 713.
	Tetrahedron **1988**, *44*, 281.
Asymmetric synthesis	*Tetrahedron* **1974**, *30*, 1503.
Carbon-sulfur bond cleavage	*Synthesis* **1990**, 89.
Chiral sulfoxides	*Synthesis* **1981**, 185.

Cleavage reactions	*Chem. Rev.* **1951**, *49*, 1.
Cycloaddition reactions	
Alkenes	*Tetrahedron* **1988**, *44*, 6755.
Dithioacetals	*Acc. Chem. Res.* **1991**, *24*, 257.
Double functional compounds	*Synthesis* **1988**, 95.
Extrusion reactions	*Tetrahedron* **1988**, *44*, 6241.
Heterocycles	
Anomeric effect	*Acc. Chem. Res.* **1989**, *22*, 357.
Ring-transformations	*Tetrahedron* **1982**, *38*, 3537.
Metal catalysis	*Chem. Soc. Rev.* **1977**, 6, 345.
Nucleophilic substitution	*Chem. Rev.* **1976**, *76*, 747.
Photochemistry	*Chem. Soc. Rev.* **1975**, *4*, 523.
Ring expansion	*Acc. Chem. Res.* **1984**, *17*, 358.
Sulfenes	*Acc. Chem. Res.* **1975**, *8*, 10.
Sulfides	*Acc. Chem. Res.* **1978**, *11*, 453.
Sulfinyl compounds	
Pummerer reaction	*Org. Reactions* **1991**, *40*, 157.
Sulfones	*Tetrahedron* **1977**, *33*, 2019.
	Tetrahedron **1987**, *43*, 1027.
Sulfoxides	*Acc. Chem. Res.* **1978**, *11*, 453.
	SYNLETT **1990**, 643.
Sulfoximines	*Chem. Soc. Rev.* **1975**, *4*, 189.
	Chem. Soc. Rev. **1980**, *9*, 477.
Sulfur ylides	*Acc. Chem. Res.* **1977**, *10*, 179.
	Angew. Chem. Int. Ed. Engl. **1983**, *22*, 516.
Sulfur-carbon multiple bonds	
Reactive intermediates	*Angew. Chem. Int. Ed. Engl.* **1991**, *30*, 361.
Sulfur-sulfur double bonds	*Chem. Rev.* **1982**, *82*, 333.
Umpolung reagents	
Carbonyl equivalents	*Synthesis* **1977**, 357.
Vinyl sulfones	*Tetrahedron* **1990**, *46*, 6951.
Conjugate additions	*Chem. Rev.* **1986**, *86*, 903.
Organotantalum chemistry	
Reaction mechanisms	*Pure & Appl. Chem,* **1980**, *52*, 729.
Organotellurium chemistry	*Acc. Chem. Res.* **1985**, *18*, 274.
Extrusion reactions	*Tetrahedron* **1988**, *44*, 6241.
Organothallium chemistry	*Acc. Chem. Res.* **1970**, *3*, 338.
	Quart. Rev. **1970**, *24*, 310.
Organothiophosphorus reagents	*Tetrahedron* **1985**, *41*, 2567.
Organotin chemistry	*Acc. Chem. Res.* **1973**, 6, 198.
	Angew. Chem. Int. Ed. Engl. **1985**, *24*, 553.
	Pure & Appl. Chem. **1981**, *53*, 2401.
	Pure & Appl. Chem. **1986**, *58*, 505.
	Tetrahedron **1989**, *45*, 909ff.

Organotin chemistry, cont'd.	
Alkoxides	*Synthesis* **1969**, 56.
Chiral compounds	*Acc. Chem. Res.* **1973**, 6, 198.
Coupling reactions	*Pure & Appl. Chem.* **1985**, *57*, 1771.
Enol triflates	*Acc. Chem. Res.* **1988**, *21*, 47.
Palladium catalysis	*Angew. Chem. Int. Ed. Engl.* **1986**, *25*, 508.
Hydrides	*Acc. Chem. Res.* **1968**, *1*, 299.
	Synthesis **1969**, 56.
	Synthesis **1987**, 665.
Reductions	*Synthesis* **1970**, 499.
Multiple bonds	*Angew. Chem. Int. Ed. Engl.* **1991**, *30*, 902.
NMR	*Pure & Appl. Chem.* **1982**, *54*, 29.
Organotin alkoxides	*Synthesis* **1969**, 56.
Organotin oxides	*Synthesis* **1969**, 56.
Stannylenes	*Chem. Rev.* **1991**, *91*, 311.
Strained molecules	*Angew. Chem. Int. Ed. Engl.* **1991**, *30*, 958.
Substitution reactions	*Acc. Chem. Res.* **1983**, *16*, 177.
Asymmetric synthesis	*Pure & Appl. Chem.* **1980**, *52*, 657.
Trialkylstannylmethyllithium	*Synthesis* **1990**, 259.
Trialkylstannyllithium	*Synthesis* **1990**, 259.
Organotitanium chemistry	*Angew. Chem. Int. Ed. Engl.* **1983**, *22*, 31.
	Pure & Appl. Chem. **1985**, *57*, 1781.
	Tetrahedron **1992**, *48*, 5557ff.
Low valent titanium	*Acc. Chem. Res.* **1974**, *7*, 281.
Nitrogen fixation	*Acc. Chem. Res.* **1970**, *3*, 361.
Organozinc chemistry	*Tetrahedron* **1987**, *43*, 2203.
Organozirconium chemistry	*Pure & Appl. Chem.* **1980**, *52*, 733.
	Synthesis **1988**, 1.
	Angew. Chem. Int. Ed. Engl. **1983**, *22*, 31.
Origin of Life	*Angew. Chem. Int. Ed. Engl.* **1981**, *20*, 500.
	Angew. Chem. Int. Ed. Engl. **1973**, *12*, 349.
Orthoesters	*Synthesis* **1977**, 73.
Carbenium ions	*Chem. Soc. Rev.* **1987**, *16*, 75.
Carboxylic acids	*Synthesis* **1974**, 153.
Hydrolysis	
Reaction mechanisms	*Chem. Rev.* **1974**, *74*, 581.
Orthometalations	
Aromatic amides	*Chem. Rev.* **1990**, *90*, 879.
Orthophosphate esters	
Hydrolysis	*Acc. Chem. Res.* **1970**, *3*, 257.
	Acc. Chem. Res. **1970**, *3*, 267.
Orthoquinodimethanes	*Tetrahedron* **1987**, *43*, 2873.
Orthosomycins	*Tetrahedron* **1979**, *35*, 1207.

Oscillating reactions	*Acc. Chem. Res.* **1987**, *20*, 186.
	Acc. Chem. Res. **1988**, *21*, 326.
	Acc. Chem. Res. **1990**, *23*, 258.
	Angew. Chem. Int. Ed. Engl. **1978**, *17*, 1.
Chaos	*Acc. Chem. Res.* **1987**, *20*, 436.
Osmium chemistry	
Carbonyl cluster compounds	*Pure & Appl. Chem.* **1982**, *54*, 97.
Macromolecules	*Pure & Appl. Chem.* **1988**, *60*, 495.
Osmium tetroxide	
Alkenes	*Chem. Rev.* **1980**, *80*, 187.
***O,S*-Sulfenyl sulfinates**	*Chem. Rev.* **1984**, *84*, 117.
Oxacarbenes	*Acc. Chem. Res.* **1975**, *8*, 209.
Oxa-di-π-methane	
Rearrangements	*Chem. Rev.* **1973**, *73*, 531.
Oxathiazinone dioxides	*Angew. Chem. Int. Ed. Engl.* **1973**, *12*, 869.
1,3-Oxazines	*Synthesis* **1972**, 333.
Oxaziridines	
Aminations	*Synthesis* **1991**, 327.
Asymmetric synthesis	*Tetrahedron* **1989**, *45*, 5703.
Oxazoles	*Chem. Rev.* **1975**, *75*, 389.
1,2-Oxazoles	*Synthesis* **1975**, 20.
Protecting groups	
Carboxylate compounds	*Chem. Rev.* **1986**, *86*, 845.
Oxazolidines	
Asymmetric synthesis	*Pure & Appl. Chem.* **1988**, *60*, 1689.
Oxazolines	*Angew. Chem. Int. Ed. Engl.* **1976**, *15*, 270.
	Chem. Rev. **1971**, *71*, 483.
Aromatic substitution	*Tetrahedron* **1985**, *41*, 837.
Asymmetric synthesis	*Acc. Chem. Res.* **1978**, *11*, 375.
Oxidation-reductions	*Russ. Chem. Rev.* **1978**, *47*, 22.
Condensation	*Angew. Chem. Int. Ed. Engl.* **1976**, *15*, 94.
Oxidations	*Pure & Appl. Chem.* **1988**, *60*, 35.
	SYNLETT **1990**, 291.
Activated dimethyl sulfoxide	*Synthesis* **1981**, 165.
Alcohols	
Activated dimethyl sulfoxide	*Synthesis* **1990**, 87.
Aldehydes	*Chem. Rev.* **1954**, *54*, 325.
Alkenes	*Acc. Chem. Res.* **1985**, *18*, 358.
	Angew. Chem. Int. Ed. Engl. **1962**, *1*, 80.
	Pure & Appl. Chem. **1981**, *53*, 2389.
	Synthesis **1971**, 527.
	Synthesis **1984**, 369.
Alkoxysulfonium ylides	*Org. Reactions* **1990**, *39*, 297.

Oxidations, cont'd.

Alkynes	*Chem. Rev.* **1970**, *70*, 267.
Amines	
Aromatic compounds	
Electrochemistry	*Acc. Chem. Res.* **1969**, *2*, 175.
Annulenes	*Chem. Rev.* **1984**, *84*, 603.
Anodic oxidations	*Tetrahedron* **1991**, *47*, 531ff.
Aryl ethers	*Chem. Rev.* **1969**, *69*, 499.
Asymmetric synthesis	*Tetrahedron: Asymmetry* **1991**, *2*, 481ff.
Benzoquinones	*Org. Reactions* **1948**, *4*, 305.
Carbohydrates	*Synthesis* **1971**, 70.
Carbonyl compounds	*Org. Reactions* **1990**, *39*, 297.
Ceric ion oxidations	*Synthesis* **1973**, 347.
Dimethyl sulfoxide	*Chem. Rev.* **1967**, *67*, 247.
	Synthesis **1990**, 87.
Dioxirane	*Acc. Chem. Res.* **1989**, *22*, 205.
Dioxygen complexes	*Pure & Appl. Chem.* **1983**, *55*, 125.
Electrochemical	*Chem. Rev.* **1968**, *68*, 449.
	SYNLETT **1990**, 301.
Glucose	*Pure & Appl. Chem.* **1991**, *63*, 1599.
Ethers	
Aryl ethers	*Chem. Rev.* **1969**, *69*, 499.
Ferricyanide	*Chem. Rev.* **1958**, *58*, 439.
Fremy's radical	*Chem. Rev.* **1971**, *71*, 229.
Haloethylenes	*Chem. Rev.* **1976**, *76*, 801.
Heterocycles	
Nitrogen aromatic heterocycles	*Pure & Appl. Chem.* **1971**, *25*, 365.
Hypervalent iodine	*Synthesis* **1990**, 431.
Indoles	*Heterocycl.* **1977**, *8*, 743.
Electrochemistry	*Chem. Rev.* **1990**, *90*, 795.
Ketones	*Synthesis* **1984**, 369.
Lead tetraacetate	*Angew. Chem.* **1958**, *70*, 173.
Manganese dioxide	*Synthesis* **1976**, 65.
	Synthesis **1976**, 133.
Methane	*Chem. Rev.* **1985**, *85*, 235.
Moffat oxidation	*Org. Reactions* **1990**, *39*, 297.
Nickel peroxide	*Chem. Rev.* **1975**, *75*, 491.
Nitrogen compounds	*Quart. Rev.* **1971**, *25*, 407.
Organoboranes	*Org. Reactions* **1953**, *13*, 22.
Organometallic chemistry	
Reaction mechanisms	*Pure & Appl. Chem.* **1984**, *56*, 35.
Oxidative condensations	
Phenolic bases	*Angew. Chem. Int. Ed. Engl.* **1964**, *3*, 192.
	Angew. Chem. Int. Ed. Engl. **1976**, *15*, 94.

Oxidative additions	*Acc. Chem. Res.* **1970**, *3*, 386.
	Acc. Chem. Res. **1977**, *10*, 434.
Alkanes	*Pure & Appl. Chem.* **1984**, *56*, 13.
Oxidative couplings	
Alkenes	*Acc. Chem. Res.* **1985**, *18*, 120.
Azo dyes	*Angew. Chem. Int. Ed. Engl.* **1962**, *1*, 640.
Methane	*Chem. Soc. Rev.* **1989**, *18*, 251.
Organocopper reagents	*Angew. Chem. Int. Ed. Engl.* **1974**, *13*, 291.
Oxidative cyclizations	
Alcohols	*Synthesis* **1970**, 209.
Carbonyl compounds	*Synthesis* **1970**, 279.
Oxidative stress	*Angew. Chem. Int. Ed. Engl.* **1986**, *25*, 1058.
Palladium catalysis	*Pure & Appl. Chem.* **1983**, *55*, 1669.
Periodic acid	*Org. Reactions* **1944**, *2*, 341.
Permanganate	*Chem. Rev.* **1958**, *58*, 403.
	Synthesis **1987**, 85.
Peroxytrifluoroacetic acid-boron trifluoride	*Acc. Chem. Res.* **1971**, *4*, 327.
	Acc. Chem. Res. **1971**, *4*, 337.
Phenols	*Angew. Chem. Int. Ed. Engl.* **1963**, *2*, 723.
Photosensitized oxidation	
Tryptophan derivatives	*Acc. Chem. Res.* **1977**, *10*, 346.
Porphyrins	*Pure & Appl. Chem.* **1989**, *61*, 1631.
Potassium nitrosodisulfonate	*Chem. Rev.* **1971**, *71*, 229.
Pyridinium chlorochromate	*Synthesis* **1982**, 245.
Selenium dioxide	*Org. Reactions* **1949**, *5*, 331.
	Org. Reactions **1976**, *24*, 261.
Sulfides	*Tetrahedron* **1986**, *42*, 5459.
Swern oxidation	*Org. Reactions* **1990**, *39*, 297.
Teuber reaction	*Chem. Rev.* **1971**, *71*, 229.
Transition metal complexes	*Acc. Chem. Res.* **1970**, *3*, 347.
Oximes	*Chem. Rev.* **1973**, *73*, 283.
Photochemistry	
NMR	
Reaction mechanisms	*Pure & Appl. Chem.* **1988**, *60*, 933.
Titanium chemistry	*Acc. Chem. Res.* **1974**, *7*, 281.
Oxiranes	*Synthesis* **1981**, 501.
	Tetrahedron **1983**, *39*, 2323.
Asymmetric synthesis	*Acc. Chem. Res.* **1973**, *6*, 341.
Oxo reaction	
Cobalt chemistry	*Acc. Chem. Res.* **1981**, *14*, 259.
3-Oxoalkanenitriles	*Synthesis* **1984**, 1.
Oxoalkylations	
Carbonyl compounds	
Nitroalkenes	*Acc. Chem. Res.* **1985**, *18*, 284.

Oxyallyl cations	
Cycloadditions	*Tetrahedron* **1986**, *42*, 4611.
3-Oxocyclopentenes	*Synthesis* **1973**, 397.
4-Oxo-2-cyclopentenyl acetate	*Angew. Chem. Int. Ed. Engl.* **1982**, *21*, 480.
α-Oxoketene acetals	*Tetrahedron* **1990**, *46*, 5423.
α-Oxoketene dithioacetals	*Tetrahedron* **1986**, *42*, 3029.
Oxycyclopropanes	*Acc. Chem. Res.* **1980**, *13*, 27.
Oxygen activation	
Phenylalanine hydroxylase	*Acc. Chem. Res.* **1988**, *21*, 101.
Oxygen binding	
Natural products	*Chem. Rev.* **1984**, *84*, 137.
Porphyrins	*Pure & Appl. Chem.* **1981**, *53*, 293.
Oxygen carriers	*Chem. Rev.* **1979**, *79*, 139.
Hemerythrin	*Acc. Chem. Res.* **1984**, *17*, 16.
Oxygen heterocycles	*Pure & Appl. Chem.* **1987**, *59*, 969.
Oxygen insertion	
Bicyclic ketones	*Tetrahedron* **1981**, *37*, 2697.
Oxygen transfer	
Metallaoxetanes	*Chem. Rev.* **1990**, *90*, 1483.
Reaction mechanisms	*Angew. Chem. Int. Ed. Engl.* **1982**, *21*, 734.
Oxygenation	
Alkanes	
Activation	*Angew. Chem. Int. Ed. Engl.* **1978**, *17*, 909.
Biometic reactions	*Tetrahedron* **1977**, *33*, 2869.
Photosensitized	*Acc. Chem. Res.* **1968**, *1*, 104.
Oxypalladation	
Alkenes	*Acc. Chem. Res.* **1990**, *23*, 49.
Oxyphosphoranes	*Synthesis* **1974**, 90.
Ozone	
Gas phase	
Reaction mechanisms	*Chem. Rev.* **1984**, *84*, 437.
Ozonides	*Acc. Chem. Res.* **1983**, *16*, 42.
Ozonization	*Chem. Rev.* **1940**, *27*, 437.
Ozonolysis	*Angew. Chem. Int. Ed. Engl.* **1975**, *14*, 745.
Alkenes	*Acc. Chem. Res.* **1983**, *16*, 42.
Reaction mechanisms	*Acc. Chem. Res.* **1968**, *1*, 313.
Reaction mechanisms	*Angew. Chem. Int. Ed. Engl.* **1975**, *14*, 745.
Palladium chemistry	*Acc. Chem. Res.* **1969**, *2*, 144.
	Pure & Appl. Chem. **1980**, *52*, 669.
	Pure & Appl. Chem. **1981**, *53*, 2371.
	Pure & Appl. Chem. **1988**, *60*, 89.
	Pure & Appl. Chem. **1990**, *62*, 1941.
	Synthesis **1990**, 739.
Addition reactions	
Butadiene	*Acc. Chem. Res.* **1973**, 6, 8.

Alkenes	*Tetrahedron* **1984**, *40*, 2415.
Alkyl halides	*Acc. Chem. Res.* **1979**, *12*, 146.
Substitution reactions	*Synthesis* **1973**, 524.
Allyl alkylations	*Acc. Chem. Res.* **1980**, *13*, 385.
Carbonyl compounds	*Acc. Chem. Res.* **1987**, *20*, 140.
π-Allyl complexes	*Tetrahedron* **1986**, *42*, 4361.
Aromatic alkenes	
Substitution reactions	*Synthesis* **1973**, 524.
Aromatic substitution	*Synthesis* **1973**, 524.
Butadiene	
Addition reactions	*Acc. Chem. Res.* **1973**, 6, 8.
Catalytic reactions	*Pure & Appl. Chem.* **1982**, *54*, 197.
Coupling reactions	*Acc. Chem. Res.* **1982**, *15*, 340.
	Pure & Appl. Chem. **1985**, *57*, 1771.
	SYNLETT **1991**, 845.
Polyenes	*Pure & Appl. Chem.* **1981**, *53*, 2323.
Enyne	*Acc. Chem. Res.* **1990**, *23*, 34.
Cycloisomerization	*Acc. Chem. Res.* **1990**, *23*, 34.
Glycosylations	*Acc. Chem. Res.* **1990**, *23*, 201.
Heterocycles	*Pure & Appl. Chem.* **1983**, *55*, 1845.
Heterogeneous catalysis	*Pure & Appl. Chem.* **1991**, *63*, 821.
Oligomerization	
Alkynes	*Acc. Chem. Res.* **1976**, 9, 93.
Organotin chemistry	*Angew. Chem. Int. Ed. Engl.* **1986**, *25*, 508.
Oxidations	*Pure & Appl. Chem.* **1983**, *55*, 1669.
Alkenes	*Synthesis* **1984**, 369.
Palytoxin	*Pure & Appl. Chem.* **1989**, *61*, 313.
Paracyclophanes	*Acc. Chem. Res.* **1980**, *13*, 65.
	Acc. Chem. Res. **1974**, *7*, 415.
D-Penicillamine	*Angew. Chem. Int. Ed. Engl.* **1975**, *14*, 330.
Penicillanic acids	
Rearrangements	*Tetrahedron* **1975**, *31*, 2321.
Penicillin	*Acc. Chem. Res.* **1973**, 6, 32.
Resistance	*Acc. Chem. Res.* **1985**, *18*, 97.
Penicillin sulfoxides	*Acc. Chem. Res.* **1973**, 6, 32.
Pentadiene	
Metal derivatives	*Acc. Chem. Res.* **1985**, *18*, 56.
Pentadiyne cation	*Angew. Chem. Int. Ed. Engl.* **1981**, *20*, 638.
Pentalene	*Acc. Chem. Res.* **1974**, *7*, 321.
Pentoses	*Acc. Chem. Res.* **1986**, *19*, 250.
Pepsin	
Active site	*Acc. Chem. Res.* **1974**, *7*, 241.
Peptidase	
Reaction mechanisms	*Acc. Chem. Res.* **1988**, *21*, 333.
Peptide hormones	*Angew. Chem. Int. Ed. Engl.* **1983**, *22*, 842.

Peptides	*Angew. Chem. Int. Ed. Engl.* **1985**, *24*, 85.
	Angew. Chem. Int. Ed. Engl. **1991**, *30*, 296.
	Angew. Chem. Int. Ed. Engl. **1991**, *30*, 1278.
	Chem. Rev. **1987**, *87*, 381.
	Pure & Appl. Chem. **1982**, *54*, 2409.
	Pure & Appl. Chem. **1984**, *56*, 979.
	Pure & Appl. Chem. **1987**, *59*, 317.
	See also retro-peptides
Alamethicin	
Transmembrane channel	*Acc. Chem. Res.* **1981**, *14*, 356.
Alkylations	
Enolates	*Angew. Chem. Int. Ed. Engl.* **1988**, *27*, 1624.
Aminoacyl transfer RNA	*Acc. Chem. Res.* **1977**, *10*, 239.
Azapeptides	*Synthesis* **1989**, 405.
Conformations	*Tetrahedron* **1988**, *44*, 661ff.
	Acc. Chem. Res. **1968**, *1*, 273.
	Acc. Chem. Res. **1975**, *8*, 306.
	Acc. Chem. Res. **1984**, *17*, 209.
	Acc. Chem. Res. **1985**, *18*, 372.
	Acc. Chem. Res. **1990**, *23*, 338.
	Chem. Rev. **1971**, *71*, 195.
Cross-linked peptides	*Angew. Chem. Int. Ed. Engl.* **1971**, *10*, 795.
Cyclic peptides	*Tetrahedron* **1975**, *31*, 2177.
Conformations	*Acc. Chem. Res.* **1972**, *5*, 193.
	Angew. Chem. Int. Ed. Engl. **1982**, *21*, 512.
Enolates	
Alkylations	*Angew. Chem. Int. Ed. Engl.* **1988**, *27*, 1624.
Enzymatic cleavage	*Acc. Chem. Res.* **1970**, *3*, 81.
	Acc. Chem. Res. **1970**, *3*, 249.
Hormones	*Acc. Chem. Res.* **1990**, *23*, 338.
Hydrolysis	*Acc. Chem. Res.* **1987**, *20*, 357.
Carboxypeptidase A	*Acc. Chem. Res.* **1972**, *5*, 219.
Ligands	
Random screeening	*Angew. Chem. Int. Ed. Engl.* **1991**, *30*, 296.
Metal complexes	*Pure & Appl. Chem.* **1983**, *55*, 23.
	Chem. Rev. **1982**, *82*, 385.
Peptide hormones	*Angew. Chem. Int. Ed. Engl.* **1983**, *22*, 842.
Poly(oxyethylene)bound peptides	
Conformations	*Acc. Chem. Res.* **1981**, *14*, 122.
Protecting groups	*Angew. Chem. Int. Ed. Engl.* **1985**, *24*, 799.
Reductions	*Chem. Rev.* **1984**, *84*, 287.

Solid phase synthesis	*Angew. Chem. Int. Ed. Engl.* **1985**, *24*, 799.
Synthesis	*Acc. Chem. Res.* **1989**, *22*, 47.
	Angew. Chem. Int. Ed. Engl. **1971**, *10*, 152.
	Angew. Chem. Int. Ed. Engl. **1991**, *30*, 1437.
	Pure & Appl. Chem. **1987**, *59*, 331.
	Pure & Appl. Chem. **1988**, *60*, 539.
	Tetrahedron **1988**, *44*, 661ff.
Azide method	*Synthesis* **1974**, 549.
Coupling agents	*Synthesis* **1972**, 453.
Mixed anhydrides	*Org. Reactions* **1962**, *12*, 157.
Side reactions	*Synthesis* **1981**, 333.
Synthetic peptides	
Enzymes	*Angew. Chem. Int. Ed. Engl.* **1985**, *24*, 719.
Peptidoglycans	*Acc. Chem. Res.* **1971**, *4*, 297.
Perfluoro compounds	
Blood substitutes	*Angew. Chem. Int. Ed. Engl.* **1978**, *17*, 621.
Sulfoxides	*Acc. Chem. Res.* **1973**, *6*, 387.
Perfluoroalkanesulfonic esters	*Synthesis* **1982**, 85.
Perfluoroaromatic compounds	*Synthesis* **1976**, 652.
Perfluorocrown ethers	*Pure & Appl. Chem.* **1988**, *60*, 473.
Perhalogenated sulfenes	*Synthesis* **1988**, 349.
Perhalogenated sulfines	*Synthesis* **1988**, 349.
Pericyclic reactions	*Acc. Chem. Res.* **1976**, *9*, 453.
	Angew. Chem. Int. Ed. Engl. **1971**, *10*, 761.
	Angew. Chem. Int. Ed. Engl. **1974**, *13*, 751.
Cation radicals	*Acc. Chem. Res.* **1987**, *20*, 371.
Photocyclization	*Acc. Chem. Res.* **1983**, *16*, 210.
Seven-membered heterocycles	*Synthesis* **1988**, 569.
Theoretical aspects	*Angew. Chem. Int. Ed. Engl.* **1974**, *13*, 751.
Thermal	*Angew. Chem. Int. Ed. Engl.* **1974**, *13*, 47.
Vinylallenes	*Acc. Chem. Res.* **1983**, *16*, 81.
Periodates	*Synthesis* **1974**, 229.
Periodic acid	*Synthesis* **1974**, 229.
	Org. Reactions **1944**, *2*, 341.
Permanganate	
Oxidations	*Chem. Rev.* **1958**, *58*, 403.
	Synthesis **1987**, 85.
Permethylscandocene derivatives	*Pure & Appl. Chem.* **1984**, *56*, 1.
Peroxidations	*Quart. Rev.* **1967**, *21*, 439.
Peroxides	*SYNLETT* **1990**, 291.
Acyloxylations	*Synthesis* **1972**, 1.
Cyclic peroxides	*Angew. Chem. Int. Ed. Engl.* **1974**, *13*, 619.
Four-membered ring peroxides	*Angew. Chem. Int. Ed. Engl.* **1983**, *22*, 529.
Organophosphorus chemistry	*Chem. Rev.* **1981**, *81*, 49.

Peroxisome	*Angew. Chem. Int. Ed. Engl.* **1974**, *13*, 706.
Peroxonium ions	
Ylides	*Chem. Soc. Rev.* **1985**, *14*, 399.
Peroxy acids	
Alkenes	*Tetrahedron* **1976**, *32*, 2855.
Epoxidations	*Org. Reactions* **1953**, *7*, 378.
Alkenes	*Org. Reactions* **1953**, *7*, 378.
Hydroxylations	*Org. Reactions* **1953**, *7*, 378.
α-Peroxy amines	*Angew. Chem. Int. Ed. Engl.* **1973**, *12*, 783.
Peroxylactones	*Acc. Chem. Res.* **1979**, *12*, 390.
Pesticides	
Enzymes	*Acc. Chem. Res.* **1988**, *21*, 348.
Petroleum refining	*Acc. Chem. Res.* **1986**, *19*, 78.
Pharmaceutical chemistry	*Angew. Chem. Int. Ed. Engl.* **1980**, *19*, 749.
Pharmaceutical industry	
Carbohydrates	*Pure & Appl. Chem.* **1989**, *61*, 1313.
Pharmaceutical proteins	*Angew. Chem. Int. Ed. Engl.* **1988**, *27*, 207.
Pharmaceuticals	*Pure & Appl. Chem.* **1986**, *58*, 365.
Phase transfer catalysis	*Angew. Chem. Int. Ed. Engl.* **1974**, *13*, 170.
	Angew. Chem. Int. Ed. Engl. **1977**, *16*, 493.
	Angew. Chem. Int. Ed. Engl. **1986**, *25*, 960.
Ion pair extractions	*Pure & Appl. Chem.* **1982**, *54*, 1769.
Magnesium chemistry	*Acc. Chem. Res.* **1988**, *21*, 261.
Reaction mechanisms	*Pure & Appl. Chem.* **1982**, *54*, 1769.
Phenanthrenes	*Chem. Rev.* **1976**, *76*, 509.
Phenolic bases	
Oxidative condensations	*Angew. Chem. Int. Ed. Engl.* **1964**, *3*, 192.
	Angew. Chem. Int. Ed. Engl. **1976**, *15*, 94.
Phenolic couplings	
Alkaloids	*Synthesis* **1972**, 657.
Phenols	*Chem. Rev.* **1986**, *86*, 821.
	Pure & Appl. Chem. **1983**, *55*, 1677.
Anionic cyclizations	*Chem. Soc. Rev.* **1983**, *12*, 213.
Biosynthesis	*Chem. Soc. Rev.* **1979**, *8*, 499.
Oxidations	*Angew. Chem. Int. Ed. Engl.* **1963**, *2*, 723.
Phenylalanine hydroxylase	
Oxygen activation	*Acc. Chem. Res.* **1988**, *21*, 101.
Phenylcarbenes	
Rearrangement reactions	*Acc. Chem. Res.* **1974**, *7*, 415.
Phenylcyclobutenediones	*Angew. Chem. Int. Ed. Engl.* **1972**, *11*, 997.
3-Phenylpropylamines	*Angew. Chem. Int. Ed. Engl.* **1980**, *19*, 184.
Pheromones	*Acc. Chem. Res.* **1988**, *21*, 294.
Structure determinations	*Angew. Chem. Int. Ed. Engl.* **1988**, *27*, 460.
Philanthotoxin-433	*Pure & Appl. Chem.* **1990**, *62*, 1223.

Philosophy of Science	*Angew. Chem. Int. Ed. Engl.* **1981**, *20*, 617.
Phosgene	*Chem. Rev.* **1973**, *73*, 75.
Phosphaalkenes	*Tetrahedron* **1989**, *45*, 6019.
Phosphaalkynes	*Angew. Chem. Int. Ed. Engl.* **1988**, *27*, 1484.
	Chem. Rev. **1990**, *90*, 191.
	Tetrahedron **1989**, *45*, 6019.
Phosphaallene ylides	*Angew. Chem. Int. Ed. Engl.* **1977**, *16*, 349.
Phosphabenzene	*Acc. Chem. Res.* **1978**, *11*, 153.
Phosphacumulene ylides	*Angew. Chem. Int. Ed. Engl.* **1977**, *16*, 349.
Phosphanediyls	*Angew. Chem. Int. Ed. Engl.* **1975**, *14*, 523.
Phosphatase	*Acc. Chem. Res.* **1983**, *16*, 244.
Phosphate esters	*Acc. Chem. Res.* **1983**, *16*, 81.
Hydrolysis	*Acc. Chem. Res.* **1968**, *1*, 70.
Phosphates	*Chem. Rev.* **1977**, *77*, 349.
Carbohydrates	
Anomerization	*Acc. Chem. Res.* **1978**, *11*, 136.
Enzymes	*Tetrahedron* **1982**, *38*, 1541.
Phosphazenes	*SYNLETT* **1990**, 651.
Friedal Crafts reaction	*Org. Reactions* **1949**, *5*, 229.
Phosphazines	*Chem. Rev.* **1972**, *72*, 315.
Phosphine ligands	
Stereoelectronic effects	*J. Organomet. Chem.* **1980**, *200*, 307.
Phosphine oxides	
Migrations	*Acc. Chem. Res.* **1978**, *11*, 401.
Phosphinic acids	*Org. Reactions* **1951**, *6*, 273.
Phosphinidenes	*Angew. Chem. Int. Ed. Engl.* **1987**, *26*, 275.
	Angew. Chem. Int. Ed. Engl. **1975**, *14*, 523.
Phosphinimides	*Synthesis* **1974**, 775.
Phosphodiesters	
Cyclic enediol phosphoryl (CEP) method	*Synthesis* **1985**, 449.
Phospholipids	*Acc. Chem. Res.* **1978**, *11*, 321.
	Angew. Chem. Int. Ed. Engl. **1984**, *23*, 257.
Membranes	*Pure & Appl. Chem.* **1982**, *54*, 2443.
NMR	*Acc. Chem. Res.* **1987**, *20*, 221.
Micelles	*Acc. Chem. Res.* **1983**, *16*, 251.
NMR	*Acc. Chem. Res.* **1987**, *20*, 221.
Sponges	*Acc. Chem. Res.* **1991**, *24*, 69.
Phosphols	*Chem. Rev.* **1988**, *88*, 429.
Phosphonates	*Chem. Rev.* **1977**, *77*, 349.
Phosphonic acids	*Org. Reactions* **1951**, *6*, 273.
Phosphonites	
Transesterification	*Russ. Chem. Rev.* **1966**, *35*, 622.

Phosphonium ylides	*Acc. Chem. Res.* **1969**, *2*, 373.
Phosphoramidite approach	
Oligonucleotides	*Tetrahedron* **1992**, *48*, 2223.
Phosphoranes	*Acc. Chem. Res.* **1979**, *12*, 257.
Phosphorus azide	*Acc. Chem. Res.* **1986**, *19*, 17.
Phosphorus esters	
Addition reactions	*Synthesis* **1979**, 81.
Cyclic	*Tetrahedron* **1980**, *36*, 2059.
Silyl esters of phosphorus	*Tetrahedron* **1989**, *45*, 2465.
Phosphorus ligands	
Steric effects	*Chem. Rev.* **1977**, *77*, 313.
Phosphorus-nitrogen heterocycles	*Synthesis* **1978**, 557.
Phosphoryl transfer enzymes	*Acc. Chem. Res.* **1971**, *4*, 214.
Phosphorylations	*Synthesis* **1977**, 737.
ATP	*Pure & Appl. Chem.* **1980**, *52*, 2213.
Enediol phosphates	*Acc. Chem. Res.* **1978**, *11*, 239.
Glycogen phosphorylase	*Acc. Chem. Res.* **1978**, *11*, 239.
Nucleotides	*Pure & Appl. Chem.* **1980**, *52*, 2213.
Phosphorylcarbenes	*Angew. Chem. Int. Ed. Engl.* **1975**, *14*, 222.
Phosphotriesters	*Tetrahedron* **1978**, *34*, 3143.
Photoadditions	
Alkenes	*Chem. Rev.* **1988**, *88*, 1453.
Aromatic compounds	*Chem. Rev.* **1987**, *87*, 811.
Enones	*Chem. Rev.* **1988**, *88*, 1453.
Photochemistry	*Angew. Chem. Int. Ed. Engl.* **1971**, *10*, 537.
	Chem. Rev. **1978**, *78*, 125.
	Pure & Appl. Chem. **1982**, *54*, 1623.
	Pure & Appl. Chem. **1986**, *58*, 1173.
	Science **1976**, *191*, 523.
	See under specific classes of compounds and reaction types
	Synthesis **1970**, 636.
	Tetrahedron **1981**, *37*, 3227ff.
Adiabatic reactions	*Angew. Chem. Int. Ed. Engl.* **1979**, *18*, 572.
Asymmetric synthesis	*Chem. Rev.* **1983**, *83*, 525.
Catalysis	*Pure & Appl. Chem.* **1984**, *56*, 1215.
Coordination compounds	*Pure & Appl. Chem.* **1990**, *62*, 1489.
Charge transfer	*Pure & Appl. Chem.* **1986**, *58*, 1285.
	Pure & Appl. Chem. **1984**, *56*, 1255.
CIDEP	*Acc. Chem. Res.* **1977**, *10*, 161.
CIDNP	
Carbenes	*Acc. Chem. Res.* **1977**, *10*, 85.

Circularly polarized light	*Angew. Chem. Int. Ed. Engl.* **1974**, *13*, 179.
Cyclizations	*Chem. Rev.* **1966**, *66*, 373.
Cycloadditions	*Acc. Chem. Res.* **1968**, *1*, 50.
	Acc. Chem. Res. **1982**, *15*, 80.
	Chem. Rev. **1969**, *69*, 845.
Cycloeliminations	*Angew. Chem. Int. Ed. Engl.* **1977**, *16*, 835.
Donor-acceptors	*Tetrahedron* **1989**, *45*, 4669ff.
Electron loss	
Reaction mechanisms	*Pure & Appl. Chem.* **1981**, *53*, 223.
Electron transfer	*Chem. Rev.* **1983**, *83*, 425.
	Pure & Appl. Chem. **1982**, *54*, 1605.
	Pure & Appl. Chem. **1982**, *54*, 1885.
	Pure & Appl. Chem. **1986**, *58*, 1279.
	Pure & Appl. Chem. **1988**, *60*, 1013.
	Pure & Appl. Chem. **1988**, *60*, 1025.
	Synthesis **1989**, 233.
Electron transport	*Pure & Appl. Chem.* **1990**, 62, 1585.
Exciplexs	*Pure & Appl. Chem,* **1982**, *54*, 1885.
	Pure & Appl. Chem. **1990**, 62, 1585.
Excited molecules	*Pure & Appl. Chem,* **1980**, *52*, 2591.
Holographic methods	*Angew. Chem. Int. Ed. Engl.* **1983**, *22*, 582.
Industrial chemistry	*Angew. Chem. Int. Ed. Engl.* **1978**, *17*, 16.
	Pure & Appl. Chem. **1986**, *58*, 1267.
Intermediates	*Pure & Appl. Chem.* **1984**, 56, 1167.
Intersystem crossing	*Pure & Appl. Chem.* **1984**, 56, 1167.
Isomerizations	*Angew. Chem. Int. Ed. Engl.* **1972**, *11*, 1072.
Cyclopentene derivatives	*Tetrahedron* **1976**, *32*, 641.
Microscopic reactors	*Pure & Appl. Chem.* **1986**, *58*, 1219.
Molecular ions	*Pure & Appl. Chem.* **1984**, 56, 1203.
Photochemical reactivity	*Pure & Appl. Chem.* **1986**, *58*, 1173.
Photoenolization	*Tetrahedron* **1976**, *32*, 405.
Photoinduction	*Pure & Appl. Chem.* **1990**, 62, 1531.
Reaction mechanisms	*Pure & Appl. Chem.* **1982**, *54*, 1885.
	Pure & Appl. Chem. **1990**, 62, 1547.
Reactive intermediates	*Pure & Appl. Chem.* **1982**, *54*, 161.
	Pure & Appl. Chem. **1986**, *58*, 1273.
Storage systems	*Angew. Chem. Int. Ed. Engl.* **1979**, *18*, 652.
Synthesis	*Synthesis* **1989**, 145.
Synthetic methods	*Synthesis* **1970**, 636.
Upper excited states	*Chem. Rev.* **1978**, *78*, 125.
Photoelectron spectroscopy	*Angew. Chem. Int. Ed. Engl.* **1983**, *22*, 283.
Photography	
Color dyes	*Angew. Chem. Int. Ed. Engl.* **1983**, *22*, 191.

Photophysics	
α,ω-Diphenylpolyene singlets	*Chem. Rev.* **1989**, *89*, 1691.
Photosynthesis	*Acc. Chem. Res.* **1973**, *6*, 177.
	Acc. Chem. Res. **1975**, *8*, 407.
	Acc. Chem. Res. **1981**, *14*, 163.
	Angew. Chem. Int. Ed. Engl. **1986**, *25*, 971.
	Angew. Chem. Int. Ed. Engl. **1991**, *30*, 1621.
	Chem. Rev. **1978**, *78*, 175.
	Chem. Rev. **1978**, *78*, 185.
	Pure & Appl. Chem. **1986**, *58*, 1171.
	Pure & Appl. Chem. **1990**, *62*, 1521.
	Tetrahedron **1989**, *45*, 4669ff.
Artifical photosynthesis	*Pure & Appl. Chem.* **1982**, *54*, 1733.
Herbicides	*Angew. Chem. Int. Ed. Engl.* **1991**, *30*, 1621.
Manganese chemistry	*Acc. Chem. Res.* **1980**, *13*, 249.
Membranes	*Acc. Chem. Res.* **1978**, *11*, 257.
Mimics	*Pure & Appl. Chem.* **1984**, *56*, 1269.
Rhodopseudomonas viridis	*Pure & Appl. Chem.* **1988**, *60*, 953.
Ribulose bisphosphate carboxylase	*Acc. Chem. Res.* **1980**, *13*, 394.
Ribulose bisphosphate oxygenase	*Acc. Chem. Res.* **1980**, *13*, 394.
Water splitting	*Angew. Chem. Int. Ed. Engl.* **1987**, *26*, 643.
Phthalimides	
Photochemistry	*Acc. Chem. Res.* **1978**, *11*, 407.
Phthalocyanine	*Pure & Appl. Chem.* **1986**, *58*, 1467.
Phycocyanin	
Reaction mechanisms	*Chem. Rev.* **1989**, *89*, 807.
Phycotoxins	*Pure & Appl. Chem.* **1980**, *52*, 165ff.
	Pure & Appl. Chem. **1989**, *61*, 7.
Physical organic chemistry	*Tetrahedron* **1988**, *44*, 7335ff.
Phytoalexins	
Chemical defense	*Angew. Chem. Int. Ed. Engl.* **1978**, *17*, 635.
Phytochrome	*Angew. Chem. Int. Ed. Engl.* **1991**, *30*, 1216.
Photochemistry	*Pure & Appl. Chem.* **1984**, *56*, 1153.
Pigments	*Angew. Chem. Int. Ed. Engl.* **1991**, *30*, 1216.
Phytohormones	
Synthesis	*Pure & Appl. Chem.* **1982**, *54*, 2501.
Phytotoxins	*Pure & Appl. Chem.* **1986**, *58*, 211ff.

Pi-sigma interactions	*Pure & Appl. Chem.* **1987**, *59*, 1585.
Pictet-Spengler synthesis	*Org. Reactions* **1951**, 6, 152.
Pig liver esterase	*Tetrahedron* **1990**, *46*, 6587.
Esters	*Org. Reactions* **1989**, *37*, 1.
Pigments	*Angew. Chem. Int. Ed. Engl.* **1975**, *14*, 688.
Fly agaric, *Amanita muscaria*	*Tetrahedron* **1979**, *35*, 2843.
Phytochrome	*Angew. Chem. Int. Ed. Engl.* **1991**, *30*, 1216.
Plant growth regulation	*Chem. Soc. Rev.* **1977**, 6, 261.
Plant tissue cultures	*Pure & Appl. Chem.* **1984**, *56*, 1011.
Anti-tumor agents	*Pure & Appl. Chem.* **1982**, *54*, 2523.
Plants	
Genetic engineering	*Angew. Chem. Int. Ed. Engl.* **1987**, *26*, 382.
Plasma flow discharges	*Angew. Chem. Int. Ed. Engl.* **1972**, *11*, 781.
Plastocyanin	*Chem. Soc. Rev.* **1985**, *14*, 283.
Platinum chemistry	*Chem. Rev.* **1986**, *86*, 451.
Decompositions	
Reaction mechanisms	*Pure & Appl. Chem.* **1981**, *53*, 287.
Macrocycles	*Pure & Appl. Chem.* **1988**, *60*, 517.
Polynucleotides	*Acc. Chem. Res.* **1978**, *11*, 211.
Unsaturated hydrocarbons	*Acc. Chem. Res.* **1973**, 6, 202.
Polar cycloadditions	*Angew. Chem. Int. Ed. Engl.* **1973**, *12*, 212.
Polynucleotides	
Metal ion complexes	*Chem. Rev.* **1980**, *80*, 365.
Polyalkalimetal derivatives	*Synthesis* **1977**, 509.
Polyalkenes	
Photodegradation	*Chem. Soc. Rev.* **1975**, *4*, 533.
Photooxidation	*Pure & Appl. Chem.* **1982**, *54*, 1667.
Poly(alkylphosphazenes)	*Chem. Rev.* **1988**, *88*, 541.
Polyamides	
Photooxidation	*Pure & Appl. Chem.* **1982**, *54*, 1667.
Polyamines	*Tetrahedron* **1992**, *48*, 4475.
Macrocycles	*Pure & Appl. Chem.* **1986**, *58*, 1461.
Natural products	*Acc. Chem. Res.* **1982**, *15*, 290.
Polyaromatic hydrocarbons	*Tetrahedron* **1988**, *44*, 2093.
Polyaromatic compounds	
Arynes	*Acc. Chem. Res.* **1978**, *11*, 283.
Polyarsanes	
Metal complexes	*Pure & Appl. Chem.* **1986**, *58*, 1429.
Polyaza compounds	
Host-guest chemistry	*Tetrahedron* **1991**, *47*, 6851.
Polyazacycloalkanes	*Pure & Appl. Chem.* **1988**, *60*, 525.
Polyazamacrocycles	*Pure & Appl. Chem.* **1989**, *61*, 1569.

Polybromoketones	
Organoiron chemistry	*Acc. Chem. Res.* **1979**, *12*, 61.
Polycarbocyclic sesquiterpenes	*Tetrahedron* **1985**, *41*, 1767.
Polycarbodiimides	*Angew. Chem. Int. Ed. Engl.* **1981**, *20*, 819.
Poly(carbon monofluoride)	*Acc. Chem. Res.* **1978**, *11*, 296.
Polycarbonates	*Angew. Chem. Int. Ed. Engl.* **1991**, *30*, 1598.
Polychloro compounds	*Synthesis* **1974**, 477.
Polycyclic aromatic hydrocarbons	*Acc. Chem. Res.* **1985**, *18*, 241.
Polycyclic aromatic ions	
Aromaticity	*Pure & Appl. Chem.* **1982**, *54*, 1005.
Polycyclic compounds	
Cage compounds	*SYNLETT* **1991**, 73.
Synthetic methods	
Aryl synthons	*SYNLETT* **1991**, 134.
Polycyclic conjugated systems	
Aromaticity	*Pure & Appl. Chem.* **1982**, *54*, 939.
Polycyclic hydroaromatic compounds	
Dehydrogenation	*Chem. Rev.* **1978**, *78*, 317.
Polycyclopentanoid chemistry	
Natural products	*Tetrahedron* **1981**, *37*, 4357ff.
Poly(dialkylsilanes)	*Pure & Appl. Chem.* **1988**, *60*, 959.
Polyenes	
Cleavage reactions	*Angew. Chem. Int. Ed. Engl.* **1978**, *17*, 150.
Coupling reactions	*Pure & Appl. Chem.* **1981**, *53*, 2323.
Cyclizations	
Biomimetic	*Angew. Chem. Int. Ed. Engl.* **1976**, *15*, 9.
Rearrangements	*Angew. Chem. Int. Ed. Engl.* **1978**, *17*, 150.
Vision	*Acc. Chem. Res.* **1986**, *19*, 42.
Polyepoxides	*Tetrahedron* **1980**, *36*, 833.
Polyethers	*Acc. Chem. Res.* **1974**, *7*, 294. *Tetrahedron* **1987**, *43*, 3309.
Biosynthesis	*Acc. Chem. Res.* **1983**, *16*, 7.
Complexes	*Angew. Chem. Int. Ed. Engl.* **1972**, *11*, 16.
Polyethylene fibrids	*Angew. Chem. Int. Ed. Engl.* **1978**, *17*, 820.
Polyfluoroaromatic compounds	*Chem. Soc. Rev.* **1986**, *15*, 261.
Polyfunctional organomagnesium compounds	*Synthesis* **1970**, 615.
Polyhalocarbonyl compounds	
Reductions	*Org. Reactions* **1983**, *29*, 163.

Polyhydroxybutyrate	
Biosynthesis	*Pure & Appl. Chem.* **1989**, *61*, 303.
Poly(iminomethylenes)	*Acc. Chem. res.* **1979**, *12*, 30.
Polyketides	
Biosynthesis	*Tetrahedron* **1977**, *33*, 2159.
***vic*-Polyketones**	*Chem. Rev.* **1975**, *75*, 177.
Polymer membranes	
Ion transport	*Pure & Appl. Chem.* **1988**, *60*, 575.
Polymer reagents	*Angew. Chem. Int. Ed. Engl.* **1974**, *13*, 99.
	Chem. Rev. **1981**, *81*, 557.
Polymer supports	*Chem. Soc. Rev.* **1974**, *3*, 65.
Protecting groups	*Tetrahedron* **1981**, *37*, 663.
Synthetic methods	*Acc. Chem. Res.* **1978**, *11*, 327.
Polymer systems	
Photochemistry	*Pure & Appl. Chem.* **1982**, *54*, 1693.
Polymerization	*Chem. Rev.* **1983**, *83*, 549.
	See also copolymerization, under specific compounds and classes of compounds.
Enzymes	*Acc. Chem. Res.* **1973**, *6*, 361.
Aromatic ketones	*Angew. Chem. Int. Ed. Engl.* **1972**, *11*, 974.
N-*Vinylcarbazole*	*Pure & Appl. Chem.* **1973**, *34*, 329.
Radical polymerization	*Pure & Appl. Chem.* **1988**, *60*, 1033.
Stereospecific	*Science* **1964**, *145*, 667.
Polymers	
Biological systems	*Pure & Appl. Chem.* **1984**, *56*, 1301ff.
NMR	*Angew. Chem. Int. Ed. Engl.* **1988**, *27*, 1468.
Photochemistry	*Chem. Rev.* **1989**, *89*, 1915.
Polymethylpolysaccharides	*Pure & Appl. Chem.* **1981**, *53*, 107.
Polynuclear metal compounds	*Pure & Appl. Chem.* **1984**, *56*, 81.
Polynucleotides	*Tetrahedron* **1978**, *34*, 3143.
NMR	*Acc. Chem. Res.* **1974**, *7*, 33.
Platinum complexes	*Acc. Chem. Res.* **1978**, *11*, 211.
Synthesis	*Angew. Chem. Int. Ed. Engl.* **1972**, *11*, 451.
1,3-Polyols	*Synthesis* **1990**, 635.
Poly(oxyethylene)bound peptides	*Acc. Chem. Res.* **1981**, *14*, 122.
Polypeptides	*Acc. Chem. Res.* **1973**, *6*, 361.
	Angew. Chem. Int. Ed. Engl. **1991**, *30*, 1051.
	Chem. Rev. **1987**, *87*, 381.
	Pure & Appl. Chem. **1984**, *56*, 1049.
Biotechnology	*Angew. Chem. Int. Ed. Engl.* **1991**, *30*, 677.
Conformations	*Acc. Chem. Res.* **1985**, *18*, 372.
	Chem. Rev. **1971**, *71*, 195.

Polypeptides, cont'd.	
Cyclic polypetides	
Conformations	*Acc. Chem. Res.* **1972**, *5*, 193.
GC-MS	*Angew. Chem. Int. Ed. Engl.* **1976**, *15*, 75.
Polyenzyme templates	*Acc. Chem. Res.* **1973**, 6, 361.
Synthesis	*Angew. Chem. Int. Ed. Engl.* **1971**, *10*, 786.
Poly(*p*-phenylene)	*Chem. Rev.* **1987**, *87*, 357.
Polyphosphanes	
Metal complexes	*Pure & Appl. Chem.* **1986**, *58*, 1429.
Polyprenoid compounds	
Palladium chemistry	*Pure & Appl. Chem.* **1988**, *60*, 89.
Polypropionates	*Pure & Appl. Chem.* **1986**, *58*, 767.
	Pure & Appl. Chem. **1987**, *59*, 345.
Polypropylene	
Polymerization	
Historical perspective	*Angew. Chem. Int. Ed. Engl.* **1980**, *19*, 857.
Polysaccharides	*Acc. Chem. Res.* **1973**, 6, 184.
	Acc. Chem. Res. **1987**, *20*, 114.
Conformations	*Acc. Chem. Res.* **1987**, *20*, 59.
Structure determination	*Pure & Appl. Chem.* **1989**, *61*, 1201.
Degradation	*Acc. Chem. Res.* **1987**, *20*, 114.
Enzymes	*Tetrahedron* **1985**, *41*, 2957.
Gels	*Angew. Chem. Int. Ed. Engl.* **1977**, *16*, 214.
Industrial chemistry	*Pure & Appl. Chem.* **1989**, *61*, 1315.
Microbial polysaccharides	*Chem. Soc. Rev.* **1990**, *19*, 29.
	Pure & Appl. Chem. **1984**, *56*, 879.
(from) *Mycobacterium phlei*	*Acc. Chem. Res.* **1968**, *1*, 366.
Shapes	*Pure & Appl. Chem.* **1981**, *53*, 1.
Sialic acid derived polysaccharides	*Pure & Appl. Chem.* **1984**, *56*, 893.
	Pure & Appl. Chem. **1984**, *56*, 907.
Structures	*Chem. Soc. Rev.* **1981**, *10*, 409.
	Pure & Appl. Chem. **1989**, *61*, 1181.
Synthesis	*Pure & Appl. Chem.* **1984**, *56*, 923.
	Tetrahedron **1987**, *43*, 2389.
Polysilanes	
Aromaticity	*Pure & Appl. Chem.* **1982**, *54*, 1041.
Polysiloxanes	
Photochemistry	*Pure & Appl. Chem.* **1990**, *62*, 1603.
Polystyrene	
Historical perspective	*Angew. Chem. Int. Ed. Engl.* **1981**, *20*, 344.
Polyunsaturated compounds	
Rearrangements	*Chem. Rev.* **1972**, *72*, 181.

Porphobilinogen	
Biosynthesis	*Acc. Chem. Res.* **1975**, *8*, 201.
	Chem. Rev. **1990**, *90*, 1261.
Porphyrins	*Acc. Chem. Res.* **1970**, *3*, 105.
	Acc. Chem. Res. **1979**, *12*, 374.
	Angew. Chem. Int. Ed. Engl. **1974**, *13*, 321.
	Chem. Rev. **1981**, *81*, 543.
	Chem. Soc. Rev. **1975**, *4*, 1.
	Pure & Appl. Chem. **1981**, *53*, 1129.
	SYNLETT **1991**, 127.
Aromaticity	
Theoretical aspects	*Chem. Rev.* **1975**, *75*, 85.
Biosynthesis	*Acc. Chem. Res.* **1975**, *8*, 201.
	Acc. Chem. Res. **1978**, *11*, 29.
	Acc. Chem. Res. **1979**, *12*, 14.
1-Bromo-19-methylbiladienes-ac	*SYNLETT* **1990**, 235.
Hemoproteins	*Acc. Chem. Res.* **1981**, *14*, 102.
Iron-porphyrins	*Pure & Appl. Chem.* **1986**, *58*, 1493.
NMR	*Pure & Appl. Chem.* **1981**, *53*, 1215.
Organometallic chemistry	*Acc. Chem. Res.* **1986**, *19*, 209.
	Pure & Appl. Chem, **1980**, *52*, 681.
	Chem. Rev. **1988**, *88*, 1121.
Oxidations	*Pure & Appl. Chem.* **1989**, *61*, 1631.
Oxygen binding	*Pure & Appl. Chem.* **1981**, *53*, 293.
Potassium *tert*-butoxide	*Chem. rev.* **1974**, *74*, 45.
Potassium nitrosodisulfonate	
Oxidations	*Chem. Rev.* **1971**, *71*, 229.
Potassium-graphite	*Pure & Appl. Chem.* **1985**, *57*, 1887.
Prelog-Djerassi lactone	*Synthesis* **1991**, 245.
Prenyl synthons	
Biosynthesis	*Acc. Chem. Res.* **1981**, *14*, 89.
Probes	
Oligonucleotides	*Angew. Chem. Int. Ed. Engl.* **1991**, *30*, 613.
Process design	
Thermodynamics	
Reaction mechanisms	*Angew. Chem. Int. Ed. Engl.* **1990**, *29*, 1246.
Propargyl compounds	
Cobalt chemistry	*Acc. Chem. Res.* **1987**, *20*, 207.
Propargylic imidic esters	
Synthetic methods	*Acc. Chem. Res.* **1980**, *13*, 218.

Propargylsilanes	
Intramolecular addition reactions	*Synthesis* **1988**, 263.
Propellanes	*Acc. Chem. Res.* **1974**, *7*, 286.
	Acc. Chem. Res. **1984**, *17*, 379.
Small ring propellanes	*Chem. Rev.* **1989**, *89*, 975.
	Acc. Chem. Res. **1972**, *5*, 249.
Propellanones	
Rearrangements	*Acc. Chem. Res.* **1974**, *7*, 106.
Propellers	*Acc. Chem. Res.* **1976**, *9*, 26.
Propene	
Ammoxidation	*Angew. Chem. Int. Ed. Engl.* **1966**, *5*, 642.
Prostacyclins	*Acc. Chem. Res.* **1987**, *20*, 364.
	Angew. Chem. Int. Ed. Engl. **1982**, *21*, 751.
	Angew. Chem. Int. Ed. Engl. **1983**, *22*, 741.
	Angew. Chem. Int. Ed. Engl. **1978**, *17*, 293.
Prostaglandin endoperoxides	*Acc. Chem. Res.* **1985**, *18*, 294.
	Angew. Chem. Int. Ed. Engl. **1978**, *17*, 293.
Prostaglandins	*Angew. Chem. Int. Ed. Engl.* **1975**, *14*, 337.
	Angew. Chem. Int. Ed. Engl. **1976**, *15*, 207.
	Angew. Chem. Int. Ed. Engl. **1983**, *22*, 805.
	Angew. Chem. Int. Ed. Engl. **1983**, *22*, 858.
	Angew. Chem. Int. Ed. Engl. **1984**, *23*, 847.
	Chem. Soc. Rev. **1973**, *2*, 29.
	Chem. Soc. Rev. **1975**, *4*, 589.
	Synthesis **1984**, 449.
	Tetrahedron **1979**, *35*, 2705.
	Tetrahedron **1980**, *36*, 2163.
Biosynthesis	*Acc. Chem. Res.* **1985**, *18*, 294.
	Chem. Soc. Rev. **1977**, *6*, 489.
Carbohydrates	
Chirons	*Pure & Appl. Chem.* **1983**, *55*, 565.
Radioimmunoassay	*Acc. Chem. Res.* **1975**, *8*, 335.
Proteases	*Angew. Chem. Int. Ed. Engl.* **1985**, *24*, 85.
Peptide synthesis	*Angew. Chem. Int. Ed. Engl.* **1991**, *30*, 1437.
Protecting groups	
Amines	*Acc. Chem. Res.* **1973**, *6*, 191.
Fluorenylmethoxycarbonyl	*Acc. Chem. Res.* **1987**, *20*, 401.
Aromatic compounds	*Synthesis* **1979**, 921.
Carboxyl group	*Tetrahedron* **1980**, *36*, 2409.
Carboxylate compounds	
Oxazoles	*Chem. Rev.* **1986**, *86*, 845.
Esters	*Tetrahedron* **1980**, *36*, 2409.

Organosilicon chemistry	*Synthesis* **1985**, 817.
Peptides	*Angew. Chem. Int. Ed. Engl.* **1985**, *24*, 799.
Photochemistry	*Synthesis* **1980**, 1.
Polymer supports	*Tetrahedron* **1981**, *37*, 663.
Removal	
Electrochemistry	*Angew. Chem. Int. Ed. Engl.* **1976**, *15*, 281.
Proteinases	*Acc. Chem. Res.* **1974**, *7*, 241.
Inhibitors	*Acc. Chem. Res.* **1974**, *7*, 315.
	Angew. Chem. Int. Ed. Engl. **1974**, *13*, 10.
Serine	*Acc. Chem. Res.* **1987**, *20*, 337.
Proteins	*Angew. Chem. Int. Ed. Engl.* **1974**, *13*, 186.
	Angew. Chem. Int. Ed. Engl. **1991**, *30*, 931.
	Chem. Rev. **1987**, *87*, 381.
	Chem. Rev. **1990**, *90*, 1359.
	Chem. Soc. Rev. **1973**, *2*, 355.
	Chem. Soc. Rev. **1980**, *9*, 325.
	Chem. Soc. Rev. **1985**, *14*, 283.
Antibodies	*Acc. Chem. Res.* **1968**, *1*, 161.
Antifreeze for fish	*Acc. Chem. Res.* **1978**, *11*, 129.
Association	*Angew. Chem. Int. Ed. Engl.* **1984**, *23*, 395.
Binuclear copper proteins	*Tetrahedron* **1989**, *45*, 3.
Biosynthesis	*Angew. Chem. Int. Ed. Engl.* **1985**, *24*, 371.
	Tetrahedron **1977**, *33*, 1671.
Codons	*Angew. Chem. Int. Ed. Engl.* **1978**, *17*, 648.
Conformations	*Acc. Chem. Res.* **1984**, *17*, 209.
	Angew. Chem. Int. Ed. Engl. **1972**, *11*, 894.
Solvent effects	*Acc. Chem. Res.* **1970**, *3*, 62.
Crystallography	*Angew. Chem. Int. Ed. Engl.* **1986**, *25*, 767.
Differentiation	
Evolution	*Angew. Chem. Int. Ed. Engl.* **1981**, *20*, 143.
DNA-protein interactions	*Acc. Chem. Res.* **1980**, *13*, 155.
Electron transfers	*Acc. Chem. Res.* **1988**, *21*, 160.
Engineering	*Angew. Chem. Int. Ed. Engl.* **1984**, *23*, 467.
Factors	
Genetic information	*Angew. Chem. Int. Ed. Engl.* **1972**, *11*, 883.
Folding	*Angew. Chem. Int. Ed. Engl.* **1984**, *23*, 395.
	Pure & Appl. Chem. **1991**, *63*, 187.
RNase A	*Acc. Chem. Res.* **1989**, *22*, 70.
Functionality	*Pure & Appl. Chem.* **1982**, *54*, 2489.
Gels	*Angew. Chem. Int. Ed. Engl.* **1979**, *18*, 599.
Green bacteriochlorophyll	*Acc. Chem. Res.* **1980**, *13*, 309.
α-Helices	
Tropomyosin	*Acc. Chem. Res.* **1982**, *15*, 224.

Proteins, cont'd.	
HPLC	*Angew. Chem. Int. Ed. Engl.* **1986**, *25*, 535.
Human plasma proteins	*Angew. Chem. Int. Ed. Engl.* **1980**, *19*, 87.
Interactions with water	*Acc. Chem. Res.* **1979**, *12*, 7.
Iron proteins	
Oxobridges dinuclear iron proteins	*Chem. Rev.* **1990**, *90*, 1447.
Iron-sulfur proteins	*Angew. Chem. Int. Ed. Engl.* **1973**, *12*, 390.
Isopeptide crosslinks	*Angew. Chem. Int. Ed. Engl.* **1974**, *13*, 514.
Membranes	
Cross-linking	*Acc. Chem. Res.* **1988**, *21*, 435.
NMR	*Acc. Chem. Res.* **1987**, *20*, 221.
Metal binding	*Chem. Soc. Rev.* **1977**, 6, 139.
Metal ions	*Pure & Appl. Chem.* **1983**, *55*, 35.
Micro-organisms	*Chem. Soc. Rev.* **1979**, *8*, 143.
NAD	*Acc. Chem. Res.* **1977**, *10*, 92.
NMR	*Acc. Chem. Res.* **1978**, *11*, 469.
	Acc. Chem. Res. **1981**, *14*, 291.
	Acc. Chem. Res. **1987**, *20*, 221.
Conformations	*Acc. Chem. Res.* **1989**, *22*, 36.
Histidine residues	*Acc. Chem. Res.* **1975**, *8*, 70.
Solid state	*Chem. Rev.* **1991**, *91*, 1307.
Nucleic acids	
Cross-coupling reactions	*Pure & Appl. Chem.* **1980**, *52*, 2717.
Oligosaccharide-protein interactions	*Chem. Soc. Rev.* **1989**, *18*, 347.
Organic fluorescent reagents	*Angew. Chem. Int. Ed. Engl.* **1977**, *16*, 137.
Organometallic chemistry	*Angew. Chem. Int. Ed. Engl.* **1991**, *30*, 931.
Pharmaceutical proteins	*Angew. Chem. Int. Ed. Engl.* **1988**, *27*, 207.
Phosphorylation	
Gene expression	*Angew. Chem. Int. Ed. Engl.* **1988**, *27*, 1040.
Photochemistry	*Acc. Chem. Res.* **1985**, *18*, 134.
	Pure & Appl. Chem, **1980**, *52*, 2705.
	Pure & Appl. Chem. **1980**, *52*, 2717.
Protein-carbohydrate interactions	*Pure & Appl. Chem.* **1987**, *59*, 1447.
	Pure & Appl. Chem. **1989**, *61*, 1293.
Protein-lipid interactions	*Angew. Chem. Int. Ed. Engl.* **1990**, *29*, 1269.
Protein-nucleic acid interactions	*Chem. Rev.* **1987**, *87*, 981.
Protein-pigment complexes	*Angew. Chem. Int. Ed. Engl.* **1988**, *27*, 79.
Protein-saccharide interactions	*Pure & Appl. Chem.* **1983**, *55*, 577.

RNA-protein interactions	*Acc. Chem. Res.* **1977**, *10*, 411.
Structures	*Acc. Chem. Res.* **1983**, *16*, 187.
	Angew. Chem. Int. Ed. Engl. **1985**, *24*, 639.
Sulfenic acids	*Acc. Chem. Res.* **1976**, *9*, 293.
Synthesis	*Angew. Chem. Int. Ed. Engl.* **1991**, *30*, 113.
Proteoglycans	*Chem. Soc. Rev.* **1973**, *2*, 355.
	Pure & Appl. Chem. **1984**, *56*, 779.
	Pure & Appl. Chem. **1991**, *63*, 545.
Proton pump	*Angew. Chem. Int. Ed. Engl.* **1976**, *15*, 17.
	Chem. Rev. **1990**, *90*, 1247.
"Proton sponges"	*Angew. Chem. Int. Ed. Engl.* **1988**, *27*, 865.
Proton transfer	
Acetal hydrolysis	
Reaction mechanisms	*Pure & Appl. Chem.* **1982**, *54*, 1853.
Reaction mechanisms	*Pure & Appl. Chem.* **1981**, *53*, 189.
	Pure & Appl. Chem, **1982**, *54*, 1837.
Protoporphyrin IX	*Acc. Chem. Res.* **1979**, *12*, 374.
Protosterols	
Biosynthesis	*Tetrahedron* **1986**, *42*, 3.
***Pseudomas fluorescens* lipase**	
Esters	*Tetrahedron: Asymmetry* **1991**, *2*, 733.
Pteridines	*Angew. Chem. Int. Ed. Engl.* **1972**, *11*, 1061.
Pulsed electron spin resonance	*Angew. Chem. Int. Ed. Engl.* **1991**, *30*, 265.
Purine bases	
Redox chemistry	*Chem. Rev.* **1989**, *89*, 503.
Purine 8-cyclonucleosides	*Acc. Chem. Res.* **1969**, *2*, 47.
Purines	
Enzymes	
Binding sites	*Acc. Chem. Res.* **1982**, *15*, 128.
Hydrogen exchange	*Chem. Soc. Rev.* **1981**, *10*, 329.
Transition metal complexes	*Acc. Chem. Res.* **1977**, *10*, 146.
Purple membrane of halobacteria	*Angew. Chem. Int. Ed. Engl.* **1976**, *15*, 187.
Pyranose derivatives	*Pure & Appl. Chem.* **1989**, *61*, 1235.
Pyranosides	
Asymmetric synthesis	*SYNLETT* **1991**, 529.
3,5-Pyrazolidinediones	*Synthesis* **1985**, 1028.
2-Pyrazolines	*Synthesis* **1985**, 1028.
Pyrethroid acids	*Angew. Chem. Int. Ed. Engl.* **1981**, *20*, 703.
Pyrethroids	*Chem. Soc. Rev.* **1978**, *7*, 473.
Pyridine nucleotides	
Enzymes	*Acc. Chem. Res.* **1973**, *6*, 289.

Pyridines *Synthesis* **1972**, 464.
Alkynes
Nitriles
Cobalt catalysis *Angew. Chem. Int. Ed. Engl.* **1978**, *17*, 505.
Macrocyles *Chem. Rev.* **1977**, *77*, 513.
Nitriles
Alkynes
Cobalt catalysis *Angew. Chem. Int. Ed. Engl.* **1978**, *17*, 505.
Nucleophilic reactions *Tetrahedron* **1981**, *37*, 3423.
Reaction mechanisms *Chem. Soc. Rev.* **1984**, *13*, 47.
Organocobalt chemistry *Angew. Chem. Int. Ed. Engl.* **1985**, *24*, 248.
Pyridinium chlorochromate
Oxidations *Synthesis* **1982**, 245.
Pyridinium salts
Michael addition *Angew. Chem. Int. Ed. Engl.* **1962**, *1*, 626.
Pyridocarcazole alkaloids *SYNLETT* **1991**, 289.
6*H*-Pyrido[4,3-*b*]carbazoles *Synthesis* **1977**, 437.
Ellipticines *Synthesis* **1984**, 289.
Pyridone
Hydroxypyridine
Tautomerism *Acc. Chem. Res.* **1977**, *10*, 186.
Pyridoxal phosphate *Angew. Chem. Int. Ed. Engl.* **1980**, *19*, 441.
Enzymes *Acc. Chem. Res.* **1980**, *13*, 455.
Pyrilium cations *Tetrahedron* **1980**, *36*, 679.
Pyrimidines *Chem. Soc. Rev.* **1977**, *6*, 43.
Substitution reactions *Acc. Chem. Res.* **1978**, *11*, 462.
Transition metal complexes *Acc. Chem. Res.* **1977**, *10*, 146.
Pyrolysis
Alkyl halides *Chem. Rev.* **1969**, *69*, 33.
Amine oxides *Org. Reactions* **1960**, *11*, 317.
Xanthates *Org. Reactions* **1962**, *12*, 57.
Pyrone
Asymmetric synthesis *Acc. Chem. Res.* **1987**, *20*, 72.
γ-Pyrones *Tetrahedron* **1977**, *33*, 3183.
Pyrrole
3-Substituted pyrroles *Synthesis* **1985**, 353.
Pyrrolenines *Synthesis* **1976**, 281.
Pyrroles *Synthesis* **1976**, 281.
Pyrrolizidine alkaloids *SYNLETT* **1990**, 433.
Biosynthesis *Chem. Soc. Rev.* **1989**, *18*, 375.
Pure & Appl. Chem. **1985**, *57*, 453.
Pyrrolizine chemistry *Synthesis* **1987**, 10.
Pyrrolopyrimidines *Synthesis* **1974**, 837.

Pyrrolo[1,4]benzodiazepine	
Biosynthesis	*Acc. Chem. Res.* **1980**, *13*, 263.
Pyruvate dehydrogenase	*Acc. Chem. Res.* **1974**, *7*, 40.
	Angew. Chem. Int. Ed. Engl. **1975**, *14*, 591.
Quaternary ammonium compounds	*Synthesis* **1973**, 441.
Rearrangement reactions	*Org. Reactions* **1970**, *18*, 403.
Quaternary carbon centers	*Tetrahedron* **1980**, *36*, 419.
Quaternary centers	*Tetrahedron* **1991**, *47*, 9503.
Quinine	
Biosynthesis	*Acc. Chem. Res.* **1969**, *2*, 59.
Quinodimethanes	*Chem. Rev.* **1989**, *89*, 1681.
	SYNLETT **1991**, 301.
Cycloadditions	*Synthesis* **1978**, 793.
Quinoid systems	*Pure & Appl. Chem.* **1990**, *62*, 395.
Quinolizidine alkaloids	
Biosynthesis	*Pure & Appl. Chem.* **1985**, *57*, 453.
Quinones	*Chem. Rev.* **1986**, *86*, 821.
Acetoxylations	*Org. Reactions* **1972**, *19*, 199.
Azulene	*Pure & Appl. Chem.* **1983**, *55*, 363.
Nucleophilic additions	*Tetrahedron* **1991**, *47*, 8043.
Thiele-Winter Acetoxylation	*Org. Reactions* **1972**, *19*, 199.
Radical reactions	*Chem. Soc. Rev.* **1973**, *2*, 397.
	Tetrahedron **1985**, *41*, 3887ff.
	Tetrahedron **1986**, *42*, 2135.
	Tetrahedron **1986**, *42*, 6097ff.
	Tetrahedron **1987**, *43*, 3541.
Acyclic compounds	
Asymmetric synthesis	*Acc. Chem. Res.* **1991**, *24*, 296.
Addition reactions	
Reaction mechanisms	*Angew. Chem. Int. Ed. Engl.* **1982**, *21*, 401.
Alcohols	
Deoxygenation	*Tetrahedron* **1983**, *39*, 2609.
Alkenes	*Acc. Chem. Res.* **1975**, *8*, 165.
	Angew. Chem. Int. Ed. Engl. **1983**, *22*, 753.
	Org. Reactions **1963**, *13*, 91.
	Org. Reactions **1963**, *13*, 150.
Addition reactions	*Tetrahedron* **1980**, *36*, 701.
Amides	*Tetrahedron* **1978**, *34*, 3241.
Aminium radicals	*Chem. Rev.* **1978**, *78*, 243.
Aminyl free radicals	*Angew. Chem. Int. Ed. Engl.* **1975**, *14*, 783.
Aromatic compounds	
Substitution reactions	*Chem. Rev.* **1956**, *56*, 77.
Ipso substitution	*Chem. Rev.* **1979**, *79*, 323.

Radical reactions, cont'd.

Asymmetric reactions	*Tetrahedron* **1981**, *37*, 3073.
Biradical intermediates	*Pure & Appl. Chem.* **1984**, *56*, 1289.
Biradicals	*Tetrahedron* **1982**, *38*, 733ff.
Carboxylic acid derivatives	*Synthesis* **1970**, 99.
Cobalt chemistry	*Chem. Soc. Rev.* **1988**, *17*, 361.
Cyclizations	*SYNLETT* **1991**, 63.
α-Oxo radicals	*Pure & Appl. Chem.* **1988**, *60*, 1645.
Spiro compounds	*Pure & Appl. Chem.* **1988**, *60*, 1645.
Stereochemistry	*Acc. Chem. Res.* **1991**, *24*, 139.
Cyclopropyl radical	*Tetrahedron* **1981**, *37*, 1625.
Diradicals	*Angew. Chem. Int. Ed. Engl.* **1972**, *11*, 92.
Electrochemistry	
Synthetic methods	*Angew. Chem. Int. Ed. Engl.* **1982**, *21*, 256.
Electron transfers	*Tetrahedron* **1990**, *46*, 6193.
ENDOR spectroscopy	*Angew. Chem. Int. Ed. Engl.* **1984**, *23*, 173.
Enzymes	
Alkanes	*Chem. Rev.* **1990**, *90*, 1343.
Free-radical clocks	*Acc. Chem. Res.* **1980**, *13*, 317.
Hydrazidinyl radicals	*Angew. Chem. Int. Ed. Engl.* **1973**, *12*, 455.
Hydroxyl radical	
Reaction mechanisms	*Chem. Rev.* **1986**, *86*, 69.
Inhibitors	*Chem. Rev.* **1987**, *87*, 1313.
Ipso substitution	*Pure & Appl. Chem.* **1981**, *53*, 239.
Kinetic studies	*Pure & Appl. Chem.* **1984**, *56*, 1767.
Lactones	*Pure & Appl. Chem.* **1988**, *60*, 1659.
Mass spectrometry	*Tetrahedron* **1980**, *36*, 2687.
N-Halamide rearrangements	*Synthesis* **1971**, 1.
Nitroxyl radicals	*Synthesis* **1971**, 190.
	Synthesis **1971**, 401.
Organoboranes	*Angew. Chem. Int. Ed. Engl.* **1972**, *11*, 692.
Organomercury compounds	*Angew. Chem. Int. Ed. Engl.* **1985**, *24*, 553.
Organotin compounds	*Angew. Chem. Int. Ed. Engl.* **1985**, *24*, 553.
Pressure effects	*Acc. Chem. Res.* **1972**, *5*, 381.
Radical anions	
Aromatic compounds	*Chem. Rev.* **1974**, *74*, 243.
Radical anions	
Historical perspective	*Angew. Chem. Int. Ed. Engl.* **1971**, *10*, 115.
Substitution reactions	*Angew. Chem. Int. Ed. Engl.* **1975**, *14*, 734.
Radical ions	
Photochemistry	*Angew. Chem. Int. Ed. Engl.* **1987**, *26*, 825.
	Pure & Appl. Chem, **1980**, *52*, 2609.
Radical pairs	*Pure & Appl. Chem*, **1982**, *54*, 1885.
Radical-radical reactions	*Chem. Rev.* **1973**, *73*, 441.

Radicophiles	
Capto-dative substituent effects	*Angew. Chem. Int. Ed. Engl.* **1979**, *18*, 917.
Substitution reactions	*Angew. Chem. Int. Ed. Engl.* **1975**, *14*, 734.
Reaction mechanisms	*Angew. Chem. Int. Ed. Engl.* **1982**, *21*, 401.
	Tetrahedron **1982**, *38*, 313.
Synthesis design	*Synthesis* **1988**, 417.
	Synthesis **1988**, 489.
Synthetic methods	*Chem. Rev.* **1991**, *91*, 1237.
	Synthesis **1973**, 1.
	Synthesis **1988**, 417.
	Synthesis **1988**, 489.
Telomerization	*Acc. Chem. Res.* **1977**, *10*, 9.
1,2,4,5-Tetrazepentenyl radicals	*Angew. Chem. Int. Ed. Engl.* **1973**, *12*, 455.
Thermochemistry	*Pure & Appl. Chem.* **1989**, *61*, 717.
Thiocarbonyl group	*Chem. Rev.* **1989**, *89*, 1413.
Unsaturated compounds	*Synthesis* **1970**, 99.
Verdazyl radicals	*Angew. Chem. Int. Ed. Engl.* **1973**, *12*, 455.
Radiochemistry	*Angew. Chem. Int. Ed. Engl.* **1973**, *12*, 712.
Radiolabelled compounds	*Chem. Rev.* **1971**, *71*, 371.
Organometallic chemistry	*Tetrahedron* **1989**, *45*, 6601.
Radiolysis	
Proteins	*Chem. Rev.* **1987**, *87*, 381.
Random screeening	*Angew. Chem. Int. Ed. Engl.* **1991**, *30*, 296.
Raney nickel	
Desulfurization	*Org. Reactions* **1962**, *12*, 356.
Reaction intermediates	
Substituent effects	*Pure & Appl. Chem.* **1987**, *59*, 1647.
Reaction mechanisms	*Chem. Soc. Rev.* **1973**, *2*, 397.
	Pure & Appl. Chem. **1989**, *61*, 613.
	See also carbocations, hyperconjugation, and under specific molecule or class of compounds
Alternative pathways	*Chem. Soc. Rev.* **1981**, *10*, 345.
Chemical oscillations	*Chem. Rev.* **1973**, *73*, 365.
Conformational analysis	*Pure & Appl. Chem.* **1987**, *59*, 1661.
Curtin-Hammett Winstein-Holness Kinetics	*Chem. Rev.* **1983**, *83*, 73.
Directionality	*Tetrahedron* **1983**, *39*, 1013.
Effective charge	*Chem. Soc. Rev.* **1986**, *15*, 125.
Electron spin alignment	*Pure & Appl. Chem.* **1987**, *59*, 1595.
Electron transfers	*Chem. Rev.* **1990**, *90*, 739.
ESR	*Chem. Soc. Rev.* **1979**, *8*, 1.
Gas phase reactions	*Tetrahedron* **1982**, *38*, 3195.

Reaction mechanisms, cont'd.	
Intramolecular reactions	
Relaxation times	*Angew. Chem. Int. Ed. Engl.* **1981**, *20*, 487.
Isotope effects	*Chem. Soc. Rev.* **1972**, *1*, 163.
	Chem. Soc. Rev. **1974**, *3*, 513.
	Chem. Soc. Rev. **1989**, *18*, 209.
Kinetic control	*Chem. Soc. Rev.* **1987**, *16*, 313.
Molecular dynamics	*Angew. Chem. Int. Ed. Engl.* **1987**, *26*, 1221.
Neighboring group effects	*Chem. Soc. Rev.* **1973**, *2*, 295.
Oscillations	*Chem. Rev.* **1990**, *90*, 355.
Pi-sigma interactions	*Pure & Appl. Chem.* **1987**, *59*, 1585.
Polar axis	
Organic solid state chemistry	*Chem. Rev.* **1981**, *81*, 525.
Polar effects	*Angew. Chem. Int. Ed. Engl.* **1976**, *15*, 569.
Reactivity	*Pure & Appl. Chem,* **1982**, *54*, 1867.
	Tetrahedron **1985**, *41*, 1613.
Selectivity	*Pure & Appl. Chem.* **1991**, *63*, 283.
Sigma-pi interactions	*Pure & Appl. Chem.* **1987**, *59*, 1637.
Solution reactions	*Tetrahedron* **1983**, *39*, 1013.
Solvent effects	*Pure & Appl. Chem,* **1982**, *54*, 1867.
Solvolysis	*Tetrahedron* **1982**, *38*, 3195.
Stereoelectronic effects	*Pure & Appl. Chem.* **1987**, *59*, 1605.
Stereoselectivity	
Theoretical aspects	*Pure & Appl. Chem.* **1989**, *61*, 643.
Strain	*Tetrahedron* **1985**, *41*, 1613.
Theoretical approach	*Chem. Soc. Rev.* **1984**, *13*, 1.
Unimolecular reactions	*Chem. Soc. Rev.* **1983**, *12*, 163.
Reactions in binary systems	
Reaction mechanisms	*Pure & Appl. Chem.* **1982**, *54*, 1797.
Reactive intermediates	*Angew. Chem. Int. Ed. Engl.* **1976**, *15*, 251.
	See also carbanions, carbenes, carbenium ions, carbocations, photochemistry, or specific intermediate
Cation radicals	*Pure & Appl. Chem.* **1991**, *63*, 223.
Dissolved electrons	*Angew. Chem. Int. Ed. Engl.* **1978**, *17*, 887.
Enzymes	*Chem. Soc. Rev.* **1984**, *13*, 97.
Flash thermolysis	*Angew. Chem. Int. Ed. Engl.* **1977**, *16*, 365.
Ion radical reactions	*Tetrahedron* **1985**, *41*, 2771.
Isotope effects	*Chem. Rev.* **1972**, *72*, 511.
Non-Born-Oppenheimer effects	*Angew. Chem. Int. Ed. Engl.* **1983**, *22*, 210.
Non-classical ions	*Tetrahedron* **1976**, *32*, 179.
Tetrahedral intermediates	*Tetrahedron* **1975**, *31*, 2463.

Reactive metal powders	*Acc. Chem. Res.* **1977**, *10*, 301.
Reactivity	*See reaction mechanisms*
Rearrangements	*Chem. Rev.* **1976**, *76*, 187.
Reaction mechanisms	*Pure & Appl. Chem.* **1989**, *61*, 661.
	Tetrahedron **1982**, *38*, 567.
Receptor sites	*Tetrahedron* **1984**, *40*, 455ff.
Receptors	
Macrocycles	*Chem. Soc. Rev.* **1989**, *18*, 409.
Siderophores	*Pure & Appl. Chem.* **1989**, *61*, 1529.
Recombinant DNA	*Angew. Chem. Int. Ed. Engl.* **1983**, *22*, 842.
Red blood pigment	
Biosynthesis	*Angew. Chem. Int. Ed. Engl.* **1982**, *21*, 343.
Red pigments	*Angew. Chem. Int. Ed. Engl.* **1974**, *13*, 305.
Redox chemistry	*Angew. Chem. Int. Ed. Engl.* **1978**, *17*, 875.
Macrocycles	*Chem. Soc. Rev.* **1990**, *19*, 239.
Purine bases	*Chem. Rev.* **1989**, *89*, 503.
Reducing agents	*Acc. Chem. Res.* **1980**, *13*, 330.
Carbonylations	*Pure & Appl. Chem.* **1985**, *57*, 1875.
Reductions	*Chem. Soc. Rev.* **1976**, *5*, 23.
	Pure & Appl. Chem. **1988**, *60*, 35.
Alkali metal reductions	*Tetrahedron* **1986**, *42*, 6351.
Alkenes	
Alkynes	
Addition reactions	*Chem. Rev.* **1942**, *31*, 77.
Alkoxyaluminohydrides	*Synthesis* **1972**, 217.
Amines	*Org. Reactions* **1944**, *2*, 262.
Annulenes	*Chem. Rev.* **1984**, *84*, 603.
Aromatic compounds	*Tetrahedron* **1989**, *45*, 1579.
Metal ammonia reductions	*Synthesis* **1970**, 161.
Aromatic nitro compounds	*Synthesis* **1969**, 9.
Asymmetric synthesis	*Pure & Appl. Chem.* **1980**, *52*, 843.
	Tetrahedron: Asymmetry **1991**, *2*, 481ff.
	Tetrahedron: Asymmetry **1991**, *2*, 843.
Azines	*Tetrahedron* **1988**, *44*, 1.
Baker's yeast	*Chem. Rev.* **1991**, *91*, 49.
Benzoylformates	
Asymmetric synthesis	*Tetrahedron: Asymmetry* **1991**, *2*, 299.
Carbonyl compounds	*Angew. Chem.* **1956**, *68*, 601.
Asymmetric synthesis	*Pure & Appl. Chem.* **1981**, *53*, 2315.
Carboxylic acid derivatives	*Org. Reactions* **1988**, *36*, 249.
Catalytic hydrogen transfer	*Chem. Rev.* **1974**, *74*, 567.
	Chem. Rev. **1985**, *85*, 129.
Complex reducing agents	*Angew. Chem. Int. Ed. Engl.* **1983**, *22*, 599.

Reductions, cont'd.

Cyclic unsaturated hydrocarbons	*Angew. Chem. Int. Ed. Engl.* **1987**, *26*, 204.
Diazo compounds	*Acc. Chem. Res.* **1988**, *21*, 400.
DIBAL	*Synthesis* **1975**, 617.
Diboranes	*Chem. Rev.* **1976**, *76*, 773.
Diimide	*Org. Reactions* **1991**, *40*, 91.
Electrolytic	*Chem. Rev.* **1962**, *62*, 19.
Enones	*Tetrahedron* **1986**, *42*, 6351.
Enzymatic reactions	
Electrochemistry	*Angew. Chem. Int. Ed. Engl.* **1985**, *24*, 539.
Enzymes	*Angew. Chem. Int. Ed. Engl.* **1974**, *13*, 569.
Hydrazine catalysis	*Chem. Rev.* **1965**, *65*, 51.
Hydrides	
Historical perspective	*Tetrahedron* **1979**, *35*, 567.
Hydrogenations	*Angew. Chem. Int. Ed. Engl.* **1988**, *27*, 679.
Asymmetric synthesis	*Angew. Chem. Int. Ed. Engl.* **1971**, *10*, 871.
Esters	*Org. Reactions* **1954**, *8*, 1.
Homogeneous catalysis	*Chem. Rev.* **1973**, *73*, 21.
	Org. Reactions **1976**, *24*, 1.
Nickel-copper alloy	*Chem. Rev.* **1980**, *80*, 417.
Ketones	*Acc. Chem. Res.* **1983**, *16*, 399.
	Acc. Chem. Res. **1984**, *17*, 338.
	Pure & Appl. Chem. **1977**, *49*, 1049.
	Tetrahedron **1986**, *42*, 6351.
Hydride reagents	*Tetrahedron* **1979**, *35*, 449.
Microbial reductions	*Angew. Chem. Int. Ed. Engl.* **1984**, *23*, 570.
Stereoselectivity	*Acc. Chem. Res.* **1984**, *17*, 338.
Steroidal ketones	*Synthesis* **1972**, 526.
Lithium aluminum hydride	*Org. Reactions* **1951**, *6*, 469.
	Synthesis **1974**, 691.
Lithium-amine reductions	*Synthesis* **1972**, 391.
Low-valent metals	*Synthesis* **1979**, 1.
Macrocycles	*Pure & Appl. Chem.* **1989**, *61*, 1555.
Magnesium	*Chem. Rev.* **1957**, *57*, 417.
Metal ammonia reductions	
Aromatic compounds	*Synthesis* **1970**, 161.
Metal hydrides	*Org. Reactions* **1985**, *34*, 1.
	Org. Reactions **1988**, *36*, 249.
Metal-amines	*Quart. Rev.* **1958**, *12*, 17.
Nickel-aluminum alloy	*Chem. Rev.* **1989**, *89*, 459.
Nitro compounds	*Acc. Chem. Res.* **1974**, *7*, 281.
	Org. Reactions **1988**, *36*, 249.
Organoboranes	
Asymmetric synthesis	*Chem. Rev.* **1989**, *89*, 1553.

Organotin hydrides	*Synthesis* **1970**, 499.
Peptides	
Sodium-ammonia	*Chem. Rev.* **1984**, *84*, 287.
Polyhalocarbonyl compounds	*Org. Reactions* **1983**, *29*, 163.
Rearrangements	
Lithium aluminum hydride	*Synthesis* **1974**, 691.
Reductive alkylations	
Amines	*Org. Reactions* **1948**, *4*, 174.
Reductive cleavage	
Nitroalkanes	*Synthesis* **1986**, 693.
Sodium cyanoborohydride	*Synthesis* **1975**, 135.
Steroidal ketones	*Synthesis* **1972**, 526.
Sulfoxides	*Tetrahedron* **1988**, *44*, 6537.
Sulfur compounds	*Org. Reactions* **1988**, *36*, 249.
Sulfurated borohydrides	*Synthesis* **1972**, 526.
Thiocarboxylic *O*-esters	*Chem. Rev.* **1984**, *84*, 17.
TIBA	*Synthesis* **1975**, 617.
Transition metal complexes	
Low valent metal complexes	*Tetrahedron* **1988**, *44*, 4295.
Trichloromethyl compounds	*Synthesis* **1983**, 773.
Unsaturated compounds	*Org. Reactions* **1976**, *23*, 1.
Replication	*Angew. Chem. Int. Ed. Engl.* **1990**, *29*, 36.
Genetic information	
Errors	*Angew. Chem. Int. Ed. Engl.* **1985**, *24*, 1015.
RNA	*Pure & Appl. Chem.* **1984**, *56*, 967.
Representation in chemistry	*Angew. Chem. Int. Ed. Engl.* **1980**, *19*, 110.
	Angew. Chem. Int. Ed. Engl. **1991**, *30*, 1.
Resolution	
Asymmetric synthesis	*Angew. Chem. Int. Ed. Engl.* **1991**, *30*, 293.
Restriction endonucleases	*Angew. Chem. Int. Ed. Engl.* **1978**, *17*, 73.
Retinal pigments	*Pure & Appl. Chem.* **1986**, *58*, 719.
Retinoids	*Acc. Chem. Res.* **1983**, *16*, 81.
Retro-peptides	*Acc. Chem. Res.* **1979**, *12*, 1.
Retrosynthetic analysis	*Angew. Chem. Int. Ed. Engl.* **1991**, *30*, 455.
	Chem. Soc. Rev. **1988**, *17*, 111.
Chirons	*Pure & Appl. Chem.* **1990**, *62*, 1887.
Computer designed synthesis	*Pure & Appl. Chem.* **1988**, *60*, 1563.
	Pure & Appl. Chem. **1988**, *60*, 1573.
Cyclizations	*SYNLETT* **1991**, 63.
Synthons	*Angew. Chem. Int. Ed. Engl.* **1991**, *30*, 455.
	Pure & Appl. Chem. **1990**, *62*, 1887.
Retroviruses	*Angew. Chem. Int. Ed. Engl.* **1990**, *29*, 707.
	Angew. Chem. Int. Ed. Engl. **1990**, *29*, 716.

Reverse micelles	
Surfactants	*Acc. Chem. Res.* **1976**, *9*, 153.
Rhenium compounds	*Pure & Appl. Chem.* **1985**, *57*, 1789.
Rheumatoid arthritis drugs	*Chem. Soc. Rev.* **1980**, *9*, 217.
Rhizopus arrhizus	
Δ^4-3-Ketosteroids	*Acc. Chem. Res.* **1984**, *17*, 398.
Rhodium chemistry	*Pure & Appl. Chem.* **1984**, *56*, 13.
	Pure & Appl. Chem. **1984**, *56*, 99.
Alkenes	*Acc. Chem. Res.* **1968**, *1*, 186.
Diazocarbonyl compounds	*Tetrahedron* **1991**, *47*, 1765.
Homogeneous catalysis	*Acc. Chem. Res.* **1978**, *11*, 301.
Hydroformylation	*Angew. Chem. Int. Ed. Engl.* **1980**, *19*, 178.
Hydrogenation	
Asymmetric synthesis	*Acc. Chem. Res.* **1983**, *16*, 106.
Hydrosilylation	
Asymmetric synthesis	*Angew. Chem. Int. Ed. Engl.* **1983**, *22*, 897.
Pentamethylcyclopentadienyl rhodium complexes	
Catalysis	*Acc. Chem. Res.* **1978**, *11*, 301.
Rhodium hydrides	
Homogeneous catalysis	*Acc. Chem. Res.* **1984**, *17*, 221.
Rhodopsins	*Pure & Appl. Chem.* **1986**, *58*, 725.
Photochemistry	*Acc. Chem. Res.* **1975**, *8*, 101.
	Acc. Chem. Res. **1975**, *8*, 407.
Ribonuclease	
Reaction mechanisms	*Acc. Chem. Res.* **1968**, *1*, 321.
Ribonuclease T1	*Angew. Chem. Int. Ed. Engl.* **1991**, *30*, 343.
Ribonucleic acids	*See RNA*
Ribonucleotides	*Angew. Chem. Int. Ed. Engl.* **1974**, *13*, 569.
Ribooligonucleotides	*Acc. Chem. Res.* **1974**, *7*, 92.
Ribose	*Acc. Chem. Res.* **1988**, *21*, 294.
Ribosomes	
Genetic information	*Angew. Chem. Int. Ed. Engl.* **1971**, *10*, 638.
Structure	*Angew. Chem. Int. Ed. Engl.* **1983**, *22*, 456.
Synthesis	*Angew. Chem. Int. Ed. Engl.* **1976**, *15*, 533.
Ribulose bisphosphate oxygenase	*Acc. Chem. Res.* **1980**, *13*, 394.
Ribulose bisphosphate carboxylase	*Acc. Chem. Res.* **1980**, *13*, 394.
Rifampcin	
Mechanism of action	*Angew. Chem. Int. Ed. Engl.* **1985**, *24*, 1009.
Rifamycin S	*Pure & Appl. Chem.* **1981**, *53*, 1163.

Ring transformations
Heterocyclic compounds
1,2-Oxazoles — *Synthesis* **1975**, 20.
Ring closures — *Acc. Chem. Res.* **1981**, *14*, 95.
Cyclizations
Carbon monoxide — *Angew. Chem. Int. Ed. Engl.* **1966**, *5*, 435.
Heterocycles
Reaction mechanisms — *Tetrahedron* **1987**, *43*, 5171.
Ring compounds — *Chem. Rev.* **1985**, *85*, 341.
Conformations — *Chem. Soc. Rev.* **1974**, *3*, 1.
Neighboring group effects — *Chem. Soc. Rev.* **1974**, *3*, 1.
Transannular reactions — *Acc. Chem. Res.* **1969**, *2*, 210.
Ring contractions
Cyclobutanes — *Acc. Chem. Res.* **1972**, *5*, 33.
Angew. Chem. Int. Ed. Engl. **1975**, *14*, 473.

Ring expansion
Bridged bicyclic ketones — *Tetrahedron* **1987**, *43*, 3.
Cyclopropylmethyl compounds — *Angew. Chem. Int. Ed. Engl.* **1975**, *14*, 473.
Macrocycles — *Tetrahedron* **1988**, *44*, 1573.
Organosulfur chemistry — *Acc. Chem. Res.* **1984**, *17*, 358.
Oxacarbenes — *Acc. Chem. Res.* **1975**, *8*, 209.
Ring fission — *Chem. Rev.* **1978**, *78*, 517.
Ring inversions
Cyclooctatetraenes — *Pure & Appl. Chem.* **1982**, *54*, 987.
Ring opening
Polymerizations — *Pure & Appl. Chem* **1976**, *48*, 257.
Risk management
Industrial chemistry — *Chem. Soc. Rev.* **1979**, *8*, 419.
RNA — *Acc. Chem. Res.* **1977**, *10*, 388.
Chem. Rev. **1990**, *90*, 1327.

Base pairing
NMR — *Acc. Chem. Res.* **1977**, *10*, 396.
Biosynthesis — *Acc. Chem. Res.* **1977**, *10*, 418.
Angew. Chem. Int. Ed. Engl. **1975**, *14*, 445.

Cleavage
RNase — *Acc. Chem. Res.* **1991**, *24*, 317.
Conformations
Pseudoknot — *Acc. Chem. Res.* **1991**, *24*, 152.
Metal ion interactions — *Chem. Rev.* **1971**, *71*, 439.
Modified nucleosides — *Acc. Chem. Res.* **1977**, *10*, 403.
NMR — *Acc. Chem. Res.* **1974**, *7*, 33.
Acc. Chem. Res. **1977**, *10*, 396.
Acc. Chem. Res. **1991**, *24*, 152.

RNA, cont'd.
Photoaffinity labels — *Acc. Chem. Res.* **1991**, *24*, 183.
Protein
Biosynthesis — *Tetrahedron* **1977**, *33*, 1671.
Replication — *Pure & Appl. Chem.* **1984**, *56*, 967.
RNA cleavage — *Angew. Chem. Int. Ed. Engl.* **1990**, *29*, 749.
RNA-protein interactions — *Acc. Chem. Res.* **1977**, *10*, 411.
Synthesis — *Pure & Appl. Chem.* **1987**, *59*, 325.
Synthetase — *Acc. Chem. Res.* **1977**, *10*, 411.
Viruses — *Acc. Chem. Res.* **1974**, *7*, 169.
RNase — *See also ribonuclease*
RNA Cleavage
Imidazole mechanism — *Acc. Chem. Res.* **1991**, *24*, 317.
RNase A — *Acc. Chem. Res.* **1989**, *22*, 70.
Rotational isomerism
1,2-Diarylethylene — *Chem. Rev.* **1991**, *91*, 1679.
Ruthenium chemistry — *SYNLETT* **1991**, 755.
Carbonyl cluster compounds — *Pure & Appl. Chem.* **1982**, *54*, 97.
Half-sandwich ruthenium complexes — *Chem. Rev.* **1987**, *87*, 761.
Macromolecules — *Pure & Appl. Chem.* **1988**, *60*, 495.
Photochemistry — *Pure & Appl. Chem.* **1988**, *60*, 973.
Water-gas shift reaction — *Acc. Chem. Res.* **1981**, *14*, 31.
Sapphyrins
Porphyrins — *SYNLETT* **1991**, 127.
Schiff bases
Macrocycles — *Pure & Appl. Chem.* **1986**, *58*, 1437.
Secoiridoid glycosides — *Angew. Chem. Int. Ed. Engl.* **1983**, *22*, 828.
Secologanin — *Angew. Chem. Int. Ed. Engl.* **1983**, *22*, 828.
Secondary metabolism — *Chem. Soc. Rev.* **1988**, *17*, 383.
Selenium chemistry — *Acc. Chem. Res.* **1979**, *12*, 22.
Selenium dioxide
Oxidations — *Org. Reactions* **1949**, *5*, 331.
Org. Reactions **1976**, *24*, 261.
Selenoethers
Dealkylations — *Synthesis* **1988**, 749.
Self-assembly
Helices — *Angew. Chem. Int. Ed. Engl.* **1991**, *30*, 407.
Macromolecules — *SYNLETT* **1991**, 445.
Self-organization — *Angew. Chem. Int. Ed. Engl.* **1972**, *11*, 798.
Serine hydroxymethyltransferase — *Chem. Rev.* **1990**, *90*, 1275.
Serine protease — *Acc. Chem. Res.* **1989**, *22*, 322.
Sesquiterpenes — *Angew. Chem. Int. Ed. Engl.* **1973**, *12*, 793.
Tetrahedron **1987**, *43*, 5467ff.

Bioconversion	*Tetrahedron* **1990**, *46*, 4109.
Biosynthesis	*Chem. Rev.* **1976**, *76*, 425.
	Pure & Appl. Chem. **1981**, *53*, 1155.
	Pure & Appl. Chem. **1981**, *53*, 1241.
Enzymes	
Biosynthesis	*Chem. Rev.* **1990**, *90*, 1089.
NMR	*Pure & Appl. Chem.* **1981**, *53*, 1241.
Polycarbocyclic sesquiterpenes	*Tetrahedron* **1985**, *41*, 1767.
Shikimic acid pathway	*Tetrahedron* **1978**, *34*, 3353.
Sialic acid	*Tetrahedron* **1990**, *46*, 5835.
Sibiromycin	
Biosynthesis	*Acc. Chem. Res.* **1980**, *13*, 263.
Siderophores	*Chem. Rev.* **1989**, *89*, 1563.
	Chem. Rev. **1984**, *84*, 587.
	Pure & Appl. Chem. **1989**, *61*, 1529.
Sigma-pi interactions	*Pure & Appl. Chem.* **1987**, *59*, 1637.
Signal transducers	*Angew. Chem. Int. Ed. Engl.* **1972**, *11*, 1.
Silenes	
Photochemistry	*Acc. Chem. Res.* **1987**, *20*, 329.
Silicon chemistry	*Angew. Chem. Int. Ed. Engl.* **1973**, *12*, 474.
Silicon-carbon bond cleavage	*Synthesis* **1984**, 991.
Siloxanes	*SYNLETT* **1990**, 651.
Siloxy dienes	*Acc. Chem. Res.* **1981**, *14*, 400.
Silver catalysis	
Rearrangements	*Acc. Chem. Res.* **1971**, *4*, 280.
	Synthesis **1975**, 347.
Silyl anions	*Acc. Chem. Res.* **1987**, *20*, 127.
Silyl enol ethers	*Synthesis* **1983**, 1.
	Synthesis **1983**, 85.
Nitroalkenes	*Acc. Chem. Res.* **1985**, *18*, 284.
Silyl enolates	*Synthesis* **1977**, 91.
Silyl esters of phosphorus	*Tetrahedron* **1989**, *45*, 2465.
Silylations	*Acc. Chem. Res.* **1970**, *3*, 299.
Trialkylsilyl perfluorosulfonates	*Synthesis* **1982**, 1.
Silyl-proton exchange	*Acc. Chem. Res.* **1970**, *3*, 299.
Singlet oxygen	*Tetrahedron* **1981**, *37*, 1825.
	Tetrahedron **1985**, *41*, 2037ff.
Addition reactions	
Alkenes	*Pure Appl. Chem.* **1971**, *27*, 635.
Alkenes	
Ene reaction	*Acc. Chem. Res.* **1980**, *13*, 419.
Reaction mechanisms	*Chem. Rev.* **1979**, *79*, 359.
Photochemistry	
Reaction mechanisms	*Pure & Appl. Chem.* **1990**, *62*, 1467.

Site-directed mutagenesis	*Chem. Rev.* **1987**, *87*, 1079.
Small molecules	
Conformations	*Chem. Soc. Rev.* **1972**, *1*, 293.
Small ring compounds	*Pure & Appl. Chem.* **1991**, *63*, 223.
	SYNLETT **1990**, 20.
Hydrogenolysis	*Chem. Rev.* **1963**, *63*, 123.
Photochemistry	*Acc. Chem. Res.* **1971**, *4*, 48.
Rearrangement reactions	*Chem. Rev.* **1976**, *76*, 461.
	Org. Photochem. **1967**, *1*, 91.
Metal catalysis	*Chem. Rev.* **1976**, *76*, 461.
Small ring propellanes	*Acc. Chem. Res.* **1972**, *5*, 249.
***S*-Nitrosations**	*Chem. Soc. Rev.* **1985**, *14*, 171.
Sodium cyanoborohydride	
Reductions	*Synthesis* **1975**, 135.
Solid supports	
Three-phase test	*Tetrahedron* **1979**, *35*, 723.
Solvent effects	
Carbocations	*Pure & Appl. Chem.* **1984**, *56*, 1797.
Cycloaddition reactions	*Pure & Appl. Chem.* **1980**, *52*, 2283.
1,3-Dipolar cycloadditions	*Synthesis* **1973**, 71.
Photochemistry	*Pure & Appl. Chem.* **1986**, *58*, 1279.
Reaction mechanisms	*Pure & Appl. Chem,* **1982**, *54*, 1867.
Sonochemistry	*Chem. Soc. Rev.* **1987**, *16*, 239.
	Chem. Soc. Rev. **1987**, *16*, 275.
	Synthesis **1989**, 787.
Spermidine	
Analogues	*Acc. Chem. Res.* **1986**, *19*, 105.
Spherands	*Angew. Chem. Int. Ed. Engl.* **1986**, *25*, 1039.
Spices	*Angew. Chem. Int. Ed. Engl.* **1978**, *17*, 710.
Spin-labelled compounds	*Synthesis* **1981**, 682.
Carbohydrates	*Chem. Rev.* **1986**, *86*, 203.
Spiro compounds	*Synthesis* **1978**, 77.
Alkylations	
Intramolecular	*Synthesis* **1974**, 383.
Carbanions	*Acc. Chem. Res.* **1972**, *5*, 354.
Rearrangements	*Synthesis* **1976**, 425.
Spiro intermediates	*Acc. Chem. Res.* **1972**, *5*, 354.
Spirodienes	
Rearrangement reactions	*Quart. Rev.* **1971**, *25*, 483.
Spiroketals	*Chem. Rev.* **1989**, *89*, 1617.
	Tetrahedron **1987**, *43*, 3309.

Sponges	*Pure & Appl. Chem.* **1986**, *58*, 357.
	Pure & Appl. Chem. **1986**, *58*, 423.
Phospholipids	*Acc. Chem. Res.* **1991**, *24*, 69.
Squalene	
Biosynthesis	*Acc. Chem. Res.* **1968**, *1*, 1.
	Acc. Chem. Res. **1968**, *1*, 111.
Stannylenes	*Chem. Rev.* **1991**, *91*, 311.
Starburst dendrimers	*Angew. Chem. Int. Ed. Engl.* **1990**, *29*, 138.
Starch	
Biosynthesis	*Pure & Appl. Chem.* **1991**, *63*, 535.
Stenhouse salts	*Tetrahedron* **1977**, *33*, 463.
Stereochemistry	
Enzyme reactions	*Pure & Appl. Chem.* **1984**, *56*, 1005.
Topology	*Chem. Soc. Rev.* **1973**, *2*, 457.
Stereoelectronic effects	*Chem. Rev.* **1987**, *87*, 1047.
	Pure & Appl. Chem. **1987**, *59*, 1605.
Stereoselectivity	*Pure & Appl. Chem.* **1989**, *61*, 643.
Steroids	*Angew. Chem. Int. Ed. Engl.* **1971**, *10*, 450.
	Angew. Chem. Int. Ed. Engl. **1991**, *30*, 1278.
	Chem. Rev. **1978**, *78*, 199.
	Pure & Appl. Chem. **1986**, *58*, 423.
	Synthesis **1970**, 74.
Asymmetric synthesis	*Acc. Chem. Res.* **1976**, *9*, 412.
Biosynthesis	*Chem. Soc. Rev.* **1981**, *10*, 435.
	Pure & Appl. Chem. **1987**, *59*, 287.
Chlorination	*Acc. Chem. Res.* **1980**, *13*, 170.
C-nor D-homosteroids	*Tetrahedron* **1979**, *35*, 911.
Cobalt catalysis	*Pure & Appl. Chem.* **1985**, *57*, 1819.
Cycloadditions	
Intramolecular cycloadditions	*Tetrahedron* **1981**, *37*, 3.
Hydroxylation	
Biosynthesis	*Chem. Soc. Rev.* **1982**, *11*, 371.
Δ^4-3-Ketosteroids	
Rhizopus arrhizus	*Acc. Chem. Res.* **1984**, *17*, 398.
Side chains	*Chem. Rev.* **1978**, *78*, 199.
Steroidal glycosides	*Pure & Appl. Chem.* **1982**, *54*, 1935.
	Pure & Appl. Chem. **1986**, *58*, 799.
Steroidal hormones	*Pure & Appl. Chem.* **1986**, *58*, 597.
Steroidal ketones	
Reductions	
Sulfurated borohydrides	*Synthesis* **1972**, 526.
Steroidal side chain	*Chem. Soc. Rev.* **1983**, *12*, 75.

Sterols	*See also natural products*
Biosynthesis	*Acc. Chem. Res.* **1991**, *24*, 371.
	Pure & Appl. Chem. **1981**, *53*, 837.
	Pure & Appl. Chem. **1981**, *53*, 1241.
	Pure & Appl. Chem. **1989**, *61*, 345.
	Tetrahedron **1986**, *42*, 3.
Stibides	*Synthesis* **1974**, 328.
Stilbenes	*Synthesis* **1983**, 341.
Cycloadditions	*Pure & Appl. Chem.* **1989**, *61*, 635.
Photochemistry	*Pure & Appl. Chem,* **1980**, *52*, 2683.
	Pure & Appl. Chem. **1982**, *54*, 1705.
Photocyclizations	*Org. Reactions* **1984**, *30*, 1.
Photoisomerization	*Chem. Rev.* **1991**, *91*, 415.
Strain	*Angew. Chem. Int. Ed. Engl.* **1986**, *25*, 312.
Reaction mechanisms	*Pure & Appl. Chem.* **1984**, *56*, 1781.
Reactivity	*Tetrahedron* **1985**, *41*, 1613.
Strain assistance	*Tetrahedron* **1989**, *45*, 2875ff.
Strained alcohols	*Chem. Rev.* **1989**, *89*, 1035.
Strained alkanes	
Electrophilic addition reactions	*Acc. Chem. Res.* **1969**, *2*, 152.
Strained cumulenes	*Chem. Rev.* **1989**, *89*, 1111.
Strained hydrocarbons	*Chem. Rev.* **1989**, *89*, 1125.
Photochemistry	*Angew. Chem. Int. Ed. Engl.* **1986**, *25*, 661.
Strained molecules	*Chem. Rev.* **1976**, *76*, 311.
Cycloaddition reactions	*Acc. Chem. Res.* **1971**, *4*, 128.
Isomerization	*Angew. Chem. Int. Ed. Engl.* **1963**, *2*, 1.
Organogermanium chemistry	*Angew. Chem. Int. Ed. Engl.* **1991**, *30*, 958.
Organosilicon chemistry	*Angew. Chem. Int. Ed. Engl.* **1991**, *30*, 958.
Organotin chemistry	*Angew. Chem. Int. Ed. Engl.* **1991**, *30*, 958.
Streptogramin antibiotics	
Synthesis	*Pure & Appl. Chem.* **1982**, *54*, 2537.
Structural chemistry	*Tetrahedron* **1986**, *42*, 6097ff.
Ab initio approaches	*Chem. Rev.* **1991**, *91*, 679.
Chirality	*Chem. Soc. Rev.* **1986**, *15*, 189.
Circular dichroism	*Chem. Rev.* **1975**, *75*, 323.
Conformational analysis	
Pentamethylene heterocycles	*Chem. Rev.* **1975**, *75*, 611.
Photoelectron spectroscopy	*Angew. Chem. Int. Ed. Engl.* **1979**, *18*, 826.
Conformations	*Chem. Soc. Rev.* **1972**, *1*, 293.
Torsional angles	*Tetrahedron* **1980**, *36*, 2809.
Isomerization	*Coordination Chem. Rev.* **1973**, *11*, 161.
Isomers	*Chem. Soc. Rev.* **1974**, *3*, 355.
Multiphoton-ionization-mass spectrometry	*Angew. Chem. Int. Ed. Engl.* **1988**, *27*, 447.

Neuralization-reionization mass spectrometry	*Angew. Chem. Int. Ed. Engl.* **1987**, *26*, 805.
Non-bonding interactions	*Angew. Chem. Int. Ed. Engl.* **1983**, *22*, 283.
Quantative mass spectrometry	*Angew. Chem. Int. Ed. Engl.* **1978**, *17*, 221.
Ring compounds	
Conformations	*Tetrahedron* **1980**, *36*, 2809.
Scanning Tunnel Microscopy	
Historical perspective	*Angew. Chem. Int. Ed. Engl.* **1987**, *26*, 606.
Topological stereochemistry	*Tetrahedron* **1985**, *41*, 3161.
Torsional angles	
Conformations	*Tetrahedron* **1980**, *36*, 2809.
Transannular interactions	*Angew. Chem. Int. Ed. Engl.* **1983**, *22*, 283.
X-ray studies	
Metal-based drugs	*Chem. Rev.* **1987**, *87*, 1027.
Structural studies	
Bio-active compounds	*Chem. Soc. Rev.* **1984**, *13*, 131.
Structure determination	
DENDRAL	*Pure & Appl. Chem.* **1982**, *54*, 2425.
Structure activity relationships	*Acc. Chem. Res.* **1969**, *2*, 232.
Free radicals	*Angew. Chem. Int. Ed. Engl.* **1970**, *9*, 830.
Strychnine	*Angew. Chem. Int. Ed. Engl.* **1985**, *24*, 365.
Substituent effects	*Pure & Appl. Chem.* **1987**, *59*, 1647.
Substitution reactions	
Acetals	*Pure & Appl. Chem.* **1988**, *60*, 49.
Aliphatic compounds	
S_N2	*Acc. Chem. Res.* **1973**, *6*, 46.
	Acc. Chem. Res. **1976**, *9*, 281.
Alkenes	
Aromatic alkenes	
Palladium chemistry	*Synthesis* **1973**, 524.
Allylic compounds	*Chem. Rev.* **1956**, *56*, 753.
Amides	
Orthometallations	*Chem. Rev.* **1990**, *90*, 879.
Aromatic compounds	
Electrochemistry	
Anodic	*Acc. Chem. Res.* **1973**, *6*, 106.
Electrophilic	*Acc. Chem. Res.* **1974**, *7*, 361.
	Chem. Soc. Rev. **1974**, *3*, 167.
	Synthesis **1971**, 455.
Ipso substitution	*Acc. Chem. Res.* **1980**, *13*, 51.
Nitration	
Ipso substitution	*Acc. Chem. Res.* **1976**, *9*, 287.
Reaction mechanisms	*Acc. Chem. Res.* **1982**, *15*, 164.
Transition metal complexes	*Acc. Chem. Res.* **1970**, *3*, 139.

Substitution reactions, cont'd.	
Aromatic substitution	
Arynes	*Angew. Chem.* **1960**, *72*, 91.
Heterolytic substitution	*Angew. Chem. Int. Ed. Engl.* **1972**, *11*, 874.
Ipso substitution	*Chem. Rev.* **1979**, *79*, 323.
Nucleophilic	*Acc. Chem. Res.* **1977**, *10*, 125.
Oxazolines	*Tetrahedron* **1985**, *41*, 837.
Palladium compounds	*Synthesis* **1973**, 524.
Radical chemistry	*Chem. Rev.* **1979**, *79*, 323.
	Pure & Appl. Chem. **1981**, *53*, 239.
Reaction mechanisms	*Angew. Chem. Int. Ed. Engl.* **1972**, *11*, 874.
Asymmetric synthesis	*Pure & Appl. Chem.* **1980**, *52*, 657.
Boronates	*Acc. Chem. Res.* **1970**, *3*, 186.
	Chem. Rev. **1954**, *54*, 1065.
Cyclophosphazenes	
Stereochemistry	*Chem. Rev.* **1991**, *91*, 119.
Electrochemical	
Aniodic	*Tetrahedron* **1976**, *32*, 2185.
Electrophilic aromatic substitution	*Acc. Chem. Res.* **1971**, *4*, 240.
	Acc. Chem. Res. **1971**, *4*, 248.
Electrophilic substitutions	
Trisdialkylaminobenzenes	*Acc. Chem. Res.* **1989**, *22*, 27.
Ethers	
Methylene groups	*Tetrahedron* **1974**, *30*, 1683.
Free radicals	
Organoboranes	*Angew. Chem. Int. Ed. Engl.* **1972**, *11*, 692.
Heteroaromatic compounds	
N-Oxides	*Acc. Chem. Res.* **1976**, *9*, 192.
Heterolytic substitution	
Aromatic substitution	*Angew. Chem. Int. Ed. Engl.* **1972**, *11*, 874.
Nucleophilic substitution	*Acc. Chem. Res.* **1987**, *20*, 282.
Acetylenes	*Acc. Chem. Res.* **1976**, *9*, 358.
Organocopper compounds	*Org. Reactions* **1975**, *22*, 253.
Organometallic compounds	*Acc. Chem. Res.* **1983**, *16*, 350.
Organophosphorus compounds	*Angew. Chem. Int. Ed. Engl.* **1973**, *12*, 91.
Organotin compounds	*Acc. Chem. Res.* **1983**, *16*, 177.
Photochemical	*Pure & Appl. Chem.* **1982**, *54*, 1633.
Metal carbonyls	*Angew. Chem. Int. Ed. Engl.* **1964**, *3*, 730.
Photosubstitutions	
Aromatic compounds	*Pure & Appl. Chem.* **1983**, *55*, 331.
Pyrimidines	*Acc. Chem. Res.* **1978**, *11*, 462.
Radical anions	*Angew. Chem. Int. Ed. Engl.* **1975**, *14*, 734.
Radical chemistry	*Angew. Chem. Int. Ed. Engl.* **1975**, *14*, 734.
Reaction mechanisms	*Angew. Chem. Int. Ed. Engl.* **1982**, *21*, 401.
	Tetrahedron **1982**, *38*, 313.

Reaction mechanisms — *Chem. Soc. Rev.* **1978**, *7*, 399.
Rearrangement reactions
S_N(ANRORC) — *Acc. Chem. Res.* **1978**, *11*, 462.
S_N2
Reaction mechanisms — *Acc. Chem. Res.* **1985**, *18*, 212.
Chem. Soc. Rev. **1990**, *19*, 133.
S_N2' — *Acc. Chem. Res.* **1970**, *3*, 281.
Chem. Rev. **1989**, *89*, 1503.
Chem. Soc. Rev. **1990**, *19*, 133.
Solvolysis — *Chem. Rev.* **1956**, *56*, 571.
Sulfenyl halides
Reaction mechanisms — *Synthesis* **1971**, 617.
Thiocyanogen — *Org. Reactions* **1946**, *3*, 240.
Succinaldehyde monoacetal — *Synthesis* **1985**, 592.
Sucrose — *Pure & Appl. Chem.* **1984**, *56*, 833.
Suicide inactivators
Enzymes — *Pure & Appl. Chem.* **1981**, *53*, 149.
Sulfamic acid derivatives — *Chem. Rev.* **1980**, *80*, 151.
Sulfate esters — *Acc. Chem. Res.* **1970**, *3*, 145.
Sulfation — *Synthesis* **1969**, 3.
Sulfenamides — *Chem. Rev.* **1989**, *89*, 689.
Inversion — *Tetrahedron* **1984**, *40*, 3345.
Sulfenes — *Acc. Chem. Res.* **1975**, *8*, 10.
Synthesis **1970**, 393.
Sulfenic acids
Cyclizations — *Synthesis* **1971**, 617.
Historical perspective — *Synthesis* **1970**, 561.
α-Sulfenyl carbonyl compounds — *Chem. Rev.* **1978**, *78*, 363.
Sulfenyl chlorides
Alkenes — *Acc. Chem. Res.* **1979**, *12*, 282.
Sulfenyl halides
Historical perspective — *Synthesis* **1970**, 561.
Substitution reactions
Reaction mechanisms — *Synthesis* **1971**, 617.
Sulfides — *Acc. Chem. Res.* **1978**, *11*, 453.
Cyclic sulfides — *Tetrahedron* **1982**, *38*, 2857.
Enzymes — *Chem. Rev.* **1988**, *88*, 473.
Oxidations — *Tetrahedron* **1986**, *42*, 5459.
Sulfoxides
Reductions — *Tetrahedron* **1988**, *44*, 6537.
Sulfilimines — *Chem. Rev.* **1977**, *77*, 409.
Elimination reactions — *Tetrahedron* **1977**, *33*, 2359.
Sulfinato complexes — *Angew. Chem. Int. Ed. Engl.* **1971**, *10*, 315.

Sulfines	
Cycloalkenones	
Asymmetric synthesis	*Acc. Chem. Res.* **1987**, *20*, 72.
α-Sulfinyl carbanions	*Pure & Appl. Chem.* **1987**, *59*, 955.
Sulfinyl dienophiles	
Cycloadditions	*Acc. Chem. Res.* **1988**, *21*, 313.
Sulfonamides	
Cleavage reactions	*Chem. Rev.* **1959**, *59*, 1077.
Rearrangement reactions	*Chem. Rev.* **1959**, *59*, 1077.
Sulfonate esters	*Acc. Chem. Res.* **1970**, *3*, 145.
Sulfonations	*Synthesis* **1969**, 3.
Aromatic compounds	*Org. Reactions* **1946**, *3*, 141.
Sulfones	*Tetrahedron* **1977**, *33*, 2019.
	Tetrahedron **1987**, *43*, 1027.
Pyrolysis	*Angew. Chem. Int. Ed. Engl.* **1979**, *18*, 514.
Sulfonyl transfer reactions	*Chem. Soc. Rev.* **1989**, *18*, 123.
Sulfoxides	*Acc. Chem. Res.* **1978**, *11*, 453.
Acids	*Acc. Chem. Res.* **1973**, *6*, 132.
Allylsulfoxides	*Acc. Chem. Res.* **1974**, *7*, 147.
Asymmetric synthesis	*SYNLETT* **1990**, 643.
Chiral sulfoxides	*Pure & Appl. Chem.* **1988**, *60*, 1699.
Enzymes	*Chem. Rev.* **1988**, *88*, 473.
Perfluoro compounds	*Acc. Chem. Res.* **1973**, *6*, 387.
Sulfides	*Tetrahedron* **1988**, *44*, 6537.
Oxidations	*Tetrahedron* **1986**, *42*, 5459.
Titanium catalyzed oxidations	*Pure & Appl. Chem.* **1985**, *57*, 1911.
Sulfoximes	*Acc. Chem. Res.* **1973**, *6*, 341.
Sulfoximines	*Chem. Soc. Rev.* **1975**, *4*, 189.
	Chem. Soc. Rev. **1980**, *9*, 477.
Sulfur chemistry	
Biosynthesis	*Tetrahedron* **1983**, *39*, 1215.
Free radicals	*Angew. Chem. Int. Ed. Engl.* **1964**, *3*, 602.
Reductions	*Org. Reactions* **1988**, *36*, 249.
Sulfur dioxide	*Synthesis* **1971**, 639.
Sulfur tetrafluoride	
Fluorination	*Org. Reactions* **1974**, *21*, 1.
Sulfur ylides	*Acc. Chem. Res.* **1977**, *10*, 179.
	Angew. Chem. Int. Ed. Engl. **1983**, *22*, 516.
Sulfur-carbon multiple bonds	*Angew. Chem. Int. Ed. Engl.* **1991**, *30*, 361.
Sulfurated borohydrides	*Synthesis* **1972**, 526.
Sulfurization	
Amides	*Angew. Chem. Int. Ed. Engl.* **1966**, *5*, 451.
Unsaturated compounds	*Chem. Rev.* **1946**, *39*, 219.

Sulfur-nitrogen compounds	*Chem. Rev.* **1985**, *85*, 341.
Sulfur-sulfur double bonds	*Chem. Rev.* **1982**, *82*, 333.
Super acids	*Acc. Chem. Res.* **1968**, *1*, 202.
	Angew. Chem. Int. Ed. Engl. **1978**, *17*, 909.
Alkanes	*Chem. Rev.* **1982**, *82*, 591.
Methane	*Acc. Chem. Res.* **1987**, *20*, 422.
Terpenes	
Rearrangements	*Acc. Chem. Res.* **1976**, *9*, 257.
Supercritical extractions	*Angew. Chem. Int. Ed. Engl.* **1978**, *17*, 710.
	Angew. Chem. Int. Ed. Engl. **1978**, *17*, 751.
Superoxide	*Acc. Chem. Res.* **1981**, *14*, 368.
	Chem. Soc. Rev. **1977**, *6*, 195.
	Tetrahedron **1979**, *35*, 1471.
Superoxide dismutase	*Acc. Chem. Res.* **1972**, *5*, 321.
	Acc. Chem. Res. **1982**, *15*, 318.
Superphane	*Acc. Chem. Res.* **1980**, *13*, 65.
Supported reagents	*Synthesis* **1979**, 402.
	Synthesis **1979**, 481.
Supramolecular chemistry	*Angew. Chem. Int. Ed. Engl.* **1988**, *27*, 89.
Surfactants	
Chiral	*Acc. Chem. Res.* **1989**, *22*, 131.
Micelles	*Acc. Chem. Res.* **1979**, *12*, 111.
	Angew. Chem. Int. Ed. Engl. **1988**, *27*, 902.
Reverse micelles	*Acc. Chem. Res.* **1976**, *9*, 153.
Sweeteners	*Chem. Soc. Rev.* **1977**, *6*, 431.
	Chem. Soc. Rev. **1985**, *14*, 357.
Oxathiazinone dioxides	*Angew. Chem. Int. Ed. Engl.* **1973**, *12*, 869.
Symmetrical compounds	*Chem. Soc. Rev.* **1990**, *19*, 1.
SYNGEN	*Pure & Appl. Chem.* **1988**, *60*, 1563.
Synthesis design	*Acc. Chem. Res.* **1986**, *19*, 274.
	Chem. Soc. Rev. **1974**, *3*, 273.
	Synthesis **1988**, 417.
	Synthesis **1988**, 489.
Computers	*Angew. Chem. Int. Ed. Engl.* **1990**, *29*, 1286.
	Angew. Chem. Int. Ed. Engl. **1979**, *18*, 111.
	Chem. Rev. **1976**, *76*, 269.
CAMD	*Angew. Chem. Int. Ed. Engl.* **1987**, *26*, 403.
Umpolung	*Angew. Chem. Int. Ed. Engl.* **1979**, *18*, 239.
Synthetase	
Isoleucyl tRNA	*Acc. Chem. Res.* **1987**, *20*, 79.
RNA	*Acc. Chem. Res.* **1977**, *10*, 411.

Synthetic methods	*Angew. Chem. Int. Ed. Engl.* **1990**, *29*, 1177.
	Pure & Appl. Chem. **1981**, *53*, 1081.
	SYNLETT **1991**, 677.
	Tetrahedron **1981**, *37*, 3871ff.
	Tetrahedron **1986**, *42*, 2777ff.
	Tetrahedron **1992**, *48*, 1959ff.
Anisotropic media	*Tetrahedron* **1987**, *43*, 1197.
Baker's yeast	*Synthesis* **1990**, 1.
Catalysts	*Acc. Chem. Res.* **1986**, *19*, 121.
1,2-Difunctionality	*Tetrahedron* **1992**, *48*, 2803.
1,3-Difunctionality	*Tetrahedron* **1992**, *48*, 2803.
Functionalized polymers	*Synthesis* **1981**, 413.
Historical perspective	*Angew. Chem. Int. Ed. Engl.* **1978**, *17*, 473.
α-Keto acids	*Chem. Rev.* **1983**, *83*, 321.
Natural products	*Pure & Appl. Chem.* **1981**, *53*, 1259.
	Pure & Appl. Chem. **1982**, *54*, 2455.
Organic "metals"	*Angew. Chem. Int. Ed. Engl.* **1991**, *30*, 1121.
Organometallic chemistry	*Top. Curr. Chem.* **1980**, *92*, 109.
Organopalladium chemistry	*Pure & Appl. Chem.* **1985**, *57*, 1799.
Oxirenes	*Chem. Rev.* **1983**, *83*, 519.
Palladium catalysts	*Pure & Appl. Chem.* **1981**, *53*, 2371.
Periodic acid oxidations	*Org. Reactions* **1944**, *2*, 341.
Radical chemistry	*Synthesis* **1973**, 1.
Solvents	*Angew. Chem. Int. Ed. Engl.* **1991**, *30*, 1306.
Strain assistance	*Tetrahedron* **1989**, *45*, 2875ff.
Strained bridgehead double bonds	*Chem. Rev.* **1989**, *89*, 1067.
Synthetic paper	*Angew. Chem. Int. Ed. Engl.* **1971**, *10*, 287.
Synthons	*Angew. Chem. Int. Ed. Engl.* **1991**, *30*, 455.
	See also aryl synthons
	SYNLETT **1990**, 20.
	SYNLETT **1990**, 441.
Allylsilanes	*Acc. Chem. Res.* **1988**, *21*, 200.
Carbohydrates	
Unsaturated carbohydrates	*Acc. Chem. Res.* **1985**, *18*, 347.
α-Metalloamines	*Chem. Rev.* **1984**, *84*, 471.
Organosilicon chemistry	*Top. Curr. Chem.* **1980**, *88*, 33.
Retrosynthetic analysis	*Pure & Appl. Chem.* **1990**, *62*, 1887.
Tannins	
Branched tannins	*Pure & Appl. Chem.* **1982**, *54*, 2465.

Tautomerism	
β-Dicarbonyl compounds	
Photochemistry	*Chem. Soc. Rev.* **1984**, *13*, 69.
Enols	*Acc. Chem. Res.* **1988**, *21*, 42.
Heterocyclic compounds	*Acc. Chem. Res.* **1977**, *10*, 186.
Reaction mechanisms	*Pure & Appl. Chem.* **1987**, *59*, 1577.
Hydroxypyridine	*Acc. Chem. Res.* **1977**, *10*, 186.
Teichoic acids	*Acc. Chem. Res.* **1970**, *3*, 98.
Tellerium reagents	*Synthesis* **1986**, 1.
	Synthesis **1991**, 793.
	Synthesis **1991**, 897.
Telomerization	
Bromo compounds	*Synthesis* **1977**, 145.
Radical chemistry	*Acc. Chem. Res.* **1977**, *10*, 9.
Unsaturated compounds	*Synthesis* **1977**, 145.
Terazolinyl radicals	*Angew. Chem. Int. Ed. Engl.* **1973**, *12*, 455.
Terpenes	*Acc. Chem. Res.* **1990**, *23*, 70.
	Pure & Appl. Chem. **1986**, *58*, 423.
Asymmetric synthesis	*Acc. Chem. Res.* **1981**, *14*, 89.
	Pure & Appl. Chem. **1990**, *62*, 1263
Biosynthesis	*Acc. Chem. Res.* **1978**, *11*, 307.
	Acc. Chem. Res. **1981**, *14*, 89.
	Chem. Rev. **1972**, *72*, 115.
	Pure & Appl. Chem. **1987**, *59*, 287.
	Pure & Appl. Chem. **1989**, *61*, 345.
Farnesyl pyrophosphate synthetase	*Acc. Chem. Res.* **1978**, *11*, 307.
Hydroxylation	
Biosynthesis	*Tetrahedron* **1984**, *40*, 3597.
Mushroom metabolites	*Tetrahedron* **1981**, *37*, 2199.
Superacids	
Rearrangements	*Acc. Chem. Res.* **1976**, *9*, 257.
Tetraalkylhydrazines	*Acc. Chem. Res.* **1981**, *14*, 131.
Tetra-*tert*-butyltetrahedrane	*Pure & Appl. Chem.* **1991**, *63*, 275.
Tetracarbonylhydridoferrates	*Chem. Rev.* **1990**, *90*, 1041.
Tetracyanoethylene	*Synthesis* **1986**, 249.
Cycloadditons	*Synthesis* **1987**, 749.
Enol ethers	
Cycloaddition reactions	*Acc. Chem. Res.* **1977**, *10*, 117.
	Acc. Chem. Res. **1977**, *10*, 199.
Organometallic chemistry	*Synthesis* **1987**, 959.
Tetracyanoquinodimethanes	*SYNLETT* **1991**, 301.

Tetracyclic isoprenoids	
Biosynthesis	*Pure & Appl. Chem.* **1990**, *62*, 1259.
Tetracyclines	*Angew. Chem. Int. Ed. Engl.* **1975**, *14*, 721.
Tetrahedrane	*Angew. Chem. Int. Ed. Engl.* **1988**, *27*, 309.
Tetrakis(trifluoromethyl)tetrahedrane	*Acc. Chem. Res.* **1981**, *14*, 76.
Tetraheterofulvalenes	*Tetrahedron* **1986**, *42*, 1209.
Tetrahydrofurans	*Acc. Chem. Res.* **1991**, *24*, 22.
	Tetrahedron **1987**, *43*, 3309.
Tetrahydropyrans	*Tetrahedron* **1987**, *43*, 3309.
Tetrahydropyridinium salts	
Nucleophilic additions	*Acc. Chem. Res.* **1984**, *17*, 289.
Tetrahydrothiophenes	*Org. Reactions* **1951**, 6, 410.
Tetrahymanol	
Biosynthesis	*Acc. Chem. Res.* **1980**, *13*, 97.
Tetrakis(trifluoromethyl)tetrahedrane	*Acc. Chem. Res.* **1981**, *14*, 76.
Tetramethyltetraselenafulvalene	*Acc. Chem. Res.* **1985**, *18*, 261.
Tetrapyrrole pigments	
Oxygen adducts	*Angew. Chem. Int. Ed. Engl.* **1976**, *15*, 648.
Tetrapyrroles	*Tetrahedron* **1983**, *39*, 1837ff.
Photochemistry	*Angew. Chem. Int. Ed. Engl.* **1983**, *22*, 656.
Tetraselenafulvenes	*Synthesis* **1976**, 489.
Tetrathiafulvenes	*Synthesis* **1976**, 489.
Thallium compounds	*Angew. Chem. Int. Ed. Engl.* **1985**, *24*, 893.
Theoretical aspects	*Angew. Chem. Int. Ed. Engl.* **1982**, *21*, 711.
Alkenes	
Pyramidalized alkenes	*Chem. Rev.* **1989**, *89*, 1095.
Allopolarization	*Angew. Chem. Int. Ed. Engl.* **1976**, *15*, 321.
Allotropic forms of carbon	*Angew. Chem. Int. Ed. Engl.* **1991**, *30*, 70.
Anions	
Ab initio calculations	*Chem. Rev.* **1987**, *87*, 535.
Aromaticity	*Angew. Chem. Int. Ed. Engl.* **1971**, *10*, 761.
	Pure & Appl. Chem. **1980**, *52*, 1397ff.
	Pure & Appl. Chem. **1982**, *54*, 1097.
	Pure & Appl. Chem. **1982**, *54*, 1129.
Berry pseudorotation	*Angew. Chem. Int. Ed. Engl.* **1971**, *10*, 687.
Binding	*Tetrahedron* **1986**, *42*, 1917.
Boron hydrides	*Pure & Appl. Chem.* **1982**, *54*, 1143.
Carbon-carbon bonds	*Angew. Chem. Int. Ed. Engl.* **1980**, *19*, 429.
Catalytic reactions	
Ab initio	*Chem. Rev.* **1991**, *91*, 823.
CH/π interactions	*Tetrahedron* **1989**, *45*, 7201.

Chirality	*Chem. Rev.* **1979**, *79*, 17.
	Chem. Soc. Rev. **1986**, *15*, 189.
Chiroptical properties	
Carbonyl compounds	*Tetrahedron* **1986**, *42*, 777.
Circular dichroism	
Alkenes	*Tetrahedron* **1976**, *32*, 2475.
Chirality	*Tetrahedron* **1981**, *37*, 4123.
Cyclohexenones	*Tetrahedron* **1982**, *38*, 3.
Computer models	
Self-organizing systems	*Angew. Chem. Int. Ed. Engl.* **1972**, *11*, 821.
Conformation analysis	*Chem. Rev.* **1974**, *74*, 519.
Electro-optical Kerr effect	*Angew. Chem. Int. Ed. Engl.* **1977**, *16*, 663.
Heterocycles	*Angew. Chem. Int. Ed. Engl.* **1972**, *11*, 739.
Conformational effects	*Tetrahedron* **1977**, *33*, 3193.
Conformations	
Hydroxylamine derivatives	*Tetrahedron* **1981**, *37*, 849.
Curtin-Hammett Winstein-Holness Kinetics	*Chem. Rev.* **1983**, *83*, 73.
Cycloaddition reactions	*Chem. Rev.* **1972**, *72*, 157.
	Pure & Appl. Chem. **1982**, *54*, 1633.
Cyclobutadiene	*Angew. Chem. Int. Ed. Engl.* **1974**, *13*, 425.
	Tetrahedron **1980**, *36*, 343.
Cyclopropane	*Angew. Chem. Int. Ed. Engl.* **1979**, *18*, 809.
Dimensional probes	*Tetrahedron* **1986**, *42*, 1917.
Donor-acceptor π-electron systems	*Angew. Chem. Int. Ed. Engl.* **1988**, *27*, 1437.
Eighteen electron rule	*Acc. Chem. Res.* **1991**, *24*, 325.
Fenestranes	*Chem. Rev.* **1987**, *87*, 399.
Flattened tetrahedral carbon	*Chem. Rev.* **1987**, *87*, 399.
Frontier orbitals	*Angew. Chem. Int. Ed. Engl.* **1982**, *21*, 801.
Hammett constants	*Chem. Rev.* **1991**, *91*, 165.
Helices	
Self-assembly	*Angew. Chem. Int. Ed. Engl.* **1991**, *30*, 1450.
Heterocycles	
Valence isomerism	*Angew. Chem. Int. Ed. Engl.* **1971**, *10*, 11.
Historical perspective	*Angew. Chem. Int. Ed. Engl.* **1975**, *14*, 507.
HSAB principle	*Chem. Rev.* **1975**, *75*, 1.
Hyperconjugation	*Chem. Rev.* **1971**, *71*, 139.
Isoinversion principle	*Angew. Chem. Int. Ed. Engl.* **1991**, *30*, 477.
Metal catalysis	
Ab initio	*Chem. Rev.* **1991**, *91*, 823.
MINDO/3	*Chem. Rev.* **1986**, *86*, 1111.

Theoretical aspects, cont'd.	
Möbius-Hückel concept	*Acc. Chem. Res.* **1971**, *4*, 272.
Molecular mechanics calculations	*Angew. Chem. Int. Ed. Engl.* **1983**, *22*, 1.
	Angew. Chem. Int. Ed. Engl. **1976**, *15*, 519.
Nitrogen inversion barrier	
Heterocycles	*Angew. Chem. Int. Ed. Engl.* **1981**, *20*, 521.
Orbital symmetry	*Top. Curr. Chem.* **1974**, *45*, 39.
Orbital symmetry rules	*Angew. Chem. Int. Ed. Engl.* **1979**, *18*, 377.
Organometallic chemistry	
Molecular orbitals	*Tetrahedron* **1982**, *38*, 1339.
Pericyclic reactions	*Angew. Chem. Int. Ed. Engl.* **1971**, *10*, 761.
Polar axis	*Chem. Rev.* **1981**, *81*, 525.
Polymetalations	
Stabilization	
Allylic compounds	
Benzylic compounds	*Tetrahedron* **1983**, *39*, 2733.
Reaction mechanisms	*Chem. Soc. Rev.* **1984**, *13*, 1.
	Pure & Appl. Chem. **1982**, *54*, 1825.
Reactivity	*Tetrahedron* **1986**, *42*, 1917.
Reactivity-selectivity principle	*Angew. Chem. Int. Ed. Engl.* **1977**, *16*, 125.
	Tetrahedron **1980**, *36*, 3461.
Representation	*Angew. Chem. Int. Ed. Engl.* **1991**, *30*, 1.
Rotational barriers	*Angew. Chem. Int. Ed. Engl.* **1976**, *15*, 87.
Rotational isomerism	
1,2-Diarylethylene	*Chem. Rev.* **1991**, *91*, 1679.
Secondary orbital overlap	*Tetrahedron* **1983**, *39*, 2095.
Selectivity	*Pure & Appl. Chem.* **1983**, *55*, 277.
Spiroconjugation	*Angew. Chem. Int. Ed. Engl.* **1978**, *17*, 559.
Steric interactions	*Angew. Chem. Int. Ed. Engl.* **1977**, *16*, 429.
Strain	*Angew. Chem. Int. Ed. Engl.* **1986**, *25*, 312.
Hydrocarbons	*Chem. Rev.* **1989**, *89*, 1125.
Third allotropic form of carbon	*Angew. Chem. Int. Ed. Engl.* **1991**, *30*, 70/
Turnstile rotation	*Angew. Chem. Int. Ed. Engl.* **1971**, *10*, 687.
Thermal reactions	
Additions	*Synthesis* **1980**, 165.
	Synthesis **1980**, 769.
Azo compounds	*Angew. Chem. Int. Ed. Engl.* **1977**, *16*, 835.
Heterocycles	
Decomposition	
Gas phase	*Chem. Rev.* **1977**, *77*, 473.
Isomerizations	*Angew. Chem. Int. Ed. Engl.* **1972**, *11*, 1072.
Organofluorine compounds	*Synthesis* **1976**, 374.
Radical reactions	*Pure & Appl. Chem.* **1989**, *61*, 717.

Thermolysin	*Acc. Chem. Res.* **1988**, *21*, 333.
Thermotropic liquid crystals	*Tetrahedron* **1988**, *44*, 3413.
Thexylborane	*Synthesis* **1974**, 77.
Thiacrown ethers	
Copper complexes	*Pure & Appl. Chem.* **1988**, *60*, 501.
Thiamin diphosphate	*Chem. Rev.* **1987**, *87*, 863.
Thiaspirane alkaloids	*Acc. Chem. Res.* **1980**, *13*, 39.
1,3-Thiazines	*Synthesis* **1972**, 333.
1,3-Thiazoinones	*Tetrahedron* **1980**, *36*, 2023.
Thiazoles	*Org. Reactions* **1951**, 6, 367.
Thiazolidinones	*Chem. Rev.* **1981**, *81*, 175.
Thiazyl fluoride	*Angew. Chem. Int. Ed. Engl.* **1980**, *19*, 883.
Thiazyl trifluoride	*Angew. Chem. Int. Ed. Engl.* **1980**, *19*, 883.
Thiele-Winter acetoxylation reaction	*Org. Reactions* **1972**, *19*, 199.
Thienopyrroles	*Synthesis* **1985**, 143.
Thiirans	
Polymerizations	*Chem. Rev.* **1966**, 66, 322.
Thin layer chromatography	
Historical perspective	*Angew. Chem. Int. Ed. Engl.* **1983**, 22, 507.
Thioacetals	
Hydrolysis	
Reaction mechanisms	*Chem. Soc. Rev.* **1990**, *19*, 55.
Thioacetylations	
Amines	*Angew. Chem. Int. Ed. Engl.* **1966**, *5*, 458.
α-Thioalkylations	*Synthesis* **1987**, 589.
Thiocarbonyl compounds	
Photochemistry	*Tetrahedron* **1985**, *41*, 5393.
Radical chemistry	*Chem. Rev.* **1989**, *89*, 1413.
Ylides	*Tetrahedron* **1976**, *32*, 2165.
Thiocarboxylic *O*-esters	
Reductions	*Chem. Rev.* **1984**, *84*, 17.
Thiocrown ethers	*Acc. Chem. Res.* **1988**, *21*, 141.
Thiocyanogen	*Org. Reactions* **1946**, *3*, 240.
Thioethers	
Cation radicals	*Acc. Chem. Res.* **1980**, *13*, 200.
Dealkylations	*Synthesis* **1988**, 749.
Metallations	*Acc. Chem. Res.* **1989**, *22*, 152.
Thiohydroxyamic acids	*Synthesis* **1971**, 111.
Thioketenes	*Tetrahedron* **1988**, *44*, 1827.
Thiol esters	
Eliminations	*Acc. Chem. Res.* **1986**, *19*, 186.
Thiollactones	*Synthesis* **1972**, 151.

Thiols	
α-Thioalkylations	*Synthesis* **1987**, 589.
Thionation reactions	
Lawesson's reagents	*Tetrahedron* **1985**, *41*, 5061.
Thiones	
Photochemistry	*Acc. Chem. Res.* **1976**, *9*, 52.
Thionyl chloride	*Synthesis* **1981**, 661.
Thiophenes	*Acc. Chem. Res.* **1975**, *8*, 139.
	Org. Reactions **1951**, 6, 410.
Macrocycles	*Chem. Rev.* **1977**, *77*, 513.
Photochemistry	*Acc. Chem. Res.* **1971**, *4*, 65.
Thiophosgene	*Synthesis* **1978**, 803.
Thiopurines	
Immunosuppressants	*Acc. Chem. Res.* **1969**, *2*, 202.
Thromboxanes	*Angew. Chem. Int. Ed. Engl.* **1978**, *17*, 293.
	Angew. Chem. Int. Ed. Engl. **1983**, *22*, 805.
	Synthesis **1984**, 449.
Biosynthesis	*Chem. Soc. Rev.* **1977**, 6, 489.
Thujane	*Chem. Rev.* **1972**, *72*, 305.
Tian Hua Fen	*Pure & Appl. Chem.* **1986**, *58*, 789.
TIBA	
Reductions	*Synthesis* **1975**, 617.
TICT compounds	*Angew. Chem. Int. Ed. Engl.* **1986**, *25*, 971.
Titanium chemistry	
Asymmetric synthesis	*Pure & Appl. Chem.* **1983**, *55*, 1807.
Coupling reactions	
Ketones	
Alkenes	*Acc. Chem. Res.* **1983**, *16*, 405.
Low-valent titanium	*Synthesis* **1989**, 883.
Oxidations	
Sulfoxides	*Pure & Appl. Chem.* **1985**, *57*, 1911.
Oximes	*Acc. Chem. Res.* **1974**, *7*, 281.
Tobacco	*Angew. Chem. Int. Ed. Engl.* **1978**, *17*, 710.
	Chem. Rev. **1977**, *77*, 295.
Nitrosamines	*Acc. Chem. Res.* **1979**, *12*, 92.
Smoke	*Chem. Rev.* **1977**, *77*, 295.
Tocopherols	
Vitamin E	*Acc. Chem. Res.* **1986**, *19*, 194.
Tomaymycin	
Biosynthesis	*Acc. Chem. Res.* **1980**, *13*, 263.
Topology	*Chem. Soc. Rev.* **1973**, *2*, 457.
Toxins	*Pure & Appl. Chem.* **1982**, *54*, 1919.
	See also natural products

Enzymes	
Inhibitors	*Acc. Chem. Res.* **1975**, *8*, 281.
Transannular reactions	*Acc. Chem. Res.* **1969**, *2*, 210.
	Pure & Appl. Chem. **1987**, *59*, 955.
7,12-Dihydropleiadenes	*Acc. Chem. Res.* **1969**, *2*, 210.
Medium rings	*Quart. Rev.* **1966**, *20*, 119.
Transcription	
Eukaryotes	*Angew. Chem. Int. Ed. Engl.* **1987**, *26*, 218.
Transition metal chemistry	*Acc. Chem. Res.* **1987**, *20*, 379.
	Angew. Chem. Int. Ed. Engl. **1975**, *14*, 195.
	Chem. Rev. **1979**, *79*, 17.
	Synthesis **1976**, 561.
	Synthesis **1988**, 833.
Addition reactions	*Acc. Chem. Res.* **1969**, *2*, 10.
Aldehydes	*Acc. Chem. Res.* **1984**, *17*, 326.
Asymmetric synthesis	*Tetrahedron* **1989**, *45*, 6901.
Carbonylations	*Synthesis* **1973**, 509.
Cyclizations	
Acetylene compounds	*Acc. Chem. Res.* **1977**, *10*, 1.
Cycloaddition reactions	
Alkynes	*Chem. Rev.* **1988**, *88*, 1081.
Cyclooligomerization	*Angew. Chem. Int. Ed. Engl.* **1973**, *12*, 975.
Decarbonylations	*Synthesis* **1969**, 157.
Dimerization	
Alkenes	*Chem. Rev.* **1986**, *86*, 353.
Hydroborations	*Chem. Rev.* **1991**, *91*, 1179.
Hydrogenation	*Adv. Organomet. Chem.* **1979**, *17*, 319.
Aromatic compounds	*Acc. Chem. Res.* **1979**, *12*, 324.
Asymmetric synthesis	*Acc. Chem. Res.* **1982**, *15*, 395.
Metal hydrides	*Angew. Chem. Int. Ed. Engl.* **1991**, *30*, 759.
Nucleosides	*Acc. Chem. Res.* **1985**, *18*, 32.
Oxidations	*Acc. Chem. res.* **1970**, *3*, 347.
Photochemistry	*Chem. Soc. Rev.* **1972**, *1*, 241.
Purines	*Acc. Chem. Res.* **1977**, *10*, 146.
Pyrimidines	*Acc. Chem. Res.* **1977**, *10*, 146.
Reductions	*Tetrahedron* **1988**, *44*, 4295.
Substitution reactions	
Aromatic compounds	*Acc. Chem. Res.* **1970**, *3*, 139.
Sulfide ligands	*Chem. Rev.* **1989**, *89*, 1.
Unsaturated carbenes	*Acc. Chem. Res.* **1980**, *13*, 327.
Transition metal dihydrogen complexes	
NMR	*Acc. Chem. Res.* **1990**, *23*, 95.

Transition metal hydrides	*Angew. Chem. Int. Ed. Engl.* **1991**, *30*, 759.
Transition states	
Hyperconjugation	*Pure & Appl. Chem.* **1984**, *56*, 1755.
Translocation	
Peptidoglycans	
Biosynthesis	*Acc. Chem. Res.* **1971**, *4*, 297.
Transmembrane channel	*Acc. Chem. Res.* **1981**, *14*, 356.
Trialkylboranes	*Acc. Chem. Res.* **1976**, *9*, 26.
Trialkylsilyl perfluorosulfonates	
Lewis acids	*Synthesis* **1982**, 1.
Trialkylstannyllithium	*Synthesis* **1990**, 259.
Trialkylstannylmethyllithium	*Synthesis* **1990**, 259.
Triazacyclononane derivatives	
Metal complexes	*Pure & Appl. Chem.* **1988**, *60*, 509.
1,2,3-Tricarbonyl compounds	*Tetrahedron* **1978**, *34*, 1285.
Tricarbonyl(diene)iron complexes	*Acc. Chem. Res.* **1980**, *13*, 463.
1,3,3-Trichloro-2-azapropene	*Synthesis* **1972**, 599.
Trichloromethyl compounds	
Reductions	*Synthesis* **1983**, 773.
Trichothecenes	*Acc. Chem. Res.* **1982**, *15*, 388.
Tricyclo[3.3.0.$0^{2,8}$]octan-3-ones	*Angew. Chem. Int. Ed. Engl.* **1982**, *21*, 820.
Tricyclo[2.1.0.$0^{2,5}$]pentane	*Chem. Rev.* **1989**, *89*, 985.
Triflates	*Acc. Chem. Res.* **1977**, *10*, 306.
	Chem. Rev. **1977**, *77*, 69.
Trifluoromethyl compounds	
Aromatic compounds	
Nucleophiles	*Acc. Chem. Res.* **1978**, *11*, 197.
Trifluoromethyl ketones	*Tetrahedron* **1991**, *47*, 3207.
Trifluoromethyl radicals	
Hydrogen abstraction	
Reaction mechanisms	*Chem. Rev.* **1971**, *71*, 247.
Trifluoromethylations	*Tetrahedron* **1992**, *48*, 6555.
Trimethylenemethane	*Acc. Chem. Res.* **1972**, *5*, 242.
Trimethylenemethane equivalents	
[3+2] Cycloadditions	*Angew. Chem. Int. Ed. Engl.* **1986**, *25*, 1.
Triosephosphate isomerase	*Acc. Chem. Res.* **1977**, *10*, 105.
Triphase catalysis	*Angew. Chem. Int. Ed. Engl.* **1979**, *18*, 421.
Trisdialkylaminobenzenes	
Electrophilic substitutions	*Acc. Chem. Res.* **1989**, *22*, 27.
Trishomocubanes	*Chem. Rev.* **1989**, *89*, 1011.
Triterpenoids	*Acc. Chem. Res.* **1975**, *8*, 152.

Trithianes	
Umpolung reagents	*Synthesis* **1969**, 17.
t-RNA	*See also RNA*
	Angew. Chem. Int. ed. Ebgl. **1985**, *24*, 371.
Tropocoronands	
Metal complexes	*Pure & Appl. Chem.* **1986**, *58*, 1477.
Tropomyosin	*Acc. Chem. Res.* **1982**, *15*, 224.
Troponin-tropomyosin complexes	
Calcium interactions	*Acc. Chem. Res.* **1980**, *13*, 185.
Tropones	*Acc. Chem. Res.* **1979**, *12*, 132.
Troponin	
Troponin-tropomyosin complexes	
Calcium interactions	*Acc. Chem. Res.* **1980**, *13*, 185.
Troponoids	
Aromaticity	*Pure & Appl. Chem.* **1982**, *54*, 975.
Tropoquinone acetals	*SYNLETT* **1990**, 301.
Tropoquinones	*SYNLETT* **1990**, 301.
Tryptophan derivatives	
Oxidation	
Photosensitized oxidation	*Acc. Chem. Res.* **1977**, *10*, 346.
Tryptophan oxygenase	
Hormonal control	*Acc. Chem. Res.* **1970**, *3*, 113.
Tumor lectinology	*Angew. Chem. Int. Ed. Engl.* **1988**, *27*, 1267.
Tumor promoters	*Pure & Appl. Chem.* **1982**, *54*, 1919.
Tungsten compounds	*Pure & Appl. Chem.* **1985**, *57*, 1789.
Turgorins	*Angew. Chem. Int. Ed. Engl.* **1983**, *22*, 695.
Turnstile rotation	*Acc. Chem. Res.* **1971**, *4*, 288.
	Angew. Chem. Int. Ed. Engl. **1971**, *10*, 687.
Twisted intramolecular charge transfer (TICT)	*Pure & Appl. Chem.* **1983**, *55*, 245.
Tyrosinase	
Mechanisms	*Acc. Chem. Res.* **1989**, *22*, 322.
Umpolung	
Acyl anion equivalents	
Cyanohydrins	*Tetrahedron* **1983**, *39*, 3207.
α-Aminoalkylations	*Angew. Chem. Int. Ed. Engl.* **1975**, *14*, 15.
Carbonyl equivalents	*Chem. Soc. Rev.* **1982**, *11*, 493.
	Synthesis **1977**, 357.
Nucleophilic acylation	*Tetrahedron* **1976**, *32*, 1943.
	Synthesis **1969**, 17.
Synthesis design	*Angew. Chem. Int. Ed. Engl.* **1979**, *18*, 239.

Unimolecular reactions	
Reaction mechanisms	*Chem. Soc. Rev.* **1983**, *12*, 163.
Unnatural products	
Buckminsterfullerene	*Angew. Chem. Int. Ed. Engl.* **1991**, *30*, 678.
	Chem. Rev. **1991**, *91*, 1213.
Cubanes	*Chem. Rev.* **1989**, *89*, 997.
Cumulenes	
Strained cumulenes	*Chem. Rev.* **1989**, *89*, 1111.
Cyclobutadiene	*Angew. Chem. Int. Ed. Engl.* **1988**, *27*, 309.
Dendralenes	*Angew. Chem. Int. Ed. Engl.* **1984**, *23*, 932.
Diamondoid hydrocarbons	*Tetrahedron* **1980**, *36*, 971.
Dodecahedrane	*Chem. Rev.* **1989**, *89*, 1051.
	Tetrahedron **1979**, *35*, 2189.
Polynitropolycyclic cage compounds	*Tetrahedron* **1988**, *44*, 2377.
Synthetic methods	*Tetrahedron* **1986**, *42*, 1549ff.
Crowded molecules	*Tetrahedron* **1978**, *34*, 1855.
Tetra-*tert*-butyl-tetrahedrane	*Pure & Appl. Chem.* **1991**, *63*, 275.
Tetrahedrane	*Angew. Chem. Int. Ed. Engl.* **1988**, *27*, 309.
Unsaturated carbohydrates	
Synthons	*Acc. Chem. Res.* **1985**, *18*, 347.
Unsaturated hydrocarbons	
Cyclic unsaturated hydrocarbons	*Angew. Chem. Int. Ed. Engl.* **1987**, *26*, 204.
Platinum chemistry	*Acc. Chem. Res.* **1973**, *6*, 202.
Unsaturated aldehydes	*Acc. Chem. Res.* **1988**, *21*, 47.
Unsaturated carbenes	*Chem. Rev.* **1978**, *78*, 383.
	Chem. Rev. **1991**, *91*, 197.
	Pure & Appl. Chem. **1983**, *55*, 369.
Transition metal complexes	*Acc. Chem. Res.* **1980**, *13*, 327.
Unsaturated carbonyl compounds	*Pure & Appl. Chem.* **1984**, *56*, 91.
Conjugate additions	*Org. Reactions* **1990**, *38*, 225.
Cyclizations	*Synthesis* **1975**, 1.
Hydrocyanation	*Org. Reactions* **1977**, *25*, 255.
α,β-Unsaturated carbonyl compounds	
Reductions	*Org. Reactions* **1976**, *23*, 1.
β,γ-Unsaturated carbonyl compounds	
Photochemistry	*Chem. Rev.* **1976**, *76*, 1.
Unsaturated carboxylic acids	
Aryl carboxylic acids	
Rearrangements	*Chem. Rev.* **1967**, *67*, 599.
Unsaturated C_{18} acids	*Acc. Chem. Res.* **1976**, *9*, 34.

Unsaturated compounds	
Addition reactions	*Synthesis* **1970**, 99.
	Synthesis **1977**, 145.
Acyl isocyanates	*Synthesis* **1982**, 433.
Free radical	*Angew. Chem. Int. Ed. Engl.* **1970**, *9*, 273.
	Chem. Rev. **1962**, *62*, 599.
	Org. Reactions **1963**, *13*, 91.
	Synthesis **1977**, 145.
Arylation	
Azonium salts	*Org. Reactions* **1976**, *24*, 225.
Bromo compounds	
Telomerizations	*Synthesis* **1977**, 145.
Cycloaddition reactions	
Photochemistry	*Chem. Rev.* **1983**, *83*, 1.
Organocuprates	*Pure & Appl. Chem.* **1988**, *60*, 57.
Polymerizations	
Photochemistry	*Chem. Rev.* **1983**, *83*, 1.
Radical reactions	*Synthesis* **1970**, 99.
Reductions	*Org. Reactions* **1976**, *23*, 1.
Sulfurization	*Chem. Rev.* **1946**, *39*, 219.
Telomerization reactions	*Synthesis* **1977**, 145.
Unsaturated esters	*Acc. Chem. Res.* **1988**, *21*, 47.
Unsaturated fatty acids	*Acc. Chem. Res.* **1969**, *2*, 196.
β,γ-Unsaturated ketones	
Rearrangements	*Acc. Chem. Res.* **1974**, *7*, 106.
Unsaturated lactones	*Chem. Rev.* **1976**, *76*, 625.
Uranium chemistry	*Acc. Chem. Res.* **1976**, *9*, 223.
Ureas	
Cyclic ureas	*Synthesis* **1973**, 243.
Uroporphyrinogens	
Biosynthesis	*Acc. Chem. Res.* **1975**, *8*, 201.
	Chem. Rev. **1990**, *90*, 1261.
Vanadium chemistry	
Bioinorganic chemistry (of)	*Angew. Chem. Int. Ed. Engl.* **1991**, *30*, 148.
Polymerization	*Angew. Chem. Int. Ed. Engl.* **1971**, *10*, 776.
Vancomycin	
Structures	*Acc. Chem. Res.* **1984**, *17*, 364.
Verdazyl radicals	*Angew. Chem. Int. Ed. Engl.* **1973**, *12*, 455.
Vesicles	
Membranes	*Acc. Chem. Res.* **1980**, *13*, 7.
Photochemistry	*Tetrahedron* **1982**, *38*, 2455.
Vinamidines	*Angew. Chem. Int. Ed. Engl.* **1976**, *15*, 459.
Vinamidinium salts	*Angew. Chem. Int. Ed. Engl.* **1976**, *15*, 459.

Vinyl trifluoromethanesulfonate esters	*Acc. Chem. Res.* **1978**, *11*, 107.
Vinyl acetals	*Synthesis* **1989**, 721.
Vinyl azides	*Angew. Chem. Int. Ed. Engl.* **1971**, *10*, 98.
	Angew. Chem. Int. Ed. Engl. **1975**, *14*, 775.
Zwitterions	*Acc. Chem. Res.* **1979**, *12*, 125.
Vinyl carbanions	*Tetrahedron* **1988**, *44*, 4653.
Vinyl cations	*Acc. Chem. Res.* **1976**, *9*, 364.
	Angew. Chem. Int. Ed. Engl. **1978**, *17*, 333.
Vinyl halides	
Solvolysis	*Acc. Chem. Res.* **1970**, *3*, 209.
Vinyl sulfonates	
Solvolysis	*Acc. Chem. Res.* **1970**, *3*, 209.
Vinyl sulfones	*Tetrahedron* **1990**, *46*, 6951.
Conjugate additions	*Chem. Rev.* **1986**, *86*, 903.
Vinyl sulfoxides	
Michael additions	
Organometallic reagents	
Asymmetric synthesis	*Pure & Appl. Chem.* **1981**, *53*, 2307.
Vinylallenes	
Diels-Alder	*SYNLETT* **1990**, 1.
Pericyclic reactions	*Acc. Chem. Res.* **1983**, *16*, 81.
Vinylcyclopropanes	
Cyclopentenes	*Org. Reactions* **1985**, *33*, 247.
Rearrangements	*Chem. Soc. Rev.* **1988**, *17*, 231.
Vinylene triheterocarbonates	*Tetrahedron* **1986**, *42*, 1209.
Vinylidene carbenes	*Chem. Rev.* **1991**, *91*, 197.
Vinyloxiranes	
Organocopper reagents	*Chem. Rev.* **1989**, *89*, 1503.
Vinylphosphine oxides	*Synthesis* **1974**, 775.
Vinylphosphonium salts	*Synthesis* **1974**, 775.
Vinylsilanes	
Cyclization reaction	*Chem. Rev.* **1986**, *86*, 857.
Electrophilic substitution	*Org. Reactions* **1989**, *37*, 57.
Viruses	*See medicinal chemistry*
Vision	*Acc. Chem. Res.* **1975**, *8*, 81.
	Acc. Chem. Res. **1975**, *8*, 92
	Acc. Chem. Res. **1975**, *8*, 101.
	Acc. Chem. Res. **1975**, *8*, 107.
	Acc. Chem. Res. **1975**, *8*, 407.
	Acc. Chem. Res. **1979**, *12*, 87.
	Angew. Chem. Int. Ed. Engl. **1987**, *26*, 170.
	Angew. Chem. Int. Ed. Engl. **1990**, *29*, 461.
Polyenes	*Acc. Chem. Res.* **1986**, *19*, 42.

Vitamin A	
Photochemistry	*Tetrahedron* **1984**, *40*, 1931.
Vision	*Angew. Chem. Int. Ed. Engl.* **1990**, *29*, 461.
Vitamin B_{12}	*Acc. Chem. Res.* **1976**, *9*, 114.
	Acc. Chem. Res. **1990**, *23*, 308.
	Angew. Chem. Int. Ed. Engl. **1976**, *15*, 417.
	Angew. Chem. Int. Ed. Engl. **1977**, *16*, 233.
	Angew. Chem. Int. Ed. Engl. **1988**, *27*, 5.
	Chem. Soc. Rev. **1980**, *9*, 125.
	Chem. Soc. Rev. **1985**, *14*, 161.
	Pure & Appl. Chem. **1983**, *55*, 1791.
Biosynthesis	*Acc. Chem. Res.* **1978**, *11*, 29.
	Acc. Chem. Res. **1986**, *19*, 147.
	Pure & Appl. Chem. **1989**, *61*, 337.
	Pure & Appl. Chem. **1990**, *62*, 1269.
	Tetrahedron **1975**, *31*, 2639.
Electrochemistry	*Pure & Appl. Chem.* **1987**, *59*, 363.
	Acc. Chem. Res. **1983**, *16*, 235.
Model compounds	*Acc. Chem. Res.* **1968**, *1*, 97.
Vitamin B_6	
Amino acids	*Acc. Chem. Res.* **1989**, *22*, 115.
Vitamin D	*Chem. Soc. Rev.* **1977**, *6*, 83.
	Chem. Soc. Rev. **1980**, *9*, 449
Vitamin E	*Acc. Chem. Res.* **1986**, *19*, 194.
Walsh rules	
MO thoery	*Acc. Chem. Res.* **1974**, *7*, 384.
Water splitting	
Photosynthesis	*Angew. Chem. Int. Ed. Engl.* **1987**, *26*, 643.
Wieland	
Historical perspective	*Angew. Chem. Int. Ed. Engl.* **1977**, *16*, 559.
Xanthates	
Pyrolysis	*Org. Reactions*, **1962**, *12*, 57.
Xanthene dye	*Acc. Chem. Res.* **1989**, *22*, 171.
Xanthonoids	*Tetrahedron* **1980**, *36*, 1465.
***o*-Xylylenes**	*Acc. Chem. Res.* **1980**, *13*, 270.
	Chem. Soc. Rev. **1980**, *9*, 41.
Photochemistry	*Pure & Appl. Chem.* **1990**, 62, 1557.
Yeast alanine tRNA	*Acc. Chem. Res.* **1984**, *17*, 393.
Ylide malonodinitriles	
Cyclizations	*Synthesis* **1976**, 705.
	Chem. Rev. **1980**, *80*, 329.
Ylides	*Chem. Rev.* **1991**, *91*, 263.
Cyclizations	*Acc. Chem. Res.* **1991**, *24*, 22.
Heterocyclic compounds	*Chem. Soc. Rev.* **1987**, *16*, 89.

Ylides, cont'd.
Organoarsinic chemistry — *Chem. Soc. Rev.* **1987**, *16*, 45.
Organophosphorus chemistry — *Acc. Chem. Res.* **1974**, *7*, 6.
Acc. Chem. Res. **1975**, *8*, 62.
Organosilicon chemistry — *Acc. Chem. Res.* **1975**, *8*, 62.
Peroxonium ions — *Chem. Soc. Rev.* **1985**, *14*, 399.
Phosphonium ylides — *Acc. Chem. Res.* **1969**, *2*, 373.
α-Silyl onium salts — *Chem. Rev.* **1986**, *86*, 941.
Ynamine — *Tetrahedron* **1976**, *32*, 1449.
Yohimbinoid alkaloids — *Pure & Appl. Chem.* **1986**, *58*, 685.
Zinc carboxypeptidase A — *Acc. Chem. Res.* **1989**, *22*, 62.
Zirconium complexes
Asymmetric synthesis — *Pure & Appl. Chem.* **1991**, *63*, 797.
Zwitterionic intermediates
Cycloadditions — *Acc. Chem. Res.* **1977**, *10*, 117.
Acc. Chem. Res. **1977**, *10*, 199.
Zwitterions
Vinyl azides — *Acc. Chem. Res.* **1979**, *12*, 125.